METHODS IN

Stream Ecology

METHODS IN

Stream Ecology

EDITED BY

F. RICHARD HAUER

FLATHEAD LAKE BIOLOGICAL STATION
UNIVERSITY OF MONTANA
POLSON, MONTANA

AND

GARY A. LAMBERTI

DEPARTMENT OF BIOLOGICAL SCIENCES
UNIVERSITY OF NOTRE DAME
NOTRE DAME, INDIANA

ACADEMIC PRESS

SAN DIEGO   LONDON   BOSTON   NEW YORK   SYDNEY   TOKYO   TORONTO

This book is printed on acid-free paper. ∞

Academic Press
a division of Harcourt Brace & Company
525 B Street, Suite 1900, San Diego, California 92101-4495, USA
http://www.apnet.com

Academic Press Limited
24-28 Oval Road, London NW1 7DX, UK
http://www.hbuk.co.uk/ap/

Library of Congress Catalog Card Number: 95-51735

International Standard Book Number: 0-12-332906-X

PRINTED IN THE UNITED STATES OF AMERICA
00 01 02 03 EB 9 8 7 6 5 4 3 2

Contents

SECTION C Stream Biota

SECTION D Community Interactions

SECTION E Ecosystem Processes

Contributors

Numbers in parentheses indicate the pages on which the authors'
contributions begin.

E. F. BENFIELD (579) Department of Biology, Virginia Polytechnic Institute and State University, Blacksburg, Virginia 24061

ARTHUR C. BENKE (557) Department of Biological Sciences, University of Alabama, Tuscaloosa, Alabama 35487

ROBERT L. BESCHTA (123) College of Forestry, Oregon State University, Corvallis, Oregon 97331

PETER A. BISSON (23) Forestry Sciences Laboratory, USDA Forest Service, Olympia, Washington 98512

THOMAS L. BOTT (533) Stroud Water Research Center, The Academy of Natural Sciences, Avondale, Pennsylvania 19311

KENNETH W. CUMMINS (453) Department of Ecosystem Research and Implementation, South Florida Water Management District, West Palm Beach, Florida 33416

CLIFFORD N. DAHM (107) Department of Biology, University of New Mexico, Albuquerque, New Mexico 87131

TERRENCE P. EHRMAN (145) Department of Biology, Virginia Polytechnic Institute and State University, Blacksburg, Virginia 24061

JACK W. FEMINELLA (409) Department of Zoology and Wildlife Science, Auburn University, Auburn, Alabama 36849

CHRISTOPHER A. FRISSELL (493) Flathead Lake Biological Station, The University of Montana, Polson, Montana 59860

FRANCES P. GELWICK (475) Department of Wildlife and Fisheries Sciences, Texas A & M University, College Station, Texas 77843

JAMES A. GORE (53) Environmental Protection Division, The Conservancy, Naples, Florida 33942

STANLEY V. GREGORY (217) Department of Fisheries and Wildlife, Oregon State University, Corvallis, Oregon 97331

NANCY B. GRIMM (625) Department of Zoology, Arizona State University, Tempe, Arizona 85287

JACK W. GRUBAUGH (191) Department of Biology, University of Memphis, Memphis, Tennessee 38152

MORGAN J. HANNAFORD (647) Department of Environmental Science, Policy and Management, University of California, Berkeley, Berkeley, California 94720

F. RICHARD HAUER (93, 339) Flathead Lake Biological Station, The University of Montana, Polson, Montana 59860

ANNE E. HERSHEY (511) Department of Biology, University of Minnesota-Duluth, Duluth, Minnesota 55812

WALTER R. HILL (93) Environmental Sciences Division, Oak Ridge National Laboratory, Oak Ridge, Tennessee 37831

MARK D. JOHNSON (233) Department of Biological Sciences, University of Alabama, Tuscaloosa, Alabama 35487

GINA D. LALIBERTE (269) Department of Biological Sciences, Bowling Green State University, Bowling Green, Ohio 43403

GARY A. LAMBERTI (217, 295, 409) Department of Biological Sciences, University of Notre Dame, Notre Dame, Indiana 46556

HIRAM W. LI (391) Oregon Cooperative Fishery Research Unit, Oregon State University, Corvallis, Oregon 97331

JUDITH L. LI (391) Department of Fisheries and Wildlife, Oregon State University, Corvallis, Oregon 97331

DAVID G. LONZARICH (493) Department of Biology, University of Wisconsin at Eau Claire, Phillips Science Hall, Eau Claire, Wisconsin 54702

REX L. LOWE (269) Department of Biological Sciences, Bowling Green State University, Bowling Green, Ohio 43403

WILLIAM J. MATTHEWS (475) University of Oklahoma Biological Station, University of Oklahoma, Kingston, Oklahoma 73439

RICHARD W. MERRITT (453) Department of Entomology, Michigan State University, East Lansing, Michigan 48824

G. WAYNE MINSHALL (591) Stream Ecology Center, Department of Biological Sciences, Idaho State University, Pocatello, Idaho 83209

DAVID R. MONTGOMERY (23) Department of Geological Sciences, University of Washington, Seattle, Washington 98195

PATRICK J. MULHOLLAND (161) Environmental Sciences Division, Oak Ridge National Laboratory, Oak Ridge, Tennessee 37831

MARILYN J. MYERS (647) Department of Environmental Science, Policy and Management, University of California, Berkeley, Berkeley, California 94720

ROBERT W. NEWBURY (75) School of Resource and Environmental Management, Simon Fraser University, Newbury Hydraulics Ltd., Gibson, British Columbia, Canada VON 1VO

MARGARET A. PALMER (315) Department of Zoology, University of Maryland, College Park, Maryland 20742

BARBARA L. PECKARSKY (431) Department of Entomology, Cornell University, Ithaca, New York 14853

BRUCE J. PETERSON (511) The Ecosystems Center, Marine Biological Laboratory, Woods Hole, Massachusetts 02543

CATHERINE M. PRINGLE (607) Institute of Ecology, University of Georgia, Athens, Georgia 30602

VINCENT H. RESH (339, 647) Department of Environmental Science, Policy and Management, University of California, Berkeley, Berkeley, California 94720

LEONARD A. SMOCK (371) Department of Biology, Virginia Commonwealth University, Richmond, Virginia 23284

JACK A. STANFORD (3) Flathead Lake Biological Station, The University of Montana, Polson, Montana 59860

ALAN D. STEINMAN (161, 295) Department of Ecosystem Restoration, South Florida Water Management District, West Palm Beach, Florida 33416

DAVID L. STRAYER (315) Institute of Ecosystem Studies, Milbrook, New York 12545

FRANK J. TRISKA (607) United States Geological Survey, Water Resources Division, Menlo Park, California 94025

H. MAURICE VALETT (107) Department of Biology, University of New Mexico, Albuquerque, New Mexico 87131

J. BRUCE WALLACE (191) Institute of Ecology, University of Georgia, Athens, Georgia 30602

AMELIA K. WARD (233) Department of Biological Sciences, University of Alabama, Tuscaloosa, Alabama 35487

JACKSON R. WEBSTER (145) Department of Biology, Virginia Polytechnic Institute and State University, Blacksburg, Virginia 24061

Preface

A fundamental feature of Earth is the presence of vast quantities of water and a hydrologic cycle that results in the deposition of rain and snow upon the landscape. Precipitation coalesces and flows downhill, as either surface or ground water. Flowing water is one of the major erosive forces on our planet and has acted upon landforms for thousands of millions of years. Streams and rivers not only affect the landscape over very long time periods, but are in turn directly affected by the valleys through which they flow. It is for these reasons that no other area of aquatic ecology requires a more interdisciplinary approach than stream ecology. Geology, geomorphology, fluid mechanics, hydrology, biogeochemistry, nutrient dynamics, microbiology, botany, invertebrate zoology, fish biology, and bioproduction are but a few of the disciplines from which stream ecology draws.

Over the past three decades, stream ecology has emerged from being a subdiscipline of limnology into an ecological discipline in its own right. Understanding of stream ecosystem structure and function has progressed rapidly and continues to be one of the most active areas of research in aquatic ecology. This is particularly evident from the growth in publication of stream ecological research. Along with the rapid increase in research activity there has been a commensurate increase in the teaching of stream ecology at the upper undergraduate and graduate levels at major colleges and universities.

With increased recognition that streams and rivers are fundamental to the human existence, as well as to global biodiversity, an appreciation for stream resources, reflected in legislation and management specifically designed for their protection or restoration, has developed. Although progress in stream management has been obvious over the past two decades, pervasive anthropogenic impacts continue to affect streams worldwide.

This book is a comprehensive and contemporary series of methods in stream ecology that can be used for teaching or conducting research. It should be useful to everyone from the stream ecology student to the most seasoned scientist. Resource managers employed in the private sector or by federal or state agencies will find this book helpful as a reference for monitoring and evaluating the efficacy of their field and laboratory techniques.

The important topics in stream ecology are presented in five major sections: Physical Processes, Material Storage and Transport, Stream Biota, Community Interactions, and Ecosystem Processes. All 31 chapters are written by leading experts in stream research and teaching. Each chapter consists of an introduction, a general design, a series of specific exercises, and finally questions. The Introduction provides background information and a literature review necessary to understand the principles of the topic. The General Design presents the conceptual approach and principles that will be used in the various exercises. A variety of exercises are presented within the Specific Exercises section, generally beginning with relatively simple goals and methods and increasing in the level of difficulty and sophistication. Each exercise provides step-by-step instructions for conducting a field or laboratory investigation. Likewise, a list of recommended supplies and equipment necessary to conduct the exercises is included at the end of each chapter. Although the methods are of research quality, it is not our intention to provide an exhaustive manual of methods for each topic. Rather, we present rigorous methods that provide sound underpinnings for both instruction and research purposes. In each case, the exercises use methods that are frequently employed by the authors in their personal research or instruction. The questions listed at the end of each chapter are intended to encourage critical evaluation of the topic and the methods that were used to address a particular stream ecology issue.

Most exercises can be completed within a few hours or an afternoon of intensive field work. However, some of the chapters include exercises that require preparatory steps such as placement of artificial substrates for colonization of algae, leaves for decomposition, or collection of invertebrates over several months for growth analysis. When this book is used for course instruction, we recommend that instructors carefully consider the chapters and exercises that they wish to use and plan carefully to budget the necessary time to complete exercise setups, sampling, and analysis. We recognize that it would be difficult to complete all the chapters and exercises in this book in an intensive one-semester stream ecology course. However, we hope that all of the chapters will enrich instruction, stimulate student interest, and enhance opportunities for research. We particularly encourage the use of this book to assist in the formulation of hypotheses, and, to that end, the chapters present sound methods for discovery for small group or graduate student research.

For course instruction, all of the field exercises are designed for moderate-sized streams from 3 to 12 m wide that are easily waded, although all the methods are adaptable to either smaller streams or larger rivers. Smaller streams should be avoided by a large class, such as 10–20 students, because of the impacts incurred on a small environment. Large rivers are limiting to class instruction because of safety concerns and the inherent difficulties associated with sampling deep, fast flowing waters.

We hope that you will find this book to be particularly "user friendly." To avoid the pitfalls of many edited books with multiple chapter authors, we attempted to present a book with a very logical flow of topics and a uniform chapter format and style. The inspiration for this book arose from our own teaching of stream ecology at the University of Montana's Flathead Lake Biological Station and at the University of Notre Dame Environmental Research Center. We thank our contributing authors for their many suggestions that improved the quality of this book and for their willingness to share their research and instructional experience to advance the field of stream ecology. Vince Resh, Jack Stanford, and Jack Webster were particularly helpful with many suggestions when the ideas for this book were in their embryonic stage. Sarah Wallace, Andrew Hauer, and Ethan Nedeau assisted with drawings in Chapter 10, 16, and 21, respectively.

We gratefully acknowledge the assistance and financial support of our home institutions, the University of Montana and the University of Notre Dame. Our graduate and undergraduate students have been a source of inspiration and encouragement to us. Finally, we thank our spouses, Brenda Hauer and Donna Lamberti, for the many hours we took from them to give to this endeavor.

F. RICHARD HAUER
GARY A. LAMBERTI

SECTION A

Physical Processes

CHAPTER 1

Landscapes and Catchment Basins

JACK A. STANFORD

Flathead Lake Biological Station
The University of Montana

I. INTRODUCTION

*S*treams, rivers, and groundwater flow pathways are the plumbing of the continents. Water coalesces and flows downhill in surface and subsurface channels in response to precipitation patterns and the complex nature of the long (geologic time) and short (decadal) term processes that determine the biophysical form of landscapes that produce runoff (*catchment basins*). Uplift of mountain ranges, caused by continental drift and volcanism, is continually countered by erosion and deposition (*sedimentation*) mediated by the forces of wind and water. The landscapes we perceive today reflect the long geologic history of the region as well as recent events such as floods, fires, and human-caused environmental disturbances (e.g., deforestation, dams, pollution, introduction of exotic species).

For a stream ecologist, a landscape view of a catchment (river) basin encompasses the entire stream network, including interconnection with groundwater flow pathways, embedded in its terrestrial setting and flowing from the highest elevation in the catchment to the point of confluence with another catchment system or with the ocean. For example, the earth's largest catchment basin, the Amazon River basin, occupies over half of the South American continent. Headwaters flow from small catchments containing glaciers and snowfields over 4300 m above sea level on the spine of the Andes Mountains to feed the major tributaries. The tributary rivers converge to form the mainstem Amazon, which flows from the base of the Andes across a virtually flat landscape to the equatorial Atlantic ocean. The elevation change is less than 200 m over the nearly 3000-km length of the larger tributaries and the mainstem river from the Andes to the ocean. Because of the enormous transport power of the massive water volume carried by the Amazon River, some channels are 100 m deep. In other places along the river corridor the channel is >5 km wide, relatively shallow, filled by sediment deposition (*alluviation*), and flood waters spread out over huge floodplains (Day and Davies 1986).

The Amazon River, like all rivers, has molded the landscape of its catchment basin over time, cutting channels into terraces in some places and building (*alluviating*) terraces on floodplains in others. Rivers drain the continents, transport sediments, nutrients, and other materials from the highlands to the lowlands and oceans, and constantly modify the biophysical character of their catchment basins. These processes occur in direct relation to a particular catchment's global position, climate, orography, and biotic character, coupled with spatial variations in bedrock and other geomorphic features of the landscape.

Within a catchment basin, stream channels usually grow in size and complexity in a downstream direction (Fig. 1.1). The smallest or first-order stream channels in the network often begin as outflows from snowfields or springs below porous substrata forming ridges dividing one catchment from another. Two first-order streams coalesce to form a second-order channel and so on to create the network (Strahler 1963). A very large river, like the Amazon, often has several large tributaries and each of those river tributaries may be fed by several to many smaller streams (Fig. 1.1). Thus, each large catchment basin has several or many subcatchments.

Erosive power generally increases with stream size. Boulders, gravel, sand, and silt are transported from one reach of the stream to the next in relation to discharge and valley geomorphometry (e.g., slope and relative resistance of substrata to erosion). Expansive deposition zones (floodplains) form between steep canyons where downcutting predominates.

All rivers feature this basic theme of alternating cut and fill alluviation. Floodplains occur like beads on a string between gradient breaks or transitions in the elevational profile of the flow pathway (Leopold *et al.* 1964). Rivers of very old geologic age have exhausted much of their erosive power; mountains are rounded, valleys are broadly U-shaped, and river channels are simple and straight, even on large alluvial floodplains. Whereas in geologically young, recently uplifted catchments, stream power and associated erosive influence on valley form is great, mountains are steep-sided, valleys narrowly V-shaped, and stream channels on alluvial floodplains are complex. Of course, no two rivers are exactly alike, but the general

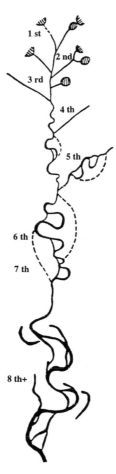

FIGURE 1.1 Hypothetical landscape view of the stream continuum of a large catchment basin. Strahler stream orders and a hierarchy of stream reaches are shown, as the river flows from headwater sources (snowfields or groundwater discharge zones in porous substrata) at high altitude to the coastal plain near oceanic confluence (modified from Stanford and Ward 1992).

longitudinal pattern of cut and fill alluviation more or less characterizes every catchment basin in the world.

The landscape at any point within the stream network is four-dimensional (Fig. 1.2). The river continuum from headwaters to ocean is the longitudinal (upstream to downstream) dimension. The second dimension is transition from the river channel laterally into the terrestrial environment of the valley uplands (aquatic to terrestrial dimension). Except where rivers

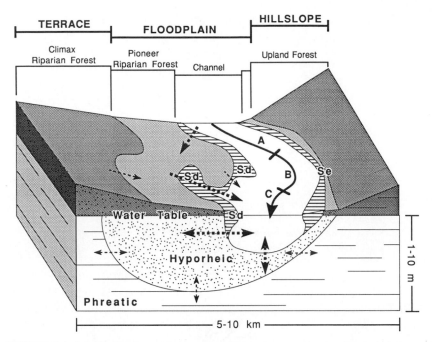

FIGURE 1.2 Major landscape features of a fifth-order montane floodplain (see Fig. 1.1), showing the three primary spatial dimensions (lateral, longitudinal or elevational, and vertical) that are dynamically molded through time (the fourth dimension) by fluvial processes. Biota may reside in all three spatial dimensions: riparos (streamside or riparian), benthos (channel), hyporheos (river influenced ground water), and phreatos (true ground water). The hatched area is the varial zone or the area of the channel that is periodically dewatered as a consequence of the average amplitude of the discharge regime. Major channel features include a run (A), riffle (B), and pool (C); Sd refers to sites of sediment deposition and Se refers to a major site of bank erosion. The heavy solid line is the thalweg and broken lines conceptualize circulation of water between benthic, hyporheic, and phreatic habitats (modified from Stanford and Ward 1992).

flow over impervious bedrock, some amount of porous alluvium is present within the channel owing to erosion at points upstream. Hence, water from the river may penetrate deeply into the substrata of the river bottom. Moreover, substrata of floodplains are composed of alluvial gravels and/ or sands and silts, which allows lateral flow of river water. Thus, interstitial flow pathways constitute a vertical dimension in the river channel and on the floodplains. All of the physical dimensions change in size over time (the fourth dimension), as floods and droughts alter hydrology, sediment transport, and distribution of vegetation and other biota (Ward 1989).

Plants and animals are distributed in relation to biophysical gradients expressed by the four-dimensional nature of the stream network within catchment basins. For example, certain species of aquatic insects reside only in the cold, rocky environs of cascading headwater streams in the high mountains (*rhithron* environments), whereas other species are found only in the much warmer waters of the often sandy, turbid, and meandering reaches of the lowlands near the ocean (*potamon* environments) (cf. Ward 1986). Hence, riverine biota have distinct preferences for specific environmental conditions that are optimal only at certain locations within longitudinal (upstream–downstream), lateral (aquatic–terrestrial), vertical (surface–ground water), and temporal (certain time) gradients that characterize lotic ecosystems (Fig. 1.2). Andrewartha and Birch (1954) observed that the essence of ecology is understanding the distribution and abundance of biota. Because environmental conditions at any point in a stream are continuously influenced by conditions at points upstream, biophysical controls on distribution and abundance of riverine biota must be examined in the context of the stream and its landscape setting (Hynes 1975).

To bring this point home, again consider the Amazon. This great river has existed for millions of years, allowing its biota to evolve highly specialized life histories and morphologies in response to long-term dynamics of the river environment. Seemingly countless aquatic and semiaquatic species coexist, each trying, and variously succeeding, to grow and reproduce in accordance with evolved life history traits and within the myriad of environmental gradients expressed by the dynamic course of the river through the massive catchment basin. For example, the adaptive radiation of Amazonian fishes is astounding, ranging from deep-water specialists that reside in the dark depths of the scoured channels to species that reproduce exclusively in the floodplain forests during floods (Lowe-McConnell 1987, Petrere 1991). Perhaps even more profound are interpretations of satellite-derived images that strongly suggest that the enormously complex and highly evolved rain forests of the Amazon Basin are composed of a mosaic of successional stages created by the river snaking its way back and forth across this huge landscape century after century (Colinvaux 1985, Salo *et al.* 1986).

We can conclude from studies on the Amazon and many other river systems (e.g., see Davies and Walker 1986) that the first task of a river ecologist is to determine the appropriate scale of study to answer the question at hand. Do I need to examine the problem in the context of the entire river continuum from headwaters to the ocean or will a particular reach or even a particular riffle or pool suffice?

Moreover, this dilemma of spatial scale is complicated by the fact that the full range of biophysical features of rivers may change suddenly as a consequence of intense, unusual events like very large floods, extended droughts, catchmentwide fires, earthquakes, volcanic eruptions, and other natural phenomena that may radically change conditions reflected by the long-term norm (Schumm *et al.* 1987). At first inspection many rivers may seem to exist in relatively stable states, especially in situations where a long period of time has passed after a major disturbance event and effects have gradually ameliorated (Schumm and Lichty 1956). In such rivers, channels remain in the same place and floodplains flood in the same general way year after year; biota are distributed rather predictably and highly evolved biotic interactions and habitat specializations are evident (e.g., resource polymorphisms: see Skulasson and Smith (1995)). In other landscapes, especially those that are geologically young and physically dynamic (e.g., glaciated landscapes of the Rocky Mountains), effects of frequent disturbances are reflected at the catchment scale by constantly changing channel positions of floodplains, perching of terraces with human artifacts and/ or relict riparian vegetation high above downcut channels, and disjunct distributions of biota. An axiom of river ecology is that sooner or later disturbance will reset successional trajectories in the myriad of biophysical gradients within catchment basins and produce significant, unpredictable variation in the distribution and abundance of biota (Reice 1994). Intermediate levels of disturbance tend to promote biodiversity by maintaining habitat complexity in time and space (Ward and Stanford 1983a, Huston 1994). So, what time period needs to be encompassed by a study in order to adequately understand the ecological significance of natural disturbance events?

Natural variation in time and space is superimposed upon environmental change induced by human activities in catchment basins. Flows in all of the larger and most of the smaller rivers in the temperate latitudes of the world now are regulated by dams and diversions (Dynesius and Nilsson 1994). Reduced volume and altered seasonality of flow radically change the natural habitat template, eliminating native species and allowing invasion of nonnatives. In many cases water from dams is discharged from the bottom of reservoirs, drastically changing temperature patterns and armoring the river bottom by flushing gravel and sand and leaving large boulders firmly paving the bottom. Problems related to flow regulation in other cases are exacerbated by pollution and channelization (Petts 1984). A wide variety of other human effects can be listed (Table 1.1). The cumulative effect is the severing or uncoupling of the complex interactive pathways that characterize the four-dimensional nature of rivers (Fig. 1.2).

As a consequence, another landscape-scale consideration in river ecological studies concerns the environmental condition of the catchment basin. How have human activities influenced environmental variables that may affect the distribution and abundance of biota? What time and space scales are involved?

TABLE 1.1
Some Pervasive Human Disturbances that Uncouple Important
Ecological Processes Linking Ecosystem Components in Large River
Basins (from Stanford and Ward 1992)

Stream regulation by dams, diversions, and revetments: *uncouples longitudinal, lateral, and vertical dimensions*
  Lotic reaches replaced by reservoirs: loss of up–downstream continuity
    Migration barrier; flood and nutrient sink; stimulates biophysical constancy in downstream environments
  Channel reconfiguration and simplification: loss of lateral connections
    Removal of woody debris, isolation of riparian and hyporheic components of floodplains
  Transcatchment water diversion: abnormal coupling of catchments
    Dewatering of channels; immigration of exotic species, import of pollutants
Water pollution: *alters flux rates of materials, uncouples food webs*
  Deposition of pollutants from airshed into catchment
    Eutrophication, acidification
  Direct and diffuse sources of waterborne waste materials from catchment
    Toxic responses, eutrophication
  Accelerated erosion related to deforestation and roading
    Sedimentation of stream bottoms, eutrophication
Food web manipulations: *induces strong interactions that alter food webs*
  Harvest of fishes and invertebrates
    Biomass and bioproduction shifts
  Introduction of exotic species
    Cascading trophic effects

The main objectives of this chapter are to provide an approach for determining the basic landscape features of a catchment basin in your region and to use that information to synthesize issues and questions about river ecology in a particular catchment basin that you select for study. Based on your examination of catchment basin features and supporting information, you will develop a list of basic and applied ecological questions about the streams and rivers in that basin. Is the hydrology of the river significantly influenced by groundwater discharge or is it predominately a snowmelt system? Is it likely that pollution from agriculture is influencing water quality in the mainstem river? Is it likely that the distribution and abundance of native fishes is compromised by dams and diversions? Questions like these should become apparent as you study maps and photos of the catchment basin and determine the basic features of the drainage network. Finally, you will have to consider what spatial scales and time periods are required to resolve your list of specific questions. The purpose of this chapter is to provide a landscape context for the other chapters in this book, which teach detailed and more site-specific analyses of river ecological processes and responses.

II. GENERAL DESIGN

The premise of this chapter is that few river research and management questions can be answered without considering landscape or ecosystem attributes and dynamics. So, how does one decide which portions of the catchment basin to include in the analysis? Almost all natural resource management questions have to be addressed in a landscape context owing to overlapping jurisdictions and the interactive nature of ecological processes that provide ecosystem goods and services that humans use. River ecosystems encompass ecological, social, and economic processes (function) that interconnect organisms (structure), including humans, with their environment within a defined catchment landscape and over some specific time period. Ecosystem boundaries are permeable with respect to energy and materials flux and often are best determined by the nature of the ecological issue or question of concern (Stanford and Ward 1992).

An example of a reasonably large catchment basin with a variety of interactive research and management problems in river ecology is the Flathead River–Lake ecosystem in northwestern Montana, USA, and southeastern British Columbia (BC), Canada (Fig. 1.3). It is a large subcatchment of the Columbia River. The catchment area is 22,241 km^2 and encompasses small urban and agricultural lands on the piedmont valley bottom, extensive national (USA) and provincial (BC) forests with forest production and wilderness management zones, and the western half of the Glacier–Waterton International Peace Park. The elevational gradient extends some 3400 m from the highest points on the watershed to Flathead Lake. Water quality in this catchment is among the best in the world but problems related to nutrient pollution and food web change have been documented (see Stanford and Ward 1992, Stanford and Hauer 1992).

For example, in the Flathead River–Lake catchment adult bull charr (*Salvelinus confluentus,* a native salmonid; Fig. 1.4) reside in the lake but migrate upstream during the spring snowmelt period into specific streams and to specific alluvial habitats (ground water upwelling zones) where they spawn. After spawning in September the adults return to the lake.

Eggs hatch after a very specific number of degree days have passed. Hatching success seems to be controlled in part by the presence of fine sediments that in some streams entrain in the spawning nests (redds). The ecology of juveniles is not well documented but it appears that they reside in nursery habitats within natal tributaries for one to several years and

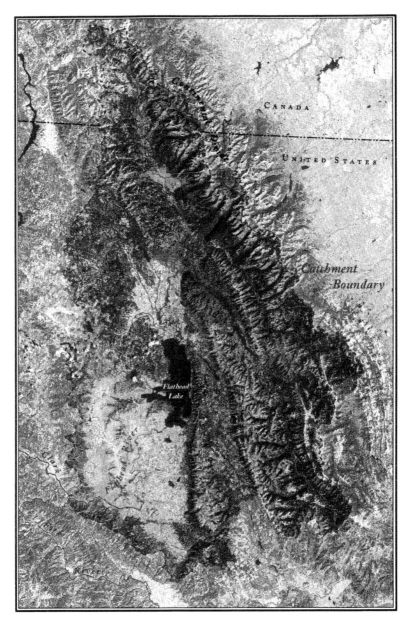

FIGURE 1.3 Topography and drainage network of the Flathead River–Lake catchment generated from spectral data obtained from Landsat satellite images. Data layer one is the topography; data layer two is the hydrography. The layers are superimposed in this figure. Spawning sites of Flathead Lake bull charr are the major tributaries on the Middle Fork and the west side of the North Fork. Note farms and urban areas in the valley bottoms up- and downstream from Flathead Lake and clear cuts on some of the forest lands, especially along the west side of the North Fork, around Hungry Horse Reservoir on the South Fork, and south of Swan Lake. The large areas with no human disturbances include Glacier National Park and the Bob Marshall–Great Bear–Scapegoat Wilderness complex (figure provided by R. Redmond and S. Holloway, Wildlife Spatial Analysis Laboratory, The University of Montana, Missoula).

FIGURE 1.4 An adult bull charr from Flathead Lake (photo, J. Stanford).

then move downstream to the lake to mature and complete their life cycle (Fraley and Shepard 1989). Bull charr in Flathead Lake are piscivorous and historic records from the 1940s and 1950s show that individuals often weighed over 15 kg, but in recent years fish larger than about 6 kg have not been observed in the lake or in the tributaries and most adults are 3 kg or less.

The bull charr is a popular sport fish in the Flathead catchment and in many respects this fish is an indicator of very clean water because throughout their range in the Pacific Northwest they are found only in cold, clean streams and lakes. Population dynamics of Flathead Lake bull charr have been monitored by the Montana Department of Fish, Wildlife, and Parks by counting redds in the tributary streams each fall. Counts were initiated in 1979 and have been done in the same way and on the same tributaries since then. The data clearly show that numbers of redds in the North and Middle Fork tributaries of the catchment are declining, precipitously so since about 1988 (Flathead Basin Commission 1994), and substantially fewer landings are now reported in Flathead Lake.

Part of the problem may be that the Flathead River–Lake system has been fragmented by the construction of hydropower dams. Note the large Hungry Horse Reservoir on the South Fork of the Flathead River in Fig. 1.3; hydropower dams also block migration at the outlet of Flathead Lake

(Kerr Dam) and on the Swan River. Separate bull charr populations now exist in Hungry Horse Reservoir and Swan Lake, apparently derived from juveniles segregated by construction of the dams. However, populations of bull charr naturally segregated by waterfalls apparently have always existed in at least one of the large lakes in the Glacier National Park portion of the catchment (Hauer *et al.* 1980, L. Marnell, Glacier National Park, West Glacier, Montana, unpublished data). In all cases the life history and habitat preferences appear to be virtually identical. In contrast to the Flathead Lake bull charr, numbers of redds and juveniles in the South Fork and the Swan River tributaries are about constant or are increasing (Flathead Basin Commission 1994).

The causes of the Flathead Lake bull charr decline are not clear, but the life history of the fish clearly encompasses a large share of the Flathead catchment. Since bull charr are voracious carnivores that grow to large size, this fish likely has a strong influence on other species in the food web. Species like the bull charr have been called "keystone" species because their behavior strongly influences ecosystem pattern and process and can be used to set ecosystem boundaries (Mills 1993). In this case the bull charr ecosystem clearly encompasses the entire Flathead River–Lake catchment.

The Flathead catchment basin is biophysically complex and human influences are apparent (Fig. 1.3). The elevation gradient is extreme and the stream network is complicated by the presence of glacial lakes and extensive foodplains. Dams have blocked fish passage and radically altered the flow and temperature patterns (Stanford and Hauer 1992); urbanization, farming, logging, and other activities also characterize the system and may be influencing the distribution and abundance of bull charr (Spencer *et al.* 1991). This example should provide a context for examining a catchment basin in your area.

III. SPECIFIC EXERCISES

A. Exercise 1: Boundaries and Hydrography of the Catchment Basin

Catchment boundaries are the ridges that separate a catchment basin from those adjacent. Technically, the catchment boundaries should be termed *watersheds*. However, in the United States *watershed* often is considered synonymous with *catchment basin*. The hydrography (spatial distribution of aquatic habitats) of a catchment basin can be conveniently examined at 1 : 24,000 scale using maps available in the United States from the United States Geological Survey.

1. Using larger scale maps of your area, determine catchment basins of various sizes for a region of at least 10,000 km². Choose one catchment of at least 100 km² area for detailed examination. Using a planimeter, determine the total area of the basin.

2. Note the stream network, shown in blue on most maps. Compare the detail of the catchment on different scale maps. The smallest streams begin at high-elevation snowfields, lakes, wetlands, or springs erupting from groundwater aquifers fed by upslope infiltration of precipitation through porous bedrock. In many cases the smallest stream channels are shown as broken lines, which indicate that surface flow is intermittent.

3. If many intermittent stream channels are shown, the catchment basin is either very dry or the substrata are very porous. In both mesic (wet) and xeric (dry) landscapes, a large amount of the runoff may follow subterranean (groundwater) pathways through porous substrata (see Stanford and Ward 1993). Differentiate intermittent and permanent stream channels in your catchment.

4. An important point to keep in mind is that the drainage network really is a geohydraulic continuum; that is, the stream corridor has both surface and groundwater components and these interactive pathways are hydrologically and ecologically interconnected (Gibert *et al.* 1990, Gibert *et al.* 1994). Water flowing at the surface at one place may be underground at another, depending on the geomorphology of the catchment basin and the volume and timing of rainfall or snowmelt. Hence, interaction zones between surface and ground waters are fundamental attributes of landscapes and are very germane to stream ecological studies. Compare topographic, geologic, and groundwater maps of your catchment and identify potential areas of near-surface ground waters that may be fed by surface waters or discharge into the stream network.

5. Stream order is determined by the coalescence pattern (Figs. 1.1, 1.3). Two first-order streams converge to form a second-order stream, two second-order tributaries form a third-order stream, and so on. Network density is related to geologic origin of the basin, time since uplift, precipitation patterns, precipitation history, types of vegetation present, and resistance of substrata to erosion and infiltration. Lay out a series of maps covering the study catchment at 1:24,000 scale. Overlay the maps with clear plastic or acetate sheets. On the plastic sheets, color code the different stream orders with markers and tabulate them on a data sheet. Measure the length of all streams within the catchment using a map wheel. Simply trace the stream corridor with the map wheel starting at zero and reading the distance on the appropriate scale of the wheel. Calculate drainage density of the catchment as total stream length divided by total area of catchment.

6. Observe the elevational gradient from highest to lowest elevation in the catchment. Carefully consider the density of topographic isopleths (lines of equal elevation). Where they converge closely adjacent to and across the stream channel, canyon segments exist. More widely spaced isopleths indicate flatter topography. Use the acetate sheets overlaying the topographic maps set up in step 5 to locate gradient breaks.

7. Identify canyons (downcutting channels confined by bedrock walls) and alluvial (aggraded, unconfined channels with wide, terraced floodplains) stream segments. In many cases alluvial deposits will be shown by special designations on the maps. Check the map key for such designations. In alluvial zones you may observe that the stream channels begin to braid, which suggests major deposition and floodplain development. On alluvial reaches of bigger streams, the general structure of the floodplains will likely be evident in the form of active zones of flood scour and terraces at higher elevations along the channel. Using elevation data from the topographic maps, plot the stream profile from highest to lowest elevation (x-axis, distance downstream; y-axis, elevation). Label the major gradient breaks and alluvial reaches.

8. Streams may flow into lakes or wetlands. In some cases wetlands may remain where lake basins have filled with sediments. In glaciated landscapes, lakes may occur in high-elevation cirques; larger, often very deep lakes may occur singly or in a series in the glaciated mountain valleys. Many lakes and wetlands are fed and drained by ground water and determination of underground flow pathways may require geohydrologic surveys. Consult Wetzel (1983) for detailed descriptions of the types of lakes and modes of origin. The main point here is to note the position and potential influence of lakes on the stream network of your catchment basin. Lakes function as sinks for fluvial sediments, nutrients, and heat. Streams flowing from lakes may well be very different than inflowing streams. Man-made reservoirs function in similar fashion, except that ecological influences on rivers below the dams will depend on the depth and mode of water release from the dams (Ward and Stanford 1983b, 1995, Petts 1984). Tabulate lakes and reservoirs in your catchments, noting elevation, area, and other available data (e.g., volume, flushing rate).

B. Exercise 2: Other Landscape Attributes of the Catchment Basin

The maps provided likely will show surficial geology, groundwater resources, broad vegetation categories, precipitation patterns, and human infrastructures (roads, pipelines, dams, railroads, urban areas, or individual buildings, etc.). Systematic summarization of these features in relation to the hydrography will provide valuable insights about potential influences on water quantity and quality and constraints on distribution and abundance of riverine biota. For example, an understanding of the general geology of the catchment basin will provide insights into discharge, water chemistry, distribution of biota, and other attributes of the catchment basin that likely will be encountered in field work. Igneous and metamorphic rocks generally do not dissolve much in water and, hence, surface waters draining such formations have low dissolved solids and little buffer capacity, whereas waters from limestone formations generally may be expected to contain high amounts of dissolved solids and be very well buffered.

Land use patterns inferred from the distribution of human infrastructures shown on the maps can be corroborated from aerial photographs. If the photos are available in a time series, changes in hydrography (e.g., channel migration on floodplains) as well as changes in land use patterns can be observed.

1. Using a map wheel and planimeter for the maps (may use digitizing tablet and computer if available) and a stereoscope for aerial photos of known scale, determine the lengths and areas of various features on the landscape of the river catchment you have chosen for study. Create a table or computer spreadsheet in which you can record the different landscape attributes identified in the steps below. Record the features by stream length, area, or other spatial measures. This will provide a basis for a general description of the study catchment and landscape attributes that may influence ecological processes and responses within the stream network.

2. Compare the catchment basins you have identified on the topographic maps with geologic maps of the region. On granitic and other "hard rock" mountains, runoff usually is dominated by surficial flow, whereas limestone and other sedimentary and volcanic formations may allow considerable infiltration and runoff may predominately follow groundwater pathways to portals back to the surface at lower elevations. Subsurface drainage networks dominate in karst (cavernous limestone) landscapes (see Mangin 1994). Tabulate the major geologic formations by type and percentage of catchment basin area. Use a planimeter to determine areas of different geologic formations.

3. Determine vegetation cover patterns within the catchment basin. At a minimum the topographic maps should show forest or grassland areas in green and exposed bedrock or other nonvegetated (e.g., clear cut forest stands) areas in white. Glaciers and wetlands likely will have special designations shown in the map key. Vegetation maps of your catchment may be available or you may be able to use aerial photos to determine the general pattern in comparison to the topographic maps. Using all available maps and photos determine at a minimum riparian (stream side), wetland, and

upland (forest and grassland) ground cover for the entire catchment basin. For montane regions it is instructive to differentiate forest types with respect to elevation (e.g., riparian, upland forest, subalpine forest, alpine). Again use the planimeter to determine areas of cover types and record percentage of basin area by type.

4. Examine the stream corridors on the topographic maps for features created by human activity, such as revetments, bridges, irrigation diversions or returns, mines, and other industrial sites. All of these may change flow patterns or otherwise influence the natural attributes of the stream corridor. On the acetate sheets, color code stream segments by type of alteration or land use. Tabulate percentage of stream corridor and/or catchment basin potentially influenced.

5. If aerial photos are available, verify all the features you have identified in the catchment basin(s) from interpretation of maps. Add notes for features more evident in the photos, such as riparian forests or stream channels. Keep in mind that the maps and photos may have been produced on very different dates and therefore show differences in the landscape features.

6. Note any discharge or precipitation gauging stations in your catchment basin. These are sometimes included on topographic maps. Prepare time series plots of available data for these stations and calculate unit area precipitation and runoff. Determination of stream flow is discussed in more detail in Chapter 3, but knowing stream flow dynamics at various points in the stream network will provide a more complete view of the catchment landscape as derived in this chapter from maps and photos.

C. Exercise 3: Computerized Spatial Analysis of Landscapes

While maps and photos are basic tools for understanding how your study basin fits into the regional landscape, digital approaches provide a means for examining landscapes in great detail. All points in any landscape can be precisely known from geodetic surveys. In fact, that is how the topographic maps used above were created. With the aid of a computer, topography can be reduced to a digital data base using algorithms that interpolate between surveyed points. Using software that is widely available, the computer operator can produce three-dimensional images of any digitized landscape. Topographic data can then be examined statistically or plotted in relation to any other spatial data bases (e.g., stream network, water quality, fish distribution).

A number of software packages that manipulate digitized data in relation to geographic references are available under the general descriptor of geographic information systems (GIS). Considerable computer sophistication is required to use a GIS properly, although most can be run on high-speed personal computers. The advantage of a GIS is that landscape data for many variables can be created in "layers" mathematically superimposed in relation to the topography (Fig. 1.3). This is a very useful way to accurately keep track of and display landscape change over time (e.g., see Johnston and Naiman 1990). For example, observed fish distributions within a catchment basin can be plotted in true spatial (geographic) context with the hydrography and, if time series data are available, changing fish distributions can be shown in spatial relation to changes in potentially controlling variables, such as land use activities. Hence, a GIS permits very large data sets to be systematically arrayed and related in time and space in a manner that facilitates interpretation of landscape pattern and process (Goodchild *et al.* 1993, Sample 1994).

Moreover, data describing landscape patterns in some cases can be derived from spectral (reflectance) data gathered from satellite or other "remote" sensors. In this case a GIS is essential to relate massive amounts of spectral data for entire landscapes to the actual topography. Different wavelengths of light are reflected by the pattern of landscape attributes on the ground. Hence, algorithms or statistical models that relate measured spatial variation for a portion of the landscape (ground truth data) to the variation in the spectral patterns recorded remotely can be derived. The algorithms then can be used to generate landscape data layers in direct relation to the topography (Barrett and Curtis 1992, Lillesand and Klefer 1993). Obviously, some landscape variables are better suited to spectral imagery than others and considerable ground truthing is needed to verify the accuracy of the remotely sensed data. For example, water bodies are easily distinguished from terrestrial environs; coniferous forests can be distinguished from glasslands. However, this technology is in a rapid state of development and should be approached with caution and a clear understanding of the research or management question.

Most research universities have spatial analysis laboratories. If a GIS is not available to demonstrate utility in landscape analysis of catchment basins, I recommend that an active spatial analysis lab be toured to clearly convey the usefulness of this technology in demonstrating pattern and process at the level of entire catchment basins.

D. Exercise 4: Identifying Ecosystem Problems at the Landscape Scale

Now that you have summarized landscape features of your catchment basins, it is important to consider what sorts of questions or problems require resolution at a landscape scale. Almost all natural resource management questions have to be addressed to some extent in a landscape context, owing to overlapping jurisdictions. For example, in the Flathead catchment described above, nearly all federal (U.S.) land management agencies (e.g., Forest Service, Environmental Protection Agency, Bureau of Reclamation, National Park Service) and a wide variety of state and tribal agencies have legislated authority for water resources. Without a landscape perspective, one agency could easily initiate a management objective that interferes with actions of another agency. Moreover, the ability of ecosystems to provide ecological goods and services (e.g., water, timber, wildlife, scenery) to humans clearly encompasses local to regional landscapes.

However, to resolve specific problems, is it necessary to study the entire catchment? If so, how big should the catchment be? Are catchment boundaries also ecosystem boundaries? All are difficult questions. Properly scaling the research and management approach perhaps is the most difficult task faced by any ecologist. However, thorough synthesis of available information about the particular problem in a landscape context is the place to start.

In the case of the Flathead Lake bull charr, the entire catchment of the Flathead River upstream from Kerr Dam (Fig. 1.3) clearly is the ecosystem that sustains this fish. In this case the catchment boundaries are the ecosystem boundaries in this very large and complex landscape. Determining the causes and consequences of the bull charr decline and implementation of a solution to sustain the fishery must involve an understanding of the habitat requirements of the various life history stages of the fish as well as a predictive understanding of the complex biophysical processes that control the quantity and quality of those habitats.

River ecosystems encompass ecological, social, and economic processes (ecosystem functions) that interconnect organisms (ecosystem structure), including humans, over some time period. The ecosystem boundaries are permeable with respect to energy and materials flux; therefore, even large systems are influenced by external events such as global climate change, pollution, national and global economies, and emigration of people and nonnative biota. This holistic view should be kept in mind as approaches to resolution of river ecological questions are considered.

1. Using the Flathead bull charr example as a general guide, list a series of river ecological questions that may be inferred from the landscape attributes of your catchment basins. For example, if your catchment is dominated by agricultural lands, what sorts of river problems might you expect?

2. Determine where in your catchment basin you would place monitoring or study sites to assemble river ecological information to solve your list of problems.

IV. QUESTIONS

A. Boundaries and Hydrography of the Catchment Basin

1. Based on your measures of the stream network, does it appear that your catchment basin is very dry or very wet?

2. How does "wetness" of the catchment basin relate to the density of the stream network?

3. Is the stream network "fragmented" in any way by natural obstructions, such as landslides, lakes, wetlands, beaver dams, or interstitial flow pathways, or is the stream profile a continuous surface flowpath?

4. Are alluvial floodplains a dominant feature of the stream corridor?

5. Is discharge from the catchment likely to be significantly controlled by reservoir storage and/or dam operations?

B. Other Landscape Attributes of the Catchment Basin

1. Is the bedrock substrata of your catchment basin likely to be very porous or will most of the precipitation run off via surficial channels?

2. Is the water chemistry of the river system likely to be well buffered or poorly buffered?

3. Is your catchment basin human dominated?

4. What man-made structures are present in the catchment basin that might change flow and channel form or obstruct migrations of biota?

5. What stream channels appear to have a closed canopy as a consequence of dense riparian forest and how might canopy cover influence ecological processes in the stream?

6. What is the predominate vegetation type in the catchment basin and how might it influence river ecology?

7. What is the predominate human land-use activity in the catchment basin and how will that activity likely affect river ecology?

C. Computerized Spatial Analyses of Landscapes

1. What sorts of problems in river ecology do not require use of a computerized GIS system?

2. What problems in river ecology might usefully be analyzed by a computerized GIS in your catchment?

3. How can a GIS be used to document landscape changes in your catchment basin?

D. Identifying Ecosystem Problems at the Landscape Scale

1. Have you included all landscape units in your catchment basin that may exert significant ecological influences on the river system?

2. Can you observe fragmentation in geohydraulic continuum caused

by human activities and, if so, how do they relate to your list of research and management problems?

4. Are your catchment boundaries also ecosystem boundaries with respect to your list of river ecological problems?

5. Based on the information about bull charr and the landscape attributes of the Flathead catchment described, answer the following questions:

    a. What processes likely are influencing the distribution and abundance of bull charr in the Flathead River–Lake catchment?

    b. State these likely processes in the form of hypotheses about the decline of Flathead Lake bull charr.

    c. What landscape information is needed to test these hypotheses?

    d. What other ecological information and data are needed to understand better the bull charr problem and in what time frame should these data be collected?

V. MATERIALS AND SUPPLIES

Aerial photo series, preferably stereo pairs
Geologic maps
GIS demonstration
Groundwater maps
Map measurement wheels
Planimeters or digitizing pads and computers
Plastic overlaps and color markers
Stereoscopes for photo interpretation
Topographic maps of the study region at various scales
Vegetation maps

REFERENCES

Andrewartha, H. G., and L. C. Birch. 1954. *The Distribution and Abundance of Animals.* Univ. of Chicago Press, Chicago, IL.

Barrett, E. C., and L. C. Curtis. 1992. *Introduction to Environmental Remote Sensing.* Chapman and Hall, New York, NY.

Colinvaux, P. A., M. C. Miller, K. Liu, M. Steinitz-Kannan, and I. Frost. 1985. Discovery of permanent Amazon lakes and hydraulic disturbance in the upper Amazon Basin. *Nature* **313:**42–45.

Davies, B. R., and K. F. Walker (Eds.). 1986. *The Ecology of River Systems.* Dr. W. Junk, Dordrecht, The Netherlands.

Day, J. A., and B. R. Davies. 1986. The Amazon River system. Pages 289–318 *in*

B. R. Davies and K. F. Walker (Eds.) *The Ecology of River Systems.* Dr. W. Junk, Dordrecht, The Netherlands.

Dynesius, M., and C. Nilsson. 1994. Fragmentation and flow regulation of river systems in the northern third of the world. *Science* **266:**753–762.

Flathead Basin Commission. 1994. *1993–1994 Biennial Report.* Flathead Basin Commission, Kalispell, MT.

Fraley, J. J., and B. B. Shepard. 1989. Life history, ecology and population status of migratory bull trout (*Salvelinus confluentus*) in the Flathead Lake and River system, Montana. *Northwest Science* **63:**133–143.

Goodchild, M. F., B. O. Parks, and L. T. Steyaert (Eds.). 1993. *Environmental Modeling with GIS.* Oxford Univ. Press, Oxford, UK.

Gibert, J., M. J. Dole-Olivier, P. Marmonier, and P. Vervier. 1990. Surface water–groundwater ecotones. Pages 199–255 *in* R. J. Naiman and H. Decamps (Eds.) *Ecology and Management of Aquatic–Terrestrial Ecotones.* Parthenon, London, UK.

Gibert, J., J. A. Stanford, M. J. Dole-Olivier, and J. V. Ward. 1994. Basic attributes of groundwater ecosystems and prospects for research. Pages 7–40 *in* J. Gibert, D. L. Danielopol, and J. A. Stanford (Eds.) *Groundwater Ecology.* Academic Press, San Diego, CA.

Hauer, F. R., E. G. Zimmerman, and J. A. Stanford. 1980. Preliminary investigations of distributional relationship of aquatic insects and genetic variation of a fish population in the Kintla Drainage, Glacier National Park, Proc. 2nd Symp. Res. National Parks. *American Institute of Biological Sciences* **2:**71–84.

Huston, M. A. 1994. *Biological Diversity: The Coexistence of Species on Changing Landscapes.* Cambridge Univ. Press, Cambridge, UK.

Hynes, H. B. N. 1975. The stream and its valley. *Verhandlungen der Internationalen Vereinigung für Theorestische und Angewandte Limnologie* **19:**1–15.

Johnston, C. A., and R. J. Naiman. 1990. The use of a geographic information system to analyze long-term landscape alteration by beaver. *Landscape Ecology* **4:**5–19.

Leopold, L. B., M. G. Wolman, and J. P. Miller. 1964. *Fluvial Processes in Geomorphology.* Freeman, San Francisco, CA.

Lillesand, T. M., and R. W. Klefer. 1993. *Remote Sensing and Image Interpretation.* Wiley, London, UK.

Lowe-McConnell, R. H. 1987. *Ecological Studies in Tropical Fish Communities.* Cambridge Univ. Press, Cambridge, UK.

Mangin, A. 1994. Karst hydrogeology. Pages 43–67 *in* J. Gibert, D. L. Danielopol, and J. A. Stanford (Eds.) *Groundwater Ecology.* Academic Press, San Diego, CA.

Mills, L. S., M. E. Soulé, and D. F. Doak. 1993. The keystone-species concept in ecology and conservation. *BioScience* **43:**219–224.

Petrere, M., Jr. 1991. Life strategies of some long-distance migratory catfish in relation to hydroelectric dams in the Amazon Basin. *Biological Conservation* **55:**339–345.

Petts, G. E. 1984. *Impounded Rivers: Perspectives for Ecological Management.* Wiley, Chichester, UK.

Reice, S. R. 1994. Nonequilibrium determinants of biological community structure. *American Scientist* **82:**424–435.

Salo, J., R. Kalliola, I. Häkkinen, Y. Mäkinen, P. Niemelä, M. Puhakka, and P. D. Coley. 1986. River dynamics and the diversity of Amazon lowland forest. *Nature* **322:**254–258.

Sample, V. A. (Ed.). 1994. *Remote Sensing and GIS in Ecosystem Management.* Island Press, Washington, DC.

Schumm, S. A., and R. W. Lichty. 1956. Time, space and causality in geomorphology. *American Journal of Science* **265:**110–119.

Schumm, S. A., M. P. Mosley, and W. E. Weaver. 1987. *Experimental Fluvial Geomorphology.* Wiley Interscience, New York, NY.

Skulasson, S., and T. B. Smith. 1995. Resource polymorphisms in vertebrates. *Trends in Ecology and Evolution* **10:**366–370.

Spencer, C. N., B. R. McClelland, and J. A. Stanford. 1991. Shrimp stocking, salmon collapse, and eagle displacement: Cascading interactions in the food web of a large aquatic ecosystem. *BioScience* **41:**14–21.

Stanford, J. A., and F. R. Hauer. 1992. Mitigating the impacts of stream and lake regulation in the Flathead River catchment, Montana, USA: An ecosystem perspective. Aquatic Conservation: *Marine and Freshwater Ecosystems* **2:**35–63.

Stanford, J. A., and J. V. Ward. 1992. Management of aquatic resources in large catchments: Recognizing interactions between ecosystem connectivity and environmental disturbance. Pages 91–124 *in* R. J. Naiman (Ed.) *Watershed Management.* Springer-Verlag, New York, NY.

Stanford, J. A., and J. V. Ward. 1993. An ecosystem perspective of alluvial rivers: Connectivity and the hyporheic corridor. *Journal of the North American Benthological Society* **12:**48–60.

Strahler, A. N. 1963. *The Earth Sciences.* Harper & Row, New York, NY.

Ward, J. V. 1986. Altitudinal zonation in a Rocky Mountain stream. *Archiv für Hydrobiologie Supplementband* **74:**133–199.

Ward, J. V. 1989. The four-dimensional nature of lotic ecosystems. *Journal of the North American Benthological Society* **8:**2–8.

Ward, J. V., and J. A. Stanford. 1983a. The intermediate disturbance hypothesis. An explanation for biotic diversity patterns in lotic ecosystems. Pages 347–356 *in* T. D. Fontaine and S. M. Bartell (Eds.) *Dynamics of Lotic Ecosystems.* Ann Arbor Scientific, Ann Arbor, MI.

Ward, J. V., and J. A. Stanford. 1983b. The serial discontinuity concept of lotic ecosystems. Pages 29–42 *in* T. D. Fontaine and S. M. Bartell (Eds.) *Dynamics of Lotic Ecosystems.* Ann Arbor Science, Ann Arbor, MI.

Ward, J. V., and J. A. Stanford. 1995. The serial discontinuity concept: Extending the model to floodplain rivers. *Regulated Rivers: Research and Management* **10:**159–168.

Wetzel, R. G. 1983. *Limnology,* 2nd ed. Saunders, Philadelphia, PA.

CHAPTER 2

Valley Segments, Stream Reaches, and Channel Units

PETER A. BISSON* AND
DAVID R. MONTGOMERY†

*Forestry Sciences Laboratory
USDA Forest Service
†Department of Geological Sciences
University of Washington

I. INTRODUCTION

Valley segments, stream reaches, and channel geomorphic units are three hierarchically nested subdivisions of the drainage network (Frissel *et al.* 1986), falling in size between landscapes and watersheds and the parameters usually measured at individual points along the stream network (see Chapters 3 and 4). Within this hierarchy of spatial scales (Fig. 2.1), valley segments, stream reaches, and channel geomorphic units represent the largest physical subdivisions that can be directly altered by human activities. As such it is useful to understand how they respond to anthropogenic disturbance, but to do so requires classification systems and quantitative assessment procedures that permit accurate, repeatable description and convey information about biophysical processes responsible for the development of current geomorphic conditions.

The location of different types of valley segments, stream reaches, and channel geomorphic units within a watershed exerts a powerful influence

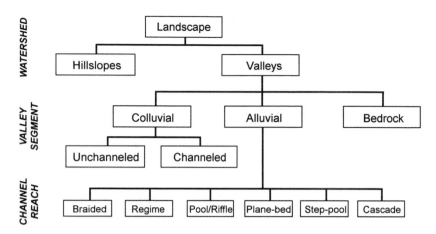

FIGURE 2.1 Hierarchical subdivision of watersheds into valley segments and stream reaches. Redrawn from Montgomery and Buffington (1993).

on the distribution and abundance of aquatic plants and animals by governing the characteristics of water flow and the capacity of streams to store sediment and transform organic material (Hynes 1970, Pennak 1979, Vannote *et al.* 1980, O'Neill *et al.* 1986, Statzner *et al.* 1988). The first biologically based classification systems were proposed for European streams. They were based on zones marked by shifts in dominant aquatic species, such as fishes, from a stream's headwaters to its mouth (Huet 1959, Illies 1961, Hawkes 1975). Recent characterizations of biologically based zones have considered the effects of physical processes and disturbance regimes on changes in faunal assemblages (Zalewski and Naiman 1985, Statzner and Higler 1986). Hydrologists and fluvial geomorphologists, whose objectives for classifying streams often differ from those of stream ecologists, have based classification of stream channels on topographic features of the landscape, substrata characteristics, and changes in patterns of water flow and sediment transport (Leopold *et al.* 1964, Shumm 1977, Richards 1982, Rosgen 1985, Montgomery and Buffington 1993). Other approaches to classifying stream types and channel units have combined hydraulic or geomorphic properties with explicit assessment of the suitability of a channel for certain types of aquatic organisms (Pennak 1971, Bovee and Cochnauer 1977, Binns and Eiserman 1979, Bisson *et al.* 1982, Beschta and Platts 1986, Sullivan *et al.* 1987, Hawkins *et al.* 1993).

There are several reasons why stream ecologists classify and measure valley segments, stream reaches, and channel geomorphic units. The first may simply be to describe physical changes in stream channels over time, whether in response to human impacts or to natural disturbances (Gordon *et al.* 1992). A second reason for stream classification may be to group sampling sites into like physical units for purposes of comparison. This is often desirable when conducting surveys of streams in different drainages. Classification of reach types and channel geomorphic units enables investigators to extrapolate results to other areas with similar features (Hankin and Reeves 1988, Dolloff *et al.* 1993). A third objective for classification may be to determine the suitability of a stream for some type of deliberate channel alteration. Habitat restoration in streams and rivers with histories of environmental degradation is currently being undertaken in many locations, and some restoration procedures may be inappropriate for certain types of stream channels (National Research Council 1992). Successful rehabilitation requires that approaches be consistent with the natural hydraulic and geomorphic conditions of different reach types (Gordon *et al.* 1992) and do not impede disturbance and recovery cycles (Reice 1994). Finally, accurate description of stream reaches and channel geomorphic units often is an important first step in describing the microhabitat requirements of aquatic organisms during their life histories or in studying the ecological processes that influence their distribution and abundance (Hynes 1970, Schlosser 1987).

Geomorphically based stream reach and channel unit classification schemes are relatively new and still undergoing refinement. Stream ecologists will do well to heed the advice of Balon (1982), who cautioned that nomenclature itself is less important than detailed descriptions of the meanings given to terms. Thus it is important for investigators to be as precise as possible when describing what is meant by the terms of the classification scheme they have chosen. Although a number of stream reach and channel unit classification systems have been put forward, none has yet been universally accepted. In this chapter we will focus on two recently proposed classification schemes that can provide stream ecologists with useful tools for characterizing aquatic habitat at intermediate landscape scales: Montgomery and Buffington (1993) for valley segments and stream reaches, and Hawkins *et al.* (1993) for channel geomorphic units. Both systems are based on hierarchies of topographic and fluvial characteristics, and both employ descriptors that are measurable and ecologically relevant. The Montgomery and Buffington (1993) classification provides a geomorphic, process-oriented method of identifying valley segments and stream reaches, while the Hawkins *et al.* (1993) classification deals with identification and measurement of different types of channel units within a given reach. The chapter begins with a laboratory examination of maps and photographs for preliminary identification of valley segments and stream reaches, and concludes with a field survey of channel geomorphic units in one or more reach types.

A. Valley Segment Classification

Hillslopes and valleys are the principal topographic subdivisions of watersheds. Valleys are areas of the landscape where water converges and where the products of erosion, sediment, and organic debris are concentrated. Valley segments are distinctive sections of the valley network that possess geomorphic properties and hydrological transport characteristics that distinguish them from adjacent segments. Montgomery and Buffington (1993) identified three terrestrial valley segment types: *colluvial, alluvial,* and *bedrock* (Fig. 2.1), although they acknowledged that a fourth type, estuarine valleys, were important transition zones between terrestrial and marine environments. Colluvial valleys were subdivided into those with and without recognizable stream channels.

Valley segment classification describes valley form based on dominant types of sediment input and transport processes. The term sediment here includes both large and small inorganic particles eroded from hillslopes. Valleys can be filled primarily with *colluvium* (sediment and organic matter delivered to the valley floor by mass wasting (landslides) from adjacent hillslopes), which is usually immobile except during rare hydrologic events, or *alluvium* (sediment transported along the valley floor by streamflow) which may be frequently moved by the stream system. A third condition includes valleys that have little soil but instead are dominated by bedrock. Valley segments distinguish portions of the valley system in which sediment inputs and outputs are transport- or supply-limited (Fig. 2.2). In transport-limited valley segments, the amount of sediment in the valley floor and its movements are controlled primarily by the frequency of high streamflows and debris flows capable of mobilizing material in the streambed. In supply-limited valley segments, sediment movements are controlled primarily by the amount of sediment delivered to the segment by inflowing water. Valley segment classification does not allow forecasting of how

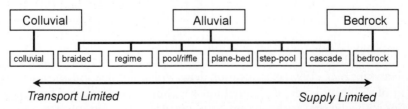

FIGURE 2.2 Arrangement of valley segment and stream reach types according to whether their substrates are limited by the supply of sediment from adjacent hillslopes or by the fluvial transport of sediment from upstream sources. Redrawn from Montgomery and Buffington (1993).

the characteristics of the valley will change in response to altered discharge or sediment supply. Reach classification, according to Montgomery and Buffington (1993), is more useful for characterizing responses to such changes.

Colluvial Valleys   Colluvial valleys serve as temporary repositories for sediment and organic matter eroded from surrounding hillslopes. In colluvial valleys, fluvial (waterborne) transport is relatively ineffective at removing materials deposited on the valley floor. These materials gradually accumulate in steep headwater valleys until they are periodically flushed by debris flows (rapidly moving slurries of water, sediment, and organic debris) or in low-gradient landscapes by periodic expansion of the alluvial channel network during episodes of very high discharge. After removal of sediment by large hydrologic disturbances, refilling processes in colluvial valleys begin again (Dietrich *et al.* 1986).

Unchanneled colluvial valleys are headwater valley segments lacking recognizable stream channels. They possess soils derived by erosional processes from adjacent hillslopes, a property which distinguishes them from steep headwater valleys of exposed bedrock (Montgomery and Buffington 1993). The depth of colluvial deposits in unchanneled colluvial valleys is related to the rate at which material is eroded from hillslopes and the time since the last major disturbances that emptied them. The cyclic process of emptying and refilling occurs at different rates in different geoclimatic regions and depends on patterns of precipitation, geological conditions, and the nature of hillslope vegetation (Dietrich *et al.* 1986). Although unchanneled colluvial valleys do not possess defined streams, they are areas where water is concentrated and have sometimes been called "zero-order" stream basins (Montgomery and Dietrich 1988). Seasonally flowing seeps and small springs serving as temporary habitat for some aquatic organisms may be present in these areas.

Channeled colluvial valleys contain low-order streams immediately downslope from unchanneled colluvial valleys. Channeled colluvial valleys may form the uppermost segments of the valley network in landscapes of low relief or they may occur where small tributaries cross floodplains of larger streams. Flow in colluvial channels tends to be shallow and ephemeral or intermittent. Because shear stresses (see Chapter 4) generated by streamflows are incapable of substantially moving and sorting deposited colluvium, channels in these valley segments tend to be characterized by a wide variety of substrata and organic matter particle sizes. Episodic scour of channeled colluvial valleys by debris flows often governs the degree of channel incision in steep terrain, and like unchanneled colluvial valleys there are cyclic patterns of sediment emptying that periodically reset the depth of colluvium. The frequency of sediment-mobilizing discharge or debris flows determines the amount of sediment stored in colluvial valleys.

Alluvial Valleys   Alluvial valleys are supplied with sediment from upstream sources, and the streams within them are capable of moving and sorting the sediments at erratic intervals. The sediment transport capacity of an alluvial valley is insufficient to scour the valley floor to bedrock, resulting in an accumulation of valley fill primarily of fluvial origin. Alluvial valleys are the most common type of valley segment in many landscapes and usually contian streams of greatest interest to aquatic ecologists. They may be *confined,* a condition in which the hillslopes narrowly constrain the valley floor with little or no floodplain development, or *unconfined,* with a developed floodplain. A variety of stream reach types (Fig. 2.1) may be associated with alluvial valleys, depending on the degree of confinement, gradient, local geology and erosional processes, and discharge regime.

Bedrock Valleys   Bedrock valleys have little valley fill material and usually possess confined channels lacking an alluvial bed. Montgomery and Buffington (1993) distinguish two types of bedrock valleys: those sufficiently steep to remain permanently bedrock floored and those associated with low order streams recently excavated to bedrock by debris flows. Bedrock channels in shallow gradient valley segments indicate that streams have enough power to maintain a high sediment transport capacity.

B. Channel Reach Classification

Channel reaches consist of relatively homogeneous associations of topographic features and channel geomorphic units, which distinguish them in certain aspects from adjoining reaches (Table 2.1). Transition zones between adjacent reaches may be gradual or sudden, and exact upstream and downstream reach boundaries may be a matter of some judgment. Colluvial and bedrock valley segments possess only colluvial and bedrock reach types (Table 2.1; Fig. 2.2), respectively, but alluvial valleys can exhibit a variety of reach types. Montgomery and Buffington (1993) hypothesized that reach differentiation in alluvial valleys was related to the supply and characteristics of sediment and to the power of the stream to mobilize its bed. Six alluvial reach types were recognized.

Cascade Reaches   This reach type is characteristic of the steepest alluvial channels. A few small pools may be present in cascade reaches, but the majority of flowing water tumbles over and around boulders and large woody debris. Waterfalls ("hydraulic jumps") of various sizes are abundant in cascade reaches. The large size of particles relative to water depth effec-

TABLE 2.1
Characteristics of Different Types of Stream Reaches (modified from Montgomery and Buffington 1993)

	Colluvial	Bedrock	Cascade	Step-pool	Plane-bed	Pool-riffle	Regime	Braided
Predominant bed material	Variable	Bedrock	Boulder	Cobble/boulder	Gravel/cobble	Gravel	Sand	Variable
Bedform pattern	Variable	Variable	None	Vertically oscillatory	None	Laterally oscillatory	Multilayered	Laterally oscillatory
Dominant roughness elements	Boulders, large woody debris	Streambed, banks	Boulders, banks	Bedforms (steps, pools) boulders, large woody debris, banks	Boulders and cobbles, banks	Bedforms (bars, pools) boulders and cobbles, large woody debris, sinuosity, banks	Sinuosity, bedforms (dunes, ripples, bars), banks	Bedforms (bars, pools)
Dominant sediment sources	Hillslope, debris flows	Fluvial, hillslope, debris flows	Fluvial, hillslope, debris flows	Fluvial, hillslope, debris flows	Fluvial, bank erosion, debris flows	Fluvial, bank erosion, inactive channels, debris flows	Fluvial, bank erosion, inactive channels	Fluvial, bank erosion, debris flows
Typical slope (%)	>20	Variable	8–30	4–8	1–4	0.1–2	<0.1	<3
Typical confinement	Strongly confined	Strongly confined	Strongly confined	Moderately confined	Variable	Unconfined	Unconfined	Unconfined
Pool spacing (channel widths)	Variable	Variable	<1	1–4	None	5–7	5–7	Variable

tively prevents substrata mobilization during typical flows. Although cascade reaches may experience debris flows, sediment movement is predominantly fluvial. The cascading nature of water movement in this reach type is usually sufficient to remove all but the largest particles of sediment (cobbles and boulders) and organic matter. What little fine sediment and organic matter occurs in cascade reaches remains trapped behind boulders and logs, or is stored in a few pockets where reduced velocity and turbulence permit deposition. The rapid flushing of fine sediment from cascade reaches during moderate to high flows suggests that transport from this reach type is limited by the supply of sediment recruited from upstream sources (Fig. 2.2).

Step-Pool Reaches Step-pool reaches possess discrete channel-spanning accumulations of boulders and logs that form a series of steps alternating with pools containing finer substrata. The diameter of the structures anchoring each step usually equals or exceeds bankfull flow depths (Montgomery and Buffington 1993). Step-pool reaches tend to be straight and have high gradients, coarse substrata, and small width to depth ratios. Pools and alternating bands of channel-spanning flow obstructions typically occur at a spacing of every one to four channel widths in step-pool reaches, although step spacing increases with decreasing channel slope (Grant *et al.* 1990). A low supply of sediment, steep gradient, infrequent flows capable of mobilizing coarse streambed material, and a heterogeneous substrata composition appear to promote the development of this reach type.

The capacity of step-pool reaches to temporarily store fine sediment and organic matter generally exceeds the sediment storage capacity of cascade reaches. Flow thresholds necessary to transport sediment and mobilize channel substrata are complex in step-pool reaches (Montgomery and Buffington 1993). Large bed-forming structures (boulders and large woody debris) are relatively stable and move only during extreme flows. In very high streamflows the channel may lose its stepped profile, but step-pool morphology becomes reestablished during the falling limb of the hydrograph (see Chapter 3, Whittaker 1987). During more frequent bankfull flow periods, fine sediment and organic matter in pools is transported over the large, stable bed-forming steps.

Plane-Bed Reaches Plane-bed stream reaches lack a stepped longitudinal profile and instead are characterized by long, relatively straight channels of uniform depth. They are usually intermediate in gradient and bed roughness (the degree to which substrata particles protrude from the streambed and impede water movement) between steep, boulder-dominated cascade and step-pool reaches and the more shallow gradient pool-riffle reaches. At low to moderate flows, plane-bed stream reaches may possess large boulders extending above the water surface, forming mid-channel eddies. However, the absence of channel-spanning structures or significant constrictions by streambanks inhibits the development of pools. Particles in the surface layer of plane-bed reaches are larger than those in subsurface layers and form an armor layer over underlying finer materials (Montgomery and Buffington 1993). This armor layer prevents transport of fine sediments except during periods when flow is sufficient to mobilize armoring particles.

Pool-Riffle Reaches   This reach type is most commonly associated with small to mid-sized streams and is a very prevalent type of reach in alluvial valleys of low to moderate gradient. Pool-riffle reaches tend to possess lower gradients than the three previous reach types and are characterized by an undulating streambed that forms riffles and pools associated with gravel bars. Also, unlike most cascade, step-pool, and plane-bed reaches, the channel shape of pool-riffle reaches is often sinuous and contains a predictable sequence of pools, riffles, and bars in the channel. Pools are topographic depressions in the stream bottom and bars form the high points of the channel. Riffles are located at cross-over areas from pools to bars. At low streamflow, riffles often travel from one side of the exposed channel to the other, although streams with sufficiently large width-to-depth ratios may have braided rather than single channels (Leopold *et al.* 1964). Pool-riffle reaches form naturally in alluvial channels of fine to moderate substrata coarseness (Leopold *et al.* 1964, Yang 1971) with single pool–riffle–bar sequences occurring every five to seven channel widths (Keller and Melhorn 1978). Large woody debris (LWD) anchors the location of pools and creates upstream sediment terraces that form riffles and bars (Lisle 1986, Bisson *et al.* 1987). Streams rich in LWD tend to have erratic and complex channel morphologies (Bryant 1980).

Channel substrata in pool-riffle reaches is mobilized annually during freshets. At bankfull flows, pools and riffles are inundated to such an extent that the channel appears to have a uniform gradient, but local pool–riffle–bar features emerge as flows recede. Movement of bed materials at bankfull flow is sporadic and discontinuous (Montgomery and Buffington 1993). As portions of the surface armor layer are mobilized, finer sediment underneath is flushed, creating pulses of scour and deposition. This process contributes to the patchy nature of pool-riffle reaches, whose streambeds are among the most spatially heterogeneous of all reach types.

Regime Reaches   Regime stream reaches consist of low gradient, meandering channels with predominantly sand substrata, although regime characteristics can occur in streams with gravel or boulder–cobble streambeds. Regime reaches occur in higher order channels within unconstrained valley segments and exhibit less turbulence than reach types with high gradients. Shallow and deep water areas are present and point bars may be present at meander bends. As current velocity increases over the fine-grained substrata of regime reaches, the streambed is molded into a predictable succession of bedforms, from small ripples to a series of large dune-like elevations and depressions (Gordon *et al.* 1992). Sediment movement occurs at all flows and is strongly correlated with discharge. The low gradient, continuous transport of sediment, and presence of ripples and dunes distinguish regime reaches from pool-riffle reaches (Montgomery and Buffington 1993).

Braided Reaches   Braided reaches usually occur in high-order streams and are characterized by numerous gravel and sand bars scattered throughout the channel (Gordon *et al.* 1992). Aside from the wide span of the active channel relative to adjacent unbraided sections of the channel network, braided reaches share many properties of regime reaches: predominantly sand and gravel substrata, easily erodible streambanks, and continuous sediment transport. In braided reaches the locations of bars change frequently and the channel containing the main flow can often move laterally.

C. Channel Geomorphic Unit Classification

Channel geomorphic units, also called channel units or habitat types, are relatively homogeneous areas of the channel that differ in depth, velocity, and substrata characteristics from adjoining areas. The most generally used channel unit terms for small to mid-sized streams are riffles and pools. Individual channel units are created by interactions between flow and roughness elements of the streambed. Definitions of channel units usually apply to conditions at low discharge. At high discharge, channel units are often indistinguishable from one another and their hydraulic properties differ greatly from those at low flows.

Different types of channel units in close proximity to one another provide organisms with a choice of habitat, particularly in small streams possessing considerable physical heterogeneity (Hawkins *et al.* 1993). Channel unit classification is therefore quite useful for developing an understanding of the distribution and abundance of aquatic plants and animals in patchy stream environments. Channel units are known to influence nutrient exchanges (Aumen *et al.* 1990), algal abundance (Tett *et al.* 1978), production of benthic invertebrates (Huryn and Wallace 1987), invertebrate diversity (Hawkins 1984), and the distribution of fishes (Bisson *et al.* 1988, Schlosser 1991). The frequency and location of different types of channel units within a reach can be affected by a variety of disturbances, including anthropogenic disturbances that remove structural roughness elements such as large woody debris (Lisle 1986, Sullivan *et al.* 1987) or impede the ability of a stream to interact naturally with its adjacent riparian zone (Beschta and Platts 1986, Pinay *et al.* 1990). Channel unit classification is therefore useful for understanding the relationships between anthropogenically induced habitat alterations and aquatic organisms.

Hawkins *et al.* (1993) modified an earlier channel unit classification system (Bisson *et al.* 1982) that had proven to have certain deficiencies, including the application of similar terms to dissimilar types of stream habitat. Hawkins *et al.* (1993) proposed a three-tiered system of classification (Fig. 2.3) in which investigators could select the level of habitat resolution appropriate to the question being addressed. The first level distinguished fast water (*riffle*) from slow water (*pool*) units. The second level distinguished turbulent from nonturbulent fast water units and slow water units formed by scour from slow water units formed by dams. The third level of classification further subdivided each type of fast and slow water unit based on unique hydraulic characteristics and the principal kind of habitat-forming structure or process.

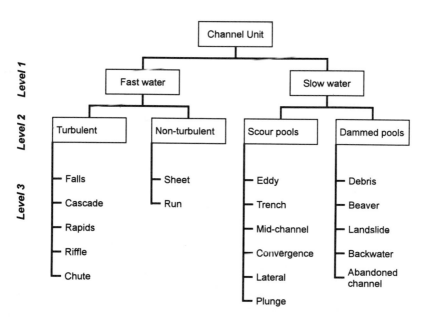

FIGURE 2.3 Hierarchical subdivision of channel units in streams. Redrawn from Hawkins *et al.* (1993).

Turbulent Fast Water Units   The term "fast water" is a relative term that describes current velocities observed at low to moderate flows and is meant only to distinguish this class of channel unit from other units in the same stream with "slow water." Most of the time, but not always, slow water units will be deeper than fast water units at a given discharge. The generic terms riffle and pool are frequently applied to fast and slow water channel units, respectively, although these terms convey limited information about geomorphic or hydraulic characteristics of a stream. Current velocity and depth are the main criteria for separating riffles from pools in low to mid-order stream channels. There are, however, no absolute values of either velocity or depth that can be used to identify riffles and pools, and in some cases the depth and velocity of certain riffles and pools may be the opposite of what is expected.

Hawkins *et al.* (1993) recognized five types of turbulent fast water channel units (Table 2.2). Channel units are classified as turbulent if they possess supercritical flow, i.e., hydraulic jumps sufficient to entrain air bubbles and create localized patches of white water (see Chapter 4). Some turbulence is present in nonturbulent channel units but it is not sufficiently strong to entrain air bubbles, and the appearance of the flow is much more uniform. Turbulent fast water channel units are listed in Table 2.2 in approximate descending order of gradient, bed roughness, current velocity, and abundance of hydraulic steps.

TABLE 2.2

Types of Turbulent and Nonturbulent Fast Water Channel Geomorphic Units and the Relative Rankings of Variables Used to Distinguish Them

	Gradient	Supercritical flow	Bed roughness	Mean velocity	Step development
Turbulent					
Falls	1	n/a	n/a	1	1
Cascade	2	1	1	2	2
Chute	3	2	4	3	5
Rapids	4	3	2	4	3
Riffle	5	4	3	5	4
Nonturbulent					
Sheet	Variable	6	6	6	5
Run	6	5	5	7	5

Note. Rankings are in descending order of magnitude where a rank of 1 denotes the highest value of a particular parameter. Step development is ranked by the abundance and size of hydraulic jumps within a channel unit. From Hawkins *et al.* (1993).

Falls are essentially vertical drops of water over a full spanning flow obstruction and are commonly found in bedrock, cascade, and step-pool stream reaches. *Cascade* channel units consist of a highly turbulent series of short falls and small scour basins, frequently characterized by very large substrata sizes and a stepped longitudinal profile; they are prominent features of bedrock and cascade reach types. *Chute* channel units are typically narrow, steep slots in bedrock. They are common in bedrock reaches and also occur in cascade and step-pool reaches. *Rapids* are moderately steep channel units with coarse substrata, but unlike cascades possess a somewhat planar (vs stepped) longitudinal profile. Rapids are the dominant fast water channel unit of plane-bed stream reaches. *Riffles* are the most common type of turbulent fast water in low gradient (<3%) alluvial channels and are found in plane-bed, pool-riffle, regime, and braided reaches. The substrata of riffles tends to be somewhat finer than that of the other turbulent fast water units and the relative abundance of white water is also reduced.

Nonturbulent Fast Water Units Two types of units were termed nonturbulent by Hawkins *et al.* (1993). *Sheet* channel units are rare in many watersheds but may be common in valley segments dominated by bedrock. Sheets occur where shallow water flows uniformly over smooth bedrock of variable gradient; they may be found in bedrock, cascade, or step-pool reaches. *Run* channel units are fast water units of shallow gradient, typically with substrata ranging in size from sand to cobbles. They are characteristically deeper than riffles and because of their smaller substrata have little if any supercritical flow, giving them a nonturbulent apperance. Runs are common in pool-riffle, regime, and braided stream reaches, i.e., mid- and higher order channels. Their average velocity tends to be lowest of the fast water units (Table 2.2).

Scour Pools There are two general classes of slow water (pool) channel units: pools created by scour that forms a depression in the streambed and pools created by the impoundment of water upstream from an obstruction to flow (Table 2.3). Scour pools can be created when discharge is sufficient to mobilize the substrata at a particular site, while dammed pools can be formed under any flow condition. Hawkins *et al.* (1993) recognized six types of scour pools.

Eddy pools are caused by the scouring action of eddies behind large flow obstructions along the edge of the stream. Eddy pools are located on the downstream side of the structure that caused the eddy and are usually proportional to the size of the obstruction. Eddy pools are often associated with large woody debris along streambanks and can be found in virtually all alluvial reach types.

TABLE 2.3
Characteristics of Slow Water Channel Geomorphic Units

	Location	Longitudinal profile	Cross-sectional profile	Substrate features	Forming constraint
Scour pools					
Eddy	Bank	Middle	Middle	Surface fines, not resistant to scour	Flow obstruction causing lateral deflection
Trench	Thalweg	Uniform	Uniform	Bedrock or sorted, resistant to scour	Bilateral resistance
Mid-channel	Thalweg	Middle	Middle	Sorted, variable resistance to scour	Constriction at upstream end
Convergence	Thalweg	Middle	Middle	Sorted, variable resistance to scour	Convergence of two streams
Lateral	Thalweg	Head or middle	Side	Sorted, variable resistance to scour	Flow obstruction causing lateral deflection
Plunge	Thalweg	Head	Upstream or middle	Sorted, variable resistance to scour	Full-spanning obstruction causing waterfall
Dammed pools					
Debris dam	Thalweg	Tail	Highly variable	Usually sorted, not resistant to scour	Large woody debris dam of fluvial origin
Beaver dam	Thalweg	Tail	Highly variable	Surface fines, not resistant to scour	Beaver dam
Landslide dam	Thalweg	Tail	Highly variable	Often unsorted, variable resistance to scour	Organic and inorganic matter delivered by mass wasting from adjacent hillslope
Backwater	Bank	Tail	Highly variable	Unsorted with surface fines, not resistant to scour	Obstruction at tail impounding water along margin of main channel
Abandoned channel	Floodplain	Highly variable	Highly variable	Unsorted with surface fines, not resistant to scour	Lateral meander bars that isolate an overflow channel from the main channel

Note. Location denotes whether the unit is likely to be associated with the thalweg of the channel (the main part of the flow) or adjacent to a bank. Longitudinal and cross sectional profiles refer to the deepest point in the unit relative to the head, middle, or tail region of the unit. Substrate characteristics refer to the extent of particle sorting (i.e., particle uniformity) and resistance to scour. The channel unit forming constraint describes the feature most likely to cause pooling. Modified from Hawkins *et al.* (1993).

Trench pools, like chutes, are usually located in tightly constrained, bedrock-dominated reaches. They are characteristically U-shaped in cross-sectional profile and possess highly resistant, nearly vertical banks. Trench pools can be among the deepest of the slow water channel units created by scour and their depth tends to be rather uniform throughout much of their length, unlike other scour pool types. Although often deep, trench pools may possess relatively high current velocities.

Mid-channel pools are formed by flow constrictions that focus scour along the main axis of flow in the middle of the stream. Unlike trench pools, mid-channel pools are deepest near the head. This type of slow water channel unit is very common in cascade, step-pool, and pool-riffle reaches. Flow constriction may be caused by laterally confined, hardened banks (bridge abutments are good examples), or by large flow obstructions such as boulders or woody debris, but an essential feature of mid-channel pools is that the direction of water movement around an obstruction is not diverted toward an opposite bank.

Convergence pools result from the confluence of two streams of some-what similar size. In many respects convergence pools resemble mid-channel pools except that there are two main water entry points, which may result in a pattern of substrata particle sorting in which fines are deposited near the head of the pool in the space between the two inflowing channels. Convergence pools can occur in any type of alluvial stream reach.

Lateral scour pools occur where the channel encounters a resistant streambank or other flow obstruction near the edge of the stream. Typical obstructions include bedrock outcrops, boulders, large woody debris, or gravel bars. Many lateral scour pools form next to or under large, relatively immovable structures such as accumulations of large woody debris or along a streambank that has been armored with rip-rap or other material that resists lateral channel migration. Water is deepest adjacent to the stream-bank containing the flow obstruction and shallowest next to the opposite bank. Lateral scour pools are very common in step-pool, pool-riffle, regime, and braided reaches. In pool-riffle and regime reaches, lateral scour pools form naturally at meander bends in gravel-bedded streams even without large roughness elements (Leopold *et al.* 1964, Yang 1971).

Plunge pools result from the vertical fall of water over a full spanning obstruction onto the streambed. The full spanning obstruction creating the plunge pool is located at the head of the pool and the waterfall can range in height from less than a meter to hundreds of meters, as long as the force of the fall is sufficient to scour the bed. A second, far less common type of plunge pool occurs in higher order channels where the stream passes over a sharp geological discontinuity such as the edge of a plateau, forming a large falls with a deep pool at the base. Depending on the height of the waterfall and the composition of the substrata, plunge pools can be quite deep. Overall, plunge pools are most abundant in small, steep headwater streams, especially those with bedrock, cascade, and step-pool reaches.

Dammed Pools   Dammed pools are created by the impoundment of water upstream from a flow obstruction, rather than by scour downstream from the obstruction. They are distinguished by the type of material causing the water impoundment and by their location in relation to the thalweg (Table 2.3). The rate at which sediment fills dammed pools depends on sediment generation from source areas and fluvial transport from upstream reaches. Due to their characteristically low current velocities, dammed pools often have more surface fines than scour pools and fill with sediment at a much more rapid rate. However, some types of dammed pools tend to possess more structure and cover for aquatic organisms than scour pools because of the complex arrangement of material forming the dam. Additionally, dammed pools can be very large, varying with the height of the dam and the extent to which it blocks the flow. Highly porous dams result in little impoundment. Well-sealed dams usually fill to the crest of the dam, creating a spill.

Hawkins *et al.* (1993) identified five types of dammed pools, three of which occur in the main channel of streams. *Debris* dam pools are typically formed at the terminus of a debris flow or where large pieces of woody debris floated downstream at high discharge lodge against a channel constriction. The characteristic structure of debris dams consists of one or a few large key pieces that hold the dam in place and that trap smaller pieces of debris and sediment that comprise the matrix.

Beaver dam pools are unlike debris dam pools in that they usually lack large key pieces but instead consist of a tightly woven smaller pieces sealed on the upstream surface with fine sediment. Some beaver dams may exceed 2 m in height, but most dams in stream systems are about ≤1 m high. In watersheds with high seasonal runoff, beaver dams may breach and be rebuilt annually. In such instances, fine sediments stored above the dam are flushed when the dam breaks.

Landslide dam pools form when a landslide from an adjacent hillslope blocks a stream, causing an impoundment. Dam material consists of a mixture of coarse and fine sediment and woody debris. When landslides take place during severe storms with high discharge, some or most of the fine sediment in the landslide deposit may be rapidly transported downstream leaving behind structures too large to be moved by the flow. Main channel landslide pools are located primarily in laterally constrained reaches of relatively small streams. They are most abundant in step-pool reaches, although some are found in pool-riffle reaches of larger order streams.

Dammed pools are nearly always less abundant than scour pools in alluvial channels, due to the rapidity with which they fill with sediment and the temporary nature of most dams.

Two types of dammed pools located away from the main channel are found only during low flows. *Backwater* pools occur along the bank of the main stream at an entrance to a blocked floodplain channel. They can be found in areas where a gravel bar or other topographic feature prevents water from the main channel from entering the secondary channel. Backwater pools often appear as a diverticulum from the main stream and possess water flowing slowly in a circular pattern. Pool-riffle, regime, and braided reaches are most likely to possess this type of channel unit.

Abandoned channel pools have no surface water connections to the main channel and are formed by bars deposited along the margin of the main stream that isolate secondary channels at low flow. Abandoned channel pools are floodplain features of pool-riffle, regime, and braided reaches that may be ephemeral or maintained by subsurface flow (see Chapter 6).

II. GENERAL DESIGN

A. Site Selection

It is generally impossible to locate examples of every type of valley segment, stream reach, and channel geomorphic unit in one watershed due to regional differences in geology and hydrologic regimes. Selection of study sites will emphasize a comparison of commonly occurring local reach types. In the laboratory, maps and photographs will be used to determine approximate reach boundaries based on stream gradients, degree of valley confinement, channel meander patterns, or significant changes in predominant rock type. The main goal of the laboratory portion of this chapter is to practice map skills and to locate two or more distinctive stream reach types.

B. General Procedures

While it is possible to infer valley segment and reach types from maps and photographs, preliminary classification should be verified by a visit to the sites. Identification of channel geomorphic units from aerial photographs, especially for small streams enclosed within a forest canopy, is virtually impossible and always requires a field survey. In the laboratory, the stream of interest can be divided into sections based on average gradient and apparent degree of valley confinement. Topographic changes in slope can provide important clues with regard to where reach boundaries might exist, but the scale of many topographic maps (including USGS 7.5-min series maps) may be too coarse to reveal key changes in stream gradient and valley confinement that mark reach transitions in very small streams. Maps may also not provide particularly accurate information on the sinuosity of the stream or the extent of channel braiding, except perhaps for maps of large rivers. However, topographic maps are essential for plotting changes in the elevational profile of a stream, as well as changes in valley confinement.

Aerial photographs are usually available from natural resource management agencies and should be used to supplement information extracted from maps. Aerial photographs can be used to accurately locate changes in channel shape in streams not obscured by forest canopies. Orthographic photographs provide a three-dimensional, if somewhat exaggerated, perspective of landscape relief but require stereoscopic map reading equipment that optically superimposes offset photos. This equipment can range from pocket stereoscopes costing $20 to mirror reflecting stereoscopes costing $2000. Low-altitude aerial photographs (1 : 12,000 scale or larger) are most useful and should be examined whenever available. Geological and soils maps of the area will help identify boundaries between geological formations, another important clue to the location of different reach types. Vegetative maps or climatological maps (e.g., rainfall or runoff), if available, provide additional information about the setting of the stream. Landsat imagery can be helpful at large landscape scales but does not usually provide the resolution needed for designation of reach boundaries in small streams.

Once the stream has been subdivided into provisional reach boundaries in the laboratory, contrasting sites are visited and all or part of the reach(es) of interest is surveyed on foot using the criteria in Tables 2.2 and 2.3 to identify channel units. This is often a time-consuming process, depending on the accessibility of the reach, its length and riparian characteristics, and the time required to conduct an inventory of channel units within the reach. Surveys of channel units in small to mid-sized streams typically involve teams of two to three people covering 1–5 km day^{-1}, and it may not be feasible for purposes of this exercise to survey an entire reach if it is a long one. Rather, representative sections of a reach can be studied, provided that the sections include examples of each type of channel unit present in the reach as a whole (Dolloff et al. 1993). A useful rule of thumb is that reach subsamples should be at least 30–50 channel widths long, for example, a survey of channel units in a reach with an average exposed channel width of 10 m should be at least 300–500 m long. During the survey the team should verify that the preliminary classification of valley segment and reach type in the laboratory was correct. Any significant changes in reach character should be noted, particularly if the stream changes from one reach type to another. The valley segment type most often surveyed by stream ecologists will be alluvial (bedrock and channeled colluvial reaches are easily recognized). Diagnostic reach characteristics are given in Table 2.1.

Surveys of channel unit composition can be used simply to determine the presence and number of each type of unit in the reach. More often, however, investigators wish to establish the percentage of total wetted area or volume in each channel unit type on the date the stream was surveyed. Simple counts of the number and type of channel unit can be completed almost as fast as it takes to walk the reach but estimates of surface area or volume can require considerable time, depending on the complexity of the channel and size of the units. Highly accurate estimates of area and volume involve many length, width, and depth measurements of each unit. Visual estimation of the surface area of individual channel units has proven to be a reasonably accurate and much less time-consuming technique (Hankin and Reeves 1988, Dolloff *et al.* 1993). However, visual estimates must be periodically calibrated by comparing them with careful measurements of the same channel units. Part of this exercise will involve performing such a comparison.

In conducting channel unit surveys the question inevitably arises "What is the relative size of the smallest possible unit to be counted?" For channels with complex topographic features and considerable hydraulic complexity, this is a very difficult question. Fast water units possess some areas of low current velocity and slow water units usually have swiftly flowing water in them at some point. Location of channel unit boundaries for survey purposes is almost always subjective. Except for waterfalls, transitions from one unit to the next are gradual. In general, an area should be counted as a separate unit if (1) its overall physical characteristics are clearly different from those of adjacent units and (2) its size is significant relative to the size of the wetted channel. A guideline for what constitutes "significant" is that the greatest dimension of the channel unit should equal or exceed the average wetted width of the reach for units in the stream's thalweg, and one-half the average wetted width of the reach for units along the stream's margin. It is quite possible (and should be expected) that channel units will not all be arranged in linear fashion along the reach but that some units will be located next to each other, depending on the presence of flow obstructions and channel braiding.

Channel unit surveys challenge investigators to balance the accuracy of characterizing stream conditions over an entire reach against the precision obtained by carefully mapping a limited subsection of the reach. The greater the desired precision, the more time will be required for the survey and the less the area that can be covered within a given time. Rapid techniques for visually estimating channel unit composition in stream reaches exist (Hankin and Reeves 1988, Dolloff *et al.* 1993) as well as precise survey methods for mapping the fine details of channel structure at a scale of one to several units (Gordon *et al.* 1992). Which technique is appropriate will be governed by the nature of the research topic. In all cases, investigators must keep in mind that discharge will strongly influence the relative abundance of different channel unit types; therefore, it is often desirable to repeat the survey at a variety of flows.

Although inventories of channel units in reaches of small streams can be conducted by one person, it is much easier and safer for surveys to be carried out by teams of at least two to three people. Because it is necessary to measure lengths and widths repeatedly, one person can be assigned to each side of the channel, while the third can record data and take additional notes. Although practiced survey crews become proficient at identifying channel unit boundaries and maximizing data gathering efficiency, it is important to work slowly and deliberately. It is far better to take the time to collect accurate data than to be in a hurry to complete the reach survey; further, the risk of accidents declines with careful planning and time management and cautious attention to detail. Work safely.

III. SPECIFIC EXERCISES

A. Exercise 1: Stream Reach Classification

Laboratory Protocols

1. Select a watershed. Assemble topographic maps, aerial photographs, and other information pertinent to the area. Within the watershed, select a stream or streams of interest.

2. Using the topographic map, construct a longitudinal profile of the channel beginning at the mouth of the stream and working toward the headwaters. Use a map wheel (also called a curvimeter or map measure) or a planimeter to measure distance along the blue line that marks the stream. If a map wheel or planimeter is not available, a finely graduated ruler may be substituted. In either case, be sure to calibrate the graduations on the map wheel, planimeter or ruler against the map scale. Record the elevation and distance from the mouth each time a contour line intersects the channel. Plot the longitudinal profile of the stream with the stream source nearest the vertical axis (Fig. 2.4).

3. Visually locate inflection points on the stream profile (Fig. 2.4). These points often mark important reach transitions. Compute the average channel slope in each segment according to the formula

$$S = \frac{E_u - E_d}{L},$$ (2.1)

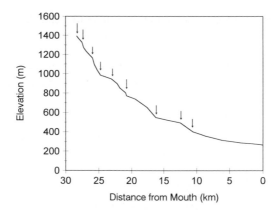

FIGURE 2.4 Hypothetical example of a stream profile constructed from a topographic map. Arrows denote changes in gradient that may mark reach boundaries.

where S represents average slope; E_u, elevation at upstream end of stream reach; E_d, elevation at downstream end of stream reach, and L, reach length. Remember to use common distance units for both numerator and denominator.

4. Examine the shape of the contour lines intersecting the stream to determine the approximate level of valley confinement in each segment. The width of the channel will not be shown on most topographic maps, but the general shape and width of the valley floor will indicate valley confinement (Fig. 2.5).

5. With the aid of a stereoscopic map reader, magnifying lens, or dissecting microscope, examine photographs of the stream segments identified on the topographic map. If it is possible to see the exposed (unvegetated)

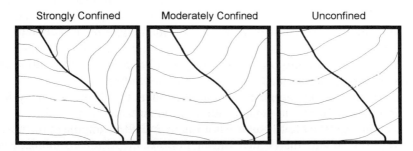

FIGURE 2.5 Appearance of strongly confined, moderately confined, and unconfined channels on topographic maps.

channel in the photographs, estimate the width of the exposed channel and compare it to the estimated width of the flat valley floor. Use the following to determine the approximate degree of confinement for the reach:

Valley Floor Width < 2 Channel Widths   Strongly Confined
Valley Floor Width = 2–4 Channel Widths   Moderately Confined
Valley Floor Width > 4 Channel Widths   Unconfined.

6. Compare average gradients and valley floor widths of each segment on the longitudinal stream profile with geological, soils, vegetation, and climatological maps of the watershed. Changes in the boundaries shown on these maps may help in more precisely locating reach boundaries and in forming hypotheses about reach conditions that can be evaluated during visits to the sites. From all available evidence, determine the most likely valley segment and reach type (or range of types) for each segment based on the features summarized in Table 2.1. Select one or more reaches for site surveys.

Field Protocols   It may be possible to combine certain aspects of the field survey in this exercise with field methods discussed in Chapters 3 and 4. One reach may be surveyed on one field trip and a second reach surveyed on a different field trip.

1. Upon arrival at the site, inspect the stream channel, adjacent valley floor, and hillslopes to verify the accuracy of preliminary valley segment and reach classification. If it is possible to do so (for example, from a vantage point that permits a panoramic view of the valley floor), locate landmarks that mark reach boundaries and that are easily visible from the stream itself.

2. If the reach is too long to complete the exercise within 2–4 h, select a representative section of the reach for the channel unit survey. Location of representative sections may be based on ease of access, but the section should typify the reach as a whole and be long enough to likely contain all types of channel units in the reach (30–50 channel widths). Use the descriptions of channel unit types in Tables 2.2 and 2.3 to identify the units.

3. If optical, electronic, or sonic rangefinders will be used to measure distances, calibrate them at the beginning of each field trip by measuring the distance between two points with a tape and adjusting the readings on the rangefinders to match the known distance. Optical rangefinders, in particular, can become misaligned if dropped and should be recalibrated frequently.

4. If surface area will be estimated visually, it may be helpful to calibrate the "eye" of the observer by placing several rectangles or circles of plastic of known area on the ground before beginning the survey. The pieces of plastic (e.g., old tarps) should approximate the sizes of typical channel units at the site.

Calculations   If channel units are measured, average width and depth are calculated according to the following formulas:

$$\text{Average width} = \frac{\text{Width measurements}}{\text{Number of measurements}} \qquad (2.2)$$

$$\text{Average depth} = \frac{\text{Depth measurements}}{\text{Number of measurements}}. \qquad (2.3)$$

Area and volume of each channel unit are calculated as follows (be sure to use common units):

$$\text{Area} = \text{Length} \times \text{Average width} \qquad (2.4)$$

$$\text{Volume} = \text{Length} \times \text{Average width} \times \text{Average depth}. \qquad (2.5)$$

The percentage of each type of channel unit in the reach, by area or volume, is

$$\% \text{ of Area} = \frac{\text{Area of channel unit type}}{\text{Total area of reach}} \times 100 \qquad (2.6)$$

$$\% \text{ of Volume} = \frac{\text{Volume of channel unit type}}{\text{Total volume of reach}} \times 100. \qquad (2.7)$$

B. Exercise 2: Visual Estimation of Channel Units

1. Most channel unit surveys progress in an upstream direction, but this is not required. It is necessary, however, to be able to recognize channel unit boundaries. These boundaries are often marked by abrupt gradient transitions, which tend to be more easily visible when looking upstream than when looking downstream. Begin at a clearly monumented starting point. This may consist of a man-made structure such as a bridge or some other permanent feature of the landscape. If semipermanent markers are used (e.g., a stake or flag tied to a tree), the location of the marker should be precisely referenced. Global positioning system (GPS) equipment has been used successfully for some reach surveys, but this technology may not work well under a heavy forest canopy in areas of high topographic relief.

2. Divide into teams of two or more individuals. Moving along the stream away from the starting point, the team should identify and record each channel unit as it is encountered (Table 2.4). Units located side-by-side relative to the thalweg (e.g., a pool in the main channel and an adjacent backwater) should be so noted.

3. Record the distance from the starting point of the reach survey to the beginning of each channel unit. This can be accomplished with a measuring tape (or hip chain), rangefinder, or GPS. Unless GPS is used, it will most likely be necessary to measure distances from intermediate reference points along the channel because bends in the channel or riparian vegetation will obscure the view of the starting point. For small streams, it may be helpful to locate intermediate distance reference points at short intervals (e.g., 50 m).

4. For each channel unit, visually estimate the wetted surface area and note it on the data form (Table 2.4). Periodically (e.g., every 10 channel units), use the techniques of Exercise 3 below to measure the length and width of a channel unit after its area has been visually estimated. Record these measurements on the data form, as they will be used to determine any systematic bias in the visual area estimates, and will make it possible to calculate a correction factor.

C. Exercise 3: Detailed Measurements of Channel Units

1. Perform steps 1–3 from Exercise 2.

2. For each channel unit, measure its greatest length in any direction and record this length on the data form (Table 2.4). Widths must be measured at right angles to the line defining the greatest length.

3. Measure the wetted width at regular intervals along the length of the channel unit. Although five width measurements are shown in Table 2.4, the number can vary at the discretion of the investigators. Geomorphically simple units require fewer width measurements than units with complex margins.

4. If the volume of each channel unit is to be estimated as well as the area, record the depth of the stream at regular intervals across the channel at each width transect. If the stream is wadeable, depths are usually measured with a telescoping fiberglass surveyor's rod, graduated wading staff, or meter stick (for very small streams). For very large streams, an electronic depthfinder operated from a boat may be appropriate. At a minimum, depth should be determined at one-third and two-thirds the distance from one side of the channel to the other at each width transect, yielding two depth measurements for each width measurement (Table 2.4). Once again, complex channel units require more depth measurements for accurate volume estimates than geomorphically simple units.

TABLE 2.4

An Example of a Field Data Form for Conducting Channel Unit Surveys

Location _____ Date _____ Surveyors _____

Stream _____ Discharge _____ _____

Quad map _____ Time _____ _____

Starting point _____ Water temp _____ _____

Channel unit	Distance from start	Area (estim.)	Greatest length	Widths					Depths									
				1	2	3	4	5	1	2	3	4	5	6	7	8	9	10

Note. Channel units can be identified by an acronym or alphanumeric designation. Modified from Dolloff *et al.* (1953).

IV. QUESTIONS

1. Were preliminary determinations of valley segment and reach types from maps and photographs correct when sites were visited in the field? What types of valley segments and stream reaches would be easy to identify from maps and aerial photographs? What types would be difficult to identify?

2. What would likely happen if each reach type were to experience a very large precipitation event, such as a flood with a 100- to 200-year recurrence interval? Would the effects be similar to other large disturbances such as inputs of massive volumes of fine sediment?

3. Give a few examples of situations where a stream reach might change from one type to another.

4. How does riparian vegetation influence the characteristics of different reach types? For one or two types, describe how alteration of the riparian plant community could affect channel features.

5. If the channel unit survey compared visual estimates of surface area with estimates derived from actual length and width measurements, was there a tendency for visual estimates to over- or underestimate area? Were errors more apparent for certain types of channel units than for others? Explain why, and suggest a way to correct for systematic bias in the visual estimates.

6. Describe several ways of displaying channel unit frequency data.

7. Describe how the properties of different types of channel units might change with increasing streamflow.

8. Based on your knowledge of the habitat preferences of a certain taxon of aquatic organism (e.g., an aquatic insect or fish species), suggest how that organism would likely be distributed among the channel units within that reach or reaches that were surveyed.

9. How would the frequency of different types of channel units in a reach likely change in response to removal of large woody debris? To extensive sediment inputs? To destruction of riparian vegetation? To a project involving channelization of the reach?

V. MATERIALS AND SUPPLIES

Field Materials

100-m fiberglass tape or hip chain
30-m fiberglass tape
Flagging

Global positioning system (GPS) instrument (optional)
Meter stick
Optical, electronic, or sonic rangefinder
Surveyor's rod or graduated wading staff

Laboratory

Aerial photographs
Geologic, soils, climate, and vegetation maps (optional)
Graph paper
Map wheel (map measure), planimeter, or digitizer
Stereoscope
Topographic maps

REFERENCES

Aumen, N. G., C. P. Hawkins, and S. V. Gregory. 1990. Influence of woody debris on nutrient retention in catastrophically disturbed streams. *Hydrobiologia* **190**:183–192.

Balon, E. K. 1982. About the courtship rituals in fishes, but also about a false sense of security given by classification schemes, comprehensive reviews and committee decisions. *Environmental Biology of Fishes* **7**:193–197.

Beschta, R. L., and W. S. Platts. 1986. Morphological features of small streams: significance and function. *Water Resources Bulletin* **22**:369–379.

Binns, N. A., and F. M. Eiserman. 1979. Quantification of fluvial trout habitat in Wyoming. *Transactions of the American Fisheries Society* **108**:215–228.

Bisson, P. A., J. L. Nielsen, R. A. Palmason, and L. E. Grove. 1982. A system of naming habitat types in small streams, with examples of habitat utilization by salmonids during low streamflow. Pages 62–73 *in* N. B. Armantrout (Ed.) *Acquisition and Utilization of Aquatic Habitat Inventory Information. Symposium Proceedings, October 28–30, 1981, Portland, Oregon.* The Hague Publishing, Billings, MT.

Bisson, P. A., R. E. Bilby, M. D. Bryant, C. A. Dolloff, G. B. Grette, R. A. House, M. L. Murphy, K. V. Koski, and J. R. Sedell. 1987. Large woody debris in forested streams in the Pacific Northwest: past, present, and future. Pages 143–190 *in* E. O. Salo and T. W. Cundy (Eds.) *Streamside Management: Forestry and Fishery Interactions.* Contribution 57, Institute of Forest Resources, University of Washington, Seattle, WA.

Bisson, P. A., K. Sullivan, and J. L. Nielsen. 1988. Channel hydraulics, habitat use, and body form of juvenile coho salmon, steelhead, and cutthroat trout in streams. *Transactions of the American Fisheries Society* **117**:262–273.

Bovee, K. D., and T. Cochnauer. 1977. *Development and Evaluation of Weighted Criteria Probability of Use Curves for Instream Flow Assessments: Fisheries.*

Instream Flow Information Paper 3, Cooperative Instream Flow Service Group, Fort Collins, CO.

Bryant, M. D. 1980. *Evolution of Large, Organic Debris after Timber Harvest: Maybeso Creek, 1949 to 1978.* United States Forest Service, General Technical Report PNW-101, Pacific Northwest Forest and Range Experiment Station, Portland, OR.

Dietrich, W. E., C. J. Wilson, and S. L. Reneau. 1986. Hollows, colluvium and landslides in soil-mantled landscapes. Pages 361–388 *in* A. D. Abrahams (Ed.) *Hillslope Processes.* Allen and Unwin, Boston, MA.

Dolloff, C. A., D. G. Hankin, and G. H. Reeves. 1993. *Basinwide Estimation of Habitat and Fish Populations in Streams.* United States Forest Service, General Technical Report SE-83, Southeastern Forest Experiment Station, Asheville, NC.

Frissell, C. A., W. J. Liss, C. E. Warren, and M. D. Hurley. 1986. A hierarchial framework for stream habitat classification: Viewing streams in a watershed context. *Environmental Management* **10:**199–214.

Gordon, N. D., T. A. McMahon, and B. L. Finlayson, 1992. *Stream Hydrology: An Introduction for Ecologists.* Wiley, Chichester, UK.

Grant, G. E., F. J. Swanson, and M. G. Wolman. 1990. Pattern and origin of stepped-bed morphology in high-gradient streams, Western Cascades, Oregon. *Geological Society of America Bulletin* **102:**340–352.

Hankin, D. G., and G. H. Reeves. 1988. Estimating total fish abundance and total habitat area in small streams based on visual estimation methods. *Canadian Journal of Fisheries and Aquatic Sciences* **45:**834–844.

Hawkes, H. A. 1975. River zonation and classification. Pages 312–374 *in* B. A. Whitton (Ed.) *River Ecology.* Blackwell Scientific, Oxford, UK.

Hawkins, C. P. 1984. Substrate associations and longitudinal distributions in species of Ephemerellidae (Ephemeroptera: Insecta) from western Oregon. *Freshwater Invertebrate Biology* **3:**181–188.

Hawkins, C. P., J. L. Kershner, P. A. Bisson, M. D. Bryant, L. M. Decker, S. V. Gregory, D. A. McCullough, C. K. Overton, G. H. Reeves, R. J. Steedman, and M. K. Young. 1993. A hierarchical approach to classifying stream habitat features. *Fisheries* **18:**3–12.

Huet, M. 1959. Profiles and biology of western European streams as related to fish management. *Transactions of the American Fisheries Society* **88:**155–163.

Huryn, A. D., and J. B. Wallace. 1987. Community structure of Trichoptera in a mountain stream: spatial patterns of production and functional organization. *Freshwater Biology* **20:**141–156.

Hynes, H. B. N. 1970. *The Ecology of Running Waters.* Univ. of Toronto Press, Toronto, Ontario, Canada.

Illies, J. 1961. Versuch einer allgemein biozönotishen Gliederung der Fliessegewässer. *Internationale Revue gesamten Hydrobiologie* **46:**205–213.

Keller, E. A., and W. N. Melhorn. 1978. Rythmic spacing and origin of pools and riffles. *Geological Society of America Bulletin* **89:**723–730.

Leopold, L. B., M. G. Wolman, and J. P. Miller. 1964. *Fluvial Processes in Geomorphology.* Freeman, San Francisco, CA.

Lisle, T. E. 1986. Effects of woody debris on anadromous salmonid habitat, Prince of Wales Island, southeast Alaska. *North American Journal of Fisheries Management* **6:**538–550.

Montgomery, D. R., and J. M. Buffington. 1993. *Channel Classification, Prediction of Channel Response, and Assessment of Channel Condition.* Washington State Timber/Fish/Wildlife Agreement. Report TFW-SH10-93-002, Department of Natural Resources, Olympia, WA.

Montgomery, D. R., and W. E. Dietrich. 1988. Where do channels begin? *Nature* **336:**232–234.

National Research Council. 1992. *Restoration of Aquatic Ecosystems.* National Academy Press, Washington, DC.

O'Neill, R. V., D. L. DeAngelis, J. B. Waide, and T. F. H. Allen. 1986. *A Hierarchical Concept of Ecosystems.* Princeton Univ. Press, Princeton, NJ.

Pennak, R. W. 1971. Toward a classification of lotic habitats. *Hydrobiologia* **38:**321–334.

Pennak, R. W. 1979. The dilemma of stream classification. Pages 59–66 *in Classification, Inventory and Analysis of Fish and Wildlife Habitat.* Report FWS/OBS-78/76, United States Fish and Wildlife Service, Biological Services Program, Washington, DC.

Pinay, G., H. Decamps, E. Chauvet, and E. Fustec. 1990. Functions of ecotones in fluvial systems. Pages 141–170 *in* R. J. Naiman and H. Decamps (Eds.) *The Ecology and Management of Aquatic–Terrestrial Ecotones.* United Nations Educational Scientific and Cultural Organization, Paris, and Parthenon, Carnforth, UK.

Reice, S. R. 1994. Nonequilibrium determinants of biological community structure. *American Scientist* **82:**424–435.

Richards, K. 1982. *Rivers: Form and Process in Alluvial Channels.* Methuen and Company, New York, NY.

Rosgen, D. L. 1985. A stream classification system. Pages 91–95 *in* R. R. Johnson, C. D. Zeibell, D. R. Patton, P. F. Folliott, and R. H. Hamre (Eds.) *Riparian Ecosystems and Their Management: Reconciling Conflicting Uses.* General Technical Report RM-20, United States Forest Service, Rocky Mountain Research Station, Denver, CO.

Schlosser, I. J. 1987. A conceptual framework for fish communities in small warmwater streams. Pages 17–24 *in* W. J. Matthews and D. C. Heins (Eds.) *Community and Evolutionary Ecology of North American Stream Fishes.* University of Oklahoma Press, Norman, OK.

Schlosser, I. J. 1991. Stream fish ecology: A landscape perspective. *BioScience* **41:**704–712.

Shumm, S. A. 1977. *The Fluvial System.* Wiley, New York, NY.

Statzner, B., and B. Higler. 1986. Stream hydraulics as a major determinant of benthic invertebrate zonation patterns. *Freshwater Biology* **16:**127–139.

Statzner, B., J. A. Gore, and V. H. Resh. 1988. Hydraulic stream ecology: Observed patterns and potential applications. *Journal of the North American Benthological Society* **7:**307–360.

Sulligan, K., T. E. Lisle, C. A. Dolloff, G. E. Grant, and L. M. Reid. 1987. Stream

channels: the link between forests and fishes. Pages 39–97 *in* E. O. Salo and
T. W. Cundy (Eds.) *Streamside Management: Forestry and Fishery Interactions.*
Contribution 57, Institute of Forest Resources, University of Washington, Seattle, WA.

Tett, P., C. Gallegos, M. G. Kelly, G. M. Hornberger, and B. J. Cosby. 1978.
Relationships among substrate, flow, and benthic microalgal pigment diversity
in Mechams River, Virginia. *Limnology and Oceanography* **23:**785–797.

Vannote, R. L., G. W. Minshall, K. W. Cummins, J. R. Sedell, and C. E. Cushing.
1980. The river continuum concept. *Canadian Journal of Fisheries and Aquatic
Sciences* **37:**130–137.

Whittaker, J. G. 1987. Sediment transport in step-pool streams. Pages 545–579 *in*
C. R. Thorne, J. C. Bathurst, and R. D. Hey (Eds.) *Sediment Transport in
Gravel-Bed Rivers.* Wiley, New York, NY.

Yang, C. T. 1971. Formation of riffles and pools. *Water Resources Research* **7:**1567–
1574.

Zalewski, M., and R. J. Naiman. 1985. The regulation of riverine fish communities
by a continuum of abiotic–biotic factors. Pages 3–9 *in* J. S. Alabaster (Ed.)
Habitat Modifications and Freshwater Fisheres. Butterworth, London, UK.

CHAPTER 3

Discharge Measurements and Streamflow Analysis

JAMES A. GORE

Environmental Protection Division
The Conservancy

I. INTRODUCTION

The most important of all the geologic processes is the force applied to land forms by running water. In the same manner, running water can have a significant effect upon the distribution of the flora and fauna in lotic ecosystems (Statzner *et al.* 1988, Gordon *et al.* 1992). The most fundamental of hydrological measurements that characterize all river and stream ecosystems is that of *discharge,* the volume of water flowing through a cross section of a stream channel per unit time. The amount of water flowing past a given point, when combined with the slope of the stream channel, yields an indication of *stream power* or the ability of the river to do work. This potential energy is dissipated as frictional heat loss on the streambed and when the stream picks up and moves material. The work performed by the stream is important to lotic ecologists because it influences the distribution of suspended sediment, bed material, particulate organic matter, and other nutrients. The distribution of these materials has substantial influence on the distribution of riverine biota (Vannote *et al.* 1980, Vannote and Minshall 1984, Statzner *et al.* 1988). In addition, discharge and stream power combine with other basin conditions to influence meander pattern and floodplain dynamics (Leopold *et al.* 1964).

Traditionally, discharge has been measured in the United States in terms of cubic feet per seconds (cfs), but, with a greater emphasis on

53

employing SI units, most ecologists prefer to express discharge as cubic meters per second (m³/s) (sometimes called "cumecs"). Because so much of the hydrological information in the United States is maintained by the United States Geological Survey (USGS), stream ecologists and hydrologists must be familiar with translating gaging records. This conversion can be expressed as 1 cfs = 0.0038 m³/s (cumecs) and 1 m³/s = 35.315 cfs.

At most gaging stations, flow is measured by recording the *stage,* or height, of the surface of the water above an arbitrary datum (or benchmark). Discharge can be calculated for different cross sections or at the same cross section at different velocities and water surface elevations (or stages). A graphical relationship (Fig. 3.1) between stage and discharge produces a rating curve so that discharges can be predicted at stages other than those measured. Often a simple *staff gage,* a piece of metal rod with measured increments representing measures of stage heights, is used at each sampling site so that stream ecologists can quickly note discharge at any observation time. Note that hydrologists often refer to "the stage at zero flow." This is not the period when the channel is dry, but rather it is the stage or water surface elevation at which the effective discharge measured across a given transect is 0.

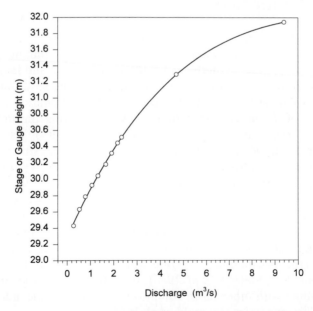

FIGURE 3.1 Stage–discharge relationship for the Olifants River (near Hexrivier Farm, Eastern Cape Province, Republic of South Africa). Based upon surveyed water surface elevations (gage height) and discharge calculated using a current meter.

An analysis of the manner in which discharge varies over time, or the *hydrograph,* allows a lotic scientist to examine the characteristics of the watershed that influence such conditions as runoff and storage. A hydrograph (Fig. 3.2) can be plotted from gaging records to display yearly, monthly, daily, or instantaneous discharges. Ecologists usually obtain gaging records from Water Supply papers published annually by the Water Resources Division of the USGS. These data can provide information on total flow (monthly and daily), mean monthly discharge, base (often groundwater maintained) flow, stage height, and periods of high and low flow.

Examination of the shape of a daily hydrograph during a storm event

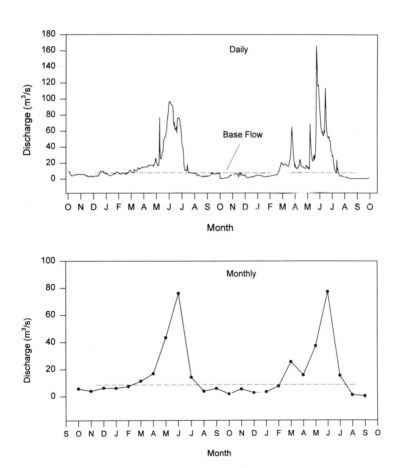

FIGURE 3.2 Hydrographs of the Tongue River (near Miles City, Montana) for water years 1965 and 1966.

can indicate the condition of the stream and its basin. The rising limb of the curve is usually concave and is an index of infiltration capacity of the catchment. In a small basin, the time from the onset of precipitation to the rise in the ascending limb of the hydrograph curve represents the time to reach soil saturation. A catchment with a large storage capacity, absorptive surface, or large channel will have a lower stage-height peak than a similar size basin with little storage (e.g., small channel, lower vegetation density, more clay in soils, greater human development, etc.). Agricultural land, for example, produces a more rapid response in the hydrograph than woodlands (Gregory and Walling 1973) because densely wooded areas restrict surface flow and enhance infiltration. The shape of the hydrograph also reflects the longitudinal profile and basin shape. A steep basin gradient is reflected in a rapid response curve, whereas a low basin gradient will produce a hydrograph with a slow and prolonged response curve. A catchment with many headwater streams but few tributaries in lower reaches will produce a hydrograph with a sharp flood peak. However, the peak is delayed from the onset of the precipitation event. An elongated catchment with many tributaries has a hydrograph that rises rapidly and falls over a longer period of time. A catchment with many subbasins often produces a hydrograph with several flood peaks depending upon distribution of rainfall in the area.

The discharge of a stream or river is also affected by conditions within the channel and the channel geometry. The location of the deepest part of the channel, the *thalweg,* is influenced by the shape of the banks, the width of the stream, the bed material, and the rate of deposition of sediment. In general, the highest stream velocities occur at or near the thalweg (see Chapter 4) and are a function of resistance to flow, usually as a result of the streambed material (i.e., *bed roughness*). The *wetted perimeter* is the cross-sectional distance along the streambed and banks where they contact water. Wetted perimeter can be the same for a deep, high-banked, narrow mountain stream and a broad, shallow lowland river, yet the same discharge through those channels will yield very different flow conditions (Lane 1937, Chow 1959). The *hydraulic radius* of the stream is the ratio of cross-sectional area to the wetted perimeter. The *hydraulic depth* is the ratio of cross-sectional area to the width of the river at the surface. In streams which are very wide in relation to their depth (e.g., greater than 20:1, width:depth), hydraulic radius and hydraulic depth are nearly equal and are approximated by the average depth of the stream. Most hydrologists and stream ecologists thus use the terms mean depth, hydraulic depth, and sometimes hydraulic radius interchangeably.

Streamflow data can also be used to produce flow-duration graphs and flood-frequency predictions. A *flow-duration curve* is a semilogarithmic plot of discharge versus the percentage of the time that a given discharge is equaled or exceeded. If the curve has an overall steep slope, the catchment has a large amount of direct runoff. If the curve is relatively flat, there is substantial storage within the catchment, either as surface or groundwater (Morisawa 1968). A frequent application of discharge records is to predict the magnitude and frequency of flood events. The *flood-frequency curve* allows hydrologists to assess the probability of a certain size of flood or greater occurring in any year. By convention, maximum discharges for each year of gaging record are ranked and plotted as a cumulative frequency curve (Fig. 3.3). A recurrence interval (the number of years within which a flood of a given magnitude or greater is likely to occur) may be calculated as an alternative way of expressing the flood frequency.

Another useful representation of flows may be obtained by plotting the cumulative discharge versus time. This allows the actual sequence and persistence of flows from month to month or year to year to be assessed.

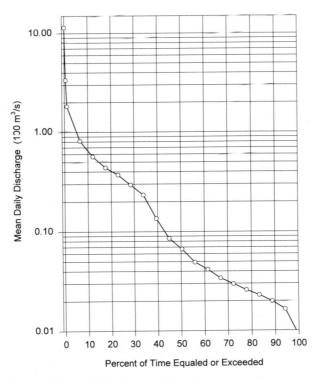

FIGURE 3.3 Flow duration curve for the Locust Fork River, at the USGS gaging station near Trafford, Alabama, for water year 1951.

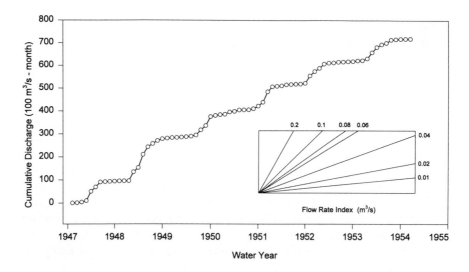

FIGURE 3.4 Discharge mass curve for the Locust Fork River, at the USGS gaging station near Trafford, Alabama, for water years 1947 through 1954.

The shape of the flow line in this form of plot (a mass curve) is equal to a rate of flow (Fig. 3.4).

In this chapter, several field methods to measure discharge are presented along with analytical techniques to produce and examine hydrographs. Some methods are appropriate only for low-order streams, but most can be adapted for larger order systems. The specific objectives are to: (1) understand methods used to pick a specific site for discharge measurement, (2) familiarize lotic researchers with the proper techniques for using current meters, calculating velocities and discharges, and producing and analyzing hydrographs, and (3) provide a better understanding of the use of discharge analysis to interpret channel form, basin shape, land-use patterns, flood conditions, and the distribution of biota in the river system.

II. GENERAL DESIGN

Discharge is usually determined by multiplying the mean velocity by the cross-sectional area of the flow. The cross-sectional area can be measured directly by stretching a measuring tape across the stream or river[1] and taking several measurements of depth with a meter stick or surveyor's stadia rod or staff. Several measurements of mean velocity must be taken

[1]This is modified in large rivers by using premeasured cables or surveying techniques.

across the stream, because flow is unevenly distributed across the stream channel. However, if the flow is very irregular, say on a meander bend or where undercut banks and boulders obstruct or alter flow, the entire velocity distribution must be measured and plotted to determine a mean. In general, stream ecologists and hydrologists try to avoid these situations because of the relative difficulty in obtaining accurate measures at these sites. Between entering tributaries, discharge should be fairly constant, but may vary with gains or losses to the stream channel.[2] Likewise, measures of true discharge may vary according to the sensitivities of the equipment and abilities of the researchers. A useful tactic is to measure the discharge across several transects in a stream reach and compare calculated flows.

A. Site Selection

The selection of the site for measurement of discharge is a critical consideration. In general, the best sites are those in which the flow appears to be relatively uniform across the width of the channel and the surface is not broken by protruding objects, which tend to alter local velocity and depth measurements. The selected section should then have uniform flow that is parallel to the banks. In small streams or those streams with very low discharge (usually with nonparallel or sinuous flow patterns), a small straight section of channel (essentially, a small weir) can be built up using large stones between which the majority of flow passes. This system can be used to measure discharge, but cannot be used to describe the pattern of flow; that is, measurements from this artificial section should not be reported as typical mean depths and velocities.

In general, volumetric analysis (Exercise 1) is most appropriate for low-order streams (first- and second-order). The velocity–area method (Exercise 2) is appropriate for any stream order, but works best on third-order and higher systems. Under unusual flow conditions (extremely shallow, low flows or bankfull or over-bank floods), the slope–area method (Exercise 3) is most useful in estimating discharge. Analysis of discharge patterns over a long period of time is best accomplished in the field by establishing a stage–discharge relationship (Exercise 4) and through graphical and mathematical analysis of published gaging records (Exercise 5).

B. Discharge, Cross-Sectional Area, and Velocity

The simplest form of discharge measurement is

$$Q = A \cdot v, \qquad (3.1)$$

---

[2] Particularly in alluvial, gravel-bed streams, a significant amount of water may be lost to or gained from the hyporheic zone along an unconfined stream reach (see Chapters 6 and 30).

where Q represents discharge (in cfs or cumecs); A, cross-sectional area of the channel at a certain transect; and v, the mean water-column velocity at a designated transect.

To measure discharge (Q), stretch a measuring tape across the stream and then divide the transect into convenient increments, or *cells*. If the flow is relatively uniform, the transect should be divided up into at least five cells of equal width. As a general rule, however, cell widths should not exceed 3 m. A stream of 30 m width, then, should have at least 10 cells to be measured. It is not necessary to make uniform width cells. If there are any hydraulic irregularities (a protruding boulder, a cascade, a pool, etc.) across the transect, a new cell should be designated at the point where the irregularity begins and a new cell designated where more uniform conditions resume. If flow is uniform, the mean velocity is measured in the middle of each cell at a height of 0.4 times the depth at that location (see Chapter 4 for the reasoning behind this choice). At a cell where the depth exceeds 60 cm, the mean velocity should be calculated as an average between velocities measured at 0.2 and 0.8 times the depth at that point.

There are a variety of velocity meters available and all are acceptable. Each requires its own special technique for use. In general, most hydrologists prefer either the (Gurley) pygmy or (Price AA) regular horizontal bucket impeller types. However, horizontal screw (Ott) or electromagnetic (Marsh–McBurney) meters are equally useful but usually require more frequent calibration and servicing.

The current meter is attached to a wading rod made of stainless steel that is marked in increments of 0.1 m or 0.1 ft. Depending upon construction, some will automatically set the current meter at the appropriate depth, using a Vernier scale, or quick field calculations must be made. In large rivers, a winch and cable in conjunction with a sonde replaces the wading rod. The winch is attached to a boat which is tethered to a cable stretched across the river at the site of the transect or attached to a "bridge board" that is suspended on a bridge over the river. The current meter is attached to the end of the cable. The winch, which usually has a depth meter attached, is calibrated to a zero point when the current meter is at the surface of the river. When using the cable and sonde, a finned weight (up to 50 kg) is attached below the current meter to maintain orientation in the current and to keep the cable vertical in each cell.

When using a mechanical current meter, a set of earphones is employed to "count clicks" produced by each revolution of the current meter. The count should last for at least 30 s and the exact number of revolutions and the time counted should be recorded for each cell. Many newer instruments have automatic timing devices and internal calculators that will provide a direct readout of the velocity. Usually some sort of rating curve is provided with each instrument to yield a velocity measured by each revolution.

The product of the width of the cell, the depth at the midpoint of the cell, and the mean velocity is calculated as the cell discharge. The sum of the cell discharges, then, is the discharge for the stream on that date at that stage height.

C. Incorporating Channel Resistance and Slope

For a variety of purposes, such as more sophisticated hydrological and ecological modeling, it may be necessary to incorporate changes in streambed roughness or changes in gradient as they affect discharge over the length of the stream. These conditions, especially roughness, can alter velocity significantly (see Chapter 4). As a result, equations that incorporate resistance to flow (*shear stress*) have been developed to get a more accurate picture of velocity along channel boundaries. Roughness is evaluated in a number of different ways, as described below. Energy slope (S) is calculated as the change in elevation over a given distance. In a stream with uniform flow, this can be taken as either the change in elevation of the water surface or the channel bottom. If the reach can be located on a topographical map, slope can be estimated in this manner. A far better, but more time consuming method is to take measures 30 m upstream and 30 m downstream of the transect using a stadia rod and a surveyor's level to estimate change in elevation over that short distance.

Chezy's Equation was developed in the 1700s and incorporates channel roughness (C) to estimate stream discharge. The equation is

$$V = C(RS)^{1/2}, \tag{3.2}$$

where R is the hydraulic radius (m) and S the energy slope (Henderson 1966). Chezy's C varies from approximately 30 for small, cobble-bottomed streams up to 90 for large, smooth sand-bottomed rivers (White 1986). Discharge is calculated as the product of cross-sectional area (A) and the calculated velocity (V) value (see Eq. (3.1)). The Chezy equation is used primarily in Europe. The details for calculating C are discussed by Chow (1959).

Manning's Equation is more commonly used for calculations of discharge where bed roughness is of great concern. It is expressed as

$$V = 1/n \, (R^{2/3} S^{1/2}) \tag{3.3}$$

or

$$Q = 1/n \, (AR^{2/3} S^{1/2}), \tag{3.4}$$

where n is an index of channel roughness known as "*Manning's n.*" The standard technique for approximating Manning's n is presented in Table 3.1. It should be noted that calculating discharge according to Manning's

TABLE 3.1
Calculation of *Manning's "n"* from Field Observation

Channel condition	Value
$n = (n_0 + n_1 + n_2 + n_3 + n_4)$ m	
Additive factors	
Material involved	n_0
Earth	0.020
Rock Cut	0.025
Fine Gravel	0.024
Coarse Gravel	0.028
Cobble	0.030–0.050
Boulder	0.040–0.070
Degree of irregularity	n_1
Smooth	0.000
Minor (slight scour)	0.015
Moderate (slumping)	0.010
Severe (eroded banks)	0.020
Variation in channel cross section (location of thalweg)	n_2
Gradual	0.000
Alternating occasionally	0.005
Alternating frequently	0.010–0.015
Effect of obstructions	n_3
Negligible	0.000
Minor (15% of area)	0.010–0.015
Appreciable (up to 50%)	0.020–0.030
Severe (>50% is turbulent)	0.040–0.060
Vegetation	n_4
None	0.000
Low (grass/weeds)	0.005–0.010
Medium (brush, none in streambed)	0.010–0.025
High (young trees)	0.025–0.050
Very high (brush in streams, full grown trees)	0.050–0.100
Multiplicative factors	
Degree of Meandering	m
Minor	1.000
Appreciable	1.150
Severe	1.300

Note. Adapted from Cowan (1956).

equation does not require direct measurement of average velocities, but instead depends upon reliable and consistent evaluations of the channel condition and an accurate measurement of the cross-sectional area, hydraulic radius, and slope.

D. Flow-Duration Curve

These curves can be prepared only if gaging records for a single location on a stream are available for a substantial period of time, usually several years. In the United States, gaging records can be obtained through local offices of the USGS. Otherwise, a gaging station or a staff gage that has been calibrated and read at regular intervals must be installed to generate the flow data.

To prepare the flow-duration curve, all flows during the given period (i.e., daily, monthly, or yearly, depending upon the analysis needed) are listed according to their magnitude. The percentage of time that each was equalled or exceeded is then calculated and plotted on a semilogarithmic plot (percentages on an arithmetic scale on the x-axis and the log of the discharge on the y-axis). Analysis of the shape of the curve provides an idea of basin or catchment characteristics. A manual of duration curve interpretations has been published by Searcy (1959).

Several duration indices have been used to compare various stream systems. For these purposes, the same period of record must be used for production of all flow duration curves. The discharge at which flows are exceeded 50% of the time is the median value, or Q_{50}. The Q_{90} is often used as a low-flow (or minimum flow) index. The ratio Q_{90}/Q_{50} is often used as an index of baseflow contribution (Gordon et al. 1992), whereas Q_{10}/Q_{50} may be used as an index of flood peaks. At the upper range of discharges, the values between Q_{30} and Q_{10} have been used to analyze the value and importance of the floodplain by the amount of time it is under water.

E. Flood-Frequency Analysis

The Weibull plotting method is the most commonly applied technique for analyzing flood conditions (Dalrymple 1967). To construct a recurrence curve, the average daily flows are most often examined. When producing a flood-frequency curve, the maximum discharge in a stream or river each year, or all discharges greater than a certain level (e.g., one that will flood a certain area, like a lowland pasture or structure, like a levee) irrespective of year, are used. Most commonly, the annual maximum discharge is used. The best flood-frequency analyses are produced by gaging records of long duration. In most cases, at least 20 years of record should be used to obtain reasonable predictions. Peak discharges are listed according to magnitude with the highest discharge first. Probability of exceedance, P, is calculated as

$$P = \left[\frac{m^*}{(n^* + 1)} \right] 100\%. \tag{3.5}$$

The recurrence interval, T (usually in years), is calculated as

$$T = \frac{n^* + 1}{m^*},\tag{3.6}$$

where n^* is the number of years of record and m^* the magnitude of the flood by its rank value ($m^* = 1$ at the highest discharge on record). Each flood discharge (y-axis) is plotted against its probability of exceedance or recurrence interval on probability paper. The points are joined to form a flood-frequency curve or exceedance curves (see Figs. 3.3 and 3.4). Even without having extremely long-term records, these sorts of curves are used to calculate the discharge for a 100-year event. In turn, that 100-year event discharge can then be compared to a rating curve for the stage–discharge relationship and an estimated height required for a levee or building to withstand that event can be estimated.

III. SPECIFIC EXERCISES

A. Exercise 1: Volummetric Analysis[3]

This exercise works well for only the lowest discharge conditions or for low-order streams.

1. Choose a container of known volume or graduated with known volumes. It should be of at least 4 liters capacity (for stream orders greater than 2 or 3, a larger volume may be required).

2. Place the container under the outflow and begin recording the time it takes to fill the container to the known volume mark.[4] A stopwatch is best for the timing and should be started at the exact time the container is placed into the flow. Be sure that the volume is sufficient that it takes at least 3 s or longer to fill the container. A more accurate measurement would be to start the timing as the level passes a certain graduation and stop it when the level passes yet another.

3. Discharge is calculated as

$$Q = \forall/t,\tag{3.7}$$

---

[3]This is the most accurate technique but can be used only in places where the flow is concentrated, e.g., the notch of a permanent weir or the outflow of a pipe or a culvert under a bridge or highway.

[4]Another handy method is to use a heavy-gage plastic garbage bag that can be held down and open on the stream bed. The rate of fill is timed and the contents of the bag are poured into a measuring container.

where Q is the discharge in m³/s (or liters/s); ∀, volume in m³ (or liters); and t, time (s).

4. Several readings should be taken to obtain a mean and variance in the measure.

B. Exercise 2: Velocity–Area Method

1. Stretch a measuring tape across the stream and divide it into at least 10 intervals or cells. In any case no individual interval or cell should exceed 3 m. Record the width (m) of each cell.

2. At the center point of each cell, measure the depth (m) and record.

Option a: Float Method

3. Measure a length of stream equal to at least 20 m to assure a travel time of at least 20 s. This is the designated reach length, L. This section should overlap one of the sections being measured for cross-sectional area. Mark the upper and lower ends of this interval with a stake or a string across the stream.

4. Choose a float that is only slightly buoyant. This will allow the object to flow with the velocity and minimize influence from air currents. An orange (peeled oranges float lower in the water), a chunk of ice, a half-filled fishing float or bobber, or water-logged branch are ideal.

5. Introduce the float a slight distance upstream of the upstream mark so that the float can reach the speed of the water before it passes the first mark. In large rivers (>10 m width), divide the stream into thirds and make several passes with the float in each third to obtain an average velocity.

6. Use a stopwatch to measure the time (t) of travel of the float between the upstream and downstream marks. Record several measurements through each section to obtain an average. Surface velocity (V_s) is calculated as

$$V_s = L/t. \tag{3.8}$$

A correction factor, k, for the roughness of the bed that affects the slope of the velocity profile must be applied to get an estimate of the mean velocity, V:

$$V = kV_s. \tag{3.9}$$

The correction factor varies between 0.8 for rough beds to 0.9 for smooth beds, but 0.85 is most commonly used unless a singularly rough or smooth bed is being measured. Go to step 7 below.

Option b: Current Meter Method

1. At each section midpoint, place the current meter into the stream, with the meter facing into the current and the researcher standing downstream of the measuring device. Make sure that eddies around legs do not disturb the activity of the current meter.

2. If depth (D) is less than 60 cm, read the velocity at $0.4 \times D$, measured upward from the streambed. If depth is greater than 60 cm, read and record velocities for $0.2 \times D$ and $0.8 \times D$. The mean velocity is the average of the two readings.

3. If the water column for the cell being measured contains large submerged objects (logs, boulders, etc.) or is disturbed by overhanging vegetation, read and record velocities at $0.2D$, $0.4D$, and $0.8D$. Calculate mean velocity as

$$V = 0.25(V_{0.2} + V_{0.8} + 2V_{0.4}).\qquad(3.10)$$

4. If velocities are extremely high or flood flows exist and it is difficult to place the current meter and wading rod (or sounding cable) into the water and maintain a vertical position, measure and record the velocity at the surface. Calculate mean velocity using Eq. (3.9), where k is usually 0.85.

5. Calculate and record the discharge for each cell (n) as

$$Q_n = w_n D_n V_n,\qquad(3.11)$$

where w_n is the width of cell (m); D_n, depth of the cell at the midpoint (m); and V_n, mean velocity of the cell at the midpoint (m/s).

6. Discharge for the transect is calculated as

$$Q = \Sigma Q_n = w_1 D_1 V_1 + w_2 D_2 V_2 + \cdots + w_n D_n V_n.\qquad(3.12)$$

C. Exercise 3: Slope–Area Method

This is an indirect method for estimating discharge when no gaging information is available. Most often this is used to estimate discharges at high flows such as bankfull flows or recent flood events. It can also be used when a current meter or float is not practical (e.g., low flows that barely cover the stream bed). However, it should be noted that the prediction of *Manning's "n"* is more difficult.

1. Choose a straight reach of stream where flows are uniform. The water slope and channel bed slope should be relatively parallel. The length of the study reach should be at least six times the mean channel width (the

average recurrence interval of pools and riffles). The important factor is that a pool and riffle pair be included for the best estimate of average slope.

2. Stretch a measuring tape across the stream and divide it into at least five cells. In any case, no cell should be wider than 3 m. Measure and record the width (m) of each cell.

3. At the center point of each cell, measure and record the depth (m).

4. Identify the water level of interest. This does not necessarily have to be the present water surface elevation. Levels such as bankfull or high-water marks (i.e., indicating the last flood) can be flagged with surveyor's tape or markers.

5. Surveys should be made for three or more typical cross sections in the reach. At each survey point, set up a surveyor's level able to swivel to see points at least 20 m upstream and downstream of the transect. In some instances this may be a clear position along the bank or a position on a mid-channel bar. Using a surveyor's level and rod, measure and record bed elevations and water surface elevations at points 20 m upstream and downstream of the transect (without moving the level). For bed elevations, the rod should be placed at or near a point equal to the average depth and close to the thalweg. For water surface elevations, the rod holder should just touch the water surface several times while elevations are recorded. The average of three or four of these readings will be acceptable as water surface elevations.

6. During a walking survey of the stream reach, estimate *Manning's* "*n*" according to values printed in Table 3.1 or Table 3.2. Strickler's (1923) estimate for deeper channels where depth of flow is at least three times greater than the median diameter (D_{50}) of streambed material projecting into the flow is calculated as

$$n = 0.04D_{50}^{1/6}. \tag{3.13}$$

7. Calculate the cross-sectional area of each cell (A_n) as the product of cell width (w_n) and cell depth (D_n).

8. Total cross-sectional area (A) is calculated as

$$A = \Sigma A_n = A_1 + A_2 + \cdots + A_n. \tag{3.14}$$

9. Calculate the mean depth as an average of the cell depths. For a wide shallow stream this approximate value may be used for the hydraulic radius (R); however, at bankfull and flood stages the calculated hydraulic radius should be used.

TABLE 3.2
Typical *Manning's* "*n*" Values for Low-Order, Natural Streams
(bankfull stage <30 m)

Channel	Typical "*n*"
Lowland and Foothill Streams	
Clean, straight, no deep pools	0.030
Clean, straight, some cobble and weeds	0.035
Clean, winding, some pools and riffles	0.040
Clean, winding, pools, riffles, some cobble and weeds	0.045
Clean, winding, pools, riffles, many cobbles	0.050
Sluggish, deep, weedy pools	0.070
Weedy reach, deep pools, riparian with stands of timber and brush	0.100
Mountain Streams	
Streambed of gravel, cobble and a few boulders	0.040
Bed of medium and large cobble and boulders	0.050

Note. Adapted from Chow (1959).

10. The energy slope[5] of the stream (S) is calculated as the difference in water surface elevations (E, in meters) between the upstream point (E_{upstream}) and the downstream point ($E_{\text{downstream}}$) divided by the distance between the points (L, in meters):

$$S = \frac{(E_{\text{upstream}} - E_{\text{downstream}})}{L}. \tag{3.15}$$

11. Calculate discharge for that transect and water surface elevation as using Eq. (3.4).

D. Exercise 4: Stage–Discharge Method

This method requires many discharge measurements at a number of different water surface elevations. It is used to construct a gaging system (i.e., rating curve) for a particular sampling site that will be visited frequently and a rapid measure of discharge is required for each sampling visit.

1. Discharge measurements must be made for at least three different water surface elevations: low flow, median flow, and high flow.

2. The section that is measured should be accessible at all water surface

---

[5]Except for purposes of examining local hydraulic conditions, bed slope is not appropriate as a substitution for power slope.

elevations and flows to be measured. Choose a straight reach of stream where flows are uniform. The water slope and channel bed slope should be relatively parallel.

3. At each flow, measure the discharge by the current meter method listed in Exercise 2, Option b.

4. Plot the dependent variable (i.e., discharge) on the x-axis and the independent variable (i.e., water surface elevation or stage) on the y-axis. The points are plotted on log–log graph paper. In most cases, this will plot the points as a straight line.

5. For most ecological studies, an approximation of the rating curve can be made by visually constructing a straight line through the points that were measured and plotted. In most instances, the line can be safely extended to a discharge 2.5 times higher than the highest discharge measured and to 0.4 times the lowest discharge measured (Bovee and Milhous 1978).

6. For a more accurate rating curve, the three flows can be fit to the equation

$$Q = a(h - z)^b, \tag{3.16}$$

where h represents gage height or water surface elevation; z, gage height at "zero flow"; and a and b are regression coefficients. The equation is fitted through simple regression techniques (see Haan 1977 or any other standard text on statistics). The regression equation is fitted with $(h - z)$ as the independent variable and Q as the dependent variable, despite the fact that the rating curve was plotted with the axes reversed. The value of z must be derived by trial and error. The true value of z is assumed to be that point at which the rating curve plots a straight line on the log–log paper. Thus, it is possible to visually estimate z by graphical extrapolation and "test" this value into the regression equation. If the z value is too small, the plotted equation will be concave downward. If the z value is too large, the plotted equation will be concave upward.

7. At the sampling site, place a staff gage into the stream. The staff gage consists of a rod[6] that has been painted a bright color for visibility and marked at appropriate intervals to match the rating curve. For example, markings can be every 0.1 m and a meter stick used to measure exact distances between major marks to get exact water surface elevation. The staff gage[7] should be placed well enough away from the bank so that the water surface will still wet the gage at the lowest flows. The staff rod should

---

[6]Reinforcing bar ("rebar") of 2.5 cm diameter works well.

[7]Commercial staff gages, already marked with appropriate intervals are also available (e.g., Ben Meadows Company, Forestry Suppliers).

extend 1 m or more below the marked section. Pound the rod into the substratum until the water surface covers the mark at the appropriate elevation for the discharge on the day the gage is installed.

8. During subsequent sampling trips, read the water surface elevation from the staff gage, compare to the derived rating curve, and record the corresponding discharge for the activities undertaken on that day.

E. Exercise 5: Hydrographs—Flood-Frequency, Flow-Duration, and Discharge-Mass

Option a: Flood-Frequency

1. Obtain a gaging record for the stream or river to be analyzed. Under optimum conditions, at least 20 years of record should be available. For predictions of a 100-year event, at least 100 years of record will provide an estimate with reasonable accuracy. Monthly or annual data are adequate for this analysis.

2. List annual peak discharges according to magnitude with the highest discharge first.

3. The recurrence interval (T) is calculated using Eq. (3.6). As an alternative, a less biased estimate of peak floods (Cunnane 1978) can be produced by calculating the recurrence interval as

$$T = \frac{[(n^* + 1) - 0.8]}{m^* - 0.4}. \tag{3.17}$$

The probability (P) that a given discharge will be exceeded (i.e., probability of exceedance) is calculated using Eq. (3.5), or as the reciprocal of T ($P = 1/T$).

4. Each flood discharge (y-axis) is plotted against its recurrence interval or probability of exceedance on log–probability paper. In theory the largest flood should plot at $P = 0$, as it will never be exceeded and the smallest at $P = 1$, since it will always be exceeded. In all situations, all of the values obtained from the calculations will plot between these two values because the numerator or denominator has been adjusted to be greater than the number of observations.

5. The points are joined to form the flood-frequency curve. In general, a curve fitted by eye can be used if the intention is to provide information on floods with a recurrence interval of less than $n^*/5$. When eye-fitting the straight line, greater emphasis should be placed on the middle and high discharge events since the primary purpose of the plot is to estimate high flow events. For recurrence intervals greater than $n^*/5$, where greater accuracy is required, a theoretical probability distribution should be fitted to

the data to obtain more reasonable estimates. The standard method applied by the USGS is the *log Pearson Type III* distribution (see Haan 1977 for specific techniques).

Option b: Flow-Duration

1. Obtain a gaging record for the stream or river to be analyzed. Under optimum conditions, several years of record should be available. If annual duration curves are the objective, then at least 20 years of record are advisable. However, daily, weekly, or monthly data can also be used to examine flow duration over shorter intervals.

2. All flows during the given period (daily, monthly, yearly, etc.) are listed according to their magnitude.

3. The range of discharges should be partitioned into 20 to 30 intervals. For example, if the total range of discharges for daily records ranged from 10 to 300 m³/s, the researcher might enter the intervals as 0–10, 11–20, 21–30, ..., 291–300.

3. The percentage of time that each was equalled or exceeded is then calculated and plotted on a semilogarithmic plot; put percentages on an arithmetic scale on the *x*-axis and the log of the discharge on the *y*-axis.

4. A manual of duration curve interpretations has been published by Searcy (1959) and can be used to analyze specific situations in which dilutions for pollution or flow durations for irrigation, hydropower, or transport of particulates or sediment are necessary.

Option c: Discharge-Mass

1. Obtain a gaging record for the stream or river to be analyzed. Under optimum conditions, several years of record should be available. If annual duration curves are the objective, then at least 20 years of record are advisable. However, daily, weekly, or monthly data can also be used to examine duration over shorter intervals. Traditionally, monthly total discharge values are used.

2. Cumulative discharge values for each month are plotted against the time intervals involved (see Fig. 3.4).

3. A flow rate index based upon critical discharge values (e.g., discharges required for incubation of eggs, spawning, instar success, or year-class strength) are compared to the slopes of the mass curve to determine the percentage of time, historically, a certain flow rate has been sustained.

4. Newbury and Gaboury (1993) have described various biological applications of mass curve analysis and Chow (1964) has provided information on the use of mass curves for setting flows in reservoir design. Newbury and Gaboury suggest that mass curves can be used to establish minimum

flows and indicate the amount of time necessary to recharge a system if those flow are not met or are exceeded. These curves can also be used to estimate flows at ungaged sites and to estimate bankfull conditions.

IV. QUESTIONS

1. Consider each of the techniques for directly measuring discharge. Where is error introduced into the calculations?

2. When choosing a sample transect for discharge calculation, what precautions must be taken in order to ensure that the best estimates of mean depth and velocity are obtained?

3. What are the difficulties that can be encountered when attempting to describe the resistance of the channel to flow, (i.e. Manning's "n")?

4. In what ways can environmental scientists and engineers use flood-frequency and stage–discharge relationships to design levees and dams yet continue to promote ecological integrity?

5. What is the value of flow-duration curves to the management and analysis of floodplains?

6. How might weeks or days of flow persistence during a sensitive spawning or incubation period be estimated using a mass curve?

7. For a measured discharge, how much variation was there in the mean velocity between riffle and pool transects? How does this affect discharge estimates?

8. Compare the hydraulic radius and the mean depth for the sample reach at high and low discharges. How would these two values alter discharge predictions?

9. Examine your estimate of roughness, Manning's n. What were the major factors that influenced it at the discharge you analyzed? What values will dominate the estimate of Manning's n at higher or lower flows?

V. MATERIALS AND SUPPLIES

Discharge Measurements

4-Liter or larger bucket or wide-mouthed container
Calculator
Current meter with wading rod (any of the standard meters; Pygmy, Price AA, Ott, Marsh–McBurney)
Float
Meter stick

Reinforcing bar (2.5 cm diameter)–2 to 3 m length
Stop watch
Surveyor's level, tripod, and stadium/rod
Tape mesure (at least 50 m)

Hydrographs

Calculator
Gaging records for local streams and rivers (in the United States, these can be obtained in Government Documents section of most major university libraries or from area/regional office of the USGS)
Log–log graph paper
Log–probability graph paper
Semilogarithmic graph paper

REFERENCES

Bovee, K. D., and R. Milhous. 1978. *Hydraulic Simulation in Instream Flow Studies: Theory and Techniques.* Instream Flow Information Paper 5, United States Fish and Wildlife Service, Office of Biological Services, Washington, DC.

Chow, V. T. 1959. *Open Channel Hydraulics.* McGraw–Hill, New York, NY.

Chow, V. T. 1964. *Handbook of Applied Hydrology.* McGraw–Hill, New York, NY.

Cowan, W. L. 1956. Estimating hydraulic roughness coefficients. *Agricultural Engineering* **37:**83–85.

Cunnane, C. 1978. Unbiased plotting positions—A review. *Journal of Hydrology* **37:**205–222.

Dalrymple, T. 1960. *Flood-Frequency Analysis.* Manual of Hydrology. 3. Flood-Flow Techniques, Water Supply Paper 1543-A, United States Geological Survey, Washington, DC.

Gordon, N. D., T. A. McMahon, and B. L. Finlayson. 1992. *Stream Hydrology: An Introduction for Ecologists.* Wiley, Chichester, UK.

Gregory, K. J., and D. E. Walling. 1973. *Drainage Basin Form and Process.* Edward & Arnold, London, UK.

Haan, C. T. 1977. *Statistical Methods in Hydrology.* Iowa State Univ. Press, Ames, IA.

Henderson, F. M. 1966. *Open Channel Flow.* MacMillan, New York, New York, USA.

Lane, E. W. 1937. Stable channels in erodible materials. *Transactions of the American Society of Civil Engineers* **102,** 123–194.

Leopold, L. B., M. G. Wolman, and J. P. Miller. 1964. *Fluvial Processes in Geomorphology.* Freeman, San Francisco, CA.

Morisawa, M. 1968. *Streams: Their Dynamics and Morphology.* McGraw–Hill, New York, NY.

Newbury, R. W., and M. N. Gaboury. 1993. *Stream Analysis and Fish Habitat Design.* Newbury Hydraulics Ltd., Gibsons, British Columbia, Canada.

Searcy, J. K. 1959. *Flow duration curves.* Manual of Hydrology. 2. Low-Flow Techniques. Water Supply Paper 1542-A, United States Geological Survey, Washington, DC.

Statzner, B., J. A. Gore, and V. H. Resh. 1988. Hydraulic stream ecology: Observed patterns and potential applications. *Journal of the North American Benthological Society* **7:**307–360.

Strickler, A. 1923. *Beitrage zur Frage der Gesewindigheits-formel und der Rauhingkeiszahlen für Strome, Kanale Geschlossen Leitungen.* Mitteilungen des Eidgenössischer Amtes für Wasserwirtschaft, No. 16, Bern, Switzerland.

Vannote, R. L., G. W. Minshall, K. W. Cummins, J. R. Sedell, and C. E. Cushing. 1980. The river continuum concept. *Canadian Journal of Fisheries and Aquatic Science* **37:**130–137.

Vannote, R. L., and G. W. Minshall. 1982. Fluvial processes and local lithology controlling abundance, structure, and composition of mussel beds. *Proceedings of the National Academy of Science* **79:**4103–4107.

White, F. M. 1986. *Fluid Mechanics,* 2nd ed. McGraw–Hill, New York, NY.

CHAPTER 4

Dynamics of Flow

ROBERT W. NEWBURY

School of Resource and Environmental Management
Simon Fraser University

I. INTRODUCTION

Rivers and streams are integrated flowing systems that create and maintain aquatic habitats within the turbulent structure of the flow as well as on and below the channel bed (Vogel 1981). In a river basin, the hydraulic habitats are nested within one another at smaller and smaller scales (Fig. 4.1). At the catchment scale (Fig. 4.1, Level I), the hydraulic condition of the flow may be generalized as uniform or gradually varying above and below interruptions in the longitudinal (long) profile of the stream (Chow 1959). Uniform flow conditions occur when the slope of the water surface and the channel bed are approximately parallel and there is little change in the cross-sectional area of the flow. The Chezy or Manning uniform-flow equations may be applied to estimate the velocity of the flow by assuming that frictional resistance on the stream bed accounts for all of the energy losses in the flow (see Chapter 3). In many natural streams, particularly at lower stages, the uniform flow equation must be applied with surrogate resistance factors to account for major obstructions on the stream bed and the turbulent energy losses created by local nonuniform flow conditions (Newbury and Gaboury 1993a). The uniform flow equations, with correction factors for natural channels and floodplains, are an integral part of the runoff and flow models used for instream flow simulation (PHABSIM, Milhous *et al.* 1989) and flood routing (Hydraulic Engineering Center, U.S. Army Corps of Engineers; Bedient and Huber 1992).

At the stream reach scale (Fig. 4.1, Level II), nonuniform flow conditions that occur in pools, riffles, and meanders can be distinguished. Regu-

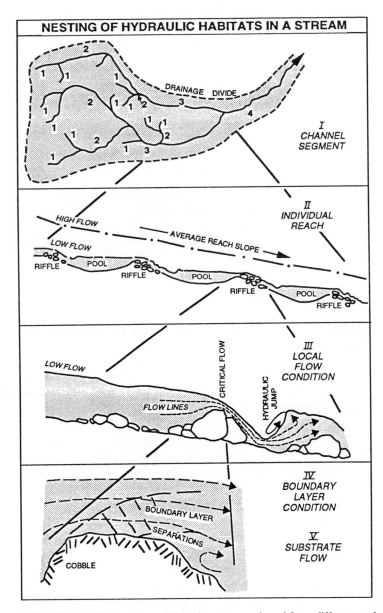

FIGURE 4.1 Hydraulic conditions in a stream viewed from different scales.

larly recurring floods erode and maintain these channel forms (Leopold *et al.* 1964). The natural wavelength of the flood flows appears to be 6 times the width of the channel, corresponding to the average spacing of pools and riffles (Fig. 4.2) found in a wide sample of rivers and channelized streams (Chang 1988, Gregory *et al.* 1994). For certain combinations of discharge, slope, and bed materials, the channel meanders with an average wavelength of 12 times the bankfull width, as there are two pool and riffle reaches in each full meander (Fig. 4.2). This channel and meander geometry for mid-latitude streams in North America has been summarized by Leopold *et al.* (1964) and by Dunne and Leopold (1978).

To distinguish the pattern of nonuniform flows, the mean depth, velocity, and direction of the flow may be mapped on sketches or surveyed plans of a reach. The channel configuration and flow condition are major components used to characterize the preferred habitats of fish, such as the "ideal" trout meanders described by Newbury and Gaboury (1993b) and the habitat suitability curves derived for different fish species by Bovee and Cochnauer (1986).

At the habitat scale (Fig. 4.1, Level III), individual stream flowlines and different states of flow can be delineated and analyzed as rapidly varying, nonuniform flow. The local velocity and depth of the flow is dominated by its momentum and gravitational forces rather than by boundary friction (Chow 1959; Part IV). In riffles or rapids, the flow is broken into segments by cobble bars and boulders that create zones of smoothly accelerating flow over and around obstacles followed by turbulent areas

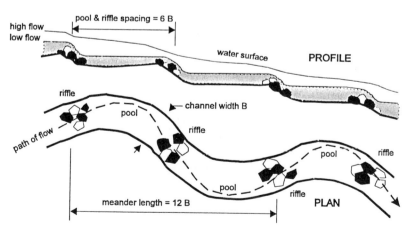

FIGURE 4.2 Average meander, pool, and riffle dimensions expressed as a ratio to the bankfull width.

of deceleration and side eddies. If the flow drops over an obstacle, or is drawn through a narrow gap between boulders, it often reaches the critical state, a condition where the velocity is maximized for the head of water (H) that exists above the obstruction (Fig. 4.3). The critical depth of flow occurs as the water passes over the obstruction and is equal to two-thirds of the upstream head. At the point of overflow, the critical velocity occurs. The critical velocity (V_c) is equal to the velocity of a wave of disturbance in the same depth of still water such that

$$V_c = (gd_c)^{1/2}, \tag{4.1}$$

where V_c represents critical velocity (m/s); g, gravitational acceleration (9.81 m/s); and d_c, critical depth of flow (m).

Smooth standing waves of disturbance are often seen slowly shifting upstream and downstream in the critical flow zone as the downstream velocity and the wave of disturbance velocity are equal, trapping the wave form in place. If the water continues to accelerate past the obstruction or through a gap, it attains supercritical velocities. The supercritical flow brings air into the water as it enters a local pocket or pool of slower-moving water downstream. This is the whitewater plume formed in a hydraulic jump where the flow rapidly rotates vertically, decelerates, and climbs to the greater depth of the water downstream. It is the source of noise in the river. The breaking bubbles of air entrained by the supercritical flow make babbling brooks and roaring rapids. Uniform flow makes no noise. The state of flow relative to the critical velocity is characterized by the Froude number (Fr),

$$\mathrm{Fr} = V_m/(gd)^{1/2}, \tag{4.2}$$

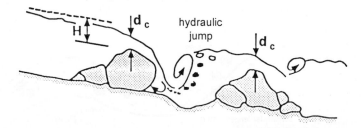

FIGURE 4.3 Rapidly varied flow conditions created by large cobbles and boulders relative to the stream depth. The critical depth at the point of overflow is two-thirds of the upstream head H.

where V_m is the mean velocity of the flow (m/s) and d the depth of flow (m). The Froude number has a value of 1 for critical flow, <1 for subcritical flow, and >1 for supercritical flow (Chow 1959, Vogel 1981).

Critical and near-critical flows are efficient habitats for net-spinning caddisflies, black flies, and other benthic insects. By locating on the tops and sides of boulders in this zone of rapid and converging flow, they are able to expand their capture nets or cephalic fans to efficiently gather particles from the flow (Wetmore et al. 1990, also see Chapters 10, 16, and 21). Nonuniform local flow patterns also create protected habitats and feeding opportunities for fish. At high flows, when the pools are subject to scouring velocities, fish move into calmer water zones created in eddies behind boulders and other obstacles. The local flow conditions are used to navigate the river, allowing fish to pass upstream through the rapids zone in short bursts between protected water pockets and eddies. Salmon will also jump over obstacles in the rapids by swimming to the top of the hydraulic jump and launching themselves from the upstream surface current (Fig. 4.3).

The diverse flow patterns and phenomena in natural rivers have not been fully analyzed. The broadest work that summarizes the general theory of open channel flow was prepared by Chow (1959). Many open channel flow studies and textbooks refer to this work (Davis and Barmuta 1989, Gordon et al. 1992). Contemporary summaries of open channel hydraulics are clearly presented in Rangaraju (1981) and Roberson and Crowe (1993). A comprehensive assessment of the application of hydraulic theory to stream ecology has been prepared by Statzner et al. (1988). Other qualitative knowledge that is useful in describing local flow conditions exists in the angling literature (Newbury 1995).

At the microhabitat scale (Fig. 4.1, Level IV), local flow conditions must be characterized indirectly as there are limited techniques available for direct field measurements. This is the scale at which near-boundary laminar flow conditions can be detected. Although the main body of the flow in streams is turbulent, as the fixed boundary of the flow is approached the velocity decreases until the viscous forces overcome the turbulence and the flow becomes laminar. In the laminar boundary layer, the water moves in parallel streams without mixing. There are several useful studies and opinions regarding its importance in this rapidly developing area of hydraulic stream ecology (Nowell and Jumars 1984, Statzner et al. 1988, Carling 1992). Near-boundary fluid habitats and flow assumptions are discussed by Vogel (1981).

The basic hydraulic habitat exercises described in this chapter focus on qualitatively identifying and mapping the flow pattern and habitats on a plan of a sample stream reach. Advanced exercises measure the flow velocities and depths at selected habitat and hydraulic sites to determine the state of the flow described by the Froude number. The same measurements may be applied to estimate the forces exerted by the flow near the stream bed that affect substratum stability and organisms that cling to the stream boundaries.

A. Streambed Stability and Shear Stress

Tractive Force   The total energy of the flow at any point in an idealized stream channel expressed in units of height above a datum consists of the elevation of the channel bed, the depth of flow, and the kinetic energy of the flow. An imaginary line may be drawn through the position of the total energy of the flow that is elevated above the water surface by the amount of the kinetic energy head (Fig. 4.4), or

$$\text{kinetic energy head} = V_m^2/2g. \tag{4.3}$$

In uniform flow equations, the slope of the energy line is assumed to be nearly parallel to the slope of the reach. If the slopes are not parallel, the energy line slope is used in flow analyses (Chow 1959). The general shear stress or tractive force exerted by the flow on the stream bed may be estimated by the relationship

$$T_G = \rho g R S, \tag{4.4}$$

where T_G is the tractive force or general bed shear stress (N/m^2); ρ, the density of water (1000 kg/m^3); g, gravitational acceleration (9.81 m/s); R, the hydraulic radius (m) (see Chapter 3); and S, the slope of the energy line.

Studies of canals and mobile-bed streams have showed that the tractive force can be related to the size of material transported and to the substratum stability (Lane 1955). This has been used to describe the effects of substra-

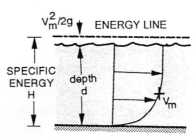

FIGURE 4.4 The depth of flow, kinetic energy, and specific energy line for uniform flow conditions.

tum instability on the density and composition of benthic insects (Cobb *et al.* 1992).

Velocity Profiles, Local Shear Stress, and Boundary Layers   The local shear stress acting on a portion of the stream bed, such as on the surface of a cobble or on an insect clinging to the cobble, must be known to describe the specific hydraulic habitat conditions. To estimate the local shear stress, measurements of the velocity profile perpendicular to the specific site are required. The relationship between the near-bed shear stress and the velocity profile in the turbulent flows of natural streams is largely empirical, with founding laboratory studies conducted by Prandtl, von Karman, and Nickuradse (Chow 1959). The velocity profile above a rough boundary has been characterized in three segments: (1) a laminar flow viscous layer immediately next to the wetted channel boundary, (2) a buffer zone or transition layer of viscous turbulent flow, and for the major part of the profile, (3) a fully developed turbulent layer in which the velocity is distributed logarithmically with the height above the stream bed (Roberson and Crowe 1993). The velocity profile in the turbulent zone predicted by the Prandtl–von Karman universal velocity distribution law may be written in the form

$$u/u_f = 5.75(\log y - \log y_0), \tag{4.5}$$

where u represents the velocity at location y in the turbulent layer (m/s); u_f, friction velocity defined as $u_f = (t_0/\rho)^{1/2}$ (m/s); y, distance above the stream bed (m); and y_0, the height at which the logarithmic velocity profile extrapolates to zero (m).

Using sand grains as roughness elements, Nickuradse demonstrated that $y_0 = k/30$, where k was the diameter of the sand grains (Chow 1959). In some stream habitat studies, this observation has been extrapolated and substituted in the velocity distribution law for y_0 by assuming that k is the mean diameter of the largest substratum paving the stream bed (m). If the depth (d) and mean velocity (V_m) are measured and assumed to be at $0.4 \times d$ from the channel bottom (see Chapter 3), the equation may be rearranged to estimate u_f as

$$u_f = \frac{V_m}{5.75(\log 12d/k)}. \tag{4.6}$$

An estimate of the local bed shear stress may be obtained also by measuring the velocity profile perpendicular to the channel boundary and plotting the relationship $u = f(\log y)$. By fitting a line to the plot, the abscissa y_0 at which the logarithmic velocity distribution is nominally zero can be found, allowing the value of u_f to be determined at the measurement site. If the slope of the regression line, $u/(\log y - \log y_0)$, is substituted in the velocity distribution law, then $u_f = \text{slope}/5.75$. The bed shear stress at the sampling site is $t_0 = \rho(u_f)^2$ (N/m^2) by definition (Chow 1959).

If the flow is transparent and shallow, the shear stress may be characterized with a graded set of shear stress testing hemispheres (FST hemispheres; Statzner and Muller 1989). The shear stress required to move the FST hemispheres was determined by measuring the velocity profile and solving the Prandtl–von Karman formula with the same assumptions as described above (Statzner et al. 1991). FST hemispheres have been used to determine the effects of shear stress on benthic insects and to characterize riverine habitats (Peckarsky et al. 1990, Gore et al. 1994). They have generated a useful discussion of stream habitat measurements and their interpretation (Carling 1992, Frutiger 1993, Frutiger and Schib 1993, Statzner 1993).

A nominal thickness of the viscous boundary layer δ_N (m) consisting of the laminar sublayer and a portion of the buffer zone has been estimated as

$$\delta_N = 11.8 \left(\frac{v}{u_f} \right), \tag{4.7}$$

where v is the kinematic viscosity of water (m^2/s). This approximation is based on sand grains and should be viewed only as an index of relatively smooth patches on the stream boundary (see "the law of the wall" (Roberson and Crowe 1993) and the discussion of Carlin (1992)).

In a natural stream, turbulent eddies and separations of high-velocity flows occur in the rapids and riffle zones that are too complex to analyze. Many researchers have addressed this problem by assuming that representative shear stress and boundary layer equations apply. One should follow Vogel's sage advice "don't perpetuate the practice of equation-grabbing predecessors ... don't use formulas unless they demonstrably apply ... and don't be intimidated by the prospect of measuring low flows in small places" (Vogel 1981).

Substrata and Stream Bed Stability   The stream bed materials or substrata may be sampled by different methods, depending upon the habitat being investigated. In studies of hyporheic habitats, all sizes of the bed materials may be sampled by excavating a portion of the bed or by withdrawing a frozen segment of the bed material (Platts and Penton 1980). In studies of the stability of the reach and the flow resistance the substrata may be subsampled for only the largest sizes of bed materials that project into the flow. This is described as a bed paving material sample in the sense that it is the largest protective fraction that must resist the bed shear stress (see Chapter 7). Methods of sampling the bed paving materials can be detailed, but for bed stability and channel roughness studies the simple random selection "pebble-count" method proposed by Wolman (1954) is adequate (see test by Statzner *et al.* 1988). The sizes of bed paving materials may be conveniently summarized in a cumulative frequency plot.

The relationship between the general bed shear stress and the size of particle that can be transported has been widely researched in river engineering. An early empirical summary of the critical tractive force T_c and the diameter of bed material at incipient motion that is often reproduced in later works was prepared by Lane (1955) (Fig. 4.5). For noncohesive bed

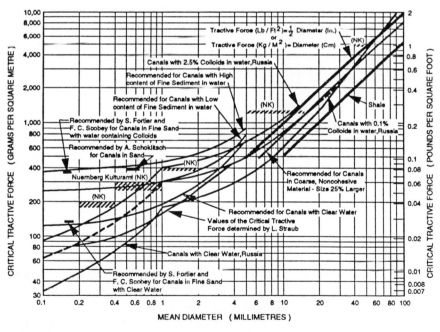

FIGURE 4.5 General relationship between the tractive force and size of bed material at incipient motion prepared from observational data gathered by Lane (1955).

materials that are greater than 5 mm in diameter, the relationship recommended by Lane was

$$T_c(\text{kg/m}^2) = \text{diameter (cm)}. \qquad (4.8)$$

To convert to Lane's units, $T_c(\text{kg/m}^2) = T_G(\text{N/m}^2)/9.81$.

The mean size of bed material that can be moved at a given discharge may be estimated with Lane's relationship by solving for the tractive force at the corresponding depth of flow in the channel. An index of the fraction of the bed paving materials that can be moved at that discharge may be obtained by examining the cumulative frequency plot of the bed paving materials (Newbury 1984). This estimate may be modified for bimodally distributed stream bed sediments. These general techniques for measuring bed shear stress and instability have been applied successfully in recent studies of the effect of bed instability on the density and species composition of stream insects (Cobb *et al.* 1992) and the shear stress in refugia (Lancaster and Hildrew 1993).

The basic stream bed stability exercises presented below focus on sampling the stream bed paving materials and predicting their stability at the bankfull flood stage using the general tractive force relationship. The advanced exercises measure the velocity distribution at selected habitat sites in the sample reach to predict the local shear stress, substrata stability, and a nominal boundary layer thickness.

II. GENERAL DESIGN

These exercises are designed to illustrate and, where possible, quantify local hydraulic conditions that occur in a natural stream. This is an area of active research and discovery and lotic ecologists should be aware that their observations of hydraulically created habitats may significantly contribute to this growing field of knowledge. Exercises have been divided into basic and advanced categories. The basic exercises classify and map hydraulic phenomena in the whole reach and determine the average conditions of streambed stability at the bankfull stage. The advanced exercises quantify the local hydraulic conditions and bed shear stress, assuming that the basic mapping of the sample reach has been completed (see Chapter 2). The analysis of the data gathered in the field requires plotting, curve fitting, and elementary algebraic calculations. Cross-section, slope, and bed material data may be available from conducting the exercises described in Chapters 2, 3, 10, and 11.

A. Site Selection

Sample reaches should be selected by examining a topographic plan of the drainage basin at a scale that distinguishes rapids and meander patterns. The sample reaches should be at least 6 times the bankfull width of the stream, and if present, include at least one full pool and riffle sequence. In smaller meandering streams, a full meander length of two pool and riffle reaches is preferred. The sample reaches may be consecutive in a single branch of the stream or selected in different stream order branches from the top to the bottom of the drainage basin. This will illustrate how the river characteristics and hydraulic conditions change with the drainage area, discharge, and average slope.

Reaches that do not meet the ideal alluvial form because of bedrock, man-made, or other intrusions should not be ignored but treated as special conditions that illustrate deviations from characteristic reaches elsewhere in the basin. The research team should visit these special reaches to understand their significance relative to their sample reach (e.g., to fish blockage, water diversion, flow regulation, or profile control).

The sample reaches should have depths of flows and velocities that can be worked in safely up to waist level. To ensure this, the sum of the depth (m) and velocity (m/s) should not exceed one. In larger streams, this may restrict the field work to periods of low or moderate flows. Photographs or video recordings of the sample sites at high flows will help to visualize the reach conditions at the bankfull stage.

III. SPECIFIC EXERCISES

A. Preparation

1. Identify the drainage basin and stream order segments on a topographic map (see Chapter 2).

2. Select sample reaches that can be reached conveniently in the basin.

3. Plot the long profile of the stream using the contour lines intersecting the stream channel and locate the sample reaches on the profile. Note special reaches that may affect the general hydraulic conditions in the sample reaches.

4. Test flow meter batteries and survey equipment.

B. Exercise 1: Hydraulic Conditions and Habitats

Option a: Basic Mapping

1. Stretch a tape along a straight baseline on one bank of the reach and sketch in the bottom and top of the channel boundaries and major physical features such as boulders, logs, debris, and typical substrata. Mea-

surements of features in the stream and floodplain may be made with a
second tape held perpendicular to the baseline tape by the map maker.
The map should be scaled to include a reach that is at least six times the
bankfull width (Fig. 4.6).

2. Select four to six typical cross sections of the channel and measure
the average depth and width at the bankfull stage. A tape may be stretched
across the stream at the bankfull stage as a reference for depth measurements. Record the present depth of flow in the sample cross section.

3. Estimate the length and total drop in the reach with a simple level
and stadia rod to determine the average slope of the streambed.

4. Locate pools, riffles, and local rapidly varying flow conditions such
as hydraulic jumps, chutes, and eddy patterns on the map approximately
to scale. Describe and locate any preferred habitat sites that have concentrations of benthic insects or fish. Detailed maps of segments of the reach

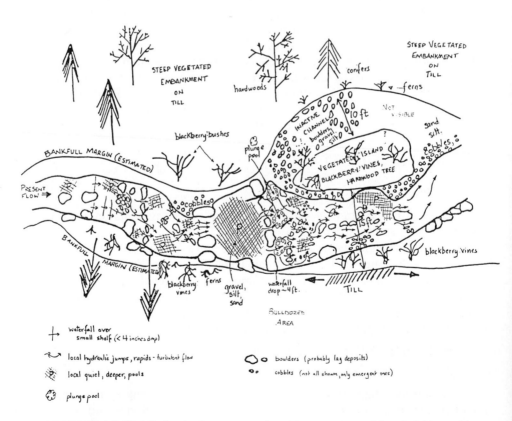

FIGURE 4.6 Sample sketch of a stream reach and features prepared in a student exercise.

drawn at a larger scale may be required to show local hydraulic conditions and habitats.

5. Release a float at the top end of the reach and sketch the pattern of the flow by following it along the bank. Repeat the exercise by releasing a float from different positions across the channel cross section and in local zones of hydraulic jumps and back eddies.

6. Redraw the reach map, eliminating unnecessary survey lines and measurement notes made in the field.

7. From the sketch and cross sections, determine the average bankfull width and depth. Measure the distance in a straight line between riffles if they are present and calculate the ratio of the riffle spacing to the bankfull width.

8. On an overlay of tracing paper, sketch the pattern of flow through the reach at the time of observation and mark in local habitats and flow phenomena. Estimate the area occupied by each flow condition and habitat.

Option b: Advanced Habitat Mapping

1. Measure the depth and mean velocity (at 0.4 depth from the bottom) at the uniform and rapidly varying local flow and habitat sites located in Exercise 1, Option a. At sites where the critical velocity occurs, measure the depth and velocity in the critical zone and in the subcritical zone upstream. Make a small sketch of the flow configuration and note the height of the water upstream above the critical depth zone.

2. Calculate the Froude number at each of the velocity and depth sampling sites and plot them on the reach map.

C. Exercise 2: Streambed Stability and Shear Stress

Option a: Basic Tractive Force and Streambed Stability

1. Measure the x, y, and z dimensions of a sample of the largest materials that pave the streambed using a meter stick. The sample should be randomly selected by walking and selecting the largest sizes projecting from the bed surface every few steps. A minimum of 50 samples should be taken by two observers in the stream with the recorder following along on the bank. If present, pools and riffles should be sampled separately.

2. Carry out steps 2 and 3 in the Basic Mapping exercise above if not done already to determine the average bankfull cross section and slope in the reach.

3. Calculate the mean diameter of the stream paving samples and plot them as a cumulative frequency curve (Fig. 4.7).

4. Using the average slope measurement and average bankfull cross

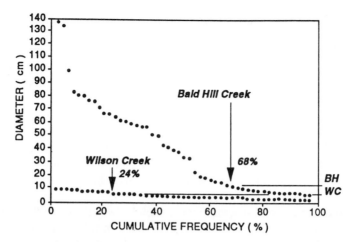

FIGURE 4.7 Cumulative frequency plot of a bed paving material sample showing the percentage stable at the bankfull stage.

section, calculate the tractive force that would be acting on the stream bed at the bankfull stage.

5. Using Lane's tractive force versus sediment size in motion plot, determine the percentage of the bed paving material that would be unstable at the bankfull stage.

Option b: Advanced Local Velocity and Shear Stress

1. At 6 to 8 sites in the reach, particularly where biotic habitats have been observed, measure the velocity of the flow at 5 to 10 increments of the depth above the streambed. Fewer intervals may be taken at shallow sites depending on the size of the flow meter. Record the size of the substrata on the streambed at the sample site.

2. If the water is clear and shallow enough, select a range of sizes of streambed cobbles and gravels and test them on the bed at the measuring site to determine the largest size that can be transported. If FST hemispheres are available, determine the highest number that is stable at the same site.

3. Plot the velocity profile at the sample sites as a function of log y (Fig. 4.8).

4. Estimate the shear velocity from a line fitted to the $u - \log y$ plot.

5. Calculate the local bed shear at the measurement site.

IV. QUESTIONS

1. Compare the channel geometry of your sample reach (the bankfull width, depth, riffle spacing, and meander length) to the average values for

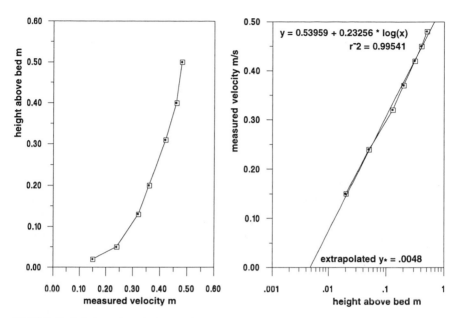

FIGURE 4.8 Sample velocity profile measurements and the velocity profile plotted as $u = f(\log y)$.

a stream in your region with the same tributary drainage area (Dunne and Leopold 1978). Are there features that you mapped in your reach that would explain deviations from the average values?

2. Measure the approximate area of each flow condition mapped in the reach. Discuss how this distribution will change at higher and lower flows.

3. What general associations of species, flow, or substratum conditions were observed? If available, compare one of the habitats to published preference curves.

4. Is there a correlation between the Froude number and the observed habitats in the reach? If available, compare one of the habitats to published Froude number preferences.

5. Estimate the volume of flow in a critical section from the cross-sectional area and velocity. From the upstream velocity measurements, estimate the cross-sectional area that is contributing flow to the critical section. Discuss the advantages and disadvantages of habitats in or near the critical flow zones.

6. If there is a rapids or riffle across the reach, what portion of the total flow would you estimate passes through critical flow zones? Discuss the impact of straightening and channelizing the sample reach into a uniform canal.

7. What was the largest size of bed material that could be transported at the bankfull stage? From your general mapping of the substrata, what areas of the channel bed would be in motion at the bankfull stage? Are these areas associated with any of the observed habitats?

8. Speculate on the strategy that biota must adopt to survive during bankfull flow events and intervening low-flow periods in the specific configuration of your sample reach.

9. (Optional, see Chapter 3) Choose a channel resistance factor and calculate the bankfull discharge from the slope and bankfull cross sections. If flow records are available, adjust them to the drainage area for the sample reach and determine the frequency of the bankfull flow. Discuss the impact of reducing the frequency of this event with flood control structures.

10. Compare the average local bed shear stress predicted from the velocity profiles with the general bed shear stress predicted by the tractive force for the reach slope and average present depth of flow noted in the sample cross sections. Discuss the reasons for variations.

11. Calculate the nominal boundary layer thickness at the velocity profile sites. Discuss the adaptations that benthic insects might have to live in high shear stress areas. Does the nominal boundary layer play a role in this strategy?

12. Compare the bed shear stress predicted by the velocity profile to the cobble test prediction, assuming Lane's relationship applies. Discuss the consistency or reasons for discrepancies in the estimates. If FST hemisphere predictions were available, compare the FST predictions to both methods and discuss the reasons for agreements or disagreements. Which method would you recommend for determining the stability of a cobble bed stream?

V. MATERIALS AND SUPPLIES

Materials

Clipboard and drawing materials
Graph paper (arithmetic and semilog 2 cycle)
Topographic map of the drainage basin containing the sample reaches
Triangular scale (1:50 to 1:250 scales)

Equipment

100-m flexible tape
30-m flexible tape
Calculator with power functions
Floats (e.g., oranges, wood blocks, water-filled balloons)

Meter sticks (2)
Round river cobbles ranging from 1 to 15 cm (may be on site)
Small velocity meter and rod
Stadia rod and small surveyors level with tripod (e.g., construction
    site type)
Stopwatch
Optional: FST hemisphere set

REFERENCES

Bedient, P. B., and W. C. Huber. 1992. *Hydrology and floodplain analysis.* Addison–Wesley, New York, NY.

Bovee, K. D., and T. Cochnauer. 1986. *Development and Evaluation of Habitat Suitability Criteria for Use in the Instream Flow Incremental Methodology.* Instream Flow Information Paper 21, Biological Report 86, United States Fish and Wildlife Service, Fort Collins, CO.

Carling, P. A. 1992. The nature of the fluid boundary layer and the selection of parameters for benthic ecology. *Freshwater Biology* **28:**273–284.

Chang, H. H. 1988. *Fluvial Processes in River Engineering.* Wiley, New York, NY.

Chow, V. T. 1959. *Open-Channel Hydraulics.* McGraw–Hill, New York, NY.

Cobb, D. G., T. D. Galloway, and J. F. Flannagan. 1992. Effects of discharge and substrate stability on density and species composition of stream insects. *Canadian Journal of Fisheries and Aquatic Sciences* **49:**1788–1795.

Davis, J. A., and L. A. Barmuta. 1989. An ecologically useful classification of mean and near-bed flows in streams and rivers. *Freshwater Biology* **21:**271–282.

Dunne, T., and L. B. Leopold. 1978. *Water in Environmental Planning.* Freeman, San Francisco, CA.

Frutiger, A. 1993. Reply to Statzner (1993). *Freshwater Biology* **30:**485–486.

Frutiger, A., and J. L. Schib. 1993. Limitations of FST hemispheres in lotic benthic research. *Freshwater Biology* **30:**463–474.

Gordon, N. D., T. A. McMahon, and B. L. Finlayson. 1992. Stream hydrology: An introduction for ecologists. Wiley, Chichester, UK.

Gore, J. A., S. Niemela, V. H. Resh, and B. Statzner. 1994. Near-substrate hydraulic conditions under artificial floods from peaking hydropower operations: A preliminary analysis of disturbance intensity and duration. *Regulated Rivers* **9:**15–34.

Gregory, K. J., A. M. Gurnell, C. T. Hill, and S. Tooth. 1994. Stability of the pool-riffle sequence in changing river channels. *Regulated Rivers* **9:**35–43.

Lancaster, J., and A. G. Hildrew. 1993. Characterizing in-stream flow refugia. *Canadian Journal of Fisheries and Aquatic Sciences* **50:**1663–1675.

Lane, E. W. 1955. Design of stable channels. *Transactions of the American Society of Civil Engineers* **120:**1234–1279.

Leopold, L. B., M. G. Wolman, and J. P. Miller. 1964. Fluvial processes in geomorphology. Freeman, San Francisco, CA.

Milhous, R. T., D. L. Wegner, and T. J. Waddle. 1989. *Users Guide to the Physical Habitat Simulation System.* FWS/OBS-81/43. United States Fish and Wildlife Service, Office of Biological Services, Washington, DC.

Newbury, R. W. 1984. Hydrologic determinants of aquatic insect habitats. Pages 323–357 *in* V. H. Resh and D. M. Rosenberg (Eds.) *The Ecology of Aquatic Insects.* Praeger, New York, NY.

Newbury, R. W., and M. N. Gaboury. 1993a. *Stream Analysis and Fish Habitat Design.* Newbury Hydraulics, Gibsons, British Columbia, Canada.

Newbury, R. W., and M. N. Gaboury. 1993b. Exploration and rehabilitation of hydraulic habitats in streams using principles of fluvial behaviour. *Freshwater Biology* **29:**195–210.

Newbury, R. W. 1995. Rivers and the art of stream restoration. Pages 137–149 *in* J. E. Costa, A. J. Miller, K. W. Potter, and P. R. Wilcock (Eds.) *Geophysical Monograph 89: Natural and Anthropomorphic Influences in Fluvial Geomorphology: The Wolman Volume.* American Geophysical Union, Washington, DC.

Nowell, A. R. M., and P. A. Jumars. 1984. Flow environments of aquatic benthos. *Annual Review of Ecology and Systematics* **15:**303–328.

Peckarsky, B. L., S. C. Horn, and B. Statzner. 1990. Stonefly predation along a hydraulic gradient: a field test of the harsh-benign hypothesis. *Freshwater Biology* **24:**181–191.

Platts, W. S., and V. E. Penton. 1980. *A New Freezing Technique for Sampling Salmonid Redds.* Forest Service Research Paper INF-248, United States Department of Agriculture, Washington, DC.

Rangaraju, K. G. 1981. *Flow through Open Channels.* McGraw–Hill, New Delhi, India.

Roberson, J. A., and C. T. Crowe. 1993. *Engineering Fluid Mechanics,* 5th ed. Houghton Mifflin, Boston, MA.

Statzner, B. 1993. Response to Frutiger and Schib (1993). *Freshwater Biology* **30:**475–483.

Statzner, B., J. A. Gore, and V. H. Resh. 1988. Hydraulic stream ecology: Observed patterns and potential applications. *Journal of the North American Benthological Society* **7:**307–360.

Statzner, B., F. Kohmann, and A. G. Hildrew. 1991. Calibration of FST hemispheres against bottom shear stress in a laboratory flume. *Freshwater Biology* **26:**227–231.

Statzner, B., and R. Muller. 1989. Standard hemispheres as indicators of flow characteristics in lotic benthos research. *Freshwater Biology* **21:**445–459.

Vogel, S., 1981. *Life in Moving Fluids.* Willard Grant, Boston, MA.

Wetmore, S. H., R. J. Mackay, and R. W. Newbury. 1990. Characterization of hydraulic habitat of *Brachycentrus occidentalis,* a filter-feeding caddisfly. *Journal of the North American Benthological Society* **9:**157–169.

Wolman, M. G. 1954. A method of sampling coarse bed material. *Transactions of the American Geophysical Union* **15:**951–956.

CHAPTER 5

Temperature, Light, and Oxygen

F. RICHARD HAUER[*]
AND WALTER R. HILL[†]

*Flathead Lake Biological Station
The University of Montana
†Environmental Sciences Division
Oak Ridge National Laboratory

I. INTRODUCTION

A. Temperature

Temperature is one of the most important variables in the biosphere. Temperature affects movement of molecules, fluid dynamics, saturation constants of dissolved gases in water, metabolic rates of organisms, and a vast array of other factors that directly or indirectly affect life on earth. Typically, the greatest source of heat in fresh water is solar radiation. This is particularly true for lakes, large rivers, or streams that are exposed to direct sunlight over most of their surface. Many small streams, however, have dense canopy cover under the riparian forest that shades the stream surface. In very heavily shaded streams, transfer of heat from the air and flows from ground water are more important than direct solar radiation in governing stream temperatures (Hauer and Stanford 1982, Stanford et al. 1988).

Generally, streams experience *diel temperature flux*. Range in daily temperatures of more than 5°C is common (Fig. 5.1). Diel temperature flux also may be very high in special environments; for example, in very small alpine streams that have direct solar radiation, afternoon temperatures in late summer may reach >20°C, whereas night temperatures approach 0°C (Hauer unpublished data). Even large rivers that have discharges in excess of 500 m³/s may experience diel temperature ranges of 3–5°C. However, because of the high latent heat of water (which means that adsorbtion or emission of a large quantity of energy is needed to change even 1°C), daily stream temperatures tend to vary much more narrowly than do air temperatures.

Annual fluctuation in stream temperature is very important to stream organisms. Critical life history variables (e.g., reproduction, growth) of lotic plants and animals (from diatoms and aquatic insects to fish and other poikilothermic vertebrates) are regulated by temperature (Ward and Stanford 1982, Sweeney 1984, Butler 1984, Hauer and Benke 1987, 1991). Many stream animals use temperature or temperature change as an environmental cue for emergence (aquatic insects) or spawning (fishes). Temperate, arctic, and montane streams and rivers often freeze during winter, which greatly

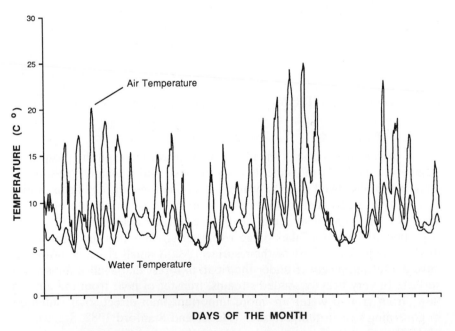

FIGURE 5.1 Hourly air and water temperatures (°C), McDonald Creek, Glacier National Park, Montana (USA) in June 1994.

affects stream discharge, light, dissolved oxygen, and many other variables, and thereby stream biota. Ice may greatly disturb or destroy stream habitats as anchor ice or large blocks of surface ice are pushed by stream flows.

A common misconception is that stream temperatures are uniform among habitats within a stream reach. On the contrary, stream temperature may be highly variable between habitats only a few meters apart. Backwater depositional areas often are much warmer than waters in the stream channel. Particularly in alluvial, gravel-bed rivers, that have high connectivity between channel waters and ground waters, habitats receiving groundwater upwelling may be several degrees colder in summer, or warmer in winter, than the main stream temperatures. Streams frequently express significant changes in temperature from small shaded headwaters to broad, open canopied river reaches. This phenomena is particularly enhanced in mountainous regions where a river may head in alpine environs, but flow through a much warmer downstream climate before confluence with other river waters, a lake, or the ocean (Stanford *et al.* 1988).

B. Light

Light, like temperature, is a critical variable in most ecosystems. In streams, as in all aquatic environments, solar radiation is necessary for photosynthesis by attached algae. It is also the medium through which all visual behavior (e.g., predation by fish) is expressed. Because streams are so closely linked with the surrounding terrestrial landscape, stream light regimes are highly influenced by terrestrial objects such as trees or geologic features. Shade created by an overhanging tree canopy restricts primary production in many streams in undisturbed forests (Hill *et al.* 1995). The longitudinal downstream change in light regime and its consequences for stream bioenergetics is an integral part of stream ecosystems (Vannote *et al.* 1980). In regions where streamside vegetation is not well-developed, shade from steep banks or canyon walls can be important (Minshall 1978). Light attenuation by dissolved organic matter or suspended organic and inorganic particles can reduce light penetration in larger streams, particularly those that carry high suspended sediment loads (see Chapter 7).

Although the wavelengths of solar radiation range from <300 to >5000 nm, the 400- to 700-nm range is of greatest interest to aquatic ecologists because this range of wavelengths is utilized for photosynthesis by autotrophs such as algae or mosses. However, recently increased UV–B radiation, as a consequence of atmospheric ozone depletion, has received interest as a possible interference to primary and secondary production (Bothwell *et al.* 1994). The wavelengths that the human eye can see also correspond roughly to the 400- to 700-nm range and are referred to as *photosynthetically active radiation* (PAR). PAR is measured with quantum sensors: specially designed photocells that quantify the number of photons in the 400- to 700-nm range falling on a specific area per unit time, called *photon flux density* (PFD). Units for this irradiance are μmol quanta m^{-2} s^{-1} or μEinsteins m^{-2} s^{-1}. The Einstein is a mole of photons, but it is not an SI unit and its use is declining. Ecological studies that integrate PAR over longer periods frequently report it as mol quanta (or Einsteins) m^{-2} day^{-1}.

Photon flux densities in stream ecosystems vary tremendously over time. During the course of a sunny day, PFDs in unshaded streams can range from 0 (before dawn or after dusk) to 2000 μmol quanta m^{-2} s^{-1} (midday). Clouds crossing in front of the sun add variability within a day, and clouds associated with passing weather fronts create significant day-to-day variability. Seasonal variation in lotic light regime is caused by changes in sun angle and day length and by phenological changes in stream-side vegetation. For example, the shading effect of vernal leaf emergence decreases PFDs at the surface of streams in deciduous forests by two orders of magnitude (W. Hill, unpublished data).

Spatial variability in lotic light regimes also is high. Variation in the amount of shade cast by streamside vegetation is responsible for much of the spatial variability of light in streams. Gaps in the tree canopy through forested streamsides create significant site-to-site differences in irradiance. One of the important impacts of humans on lotic ecosystems is the altered light regimes caused by removal of streamside vegetation. Streamside vegetation also plays an important role in the longitudinal gradient of light regime in undisturbed stream systems. As stream size increases progressively downstream, riparian trees and bushes shade proportionally less of the stream, allowing more diffuse and direct sunlight to reach the streambed.

C. Oxygen

Dissolved oxygen (DO) greatly affects aquatic life (Eriksen *et al.* 1996) as well as biogeochemical processes (Vallett *et al.* 1990). In most unpolluted streams and rivers, DO concentrations remain well above 80% saturation. Solubility of oxygen increases nonlinearly as temperature decreases and decreases with decreasing atmospheric pressure associated with different elevations or barometric change of weather.

Dissolved oxygen concentrations are not uniform within or between stream reaches. Upwelling of interstitial waters of the hyporheic zone (see Chapters 6 and 30) or side flow of ground waters may create patches of stream bottom where DO is significantly less than that of surrounding waters (Stanford and Ward 1988, 1993). Nearly all stream organisms are sensitive to oxygen concentration. Organic pollution, such as that associated with municipal sewage treatment discharge or industrial wastes, may significantly reduce DO concentrations in entire stream reaches as microbial processes consume the oxygen from the water; this is generally referred to as *biochemical oxygen demand* (BOD). In unpolluted running waters, oxygen concentration may also change dramatically between habitats. Microbial activity within leaf packs and debris dams may reduce oxygen concentrations at the microhabitat level. Streams and rivers that support luxuriant algal growth may experience broad daily ranges in DO as photosynthesis increases oxygen concentration during the day and respiration reduces oxygen concentration at night. Gross primary productivity (GPP), net primary productivity (NPP), and respiration (R) along with the ecological significance of P : R ratios are discussed in Chapter 25.

The objective of this chapter is to present methods of quantifying spatial and temporal variability in temperature, light, and oxygen. Exercise 1 is designed to familiarize the stream researcher with the spatial heterogeneity of these three parameters. The second exercise is designed to demonstrate temporal variability of each parameter, is more involved, and requires an extended field time.

II. GENERAL DESIGN

A. Temperature

Temperature is most simply measured with standard mercury thermometers. Thermometers can be hand-held in most stream habitats, with the obvious exception being habitats within the substratum that present very specific problems in sampling (see Chapter 6). When working in small, clear streams, temperatures may be read directly on the thermometer, even as it is underwater. Inexpensive thermometers are readily available from most scientific supply companies. Some suppliers offer field thermometers with metal or plastic jackets that protect the glass rod and thus reduce breakage. Unfortunately, inexpensive thermometers lack both precision and accuracy; generally measurement is only to within 0.5°C and often several degrees in error. For precise thermographic work, thermometers should have at least a minimum scale marking of 0.1°C. The thermometer should be calibrated regularly against a precision thermometer certified by the National Institute of Standards and Technology (NIST, formerly the National Bureau of Standards).

Temperature is also commonly measured in streams using electronic thermistors. The probe portion of the instrument is submerged at the location that the temperature reading is desired. The probe generally consists of a plastic covered metal that decreases in electrical resistance with increasing temperature. The probe is connected to a battery-operated analog or digital recorder that displays the temperature. Thermistor-type thermometers are generally very accurate ($\pm 0.1°C$), but like mercury field thermometers they should be calibrated regularly against a NIST-certified thermometer.

Dataloggers for monitoring stream temperatures have been available for several decades; however, with the relatively recent advent of microcomputerization, what cost >$2000 a few years ago can now be obtained for <$200. These field instruments use thermistor-type thermometers and record temperature at programmed time intervals. In all cases, the current dataloggers permit downloading of the data into easily handled computer data files.

B. Light

Light sensing devices range in sophistication from ozalid paper meters to spectroradiometers costing over $15,000. For most ecological investigations, quantum sensors are the most appropriate measuring device. Quantum sensors and their complementary meters can be obtained for ~$700–$800. Other photocell sensors that measure irradiance instantaneously, but at different wavelengths, include pyranometers and photometers. Pyranometers measure a broader range of wavelengths (400–1100 nm) than quantum sensors; pyranometer measurements are reported in energy units per area, usually W/m^2, where W is watts (Li-Cor 1992). Photometers measure approximately the same range of wavelengths as quantum sensors, but their sensitivity is weighted to match that of the human eye, which is most responsive to green-yellow wavelengths (ca. 500–600 nm). Photometer measurements are reported in footcandles or lux (SI units). Although quantum sensors are preferred, pyranometers or photometers may be used for instantaneous measurements if quantum sensors are not available.

Pyrheliometers and Ozalid paper meters can be used when an instantaneous measurement of irradiance is not needed. Pyrheliometers measure the light energy absorbed by a black metallic surface, and their measurements, similar to those recorded by the pyranometer, are expressed in energy units $(W/m)^2$. They have been used extensively in limnology and oceanography, but measure a very broad range of wavelengths (300–5000 nm), and have a slower response time than quantum sensors, pyranometers, and photometers. Pyrheliometers are best suited for monitoring solar energy at a single location. Ozalid paper meters provide an estimate of time-integrated photon flux. They are constructed by the investigator from plastic petri dishes and light-sensitive blueprint (Ozalid) paper, which strongly absorbs wavelengths around 410 nm. A stack of photosensitive paper is exposed for periods up to 24 h, and the amount of light energy received during this time is estimated from the number of layers bleached (Friend 1961). Although the Ozalid paper meter technique is sensitive to only a small portion of the PAR range, it can be calibrated against quantum sensors and provides a relatively accurate estimate of PAR, as long as the radiation spectrum does not vary substantially between measurements. The Ozalid papers are small and the technique is very inexpensive and well-suited for synoptic studies where integrated light measurements are made simultaneously at many locations over several hours. However, the overall usefulness of Ozalid paper measures is constrained by the relatively low sensitivity and poor precision. The best choice for obtaining integrated light measurements is a quantum sensor attached to a datalogger programmed to record readings at intervals appropriate for the particular study questions. For general monitoring, we have found recording of the mean, maximum, and minimum PAR for 1 h time intervals taken from readings every 5 min is sufficient for most stream ecology studies.

C. Oxygen

Dissolved oxygen is generally measured using either of two methods; the Winkler method or the membrane–electrode method (APHA *et al.* 1992). Each method has specific advantages and disadvantages. The advantages of the Winkler method are: (1) when performed by experienced persons it can very accurately measure DO with great precision, and (2) it is relatively inexpensive to acquire the necessary titration burrettes, sample bottles, and chemicals. The primary disadvantages of the Winkler method are: (1) one cannot continuously monitor change in DO, but rather must rely on discrete measures, and (2) reducing or oxidizing materials dissolved in the water can interfere with accurate measurement of DO concentration. The advantages of the membrane–electrode method are: (1) ease of use, and (2) one can continuously monitor change in DO, especially in running waters that move water across the probe membrane. The primary disadvantages of the membrane–electrode method are: (1) difficulties associated with instrument maintanence and calibration, and (2) the expense of the recording device and probe (e.g., YSI Model 58 digital recorder, 0–20 mg/liter ± 0.03, ~$1500; YSI O2 and temperature probe with cable, ~$340). However, these disadvantages are negligable if the instrument is used frequently and maintained on a regular schedule.

III. SPECIFIC EXERCISES

The two exercises presented in this chapter combine investigation for the three variables: temperature, light, and DO. Often temperature, light, and dissolved oxygen are related in interesting ways that are not always intuitive. By obtaining readings of each of these variables, both spatially and temporally, researchers can increase their understanding of these relationships.

A. Protocol for Winkler Method Determination of Dissolved Oxygen Concentration in Freshwater Laboratory Preparation of Reagents

Manganous sulfate solution

1. Dissolve 364 g $MnSO_4 \cdot H_2O$ in distilled water, filter through a 1.0-μm glass-fiber filter, and dilute to 1 liter.

Alkali–iodide–azide reagent

1. Dissolve 500 g NaOH (or 700 g KOH) and 135 g NaI (or 150 g KI) in distilled water and dilute to 1 liter.
2. Dissolve 10 g NaN_3 in 40 ml distilled water and add to NaOH and NaI solution.

Sulfuric acid

1. Fill a small glass bottle with concentrated H_2SO_4.

Starch

1. Dissolve 2 g laboratory grade soluble starch in 100 ml hot distilled water.
2. Add 0.2 g salicylic acid as a preservative if the starch solution will be kept for more than 48 h.

Standard sodium thiosulfate titrant (0.025 M $Na_2S_2O_3$)

1. Dissolve 6.205 g $Na_2S_2O_3 \cdot 5H_2O$ in ~900 ml distilled water.
2. Add 1.5 ml 6 N NaOH or 0.4 g solid NaOH and dilute to 1000 ml.

Collection of Sample

1. Collect samples very carefully in narrow-mouth, glass-stoppered, 300-ml BOD bottles.
2. Avoid entraining or dissolving atmospheric gases during sampling. This can be accomplished by using a large-mouth beaker or if a particular microhabitat is to be sampled, one may use a large (>100 ml) plastic syringe to draw water from a specific location in the stream.
3. Fill the BOD bottle to overflowing by 2–3X the volume of the bottle. Prevent turbulence and bubbles during filling of either the sampling device or the BOD bottle.
4. Stopper the BOD bottle by carefully tipping the bottle slightly and

inserting the glass stopper making certain no gas bubbles are entrained in the bottle and immediately proceed to the analysis.

Analysis Procedure

1. Add 1 ml $MnSO_4$ solution to a 300-ml BOD bottle filled with sample water using a glass pipet.

2. Immediately follow by adding 1 ml of the alkali–iodide–azide reagent using separate pipet.

3. Stopper carefully to exclude air bubbles.

4. Mix the sample and reagents by inverting the sample bottle several times.

5. A brown precipitate $[MnO(OH)_2]$ will form in the presence of dissolved oxygen in the sample water:

$$MnSO_4 + 2NaOH \rightarrow Mn(OH)_2 + Na_2SO_4$$

$$2Mn(OH)_2 + O_2 \rightarrow 2MnO(OH)_2.$$

6. When the precipitate has settle to the bottom 1/3 of the bottle, add 1 ml concentrated sulfuric acid with a glass pipet.

7. Restopper the BOD bottle and mix by inverting several times until the precipitate is completely dissolved resulting in the liberation of iodine in direct proportion to the concentration of dissolved oxygen:

$$MnO(OH)_2 + 2H_2SO_4 \rightarrow Mn(SO_4)_2 + 3H_2O$$

$$Mn(SO_4)_2 + 2NaI \rightarrow Mn(SO_4) + Na_2SO_4 + I_2.$$

8. The quantity of iodine present is determined by titrating 200 ml of the sample with 0.025 M $Na_2S_2O_3$. Titrate a volume corresponding to 200 ml of the original sample by correcting for losses as a result of addition of reagents. Thus, for a total of 1 ml $MnSO_4$ and 1 ml alkali–iodide–azide reagents added to a 300-ml sample, titrate:

$$200 \times 300/(300 - 2) = 201.3 \text{ ml.}$$

9. Titrate to a pale straw color. Add a few drops of starch solution forming a blue color. Continue to titrate carefully and slowly to the first disappearence of blue color. Record the volume of titrant used. Note that after 1 min or so a pale blue color may return; however, this should not be titrated.

Calculation

1. For titration of 200 ml of sample (201.3 ml of end product), 1 ml of 0.025 M $Na_2S_2O_3$ = 1 mg/liter dissolved oxygen.

B. Exercise 1: Spatial Variation of Temperature, Light, and Dissolved Oxygen

1. Choose two sections of the study stream that have readily apparent differences in light regime. Ideally, one section would be relatively open and the other heavily shaded by streamside vegetation or by geophysical features. Each stream section should consist of at least one riffle–pool–run sequence, if possible. Within each stream section select and mark five cross-stream transects. The distance between transects will depend upon the length of each section, but transects should be at least several meters apart and intersect different and representative habitat types.

2. Measure temperature, light, and oxygen at nine equidistant points across each transect, including points at both edges of the stream. Measure water temperature at the surface and as close to the stream substratum as possible. While measuring light, if using a quantum sensor or other instantaneous sensor (i.e., pyranometer, photometer), measure and record the photon flux just above the water surface at each point on the transect. For this exercise, oxygen is most easily measured with an oxygen probe. As with temperature, DO should be measured near the surface and at the stream bottom.

3. Begin measurements at the downstream transect and work across each transect before moving to the next upstream transect. Working in research teams of two to four, one or more persons should hold the sensor(s) and meter(s) while one person records the data. Measure and record as quickly as possible to reduce the confounding effects of temporal variation. Make note of changing cloud cover during the measurements.[1] If an underwater sensor is available, measure light at 10- or 20-cm depth intervals at a single deep site to obtain an estimate of light attenuation with depth.

4. For each transect, graph temperature, light, and DO versus transect position (m) for each data point. Conduct an analysis of variance (ANOVA) to determine whether there is greater variation between points within

---

[1]If Ozalid paper light meters are used instead of instantaneous meters, then glue (gel superglue or silicon sealant) individually labeled ozalid meters on the tops of steel rods (e.g., 3/8 to 1/2 inch rebar) driven into the stream bottom at each point on each transect. Uncover the aperture of the meters in timed sequence; the same timed sequence should be followed when the meters are collected later. Allow at least 1–2 h for Ozalid paper exposure, and be sure to retain the identity (section, transect, point) of each meter when developing and reading the meters in the laboratory.

transects or between transects for each section for each of the three variables.

5. Stratify each site of data point collection across each transect into
habitats (e.g., riffle, pool, thalweg, bank margin). Combine temperature
data from each habitat type. Calculate mean, standard error, and coefficient
of variation for each habitat type; compare habitats. Do the same for
DO data.

6. Combine light data from each section into paired frequency histograms labeled open and shaded; use a doubling scale for the x-axis, (e.g.,
0–10, 11–20, 21–40, 41–80, 81–160, 161–320, and >321 μmol quanta m^{-2}
s^{-1}) (see example, Fig. 5.2). Calculate mean, standard error, and coefficient
of variation for each section; compare open canopy and shaded stream
sections. If underwater light measurements were made, calculate the vertical
attenuation coefficient k in the exponential equation

$$E(z) = E(0)e^{-kz}, \tag{5.1}$$

where $E(z)$ and $E(0)$ are the irradiances at a depth of z meters and just
below the surface, respectively (Kirk 1983).

C. Exercise 2: Temporal Variation of Temperature, Light, and
Dissolved Oxygen

1. Locate a representative site on a reasonably homogenous stream
section.

2. Record water temperature, irradiance, and DO at regular time intervals. Intervals of 5, 10, or 15 min are preferred, but 30-min intervals may
suffice if skies are clear.

3. There are numerous manufactures of electronic data logging equipment. Most data loggers can be equipped with temperature probes, quantum
sensors, and DO meters. If available, use a datalogger to record these
variables at 5-min intervals.

4. Plot stream temperature, irradiance, and DO versus time of day.
Plot temperature versus light, oxygen versus temperature, and oxygen versus light.

IV. QUESTIONS

1. Identify the spatial variation in temperature, light, and DO in the
study stream. What appears to be the sources of variation? Are these
sources different between these three variables? Are the sources random
or predictable?

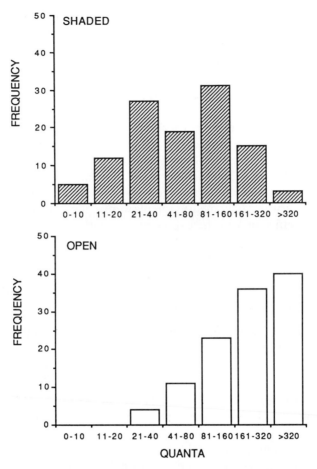

FIGURE 5.2 Example of paired frequency histograms of light intensity (μmol quanta $m^{-2} s^{-1}$) in relatively shaded and open canopy stream reaches.

2. Many investigations characterize stream temperature, light regime, and concentration of DO at a particular site in a stream based on measurements taken at a single point. Based on your data, how accurately would a measurement at a single point reflect these conditions for the stream section(s) you chose? If you measure light at only one time, when should it be done?

3. Assume that photosynthesis by stream algae is limited by insufficient light below 200 μmol quanta $m^{-2} s^{-1}$. At what percentage of sites (Exercise 1) or times (Exercise 2) is photosynthesis light-limited? What if photosaturation irradiance was reduced to 100 μmol quanta $m^{-2} s^{-1}$?

4. Is there a relationship between temperature and oxygen? What about light and oxygen? Is the relationship linear? Why or why not?

5. What are the depths at which light is attenuated to 50, 10, and 1% of surface intensity? Do you think that light attenuation with depth affects algal photosynthesis at your study site? Why or why not?

V. MATERIALS AND SUPPLIES

Field equipment for stream transects

Field notebook
Measuring tapes
Transect markers (rebar, plastic flagging, etc.)

Temperature

Electronic thermistor (±0.1°C)
Mercury thermometers (±0.1°C), or

Light

Quantum sensor (preferred), pyranometer, or Ozalid paper

Oxygen

Winkler method
    300-ml BOD bottle(s)
    500-ml beaker
    Burrett stand
    Pipettes with which to dispense reagents
    Reagents
    Starch bottle with eye dropper
    Titration burrett
Probe method
    Oxygen meter with probe

REFERENCES

APHA, AWWA, and WEF. 1992. *Standard Methods for the Examination of Water and Wastewater,* 18th ed. American Public Health Association, Washington, DC.

Butler, M. G. 1984. Life histories of aquatic insects. Pages 24–55 *in* V. H. Resh and D. M. Rosenberg (Eds.) *The Ecology of Aquatic Insects.* Praeger, New York, NY.

Bothwell, M. L., D. M. J. Sherbot, and C. M. Pollock. 1994. Ecosystem response to solar Ultraviolet-B radiation: Influence of trophic-level interactions. *Science* **265**:97–100.

Eriksen, C. H., V. H. Resh, and G. A. Lamberti. 1996. Aquatic insect respiration. Pages 29–40 *in* R. W. Merritt and K. W. Cummins (Eds.) *An Introduction to the Aquatic Insects of North America,* 3rd ed. Kendall/Hunt, Dubuque, IA.

Friend, D. T. C. 1961. A simple method of measuring integrated light values in the field. *Ecology* **42**:577–580.

Hauer, F. R., and A. C. Benke. 1987. Influence of temperature and river hydrograph on blackfly growth rates in a subtropical blackwater river. *Journal of the North American Benthological Society* **6**:251–261.

Hauer, F. R., and A. C. Benke. 1991. Rapid growth of snag-dwelling chironomids in a blackwater river: The influence of temperature and discharge. *Journal of the North American Benthological Society* **10**:154–164.

Hauer, F. R., and J. A. Stanford. 1982. Ecology and life histories of three net-spinning caddisfly species (Hydropsychidae: *Hydropsyche*) in the Flathead River, Montana. *Freshwater Invertebrate Biology* **1**:18–29.

Hill, W. R., M. G. Ryon, and E. M. Schilling. 1995. Light limitation in a stream ecosystem: Responses by primary producers and consumers. *Ecology* **76**:1297–1309.

Minshall, G. W. 1978. Autotrophy in stream ecosystems. *BioScience* **28**:767–771.

Stanford, J. A., F. R. Hauer, and J. V. Ward. 1988. Serial discontinuity in a large river system. *Verhandlungen der Internationalen Vereinigung für Theoretische und Angewandte Limnologie* **23**:1114–1118.

Stanford, J. A., and J. V. Ward. 1988. The hyporheic habitat of river ecosystems. *Nature* **335**:64–66.

Stanford, J. A., and J. V. Ward. 1993. An ecosystem perspective of alluvial rivers: connectivity and the hyporheic corridor. *Journal of the North American Benthological Society* **12**:48–60.

Sweeney, B. W. 1984. Factors influencing life history patterns of aquatic insects. Pages 56–100 *in* V. H. Resh and D. M. Rosenberg (Eds.) *The Ecology of Aquatic Insects.* Praeger, New York, NY.

Valett, H. M., S. G. Fisher and E. H. Stanley. 1990. Physical and chemical characteristics of the hyporheic zone of a Sonoran Desert stream. *Journal of the North American Benthological Society* **9**:201–215.

Vannote, R. L., G. W. Minshall, K. W. Cummins, J. R. Sedell, and C. E. Cushing. 1980. The river continuum concept. *Canadian Journal of Fisheries and Aquatic Sciences* **37**:130–137.

Ward, J. V., and J. A. Stanford. 1982. Thermal responses in the evolutionary ecology of aquatic insects. *Annual Review of Entomology* **27**:97–117.

CHAPTER 6

Hyporheic Zones

CLIFFORD N. DAHM AND
H. MAURICE VALETT

Department of Biology
University of New Mexico

I. INTRODUCTION

The *hyporheic zone* is a portion of the groundwater interface in streams where a mixture of surface water and groundwater can be found. Original use of the term can be found in the work of Orghidan (1959), who described the interface as a new groundwater environment containing a distinctive biota. The word hyporheic derives from the Greek words for flow or current (*rheo*) and under (*hypo*). Hyporheic zone waters can be found both beneath the active channel and within the riparian zone of most streams and rivers. Interest in this dynamic interface has grown substantially of late (e.g., Stanford and Simons 1992, Valett *et al.* 1993), after Danielopol (1980) and Hynes (1983) argued forcefully for better integration of groundwater and stream research. In a broad sense, the hyporheic zone can be defined as the subsurface region of streams that exchanges water with the surface (Valett *et al.* 1993). Triska *et al.* (1989) provide an empirical perspective of this interstitial environment by recognizing a surface hyporheic zone with >98% stream water and an interactive hyporheic zone where there is >10% but <98% channel water. Vervier *et al.* (1992) prefer a definition that emphasizes the ecotonal nature of the hyporheic zone, in which they stress that the hyporheic zone is a surface water–groundwater ecotone where boundaries are spatially and temporally dynamic. With this perspective, important attributes of the hyporheic zone are (1) the interface of groundwater (flow through porous medium) and channel water (free flow) and (2) the associated gradients in such variables as redox potential (E_h), organic matter content, microbial numbers and activity, and availability of nutrients and light. In general, key components of the hyporheic zone from the various definitions are the character of surface water–groundwater interfaces and the spatial and temporal dynamics of this interface.

It is important for stream ecologists to consider the hyporheic zone when studying streams and rivers. One reason is that this interface is an important habitat for numerous aquatic organisms. Hyporheic zones contain a wide variety of subterranean fauna and zoobenthos, either at various stages of their lives or throughout their life histories (e.g., Coleman and Hynes 1970, Stanford and Gaufin 1974, Williams 1984, Stanford and Ward 1988, Williams 1989, Boulton et al. 1992, Smock et al. 1992, Stanley and Boulton 1993). Much of this fauna is inadequately described and identified, and new organisms and adaptations to subterranean life are frequently being found. In addition, early research on the hyporheic zone focused on fish reproduction, as fish eggs are commonly incubated in this environment (e.g., Pollard 1955, Hansen 1975, Johnson 1980). The recent advances in understanding the role of the hyporheic zone adds significantly to our fundamental understanding of stream ecology and greatly expands the documented physical space that aquatic organisms inhabit, and the region where biotic interactions and production occur. For many streams and rivers, subterranean invertebrate production in the hyporheic zone rivals or exceeds that of the benthos (e.g., Stanford and Ward 1988, Smock et al. 1992). Although difficult to access (Palmer 1993), hyporheic zones hold fascinating biota and surprisingly complex food webs (see Chapters 12, 15, 16, and 30).

A second reason for including the hyporheic zone in studies of stream and river ecosystems is the impact that hydrologic exchange with this zone has on surface stream biota (Chapter 30, Boulton 1993). Hyporheic zone sediments and waters are metabolically active with complex patterns of nutrient cycling which vary spatially and temporally (e.g., Grimm and Fisher 1984, McDowell et al. 1992, McClain et al. 1994). Upwelling waters from the hyporheic zone can deliver limiting nutrients to the stream channel that influence rates of algal primary production, the composition of benthic algal assemblages, and the recovery of stream reaches after disturbance (Valett et al. 1990, 1994, Coleman and Dahm 1990). Stanford and Ward (1993) have described how discrete localized zones of upwelling of hyporheic waters can produce patches of increased biotic productivity within oligotrophic riverine–floodplain ecosystems. Hendricks and White (1988) and Fortner and White (1988) have pointed out how convective water movement in the hyporheic zone affects the distribution of aquatic macrophytes in some streams. Interchange of waters between groundwater and surface water can play a major role in the structure and function of the benthic interface in streams and rivers.

In this chapter, we describe field methods for sampling the hydrology, chemistry, and biota of the hyporheic zone, with emphasis placed on understanding hyporheic zone hydrology. A variety of levels of sophistication are presented, ranging from excavating pits in alluvial sediment to installing a permanent well field. Installation of minipiezometers or PVC wells is a vital part of these protocols and will need to be carried out in advance of the specific procedures described in this chapter. These protocols emphasize that access to the hyporheic zone remains one of the major challenges in studying this interface.

Throughout the rest of this chapter, we use the term groundwater to describe subsurface water that will be sampled. Much of this water will be hyporheic zone groundwater, but in some locations the sampling wells may access groundwater without a surface stream water contribution. Therefore, the more general term of groundwater will be used to identify the water samples collected from the saturated zones of stream sediments.

Specific objectives of this chapter are to: (1) determine the direction and velocity of groundwater flow in the hyporheic zone, (2) measure vertical hydraulic gradients within the groundwater to characterize the direction of vertical flow, and (3) describe sampling protocols for collecting samples for physical, chemical, and biological variables. Choice of which of the three exercises to use in the field will depend on characteristics of the stream, the availability of sampling points to access groundwater in the hyporheic zone, and the magnitude of effort appropriate for research or educational goals.

II. GENERAL DESIGN

Each of the various exercises requires a means to sample groundwater in the region adjacent to the stream or below the active channel. In all cases, a strong back and a stout heart are needed to prepare for sampling of these subsurface environments. Site selection should consider local geomorphology; stream reaches with considerable bedrock exposure at the surface or dominated by large boulders should be avoided. In general, unconstrained reaches (*sensu* Gregory *et al.* 1991) of stream are best to consider for these exercises. Convex bedform where lower gradient segments of stream begin to steepen are generally areas of surface water recharge into groundwater (downwelling), while concave bedform where higher gradient reaches change to lower gradient sections are commonly zones of groundwater discharge (upwelling) (White *et al.* 1987, Harvey and

Bencala 1993). A conceptual view of this interaction between bedform and hyporheic zone flowpaths is shown in Fig. 6.1. An unconstrained reach of stream with a riffle–pool–riffle sequence is a good site for these procedures and, once a proper reach is identified, properties of the hyporheic zone may be investigated with the following techniques.

A. Sampling Pits

A straightforward way of sampling the hyporheic zone is to dig a hole with a shovel and crowbar into the floodplain near the active channel. This is a technique that was utilized by pioneering researchers interested in the interstitial biota. One advantage of this procedure is that it allows an accurate determination of the height of the water table. The elevation of the top of the saturated zone determines the location of the water table at that location. Continue to dig the hole to a depth of 30–50 cm below the water table. The pits may then be used to sample hyporheic water, sediments, chemistry, and biota. Additional holes or a trench should be dug close to the first pit for use in tracing the direction and velocity of groundwater flow. The distance from the main pit to the secondary sampling locations

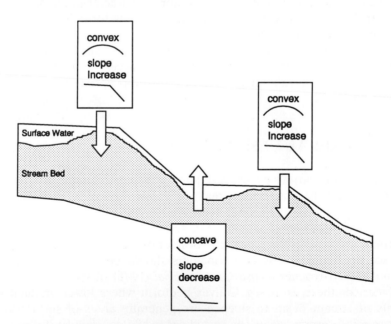

FIGURE 6.1 A schematic representation of changes in the direction of vertical hydrologic exchange (arrows) in response to alterations in stream bed morphometry (after Vaux 1968, White *et al.* 1987).

should consider the texture of the alluvium. Fine-grained alluvium calls for secondary sampling sites within 10–30 cm, whereas pits in coarse-grained alluvium may be placed 50–100 cm distant.

B. Minipiezometers

The vertical direction of groundwater movement can be measured with minipiezometers (Lee and Cherry 1978). Preparation and emplacement of minipiezometers is relatively easy with the proper tools. Minipiezometers also may be connected to a hydraulic potentiomanometer to measure differences in hydraulic head relative to surface water (Winter *et al.* 1988, Boulton 1993). Vertical hydraulic gradient (VHG) measures pressure differentials between piezometers installed to different depths. A positive VHG (greater pressure in the deeper piezometer) indicates upwelling and a negative VHG (lower pressure in the deeper piezometer) indicates downwelling. Similarly, VHG across the groundwater–benthic interface can be determined by comparing pressure differentials existing between a piezometer established beneath the active channel and ambient surface water pressure (Lee and Cherry 1978).

We suggest that minipiezometers be made of 1/2″ (12 mm) PVC pipe. The minipiezometers are inserted into the sediments by placing the plastic PVC pipe around a steel T-bar with a sharpened tip and reinforced handle. A small sledge hammer is used to drive the T-bar into the sediments. In the active channel, minipiezometers are commonly inserted to a depth of 30 cm below the sediment–water interface. While holding the PVC pipe firmly, withdraw the T-bar carefully. Using a 60-ml syringe with the tip extended with Tygon tubing, remove the water from inside the minipiezometer. Check to see if the minipiezometer refills quickly after bailing to assure that it is not clogged. The minipiezometer can now be used for measurement of VHG and for hyporheic water sampling. Minipiezometers can be placed in transects across the active channel and into lateral alluvial sediments or on an upstream–downstream transect to measure either lateral or longitudinal patterns in VHG.

C. Wells

PVC wells of 2″ (51 mm) diameter can be installed to provide permanent access points for sampling the hyporheic zone. Wells can be placed either beneath the active channel or in the floodplain. Although the initial effort of well installation may be laborious and time-consuming, the wells provide sampling opportunities for many years with minimal maintenance. It is best to install wells at times of low discharge. Wells can be emplaced using crowbars and shovels, hand augers, power augers, or drilling equipment. Well depth should extend at least 50 cm below the water table during base flow discharge. Commercially available slotted screen with known slot size is attached with the use of collars to the bottom or segments of the PVC well. The length of slotted screen determines from what depth water will be sampled. For example if the bottom 50 cm of the well is slotted, groundwater will be drawn from that section of the well. The location and length of slotted well pipe which is used can be varied if samples are desired from specific regions of the groundwater. Before inserting the completed well (consisting of a bottom cap, slotted screen, and solid PVC pipe), the base of the excavated hole should be packed with a few centimeters of coarse silica sand (commercially available). The well can then be inserted into the hole and additional sand should be added to cover the length of the slotted screen. This prevents clogging of the well and efficient groundwater flow into the well. The remainder of the hole is then refilled with alluvial material removed during excavation. A surface seal of pelletized bentonite at the top of the well ensures that surface flow will not enter the well from above. Pelletized bentonite is a swelling clay which expands when wetted. Packing a few centimeters of bentonite around the well at the ground surface will prevent water from the surface to flow vertically along the outside of the pipe. Supplies for well installation (PVC pipe, slotted screen, connectors, bentonite, silica sand, etc.) are available from most well-drilling supply stores at reasonable prices.

III. SPECIFIC EXERCISES

A. Exercise 1: Dye Injection—Measuring Groundwater Flow and Velocity

1. This experiment will make use of the hand-dug sampling pits, the minipiezometer network, or the well field. The direction and velocity of groundwater flow will be measured.

2. Inject a dye such as fluorescein or rhodamine WT into a center well or pit and record the time of injection.

3. Sample nearby wells, minipiezometers, or pits for appearance of the dye. Sampling times should be every 5–10 min except in very coarse sediments where sampling may need to be done every minute. Note the time that dye is detected in any of the adjacent sampling locations.

4. Measure the direction and distance of groundwater flow. Calculate the velocity in the appropriate distance and time (e.g., cm/day), and note the direction of flow.

5. Be sure to refill pits at the end of the experiment for the safety of wildlife and people visiting the stream.

B. Exercise 2: Measuring VHG in Minipiezometers

1. Carry out these measurements on all minipiezometers within the study reach. Sampling the height of the water within the minipiezometer can be done using: (a) a calibrated wooden dowel (cm) coated with chalk dust, (b) a voltmeter with leads attached to the base of a calibrated wooden dowel (cm), or (c) a commercial water level recorder.

2. For minipiezometers installed in the active channel, measure the distance from the top of the minipiezometer to the stream surface with a tape measure (cm).

3. Measure the distance from the top of the minipiezometer to the water level inside the minipiezometer. Repeat steps 2 and 3 for all minipiezometers at the study site.

4. For minipiezometers in the stream, calculate VHG for each minipiezometer,

$$\text{VHG} = \frac{h_s - h_p}{L},\tag{6.1}$$

where h_s represents the height of the top of the minipiezometer above stream surface (cm); h_p, height from top of minipiezometer to water in minipiezometer (cm); and L, depth of minipiezometer into sediment (cm). The resulting unitless ratio will be positive in upwelling (groundwater discharge) and negative in downwelling (groundwater recharge) zones.

5. Map out the pattern of upwelling and downwelling zones in the study section of stream. An example of such a map is shown in Fig. 6.2. Figure 6.2A shows the stream bed and surface water elevations, and Fig. 6.2B shows regions of upwelling and downwelling as measured by VHG (cm/cm).

C. Exercise 3: Sampling a Well Field

1. This exercise is designed for those sites where a well field has been installed. A survey map with well locations and elevations is required for the exercise.

2. Measure the height of the water table in each well. This can be done using a commercial water level detector (e.g., Solinst), with a calibrated dowel with leads at the base connected to a voltmeter, or with chalk and a measuring line. Be sure to also measure the stage height of the stream at various locations within the well field.

3. Bail all the wells and allow them to recharge. After recharging, the wells can be sampled for temperature, conductivity, pH, and dissolved oxygen. Measurements are best made with portable field probes inserted

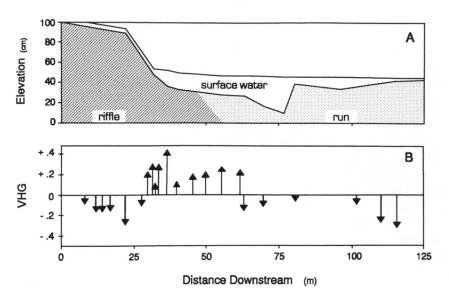

FIGURE 6.2 (A) Longitudinal section of a 125-m study reach at Sycamore Creek, Arizona, illustrating downstream changes in stream bed elevation and morphometry. Riffles are steeper gradient sections where substrata are cobbles/boulders. Runs are lower gradient reaches where substrata are predominantly sand and gravel. (B) Magnitude of vertical hydraulic gradient (cm/cm) along a mid-stream longitudinal axis of the same study reach.

into the groundwater within the wells. If field probes are not available, samples can be drawn into large gas-tight syringes (60 ml) with Tygon extensions and measurements made using standard methods (e.g., APHA *et al.* 1992). The number of physical and chemical measurements should be guided by available instrumentation and questions being asked about the characteristics of hyporheic zone groundwaters.

   4. Hydraulic conductivity of the sediments can be determined with the Hvorslev method (Hvorslev 1951, Fetter 1988). Water is withdrawn by bailing the well with a bailer. Water elevations then need to be taken at various time intervals after bailing. Sampling intervals will need to be short (seconds) for high hydraulic conductivity substrata (i.e., coarse sand, gravel, and cobble) and longer for lower hydraulic conductivity material (i.e., fine sand, silts, and clays). Compute the ratio h/h_0, where h_0 is the height the water level dropped below the static water level immediately after bailing and h is the water level drop at some time t. Plot the ratio h/h_0 versus time on semilogarithmic paper; the time-recharge data should plot on a straight line. Hydraulic conductivity (K) can be calculated with the following formula provided that the length of well screen is greater than 8 times the

well screen radius ($L/R > 8$). For example with a 5-cm-diameter (2.5-cm-radius) well, the length of screened well should be at least 20 cm,

$$K = \frac{r^2 \ln (L/R)}{2LT_0}, \qquad (6.2)$$

where K is the hydraulic conductivity (cm/s); r, radius of well casing (cm); R, radius of well screen (cm); L, length of well screen (cm); and T_0, time for water level to rise to 37% of initial change after bailing.

5. Chemical and invertebrate sampling can also be carried out from groundwater wells. Bailers or syringes can be used to collect samples for chemical analyses. Standard methods for sample preparation and storage should be employed after samples are withdrawn (APHA *et al.* 1992). Special care needs to be given to avoid contact with the atmosphere for samples collected for gas analyses. Sampling for invertebrates requires unscreened wells with open bottoms and sidewall slots, a high-volume diaphram water pump, and a fine-meshed net (e.g., 45-μm plankton net) to collect the organisms exiting the pump. Sampling methods for hyporheic zone invertebrates from wells are presented in more detail by Stanford and Ward (1988) and Hakenkamp and Palmer (1992).

IV. QUESTIONS

1. What is the direction of flow for groundwater at your sampling site? What is the groundwater velocity?

2. What is the pattern of VHG values for the minipiezometers beneath your stream? Is this section of channel gaining or losing surface water due to exchanges with groundwater? Do the locations for upwelling (groundwater discharge) and downwelling (groundwater recharge) conform to the predicted relationship with bedform concavity and convexity?

3. How does VHG change with increasing stream discharge? What happens to the water table along the edges of the channel?

4. Are there measurable changes in water level in groundwater of the hyporheic zone during a diurnal cycle with no precipitation during that period of time?

5. Darcy's Law calculates the flux of groundwater (volume per unit time) with the equation

$$Q = -KA(d_h/d_l), \qquad (6.3)$$

where Q is the flux of groundwater (m³/day); K, hydraulic conductivity (m/day); A, area through which flow occurs (m²); and d_h/d_l, vertical hydraulic gradient (VHG is unitless). Given an area of 1 m², VHG measured with your minipiezometers, and K estimated by the Hvorslev method, calculate the vertical flux of groundwater through the sediment–water interface of your stream. How does this value compare to total surface water discharge through the reach of stream?

6. Are the groundwaters of the hyporheic zone supersaturated, saturated, or undersaturated with dissolved oxygen? Why?

V. MATERIALS AND SUPPLIES

Installation of Sampling Pits or Minipiezometers

1/2″ PVC cut to required length (for minipiezometers)
Shovels, crowbar, work gloves
Sledge hammer
T-bar with reinforced handle for hammering

Sampling

Bailer
Biodegradable dyes
Hydraulic potentiomanometer or water level reader (e.g., chalk and calibrated line)
Sample preservatives/filtration apparatus
Sampling bottles and nets
Stopwatch
Syringe sampler
Tape measure
Temperature, conductivity, pH, and dissolved O_2 portable field probes

REFERENCES

APHA, AWWA, and WEF. 1992. *Standard Methods for the Examination of Water and Wastewater,* 18th ed. American Public Health Association, Washington, DC.

Boulton, A. J. 1993. Stream ecology and surface-hyporheic hydrologic exchange: Implications, techniques and limitations. *Australian Journal of Marine and Freshwater Research* **44:**553–564.

Boulton, A. J., H. M. Valett, and S. G. Fisher. 1992. Spatial distribution and taxonomic composition of the hyporheos of several Sonoran Desert streams. *Archiv für Hydrobiologie* **125**:37–61.

Coleman, M. J., and H. B. N. Hynes. 1970. The vertical distribution of the invertebrate fauna in the bed of a stream. *Limnology and Oceanography* **15**:31–40.

Coleman, R. L., and C. N. Dahm. 1990. Stream geomorphology: effects on periphyton standing crop and primary production. *Journal of the North American Benthological Society* **9**:293–302.

Danielopol, D. L. 1980. The role of the limnologist in groundwater studies. *Internationale Revue der gesamten Hydrobiologie* **65**:777–791.

Fetter, C. W. 1988. *Applied Hydrogeology,* 2nd ed. Macmillan, New York, NY.

Fortner, S. L., and D. S. White. 1988. Interstitial water patterns: A factor influencing the distribution of some lotic aquatic vascular macrophytes. *Aquatic Botany* **31**:1–12.

Gregory, S. V., F. J. Swanson, W. A. McKee, and K. W. Cummins. 1991. An ecosystem perspective of riparian zones. *Bioscience* **41**:540–551.

Grimm, N. B., and S. G. Fisher. 1984. Exchange between interstitial and surface water: Implications for stream metabolism and nutrient cycling. *Hydrobiologia* **111**:219–228.

Hakenkamp, C. C., and M. A. Palmer. 1992. Problems associated with quantitative sampling of shallow groundwater invertebrates. Pages 101–110 *in* J. A. Stanford and J. J. Simons (Eds.) *Proceedings of the First Internatioanl Conference on Ground Water Ecology.* American Water Resources Association, Bethesda, MD.

Hansen, E. A. 1975. Some effects of groundwater on brown trout reds. *Transactions of the American Fisheries Society* **104**:100–110.

Harvey, J. W., and K. E. Bencala. 1993. The effect of streambed topography on surface–subsurface water exchange in mountain catchments. *Water Resources Research* **29**:89–98.

Hendricks, S. P., and D. S. White. 1988. Hummocking by lotic *Chara:* Observations on alterations of hyporheic temperature patterns. *Aquatic Botany* **31**:13–22.

Hvorslev, M. J. 1951. *Time Lag and Soil Permeability in Ground Water Observations.* Bulletin 36, United States Army Corps of Engineers, Waterways Experimentation Station, Vicksburg, MS.

Hynes, H. B. N. 1983. Groundwater and stream ecology. *Hydrobiologia* **100**:93–99.

Johnson, R. A. 1980. Oxygen transport in salmon spawning gravels. *Canadian Journal of Fisheries and Aquatic Science* **37**:155–162.

Lee, D. R., and J. Cherry. 1978. A field exercise on groundwater flow using seepage meters and mini-piezometers. *Journal of Geological Education* **27**:6–10.

McClain, M. E., J. E. Richey, and T. P. Pimentel. 1994. Groundwater nitrogen dynamics at the terrestrial–lotic interface of a small catchment in the Central Amazon Basin. *Biogeochemistry* **27**:113–127.

McDowell, W. H., W. B. Bowden, and C. E. Asbury. 1992. Riparian nitrogen dynamics in two geomorphologically distinct tropical forest watersheds: Subsurface solute patterns. *Biogeochemistry* **18**:53–76.

Orghidan, T. 1959. Ein neuer Lebensraum des unterirdischen Wassers, der hypo-rheische Biotop. *Archiv für Hydrobiologie* **55:**392–414.

Palmer, M. A. 1993. Experimentation in the hyporheic zone: challenges and prospectus. *Journal of the North American Benthological Society* **12:**84–93.

Pollard, R. A. 1955. Measuring seepage through salmon spawning gravel. *Journal of the Fisheries Research Board of Canada* **12:**706–741.

Smock, L. A., J. E. Gladden, J. L. Riekenberg, L. C. Smith, and C. R. Black. 1992. Lotic macroinvertebrate production in three dimensions: channel surface, hyporheic, and floodplain environments. *Ecology* **73:**876–886.

Stanford, J. A., and A. R. Gaufin. 1974. Hyporheic communities of two Montana rivers. *Science* **185:**700–702.

Stanford, J. A., and J. J. Simons (Eds.). 1992. *Proceedings of the First International Conference on Ground Water Ecology.* American Water Resources Association, Bethesda, MD.

Stanford, J. A., and J. V. Ward. 1988. The hyporheic habitat of river ecosystems. *Nature* **335:**64–66.

Stanford, J. A., and J. V. Ward. 1993. An ecosystem perspective of alluvial rivers: Connectivity and the hyporheic corridor. *Journal of the North American Benthological Society* **12:**48–60.

Stanley, E. H., and A. J. Boulton. 1993. Hydrology and the distribution of hyporheos: Perspectives from a mesic river and a desert stream. *Journal of the North American Benthological Society* **12:**79–83.

Triska, F. J., V. C. Kennedy, R. J. Avanzino, G. W. Zellweger, and K. E. Bencala. 1989. Retention and transport of nutrients in a third-order stream in northwestern California: Hyporheic processes. *Ecology* **70:**1893–1905.

Valett, H. M., S. G. Fisher, N. B. Grimm, and P. Camill. 1994. Vertical hydrologic exchange and ecological stability of a desert stream ecosystem. *Ecology* **75:**548–560.

Valett, H. M., C. C. Hakenkamp, and A. J. Boulton. 1993. Perspectives on the hyporheic zone: integrating hydrology and biology: Introduction. *Journal of the North American Benthological Society* **12:**40–43.

Valett, H. M., S. G. Fisher, and E. H. Stanley. 1990. Physical and chemical characteristics of the hyporheic zone of a Sonoran Desert stream. *Journal of the North American Benthological Society* **9:**201–215.

Vaux, W. G. 1968. Intragravel flow in interchange in a streambed. *United States Fish and Wildlife Service, Fisheries Bulletin* **66:**479–489.

Vervier, P., J. Gibert, P. Marmonier, and M. J. Dole-Olivier. 1992. A perspective on the permeability of the surface freshwater–groundwater ecotone. *Journal of the North American Benthological Society* **11:**93–102.

White, D. S., C. H. Elzinga, and S. P. Hendricks. 1987. Temperature patterns within the hyporheic zone of a northern Michigan river. *Journal of the North American Benthological Society* **6:**85–91.

Williams, D. D. 1984. The hyporheic zone as a habitat for aquatic insects and associated arthropods. Pages 430–455 in V. H. Resh and D. R. Rosenberg (Eds.) *The Ecology of Aquatic Insects.* Praeger, New York, NY.

Williams, D. D. 1989. Towards a biological and chemical definition of the hyporheic zone in two Canadian rivers. *Freshwater Biology* **22:**189–208.

Winter, T. C., J. W. LaBaugh, and D. O. Rosenberry. 1988. The design and use of a hydraulic potentiomanometer for direct measurement of differences in hydraulic head between groundwater and surface water. *Limnology and Oceanography* **33:**1209–1214.

SECTION B

Material Storage and Transport

CHAPTER 7

Suspended Sediment and Bedload

ROBERT L. BESCHTA

College of Forestry
Oregon State University

I. INTRODUCTION

For many watersheds, the movement of sediment into stream systems generally occurs by two major processes: surface erosion and mass wasting. *Surface erosion* by water is perhaps most common in arid areas, agricultural areas, or where watersheds have relatively little vegetation or organic matter (litter) on the soil surface. When bare soils occur, rainfall can detach inorganic soil particles and transport them to a channel network by overland flow; additional sediment can be entrained by surface runoff as sheet and rill erosion. *Mass wasting,* or landslides, represent the *en mass* movement of soil, rock, and organic debris downslope by gravity. Various types of mass failures can occur including rockfalls, debris avalanches, debris flows, slumps, soil creep, and others. Each type of failure can deliver a wide variety of particle sizes to a stream system. Landslides and other forms of mass failures usually are the predominate form of erosion in areas with relatively steep terrain. Furthermore, where soil, rock, or previously deposited alluvium are being eroded by a stream or river system, these materials can represent another important source of sediment to aquatic systems.

The amount of sediment that moves into a stream network from hillslopes or other land surfaces or is eroded by fluvial systems can vary greatly among watersheds because of the numerous factors involved in erosional processes. These factors include climate (precipitation and temperature regimes), topography (terrain steepness, aspect), vegetation (type and density of vegetation), soils (particle sizes and erodibility), and geology (characteristics of parent material and bedrock). In addition, human perturbations and management practices that affect watersheds and stream systems can greatly augment natural rates of erosion and sediment yield.

The inorganic soil or rock particles of various sizes that reach a channel may be deposited in the stream or stored along the banks. In other instances, such particles may be transported by the stream or river and be deposited at some downstream location (e.g., a floodplain, channel, lake, or estuary). The detachment, transport, and deposition of inorganic particles by water collectively represent the process of *sedimentation*; inorganic particles that have experienced entrainment, transport, and deposition by flowing water are referred to as sediment.

While the term "sediment" generally refers to only inorganic particles, some sediment deposits may be organic-rich. Similarly, particulate organic matter transport (see Chapters 10 and 11) often occurs simultaneously with the transport of inorganic particles. However, because the quantity of particulate organic matter in a sample, on a mass basis, often is relatively small in comparison to the amount of inorganic particles, and because additional analytical steps are needed to determine the relative amounts of each, from a practical standpoint the amount of organic matter within a "suspended sediment sample" is seldom determined unless it is specifically a portion of the research effort. For example, for streams draining forested watersheds or streams with high densities of riparian vegetation, the amount of organic matter in transport can be relatively high at certain times of the year and thus both organic and inorganic constituents of a "suspended sediment sample" should be determined.

The capacity of a stream or river to transport sediment depends upon a variety of factors (e.g., particle size and availability, flow conditions, channel morphology). Although particle sizes and flow conditions can vary greatly for streams and rivers, inorganic sediments are typically characterized by two primary modes of transport: (1) *suspended sediment* or (2) *bedload sediment*. Each of these categories delineates relatively different groups of particle sizes with different implications for the morphology and ecology of a stream system, and in sampling and analytical techniques.

A. Suspended Sediment

Suspended sediment concentration is defined as the mass of suspended sediment per unit volume of sample and is expressed in units of mg/liter. Sediment particles transported in suspension by a stream are typically less than 0.1 mm in diameter and consist mostly of silt- and clay-sized particles.

Once entrained, the turbulence of flowing water is sufficient to maintain these particles in suspension. Suspended sediment particles are transported downstream at essentially the same velocity as the flowing water.

Although suspended sediment concentrations are naturally variable in many stream systems, several patterns may occur during storm events. For example, suspended sediment concentrations typically increase during periods of increasing flow from rainfall or snowmelt. Concentrations tend to attain their maximum before or near the peak of a storm or snowmelt hydrograph, and then decrease relatively rapidly during the recession limb of the hydrograph (Beschta 1987). During periods of low flow or between storm hydrographs, concentrations typically are low.

When suspended sediment concentrations are plotted as a function of stream discharge over a wide range of flows, a sediment rating curve can be developed for a particular stream or river (Fig. 7.1). Considerable scatter in the data usually occurs around the sediment rating curve; this scatter reflects the high degree of variability of sediment sources, rainfall intensities, snowmelt effects, and other factors. Once a sediment rating curve has been developed, it provides a means of estimating concentrations for flows and periods when suspended sediment samples have not been collected.

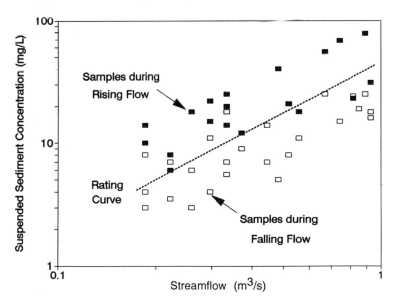

FIGURE 7.1 Relationship between suspended sediment concentration and streamflow (i.e., a sediment rating curve).

Changes or shifting of sediment rating curves also can be used to demonstrate the effects of changing landuses.[1]

Suspended sediment yields also can be accumulated over various time periods. For example, an annual suspended sediment yield represents the total mass (or volume) of sediment exported from a particular watershed or basin over the course of a year.

B. Bedload Sediment

Bedload sediment consists of relatively large inorganic particles that are transported by water along the bed of the stream. Bedload transport discharge (Q_b) is defined as the mass of bedload particles (M_b) per unit time moving past a given stream cross section. Bedload discharge is an instantaneous rate of transport and can be expressed in the same units used for suspended sediment discharge (e.g., kg/s). Similarly, total yields of bedload sediments can be determined over various periods of time (e.g., kg/day, mt/day, mt/year; mt represents metric tons). Although the transport of bedload sediments is not expressed as a simple concentration nor used as an indicator of water quality, these sediments have important implications to aquatic organisms because of their influence on the character of stream substrates and channel morphology.

Sediment particles transported as bedload are relatively large (generally greater than 1 mm in diameter) and consist mostly of coarse sands, gravels, cobbles, or larger. Because of their size and density, the turbulent forces of flowing water are unable to continuously sustain these larger particles in suspension or partial suspension and they tend to move downstream by rolling, sliding, and saltating near the streambed. Saltation consists of a bouncing motion whereby the flow briefly lifts particles into the water column only to have them quickly settle back to the streambed a short distance downstream. Because bedload particles remain largely in contact with the streambed during periods of transport, frictional resistance slows their transport velocities relative to that of the water. Furthermore, because these larger particles tend to move only during periods of high flow, on an annual basis the typical distance that these particles move downstream ranges from only a few meters to several tenths of a kilometer.

Particle sizes in the range of fine to medium sands (i.e., approximately 0.1 to 1 mm in diameter) can be transported either as suspended sediment or as bedload sediment, depending upon flow conditions. At higher flows, these particles tend to be transported as suspended load; at lower flows they would tend to move predominantly near the bed of a channel.

---

[1]Sediment rating "curves" tend to be linear over a wide range of flows and concentrations when plotted on log–log scale graph paper.

As flows increase during a period of rainfall or snowmelt, bedload transport ultimately commences and continues to increase with increasing flows. Similar to suspended sediment, a relationship between bedload discharge and stream discharge (i.e., a bedload rating curve) can be established. However, it is common for such relationships to be even more variable than those associated with suspended sediment rating curves.

Sediments transported as bedload are important for maintaining spawning gravels and channel morphology (pools, riffles, runs, cascades; see Chapter 2) of streams and rivers. Large amounts of bedload transport may scour benthic plants and organisms, fill pools, bury spawning gravels, or cause relatively rapid channel adjustments (e.g., shifts in channel locations, sinuosity, or morphology). In contrast, where bedload transport is reduced below dams, long-term channel downcutting may occur because bedload supplies from upstream sources have been eliminated.

II. GENERAL DESIGN

Measurements of sediment concentration and bedload provide important information about stream systems that has direct significance for aquatic biota. Such data have often been used to assess the effects of human disturbances and alterations to stream channels, riparian areas, or watersheds. The purpose of the following exercises is to provide an understanding of the respective methodologies associated with sampling and measurement of suspended sediment concentration and bedload discharge in streams.

A. Site Selection

Suspended sediment and bedload transport theoretically can be sampled anywhere along a stream or river. However, the sampling methodologies, the types of sampling equipment employed, and the general need for concurrent flow data mean that certain types of channel locations are preferred. Companion measurements of streamflow are required to allow calculation of sediment discharges, develop rating curves, or determine yields for sediment budgets. Because of the desire to have reasonably uniform flow patterns and for a stream to be a wadeable for streamflow determinations (unless a bridge or cable system over the stream is used), measurements are generally obtained in relatively straight and shallow channel reaches that lack large flow obstructions (see Chapter 3). Thus, field sites for streamflow measurements are typically located in riffles that are uniformly shaped. These locations also provide suitable sites for sediment sampling.

B. Suspended Sediment Sampling

Grab Samples   A grab sample, consisting of holding a clean open container below the water surface until it fills, is a simple means of acquiring a water sample for suspended sediment analysis. This procedure may provide an appropriate sample if the stream is relatively shallow (≤ 0.5 m) and well mixed. However, grab samples are generally considered inadequate for accurate determinations of suspended sediment concentration in most streams and rivers. This is true because different sized particles have different settling velocities and thus the concentration of suspended sediment tends to increase with depth. Furthermore, flow velocities are generally much greater near the center of a stream than near the banks and edges of the channel and these flow patterns also influence concentrations.

Depth-Integrated Samples   To help account for changes in suspended sediment concentrations that occur with depth, depth-integrating samplers are employed. A commonly used depth-integrating sampler is the DH-48 sampler (where DH-48 identifies it as a depth-integrating, hand-held sampler developed in 1948). The sampler consists of a 0.47-liter container, mounted at an angle in a holder that is attached to a wading rod (Fig. 7.2). As the sampler is lowered and raised in the water column, a water–sediment mixture flows into the nozzle and container and displaces air that escapes through a small exhaust vent. The sampler's 6.4-mm-diameter intake nozzle is designed to remain 7–8 cm above the streambed so that when the sampler is fully lowered into a stream and is touching the bed, sediments moving as bedload are not collected by the sampler.

Pumping Samplers   Both grab and depth-integrated sampling techniques require personnel to be present in the field during sampling operations. When frequent sampling is required over a specific period of time or when multiple samples are required over periods of hours, days, or weeks, pumping samplers can provide an efficient means of obtaining the samples. A pumping sampler consists of an intake system, a pump, an electronics package designed to control sampling frequency, and a power source. Pumping samplers are generally self-contained, field-portable units that can greatly increase the effectiveness of a sampling program. Various types of pumping samplers are commercially available, ranging from relatively simple units that combine or composite all samples into a single large container to those that pump individual samples into separate containers.

The placement of the pumping sampler's intake is an important concern as it may become quickly fouled by floatable debris (e.g., leaves, twigs, branches, or other materials), particularly when the intake is mounted in

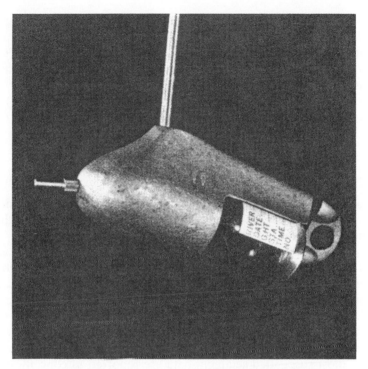

FIGURE 7.2 DH-48 depth-integrating sampler, 0.47-liter container, and wading rod.

a fixed position. However, by mounting the nozzle on a free-swinging rod supported from above the water surface (from a bridge or cable across the stream), the potential for intake plugging can be greatly reduced or alleviated (Beschta 1980).

Sample Handling Each suspended sediment sample or subsample should be clearly labeled in the field such that it has a unique identity and cannot be confused with any other sample when taken to a laboratory. For laboratory analysis, a sample volume of 100 to 500 ml is usually adequate for accurately determining suspended sediment concentration. If a sample cannot be analyzed within 24 h, it should be stored in a dark, cool environment (to prevent microbial growth) and analyzed within 7 days.

C. Suspended Sediment Analysis

The determination of a sample's suspended sediment concentration is a gravimetric procedure requiring the separation of sediment from water and is typically done in a laboratory. A measured volume of sample (V_s)

is drawn via a vacuum through a preweighed, glass-fiber filter. Water and dissolved materials pass through the filter while the sediment is retained on its surface. The filtered sediment and filter are oven-dried at 105°C for 24 h and weighed. The amount of suspended sediment (M_s) in the filtered sample is found by subtracting the original weight of the filter paper. The suspended sediment concentration (C) is calculated from Eq (7.1) and expressed in mg/l:

$$C = \frac{M_s}{V_s}. \tag{7.1}$$

If several samples were collected from a transect across the stream, the average suspended sediment concentration would be calculated as arithmetic mean of the sample concentrations.

For streams draining forested environments or where the riparian vegetation provides high loadings of leaves or needles, a significant portion of the calculated "suspended sediment concentration" may be composed of organic matter. The relative proportions of inorganic and organic fractions can be determined by combusting the previously dried sample at 550°C for 24 h, reweighing the sample ash, and calculating the relative proportions of organic and inorganic solids (see Chapter 10 for detailed methods).

While most suspended sediment data are reported as concentrations, this information also can be useful for other determinations. For example, suspended sediment discharge (Q_s) represents the mass of sediment transported past a given stream cross section per unit time and is calculated as

$$Q_s = C \times Q = \frac{M_s}{V_s} \times \frac{V}{T}, \tag{7.2}$$

where C represents sediment concentration; Q, stream discharge; M_s, suspended sediment mass; V_s, volume of sample; V, volume of water; and T, time. When average concentrations and streamflows are known for a particular period of time (T_p), suspended sediment yields (Y_s) can be determined by

$$Y_s = C \times Q \times T_p = \frac{M_s}{V_s} \times \frac{V}{T} \times T_p. \tag{7.3}$$

Suspended sediment yields can be expressed in units of kg/day, mt/year, or other comparable units. Whereas suspended sediment discharge provides a measure of the instantaneous level of sediment in transport, suspended sediment yields integrate the effects of changing flows and concentrations over time and thus indicate the total mass of suspended sediment exported from a watershed during the time period of interest.

Suspended sediment concentrations often are used as a general indicator of water quality for domestic, municipal, and industrial users. High suspended sediment levels also have implications to stream organisms in that visual feeders are less efficient, gill damage to fish may result, and the siltation of spawning gravels and streambeds may occur. Both sediment concentration and sediment yield data are used for evaluating the effects of landuses and other watershed disturbances on stream systems.

D. Bedload Sediment Sampling

Hand-Held Sampler Because of variable hydraulic conditions in streams, a varying composition of material in transport along the streambed, and the effects of bedforms and large roughness elements along the channel (see Chapter 2), bedload transport can vary substantially at a given cross section and from one location to another across the bed of a stream. Thus, to adequately characterize bedload transport, a number of bedload subsamples should be obtained systematically at equally spaced points across the channel. These subsamples are then combined to form a single composite sample. The entire sampling procedure should be repeated several times to help ensure that a reasonable representation of a stream's bedload transport has been obtained for a given flow condition.

The Helley–Smith pressure differential sampler is a hand-held sampler commonly used for bedload measurements. The sampler has a 0.076-m square orifice, a flared body that develops a pressure differential at the orifice and ensures that bed material in transport will continue into the sampler, and a 0.2-mm mesh catch bag attached to the downstream end (Helley and Smith 1971). This sampler originally was developed with a triangularly shaped catch bag approximately 45 cm long and with a surface area of 2000 cm². However, a cylindrically shaped catch bag approximately 0.9 m long with a surface area of 6000 cm² (Fig. 7.3) greatly reduces the potential of the mesh clogging during use. If the mesh becomes clogged, sampling efficiency is reduced drastically and bedload discharge rates are underestimated (Beschta 1981).

With a Helley–Smith sampler, bedload samples are collected from multiple locations across the channel. Depending upon stream size, as few as 5 or as many as 20 subsamples may be needed to adequately sample a given cross section. When the entire cross section has been sampled, a single composite bedload sample will have been obtained. This composite sample can then be processed in the field or later in the laboratory.

Sample Handling For bedload samples, as with suspended sediment samples, completeness of labeling is of critical importance. Additional infor-

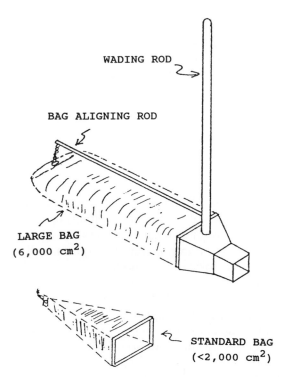

FIGURE 7.3 Helley–Smith bedload sampler, wading rod, and an enlarged sampling bag for improved sampler efficiency in streams with high sand or organic matter transport. For comparative purposes, a "standard" bag is also shown.

mation required of bedload samples includes the duration of sampling, number of subsamples in a cross section, and wetted width of the stream. If bedload samples are not processed in the field, they should be labeled and stored in durable bags or other containers for transport and storage.

Other Bedload Measurements   Bedload transport is a complicated and highly variable phenomenon. Although differing types of instream sampling devices have been developed (i.e., basket, box, pan, and pit samplers (Graf 1971, Dejia *et al.* 1981)), each has various advantages and limitations. An estimate of sediment yield (potentially including both suspended and bedload sediments) sometimes can be obtained by measuring the volume of material deposited in sediment basins, lakes, or reservoirs. Repeated topographic surveys over time can be used to provide a measure of the change in sediment storage, which in turn can be used to calculate sediment yields from a drainage.

E. Bedload Analysis in the Field and Laboratory

Volume and Weight Bedload samples can be heavy and bulky, thus making them difficult to transport. As a result, they are somtimes processed in the field. The total mass of bedload sediment (M_b, in kg) in a sample can be determined by direct weighing of the samples using a balance or scale with sufficient capacity. Once the sample has been weighed, the instantaneous bedload transport rate (Q_b, in kg/s) can be calculated from the subsample duration (T, in s), the number of subsamples (N), and the wetted width (W, in m) of the channel

$$Q_b = \frac{M_b}{T} \times \frac{1}{N} \times \frac{W}{0.076\,\text{m}}. \tag{7.4}$$

If it is not possible to weigh the sample, the amount of sediment may alternatively be estimated using volumetric techniques. For example, the sample can be placed into a graduated container and the total volume (V_b, in liters) determined. By assuming a specific weight (SW_b, in kg/m^3) that is characteristic for unsorted sediments (Table 7.1), the total mass of bedload sediment (M_b, in kg) can be estimated as

$$M_b = V_b \times SW_b. \tag{7.5}$$

Wet Sieving If information concerning particle sizes of bedload in transport is desired, samples can be wet-sieved in the field. Sieving of samples provides additional information about the types of material in transport and often is summarized in graphical form (e.g., Fig. 7.4). Although weighing of samples often can be accomplished efficiently in the field, wet-sieving is more time consuming and may slow the collection of bedload samples. Thus, in many instances bedload samples are simply labeled, placed in storage bags or containers, and transported to a laboratory for subsequent analysis.

TABLE 7.1
Ranges of Specific Weights for Unsorted Sediments (from Gottschalk, 1964)

Particle size	Specific weight (kg/m^3)
Sand	1350–1600
Gravel	1360–2000
Sand and gravel mixture	1520–2080

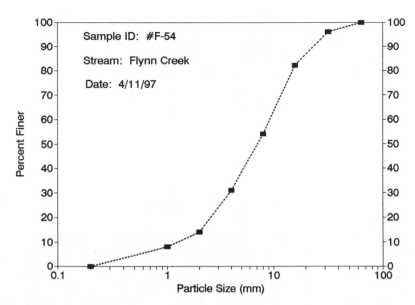

FIGURE 7.4 Particle size distribution (on a weight basis) for sieved bedload sediments.

Laboratory Analysis Bedload samples brought in from the field should be oven-dried (105°C for 24 h) and then weighed. If there is a significant amount of organic matter in the sample that cannot be removed easily by hand, the sample should be combusted before continuing. Placing a sample in a muffle furnace (550°C for 24 h) generally is sufficient to oxidize all organic matter. After cooling to room temperature, the ashed sample is weighed again; this weight represents the total mass of inorganic sediments (M_b) in the sample. Equation (7.4) then can be used to calculate the bedload transport rate. Should information be desired regarding the particle size distribution of bed material in transport, a series of nested sieves is used to separate various size classes. The amount of sediment retained on each sieve is weighed and a cumulative distribution, on a mass basis, is developed for the sample (Fig. 7.4).

III. SPECIFIC EXERCISES

A. Exercise 1: Suspended Sediment

Sampling Protocol A riffle typically is chosen for sampling because of the relatively shallow depths encountered and because flow streamlines are generally parallel. If sampling is undertaken during periods of high flow, when suspended sediment levels typically are high, caution must be exercised in traversing the stream. Not only is it difficult to see the bottom at high flows for stepping onto stable substrata but water velocities may make it difficult to move safely about.

    1. Stretch the measuring tape across the stream and anchor it on each streambank. Determine the wetted width of the stream and subdivide this width into several equal-width subsections (approximately three to five subsections will be adequate for most small streams although larger streams and rivers will require more). With a DH-48 depth-integrating sampler, collect a suspended sediment subsample at the mid-point of each subsection.

    2. At each sampling location orient the intake nozzle of the DH-48 sampler directly upstream as the sampler is lowered at a uniform rate from the water surface to the bed of the stream and back to the surface again. The sample container should be approximately 2/3 to 3/4 full when removed from the water.[2] Individual subsamples from a given cross section can be analyzed separately in the laboratory and used to calculate an average suspended sediment concentration for the cross section. However, if the subsamples are to be combined into a single composite sample for analysis, it is particularly important that the DH-48 sampler be lowered and raised at the same uniform rate for all subsamples in a given cross section. This latter sampling procedure is often referred to as the equal-width increment (EWI) or equal transit rate (ETR) method (USGS 1977).

    3. Label the sample and either (a) store it in a separate container or (b) combine it with other subsamples in the cross section to form a single composite sample. Samples should be sequentially numbered and then labeled immediately as to date, time, stream, station number or reach location, stage (gage height) or streamflow, and initials of the person doing the sampling; each sample should have a unique identity.

Laboratory Analysis Protocol   Laboratory analysis of suspended sediment is a gravimetric procedure requiring the separation of the sediment from the sample.[3] It is accomplished by doing the following:

---

[2]Caution must be taken to not overfill the sampler.

[3]If suspended sediment samples are not available or cannot be obtained for laboratory analysis (e.g., because of seasonally low streamflow), samples suitable for analysis can be improvised. Analysis of such samples provides an opportunity for researchers to develop experience in sample processing as well as establishing the reproducibility of analytical results. To create such samples, a volume of soil needs to be sieved through a 0.1-mm mesh, or finer, soil sieve. The particles that pass through the fine-mesh sieve are added to distilled water to create suspensions of various concentrations.

1. The desired amount of sample needed for filtration purposes generally is 100 ml. Thus, if field samples are considerably larger than this or have relatively high sediment concentrations, only a portion of the sample may be used. However, before splitting or removing part of the sample for analysis, thoroughly resuspend the sediment in the field sample by vigorously stirring or shaking. Then, before suspended particles can settle, withdraw approximately 100 ml of suspension by pouring into a clean graduated cylinder. In some cases, it may be more convenient to use large pipettes and syringes to obtain approximately 100 ml of suspension for analysis. Visually determine the exact volume of sample (V_s, in ml) in the graduated cylinder and record the amount.

2. Previously oven-dried (105°C for 24 h) filters need to be available for filtering the sample. However, before beginning the filtering process, weigh the filter (in mg) and record its mass.

3. Place the filter onto a filtration funnel and initiate a vacuum. Pour the suspended sediment subsample from the graduated cylinder into the filtration funnel and continue filtration. Use a wash bottle to rinse residual particles from the cylinder into the funnel; near the end of the filtration use the wash bottle to rinse any sediment adhering to the sides of the funnel onto the filter paper. After all the sample's sediment has been trapped on the filter, stop the filtering process by releasing the vacuum. Carefully remove the filter and its attendant sediment, lay it on a small metal or porcelain holder, and place the holder into an oven for drying. After drying (105°C for 24 h), carefully remove the holder from oven, reweigh the filter with its attached sediment and record the combined mass.

4. Subtract the original dry weight of the filter from the combined dry weight of the filter and the sediment. This weight represents the amount of suspended sediment (M_s) in the sample.

5. From the mass of sediment (M_s, in mg) and the volume of sample (V_s, in ml), calculate and record the suspended sediment concentration (C, mg/liter) of the sample using Eq. (7.1).

B. Bedload

Bedload Sampling and Field Analysis Protocol Generally, a riffle of relatively uniform channel dimensions is selected for bedload sampling. However, because bedload transport normally occurs during periods of high flow when deep water and swift velocities are common, care must be exercised regarding the safety of personnel when sampling under such conditions. Bedload samples can be safely obtained during high flows in small streams by carefully wading in the channel, but for larger streams a small temporary bridge may be needed from which the sampler can be lowered to the bed of the stream. For yet larger streams and rivers, bedload sampling from cableways or bridges may be required. Once at a stream cross section, a bedload sample can be obtained as follows:

1. Stretch the measuring tape across the stream and anchor the tape at both ends. Determine the wetted width of the stream and subdivide the width into a minimum of five or more equal-width subsections.

2. At the middle of each subsection, collect a subsample with the Helley–Smith bedload sampler. Lower the sampler quickly through the water column and place firmly on the streambed with the orifice oriented directly upstream. Care must be taken not to scoop bed material into the sampler as it becomes positioned on the bed of the stream. After leaving the sampler on the bed for a specific length of time, raise the sampler quickly and remove from the water. The length of time the bedload sampler remains on the bed should be the same for each subsample in a given cross section and needs to be recorded for later use. A sampling time of 30 s or less may provide an adequate sample of bedload sediments, but actual duration will depend upon the amount of material in transport at the time. The sample bag should be no more than 20–25% full of bedload sediments when it is retrieved from the stream.

3. Once the sampler has been removed from the water, carry it to the streambank and empty it into a pan or other open container. Pour stream water onto the outside of the catch bag to rinse all sediments from the sampler. Once the bag is empty of sediment, return the sampler to the next location in the cross section to obtain another subsample. Repeat this procedure until the desired subsamples have been obtained from across the channel. If desired, combine these subsamples to form a single, composite sample.

4. As with suspended sediment samples, bedload samples should be sequentially numbered and labeled immediately as to date, time, stream, station number or reach location, stage (gage height) or streamflow, and initials of the person doing the sampling. In addition, the length of time that the sampler was on the bed of the stream for a given subsample (T, in s), the number of subsamples in a cross section (N), and the wetted width (W, in m) of stream need to be recorded for each composite sample. If bedload samples are to be analyzed in a laboratory, they should be stored in durable bags or other containers for transport and storage. For samples that are to be processed in the field, procedures may vary somewhat depending upon the availability of equipment.

5. If the objective is to quantify the total amount of bedload sediment and its grain-size distribution remove all readily visible organic matter (e.g.,

twigs, leaves) from the sample. Retaining organic matter in the sample not only can cause an overestimate of sample weight, but can also hamper subsequent sieving operations. After the organic matter has been removed, the sample can be simply weighed with a balance or scale, and its total mass recorded as wet weight (in kg). Because sediment particles will retain moisture on their surfaces, the wet weight of the bedload sample will be higher than that for a sample that has been dried. Thus, a number of the samples weighed in the field should be retained and later dried in a laboratory oven (105°C for 24 h) to determine "dry weight." The ratio of average dry weights to average wet weights is used to adjust field-measured wet weights to obtain the total mass (M_b, in kg) of bedload in individual samples. As an alternate field procedure for determining sample mass, place the bedload sample (after any organic matter has been removed) into a graduated container and record the total volume (V_b, in liters). By assuming, or determining in the laboratory, a specific weight (SW_b, in kg/m^3) for the bedload sediments being evaluated (see Table 7.1), the total mass (M_b, in kg) of bedload sample can be determined using Eq. (7.5).

6. After the mass of bedload associated with a particular sample has been determined, the bedload transport rate (Q_b, in kg/s) for the stream cross section can be calculated in conjunction with information on subsample duration (T, in s), number of subsamples (N), and channel wetted width (W, in m) using Eq. (7.4).

7. Additional information regarding particle sizes in transport can be obtained by wet-sieving the bedload sample. Place a bedload sample onto the top of a series of sieves (grading from the largest sieve on top to the smallest on the bottom) and wash the sediment particles down through the sieves with stream water. Weigh and record the amount of sediment retained on each sieve. From the individual sieve weighs, the percentage of the total sample represented by various ranges of particle sizes and cumulative percent finer can be calculated and graphed (Fig. 7.4). If sieved weights cannot be determined in the field, the volume of material retained on each sieve can be determined and used as a basis for estimating the relative amount of material in each size class.

Protocol for Laboratory Analysis

1. If a significant amount of organic matter is readily visible, it should be removed by hand prior to weighing the sample. However, if the amount of organic matter in transport along the bed of the stream is of interest, the organic material should be oven-dried (105°C for 24 h). After drying, record the weight (M_o, in kg) of this material. The remaining sample of inorganic sediment is oven-dried (105°C for 24 h) and weighed. This weight,

assuming the organic matter content of the sample is negligible, represents the total mass of bedload (M_b) for the sample. For many streams that drain forested watersheds, the amount of organic matter (e.g., partially decomposed leaves, twigs, cones) can be large and is not easily removed by hand. Including this material in the dry weight of a bedload sample will cause an overestimate of sediment in transport. However, the amount of organic and inorganic sediment can be accurately determined. After weighing, the oven-dried sample is placed in a muffle furnace (550°C for 24 h) to oxidize the organic matter. The furnace needs to have a vented hood so that the products of combustion are removed from the laboratory. The combusted sample is removed from the oven, allowed to cool, and weighed. The amount of organic matter (M_o, in kg) that was in the bedload sample can be calculated as the difference between the weight of the original sample (M_{total}, in kg) and the weight of the inorganic sediment (M_b).

2. The amount of bedload sediment determined in the previous step is used, in conjunction with previously recorded field information (i.e., length of subsample time (T, in s), number (N) of subsamples per cross section, and wetted width (W, in m) of stream cross section), to calculate

TABLE 7.2

Example Form and Sample Data Associated with Wet- or Dry-Sieving of Bedload Sediment Sample

Sample ID _____   DATE _____
Oven-dry Wt. _____   SAMPLE ANALYZED BY _____
COMMENTS _____
_____

Sieve size	Sieve number	Mesh size		Weight retained (kg)	Percent of total weight	Cumulative percent finer
Smallest	1	1	mm	_____	_____	_____
	2	2	mm	_____	_____	_____
	3	4	mm	_____	_____	_____
	4	8	mm	_____	_____	_____
	5	16	mm	_____	_____	_____
	6	32	mm	_____	_____	_____
Largest	7	64	mm	_____	_____	_____
		Totals =		_____	100%	

a total bedload transport rate (Q_b, in kg/s) for the stream cross section by using Eq. (7.4). Similarly, a transport rate for particulate organic matter moving along the stream bed also can be calculated.

3. Once the organic matter in a sample has been removed by hand or has been combusted, further processing of the sample can provide particle size information of the bedload sediment. In contrast to the wet-sieving procedures for particle size analysis utilized in the field, dry-sieving techniques in the laboratory are more accurate and generally can be accomplished more efficiently. The bedload sample, which has either been oven-dried with organic material removed by hand or has been combusted in a high-temperature oven, is placed onto the top of a series of nested sieves that are then placed onto a mechanical shaker. The sample is shaken for a specific length of time (usually several minutes is sufficient) after which the sediment trapped on each sieve is individually weighed and the weights recorded (Table 7.2). From the individual sieve weights, the percentage of the total sample represented by various particle sizes and cumulative percentage finer can be calculated in Table 7.2 and graphed (e.g., Fig. 7.4).

IV. QUESTIONS

1. Assume that an analysis of suspended sediment concentration is underway and 215 ml of sample have been filtered. After oven-drying and subtracting the weight of the filter, the weight of the sediment on the filter is 365 mg. The sample was then placed in a high-temperature oven (550°C for 24 h) where it experienced a 24-mg weight loss. What is the concentration (mg/liter) of suspended sediment for this sample? What proportion (%) of the original sample was organic matter?

2. Assume that on 3 consecutive days, average streamflow (Q) and suspended sediment concentrations (C) were, respectively, 3.4 m³/s and 189 mg/liter, 4.8 m³/s and 483 mg/liter, and 4.1 m³/s and 152 mg/liter. Determine the suspended sediment discharge (kg/s) for each day. Also determine the suspended sediment yield (kg) for the 3-day period.

3. A bedload sample was obtained at a cross section of Deer Creek when streamflow was at 2.3 m³/s. A Helley–Smith bedload sampler was used to obtain the sample under the following conditions: each subsample was on the streambed for a duration (T) of 30 s, the total number of subsamples (N) obtained across the channel was 6, and the wetted width (W) of the stream was 4.8 m. The subsamples were combined in the field and the resultant composite sample was processed in a laboratory. The

oven-dry (105°C for 24 h) weight of the sample was 850 g. Sieving the sample yielded the following:

Sieve size (mm)	Weight retained (g)
2	30
4	110
8	145
16	320
32	130
64	95

Assuming that organic matter was not present in the sample (or had been effectively removed before sample processing), (1) calculate the bedload discharge (kg/s) represented by the sample and (2) plot the sample's particle size distribution.

4. Suspended sediment data and bedload data are seldom available for most streams and rivers in the United States. Identify and discuss several reasons why this situation occurs.

5. Identify potential biological impacts associated with each of the following:

(a) increased concentrations of suspended sediment.
(b) increased levels of bedload transport

Which impacts are similar and which are different? Which impacts are likely to be short-lived? Which are not? How can various land uses (e.g., agriculture, forestry, grazing, mining, urbanization, or others) affect suspended sediment concentrations and bedload discharge? Describe a possible monitoring plan that could be used to assess the effects of a particular land use pattern on suspended sediment transport and on bedload transport.

V. MATERIALS AND SUPPLIES

Because streamflow is such an important factor relative to the collection and interpretation of both suspended and bedload sediment transport, concurrent measurements of streamflow (or at a minimum stage) should be undertaken at field sampling sites. However, stream-gaging equipment is not specifically included in the following listing of materials and supplies (but see Chapter 3).

Suspended Sediment Sampling

DH-48 depth-integrating sampler, collection bottle, and attached wading rod
Measuring tape of sufficient length to reach across stream
Sample labels and recording forms
Several 500-ml glass bottles or other clean containers for storing samples.

Laboratory Analysis of Suspended Sediment

100-ml graduated cylinders
100-ml syringes or pipettes
Drying oven (set at 105°C)
Filters (glass fiber)
Glassware (numerous 500-ml to 1 liter beakers)
Metal weigh boats (for holding filter papers and sediment during drying)
Precision balance (±0.1 mg)
Recording forms
Thermally padded gloves (for handling high-temperature samples)
Vacuum pump and filtration apparatus (funnels)
Wash bottles

Bedload Sampling and Field Analysis

Containers (for holding sieved fractions)
Durable bags or containers (for storing bedload samples)
Field-portable balance or spring scale
Helley–Smith hand-held bedload sampler
Measuring tapes
Sample labels and recording forms
Series of sieves (e.g., 2-, 4-, 8-, 16-, 32-, and 64-mm sieves)
Water containers (for washing sediments through nested sieves)

Laboratory Analysis of Bedload Samples

Drying oven (high-capacity set at 105°C)
Mechanical shaker (for use with sieves)
Muffle furnace (high-capacity set at 550°C)
Recording forms
Rugged laboratory balance
Series of sieves (e.g., 2-, 4-, 8-, 16-, 32-, and 64-mm mesh size)
Thermally padded gloves (for handling high-temperature samples)

REFERENCES

Beschta, R. L. 1980. Modifying automated pumping samplers for use in small mountain streams. *Water Resources Bulletin* **16:**137–138.

Beschta, R. L. 1981. Increased bag size improves Helley-Smith bedload sampler for use in streams with high sand and organic matter transport. Pages 17–25 *in Erosion and Sediment Transport Measurement Symposium Proceedings.* International Association of Hydrologic Sciences, Publication 133, Washington, DC.

Beschta, R. L. 1987. Conceptual models of sediment transport in streams. Pages 387–419 *in* C. R. Thorne, J. C. Bathurst, and R. D. Hey (Eds.) *Sediment Transport in Gravel-Bed Rivers.* Wiley, New York, NY.

Dejia, Z., L. Daorong, and G. Hoachuan. 1981. The development of a sand bed load sampler for the Yangtze River. Pages 35–46 *in Erosion and Sediment Transport Measurement Symposium Proceedings.* International Association of Hydrologic Sciences, Publication Number 133, Washington, DC.

Gottschalk, L. C. 1964. Reservoir sedimentation. Pages 1–34 *in* V. T. Chow (Ed.). *Handbook of Applied Hydrology.* McGraw–Hill, New York, NY.

Graf, W. H. 1971. *Hydraulics of Sediment Transport.* McGraw–Hill, New York, NY.

Helley, E. J., and W. Smith. 1971. *Development and Calibration of a Pressure-Difference Bedload Sampler.* United States Geological Survey, Open-File Report, Menlo Park, CA.

United States Geological Survey (USGS). 1977. *National Handbook of Recommended Methods for Water-Data Acquisition.* United States Geological Survey, Reston, VA.

CHAPTER 8

Solute Dynamics

JACKSON R. WEBSTER AND
TERRENCE P. EHRMAN[1]

Department of Biology
Virginia Polytechnic Institute and State University

I. INTRODUCTION

The term *solute* is used for materials that are chemically dissolved in water. This includes materials such as calcium, chloride, sodium, potassium, magnesium, silica, and carbonate, which are often in relatively large concentrations. More biologically important solutes such as phosphate and nitrate are normally at very low concentrations. These solutes enter streams from three natural sources (e.g., Webb and Walling 1992). First, the atmosphere (i.e., rainwater) is often the major source of chloride, sodium, and sulfate. Second, other solutes come from soil and rock weathering, including calcium, phosphate, silica, and magnesium. Third, biological processes may be important. For example, nitrate may enter from the atmosphere or from weathering, or it may also come from biological fixation by blue-green algae. Also, inorganic carbon (CO_2, bicarbonate, or carbonate) comes from the atmosphere, weathering, or respiration by soil and stream organisms. Point (i.e., pipes) and nonpoint (e.g., agricultural runoff) are often major sources of solutes.

Solute dynamics refers to the spatial and temporal patterns of solute transport and transfer (Stream Solute Workshop 1990). These processes are tightly coupled to the physical movement of water in all ecosystems, but in streams this coupling is especially important. As the materials cycle

---

[1]Current address: Moreau Seminary, Notre Dame, IN 46556.

between biotic and abiotic components of the stream ecosystem, they are continuously or periodically transported downstream. Thus the cycles are longitudinally drawn out in spirals (Webster and Patten 1979, Newbold 1992). While the dynamics of many solutes are determined primarily by biogeochemical and hydrologic interactions occurring in the whole water-shed (Webb and Walling 1992), important in-stream dynamics also occur. Studies of solute dynamics in streams provide two types of information. First, they provide information on the rates of transport and transformation of the solutes themselves, which is important to the understanding of the availability (or impact) of the solutes. Second, they can be used to quantify various hydrologic properties of a stream. In this chapter, we will investigate solute dynamics from both perspectives.

Solutes in streams can be classified in various ways (Stream Solute Workshop 1990). Nutrients are those solutes that are essential to the growth and reproduction of some organisms. Nutrients may be limiting if their concentration is too low to meet biological demand. Other substances such as heavy metals may be inhibitory or toxic to stream organisms. Stream solutes also can be classified according to their biological and chemical reactivity. If their concentration is changed by biotic or abiotic processes, they are referred to as *nonconservative.* On the other hand, if their concentration is not changed by in-stream processes, they are called *conservative* solutes. Conservative solutes include things that are not nutrients and do not react chemically with water or the stream substrate, such as lithium (e.g., Bencala *et al.* 1991). Also, some nutrients may be so abundant that biotic and abiotic exchanges are very small relative to the stream concentration, and the solute may appear to be conservative and may in fact be treated as a conservative solute. Chloride is an example of a biologically essential solute that exists in most streams in concentrations that far exceed biological need. Chloride is often used as a conservative solute in stream studies (e.g., Triska *et al.* 1989).

The dynamics of a conservative solute are primarily driven by two processes; *advection* and *dispersion.* Advection is the downstream transport at the water velocity. Dispersion can occur by molecular diffusion, but in streams is primarily caused by turbulence. The two processes are expressed in the partial differential equation

$$\frac{\partial C}{\partial t} = -u\frac{\partial C}{\partial x} + D\frac{\partial^2 C}{\partial x^2}, \tag{8.1}$$

where C represents solute concentration; t, time; x, distance in the down-stream direction; u, water velocity; and D, a dispersion coefficient. However, this equation applies only to conservative solutes in uniform channels with constant discharge. Other terms can be added to this equation to include variable stream morphology, groundwater and tributary inputs, and transient storage. *Transient storage* refers to the temporary storage of solutes in water that is moving more slowly than the main body of water, such as pools, backwaters, and hyporheic water (Bencala and Walters 1983). Including these factors, the equation becomes:

$$\frac{\partial C}{\partial t} = \frac{-Q}{A}\frac{\partial C}{\partial x} + \frac{1}{A}\frac{\partial}{\partial x}\left[\frac{AD\partial C}{\partial x}\right] + \frac{Q_L}{A}(C_L - C) + \alpha(C_S - C) \quad (8.2)$$

and

$$\frac{\partial C_S}{\partial t} = -\alpha\frac{A}{A_S}(C_S - C),$$

where Q is discharge; A, the cross-sectional area of the stream; Q_L, the lateral inflow from groundwater or tributaries; C_L, the solute concentration of the lateral inflow; α, a coefficient for exchange with the transient storage zones; A_S, the size of the transient storage zones, and C_S, the concentration of solute in the transient storage zone.

Dynamics of nonconservative solutes are more complicated because of the exchanges between solute in the water column and on the stream substrate. These exchanges include abiotic processes, such as adsorption, desorption, precipitation, and dissolution. There are also many important biotic exchanges. Examples of biotic exchanges include heterotrophic (i.e., microbial) uptake, plant uptake, leaching, and mineralization. In general, abiotic or biotic process that remove solutes from the water column are called *immobilization*. In streams the most important immobilization processes for biologically important solutes (i.e., nutrients) are adsorption (especially for phosphate), heterotrophic uptake, and plant uptake. Ignoring the complications we just added in Eq. (8.2), the dynamics of a nonconservative solute can be expressed as

$$\frac{\partial C}{\partial t} = -u\frac{\partial C}{\partial x} + D\frac{\partial^2 C}{\partial x^2} - k_C C, \quad (8.3)$$

where k_C is the overall uptake rate. Of course, nutrients that are immobilized may eventually be returned to the water column. This can be most simply expressed by adding another term to Eq. (8.3) and adding another equation for the immobilized nutrient,

$$\frac{\partial C}{\partial t} = -u\frac{\partial C}{\partial x} + D\frac{\partial^2 C}{\partial x^2} - k_C C + \frac{1}{h}k_B C_B \qquad (8.4)$$

and

$$\frac{\partial C_B}{\partial t} = hk_C C - k_B C_B,$$

where C_B is the immobilized (i.e., benthic) nutrient concentration and k_B is the rate of remobilization.

These equations (or models) of solute dynamics can get much more complex. This description was adapted from the presentation by Stream Solute Workshop (1990), and a more complete description is given there. The very simplest equation (Eq. (8.1)) can be solved analytically, but the other equations can be solved only by using computers and numerical solution techniques.

As noted above, nutrients cycle (in the standard ecological sense) between abiotic and biotic forms, but in streams this cycling is constantly subject to downstream displacement, resulting in a pattern described as *spiralling*. Another way of looking at nutrient dynamics (i.e., the dynamics of nonconservative, biologically important solutes) is in terms of *spiralling length;* the distance a nutrient atom travels while completing a cycle (i.e., while going from abiotic form to biotic and back to abiotic (e.g., Elwood *et al.* 1983, Newbold 1992)). Spiralling length has two components: (1) the distance traveled while in abiotic form (dissolved in the water column) before being immobilized, called the *uptake length,* and (2) the distance traveled before being remobilized and returned to the water column, called the *turnover length.* Thus,

$$S = S_W + S_B, \qquad (8.5)$$

where S represents spiralling length; S_W, uptake length; and S_B, turnover length. Uptake length can be related back to the previous equations because it is the inverse of the uptake rate:

$$S_W = 1/k_C. \qquad (8.6)$$

As we will see in this chapter, uptake length can be determined with fairly simple experimental techniques. However, determination of turnover length is much more difficult and has been done only using radioactive tracers (e.g., Newbold *et al.* 1983). In most cases uptake length is the major component of spiralling length. When a nutrient atom is dissolved in the water column it is free to travel with the water (i.e., it is mobile), but after it is immobilized it is attached to or part of a particle and its downstream velocity is much slower.

The objective of the experiments described in this chapter is to examine the dynamics of both a conservative solute and a nonconservative solute in a stream or in a variety of streams. Because of the variability of equipment that might be available and the highly variable nature of stream chemistry, we have provided a number of procedural and experimental options. At a minimum, you should be able to determine discharge, velocity, and the importance of transient storage.

II. GENERAL DESIGN

The general design of these experiments is that a known concentration of solute is released at a constant rate into a stream for one to several hours. Measurements are made downstream to determine the concentration and timing of the passage of the solute pulse.

A. Site Selection

Most solute studies have been done on first- to fourth-order streams that range in discharge from <1 up to 250 liters/s. Streams this size allow wadeable access for physical measurements and sampling. Stream flows greater than this may require scaling up of the release apparatus and modification of sampling design and execution. It may be necessary to calculate discharge prior to the experiment either with a "quick and dirty" dye release or with physical measurements (see Chapter 3).

Choice of a stream or section of stream will depend on the question posed (e.g., single reach or comparison of multiple reaches—see options). Ideally, a stream or set of streams should be selected that provide a range of physical and biological conditions. A comparison of hydraulic properties between two reaches should encompass one simple reach (e.g., a straight channel with homogeneous substrate and low amount of wood) and one more complex reach (e.g., sinuous channel, heterogeneous substrate, high amount of wood). Try to avoid reaches with tributary input. The length of experimental reaches will vary with flow, but minimally must be long enough for mixing and dispersion of released solute (a preliminary dye release may be in order). Typical lengths range from 50 m in fairly small streams to several hundred meters in larger streams.

B. Choice of Solutes

Selection of a conservative solute tracer is a function of local geology, ambient levels of solute in the stream, research budget, and analytical equipment available. It is desirable to raise stream concentration of the solute 5- to 10-fold over background levels. Typical conservative solutes used are salts of Cl, Na, Li, K, and Mg. Of these, Cl, either as NaCl or LiCl, is the most common. Cl can easily be obtained as NaCl (available at the local grocery store—make sure it is noniodized) and can be measured several ways. The most convenient way is with a portable ion-specific probe, which eliminates any laboratory analysis. Sodium-specific probes are also available, but sodium loses 5–10% by mass through sorption to stream bottom materials compared to almost no loss of chloride (Bencala 1985). Salt concentration also can be measured with a high-quality conductivity meter (e.g., Mulholland *et al.* 1994). If portable instruments are not available, samples can be collected in the field and analyzed in a laboratory by various spectrographic means.

C. Mariotte Bottle

A simple, inexpensive, reliable, and nonelectrical release apparatus is the Mariotte bottle (Fig. 8.1). Named for its 17th century creator, Edme Mariotte, the "bottle" allows for delivery of solute solution at a constant release rate, despite the change in head of the reservoir. Its parts include only a carboy with volume of approximately 12 liters sealed at the top with a rubber stopper. A rigid plastic tube extends through a hole in the stopper to just above the bottom of the carboy. As long as the tube remains below the liquid level, the solution drains at a constant rate through a spigot at the bottom. Calibrated tips (50-μl automatic pipet tips) connected to the spigot by rubber tubing allow variable release rates. In the lab, one can calibrate tips cut to various aperture sizes to obtain various delivery rates. The delivery rate can also be adjusted by connecting the tip to the carboy spigot with flexible tubing and changing the height of the dispensing tip. Release rate may fluctuate with changes in barometric pressure or elevation.

D. Optional Exercises

Several optional experiments are presented in this chapter. Beyond the single reach release, solute dynamics can be compared spatially among the reaches of one to several streams, before and after a manipulation, and over time at different flows. For each solute release, a computer model can be used to simulate the actual release data and calculate hydraulic parameters such as dispersion and transient storage zone retention. Nonconservative (nutrient) releases can also be run simultaneously with the

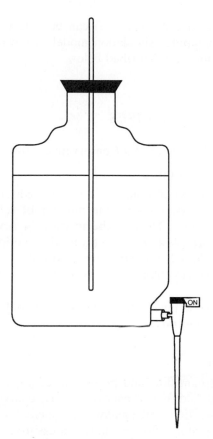

FIGURE 8.1 Mariotte bottle.

conservative tracer. A computer simulation of the nonconservative solute dynamics also can be run and uptake length calculated.

E. Data Analysis

Necessary physical measurements include discharge, average water depth, and average wetted-channel width for the stream reach over which the release is being conducted. Measurements of thalweg velocity, gradient, and large woody debris area or volume are optional. One can calculate hydraulic characteristics (discharge, nominal transport time (NTT)) from a graph of conservative solute concentration versus time, and uptake length and rate can be calculated from nutrient data fit to a negative exponential model. Further hydraulic properties of the reach (dispersion, transient storage zone area and exchange rate) can be determined by subjective curve fitting of a computer simulation model to the conservative solute data. These techniques are described below.

III. SPECIFIC EXERCISES

A. Exercise 1: Dynamics of a Conservative Solute

Laboratory Preparation

1. Mix stock solution of sodium chloride (238 g/liter) in distilled water. Total volume needed depends on the number of releases, duration of release, and release rate. Heating the mixture in a water bath aids in dissolution. Mix vigorously and repeatedly for the solution is close to saturation. Make certain the salt is completely dissolved.

2. Prepare a series of chloride standards (1–20 mg/liter) for calibrating the probe.

3. Calibrate several pipet tips for the Mariotte bottle to cover a broad span of possible release rates.

Field Prerelease

1. Calculate stream flow and necessary release rate to raise stream concentration 5–10 fold that of background. Discharge can be estimated quickly from cross-sectional area and water velocity (determined by timing a buoyant piece of material floating down a measured reach) or with a dye release (see Chapters 3 and 11). Release rate (Q_I) is calculated as

$$Q_I = Q \times C_S/C_I, \tag{8.7}$$

where Q is discharge; C_S, target stream concentration of solute; and C_I, the concentration of solute in the release solution (238 g/liter). Select a tip to deliver solute at this rate.

2. Use a tape measure to delimit the extent of the experimental reach. Mark every 5 m (for a 100-m reach) within the reach with labeled flagging tape.

3. At each 5-m cross section, measure wetted channel width, depth across the stream (every 0.5 m), and thalweg velocity (optional). Stream temperature and gradient (optional) should also be measured.

4. Calibrate the Cl probe with the standards. The standards should be placed in the stream until they equilibrate with ambient stream temperature.

Field Release

1. Collect a series of background water samples in mid-stream at 10-m intervals over the reach. Take three replicates at each site. These samples can be taken in any type of clean container. We use disposable urine cups. Work from downstream up and avoid unnecessary stomping in the stream.

2. Position chloride probe and recorders at the downstream site. Place probe securely in a well-mixed area.

3. Add solute solution to Mariotte bottle and seal with rubber stopper. Position the Mariotte bottle on a stand directly in the stream (if shallow and stable enough) or on bank (with sufficiently long tubing to reach stream) such that the solution will enter a turbulent, well-mixed zone. Do not attach tip to Mariotte bottle at this time.

4. With a bucket under the spigot, open to full and allow Mariotte bottle to equilibrate; you will hear a glug-glug-glug sound as air comes down through the tube. Turn off spigot. Do not break the seal at top or this step will have to be repeated.

5. Connect tubing with appropriate tip. Place bucket under tip and open to flush out any air bubbles. Measure the release rate with a graduated cylinder and stopwatch. Keep the bucket under the tip to avoid any premature addition to the stream. If the release rate is unacceptably higher or lower than expected, a new tip should be used. During the release, periodically recheck the release rate, emptying solute in the graduated cylinder into the stream. (Caution: do not do this prior to the release, rather empty the graduated cylinder into the bucket.)

6. Synchronize stopwatches and open spigot to commence release.

7. Frequency of chloride readings at downstream site depends upon rate at which the concentration changes in the stream. Record probe readings every 1–5 min (flow dependent) until pulse arrives and then measure every 15–30 s as chloride concentration increases rapidly.

8. At plateau (10 min to several hours after commencing release), working from downstream to upstream, take three samples from mid-stream at 10-m intervals (see step 1 above). Again, avoid unnecessary stomping in the stream. Shut off the Mariotte bottle once samples have been collected from all sites. Record the total time of release.

9. Continue recording chloride concentration until stream levels return to prerelease levels. Once measurement in the stream has been terminated, use the probe to measure chloride concentrations of the background and plateau samples. Recalibrate probe, for it may experience electronic drift during the release.

10. If no probe is available, water samples at the downstream site

should be taken before the release and every 1–5 min throughout the release. The volume of sample taken will depend on the laboratory method of measuring solute concentration.

Data Analysis

1. Summarize physical parameters: mean width and mean depth at each cross section and over the whole reach, mean velocity (optional), and gradient (optional).

2. Graph conservative solute concentration versus time at the downstream end of the reach (Fig. 8.2).

3. From this graph calculate discharge, Q, from plateau concentrations,

$$Q = (C_I - C_b) \times Q_I/(C_p - C_b), \tag{8.8}$$

where Q_I is release rate; C_I, the solute concentration of the release solution; C_p, the plateau solute concentration; and C_b, background (i.e., prerelease) concentration. Compare this measurement of discharge with direct measurements.

4. A useful measure of hydraulic retention is the nominal transport time (NTT), which is the time required for 50% of the chloride (or other conservative solute) to pass out of the stream reach (Triska *et al.* 1989). This can be determined by integration of the chloride curve (digitizer or computer approximation). Dividing the length of the reach by NTT gives the average stream velocity, which can be compared with direct measure-

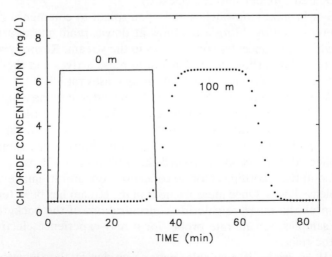

FIGURE 8.2 Chloride concentration versus time for a small stream with very little transient storage and no increase in flow over the reach.

ments of thalweg velocity. For example, in Fig. 8.2 the solute release began at 12:04 and lasted 30 min. One hundred meters downstream, the solute pulse came by between 12:10 and 1:10. By integrating the curve, we determined that half of the added solute had passed the 100-m point by 12:35. Since one-half of the 30-min release was completed by 12:19, NTT was 16 min (12:35 to 12:19). One hundred meters divided by 16 min is 10.4 cm/s. (This example was for a very small stream with a very low gradient.)

5. Similarly, you can calculate discharge along the reach by using the plateau concentrations (Fig. 8.3). Graph discharge versus distance to see if there is evidence of groundwater input. If there is evidence of flow increase at a specific point (or points), go back to the stream and see if you can identify landscape features associated with this subsurface input.

6. Comparison of your data to the curves in Figs. 8.2 and 8.4 should give you some idea of the transient storage in your experimental reach. A reach with little or no transient storage will have a nearly rectangular graph (Fig. 8.2). If there is lots of transient storage, the uptake arm of the curve will be sloped, instead of a plateau there will be a period of slowly rising concentration, and the falling side of the graph will have a long tail (Fig. 8.4).

B. Exercise 2: Dynamics of a Nonconservative Solute

Simultaneously with the conservative solute, a nonconservative solute may be released to determine nutrient uptake. Samples should be taken

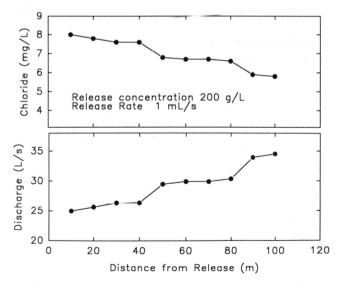

FIGURE 8.3 Plateau concentrations versus distance and calculated discharge versus distance for a stream with significant groundwater input over the reach.

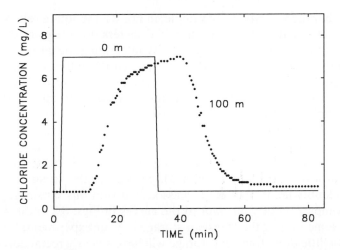

FIGURE 8.4 Chloride concentration versus time for a stream with considerable transient storage.

before the release and at the plateau of the release as with the chloride samples. Choice of appropriate nutrient depends upon local geology and which nutrient is possibly limiting in the stream. Be sure not to pick a nutrient that precipitates with the conservative solute. For example, calcium and phosphate cannot be used together because they form a highly insoluble salt.

Graph normalized nonconservative solute (nutrient) concentration versus distance and calculate uptake rate (k_C) and uptake length (S_W) (Fig. 8.5). Nutrient concentrations of the samples collected at plateau must be corrected for background levels (C_b) in order to get the added nutrient level. Then calculate normalized concentrations (C_N) by dividing the nutrient concentrations at a specific site (C_t) by the conservative solute (C_0, corrected for background) concentrations at the site:

$$C_N = (C_t - C_b)/C_0. \tag{8.9}$$

By doing this you avoid the necessity of correcting for possible increase in flow over the reach. For steady conditions (e.g., at plateau) the solution of Eq. (3) is a negative exponential,

$$C_N = C_{N0}e^{-k_c x}, \tag{8.10}$$

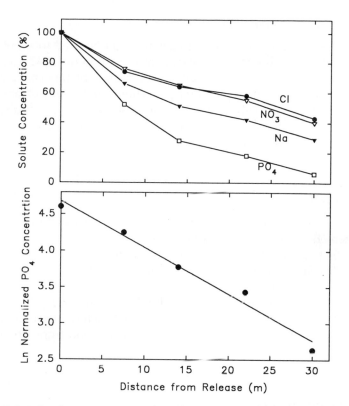

FIGURE 8.5 (Top) Plateau concentrations of solutes versus distance expressed as percentage of upstream concentrations. In this stream NO$_3$ is relatively abundant and behaves like
a conservative solute. PO$_4$ is taken up from solution. (Bottom) Semi-log plot of PO$_4$ concentration versus distance. The slope of this line is the PO$_4$ uptake rate.

where C_{N0} is the theoretical concentration at the release site and x is
distance downstream from the release site. Taking the logarithm of both
sides of this equation gives

$$\ln(C_N) = \ln(C_{N0}) - k_C x. \qquad (8.11)$$

This is the equation for a straight line with intercept of $\ln(C_{N0})$ and a slope
of k_C. So if you use your data to run a regression (or draw a graph) of
$\ln(C_N)$ versus x, the slope (k_C) will be an estimate of the uptake rate
per unit distance. However, be aware that there are lots of simplifying
assumptions in doing this, for example, first-order (linear) uptake, no remin-
eralization of added nutrient, and no saturation of uptake processes.

C. Computer Simulation

There are various computer models that can be used to simulate the results of your experiment. A fairly straightforward FORTRAN model can be obtained for free by writing the senior author of this chapter.[2] This model should run on most DOS-type machines and with just about any FORTRAN compiler. This simulation will allow you to calculate dispersion (D), the rate of solute exchange with the transient storage zone (α), and the size of the transient storage zone (A_S).

IV. QUESTIONS

1. What are causes of hydraulic retention in a stream? (That is, what causes temporary retention of conservative solutes?)
2. What stream features affect retention of solutes?
3. What factors determine the usefulness of various conservative and nonconservative solutes?
4. How does stream size affect hydraulic parameters?
5. What is the significance of wood in streams in terms of solute dynamics? How do you think the historical removal of wood from streams and rivers has affected solute dynamics?

V. MATERIALS AND SUPPLIES

Laboratory Materials

Beaker (for weighing salt)
Carboy for stock solution of solutes
Conservative solute (noniodized table salt)
Containers for standards
Distilled water
Graduated cylinders (100 ml and 1000 ml)
Nonconservative solute

Lab Equipment

Analytical instruments (for measuring solute concentrations)
Computer (optional)
Electronic balance (± 0.01 g)

[2]Dr. J. R. Webster. Please send a blank floppy disk.

Field Materials

Bucket
Calibrated pipet tips
Flagging tape
Graduated cylinder (100 ml)
Mariotte bottle: carboy, rubber stopper, plastic tube, rubber tubing,
    plastic connectors
Meter stick
Permanent marking pen
Sample bottles
Squirt bottle with distilled water
Stand for Mariotte bottle
Stopwatches
Tape measure (50–100 m)
Thermometer
Velocity meter (optional)
Water-resistant paper or notebook, pencils

REFERENCES

Bencala, K. E. 1985. *Performance of Sodium as a Transport Tracer: Experimental and Simulation Analysis.* Water Supply Paper 2270:83–89, United States Geological Survey, Reston, VA.

Bencala, K. E., and R. A. Walters. 1983. Simulation of solute transport in a mountain pool-and-riffle stream: A transient storage model. *Water Resources Research* **19:**718–724.

Bencala, K. E., B. A. Kimball, and D. M. McKnight. 1991. Use of variation in solute concentration to identify interactions of the substream zone with instream transport. Pages 377–379 *in* G. E. Mallard and D. A. Aronson (Eds.) U.S. *Geological Survey Toxic Substances Hydrology Program, Water-Resources Investigations Report 91-4034.* United States Geological Survey, Reston, VA.

Elwood, J. W., J. D. Newbold, R. V. O'Neill, and W. VanWinkle. 1983. Resource spiralling: an operational paradigm for analyzing lotic ecosystems. Pages 3–27 *in* T. D. Fontain III and S. M. Bartell (Eds.) *Dynamics of Lotic Ecosystems.* Ann Arbor Science, Ann Arbor, MI.

Mulholland, P. J., A. D. Steinman, E. R. Marzolf, D. R. Hart, and D. L. DeAngelis. 1994. Effect of periphyton biomass on hydraulic characteristics and nutrient cycling in streams. Oecologia **98:**40–47.

Newbold, J. D. 1992. Cycles and spirals of nutrients. Pages 370–408 *in* P. Calow and G. E. Petts (Eds.) The Rivers Handbook. Blackwell Scientific, Oxford, UK.

Newbold, J. D., J. W. Elwood, R. V. O'Neill, and A. L. Sheldon. 1983. Phosphorus dynamics in a woodland stream ecosystem: A study of nutrient spiralling. Ecology **64:**1249–1265.

Stream Solute Workshop. 1990. Concepts and methods for assessing solute dynamics in stream ecosystems. *Journal of the North American Benthological Society* **9:**95–119.

Triska, F. J., V. C. Kennedy, R. J. Avanzino, G. W. Zellweger, and K. E. Bencala. 1989. Retention and transport of nutrients in a third-order stream in northwestern California: Hyporheic processes. Ecology **70:**1893–1905.

Webb, B. W., and D. E. Walling. 1992. Water quality. Chemical characteristics. Pages 73–100 *in* P. Calow and G. E. Petts (Eds.) *The Rivers Handbook.* Blackwell Scientific, Oxford, UK.

Webster, J. R., and B. C. Patten. 1979. Effects of watershed perturbation on stream potassium and calcium dynamics. *Ecological Monographs* **49:**51–72.

CHAPTER 9

Phosphorus Limitation, Uptake, and Turnover in Stream Algae

ALAN D. STEINMAN[*] AND
PATRICK J. MULHOLLAND[†]

*Department of Ecosystem Restoration
South Florida Water Management District
†Environmental Sciences Division
Oak Ridge National Laboratory

I. INTRODUCTION

Increased loading of nutrients into streams and lakes has become one of the major environmental problems facing society today. Indeed, greater attention is now being devoted to determining a watershed's capacity to absorb these nutrients before its ecological integrity is threatened (i.e., its so-called "nutrient assimilative capacity"). The ability of an ecosystem to assimilate a nutrient load depends largely on its biology, chemistry, and geomorphology. In stream ecosystems, benthic plants represent a potentially important biotic sink for nutrients. Determining the rates at which nutrients are taken up by plants and how quickly they are released can provide important information in assessing how large a nutrient load an ecosystem can absorb before its integrity is negatively impacted.

The nutrient that we focus on in this chapter is phosphorus. Inorganic phosphorus is commonly considered to be the element most likely to limit primary production in freshwater ecosystems (Schindler 1977, Hecky and

Kilham 1988). This follows from the fact that algae require elements in relatively fixed proportions to grow and reproduce. When these proportions are compared to ambient levels in streams and rivers, most elements in the ambient water are present in a much greater concentration than needed by algae (Hecky and Kilham 1988), and thus they are not likely to limit algal growth. Hecky and Kilham concluded that the only elements found in similar proportions between river water and algae, and thus may be found in low enough concentrations to limit algal production, are phosphorus, iron, and cobalt. However, their analysis was performed on a small set of streams and analyzed phytoplankton, not benthic algae. Although a significant number of stream studies have indicated that phosphorus limits the growth of benthic algae (Stockner and Shortreed 1978, Elwood *et al.* 1981, Peterson *et al.* 1983, Bothwell 1989, and others), it is by no means the only limiting nutrient in lotic ecosystems. In some streams in the western United States, where watersheds may be relatively rich in geologic sources of P, nitrogen can be the limiting nutrient in streams (Grimm and Fisher 1986, Hill and Knight 1988, Lohman and Priscu 1992). In addition, Pringle *et al.* (1986) reported that micronutrients (e.g., iron, boron, manganese, zinc, cobalt, molybdenum) limited algal growth more than nitrogen or phosphorus in a Costa Rican stream.

Although phosphorus concentrations in healthy plants are relatively low, usually ranging from 0.1 to 0.8% of dry mass (Raven *et al.* 1981), P is an essential element. Some of the more important functions played by phosphorus in plants include being a structural component of "high-energy" phosphate compounds (e.g., ADP and ATP), nucleic acids, several essential coenzymes, and phospholipids, as well as being involved in the phosphorylation of sugars.

In this chapter, three different aspects of phosphorus utilization by benthic algae will be covered: assessment of P limitation, measurement of P uptake rates, and the determination of the release rate of P (expressed as the turnover rate). Although we use the term *benthic algae* throughout the chapter, it should be noted that the benthic algae attached to submerged substrata in streams usually exist as part of a complex assemblage variously referred to as *periphyton, aufwuchs,* or *biofilm.* This assemblage usually consists of algae, bacteria, fungi, and meiofauna (see Chapter 15) that are held within a polysaccharide matrix (Lock *et al.* 1984).

A. Assessment of P Limitation

Nutrient limitation in algae can be assessed in several different ways, including chemical composition of biomass, nutrient enrichment bioassays, enzymatic activities, and physiological responses. Chemical composition has the potential to be an indicator of nutrient limitation because the proportions of carbon, nitrogen, and phosphorus, while confined to a relatively narrow range in algae, nonetheless vary in response to nutrient availability in the water. Cells growing near nutrient-saturated growth rates (i.e., not nutrient-limited) typically contain the elements C, N, and P in a molar ratio of 106:16:1, the so-called *Redfield ratio* (Redfield 1958). Although Redfield based this ratio on phytoplankton, it also may be applicable for benthic algae. When one of these elements becomes limiting in the environment, this can be reflected in a slightly lower level of nutrient present in the algal cell. For example, if P concentration becomes limiting in a stream, tissue C:P and N:P ratios would be expected to increase because the algae make more efficient use of the P incorporated into cells. Hecky *et al.* (1993) suggested that C:P values in excess of 129 (as opposed to the Redfield ratio of 106) and N:P values in excess of 22 (as opposed to the Redfield ratio of 16) indicate at least moderate phosphorus deficiency in algae.

Nutrient enrichment bioassays involve the addition of nutrients to a stream, either in the form of diffusing substrata (see Chapter 29), powdered fertilizers, or solute injections (see Chapter 8). The enrichment would continue for some designated period of time, and its effect would be evaluated by change in algal biomass (see Chapter 14) or productivity (see Chapter 25) compared to that seen in unenriched algae.

An enzymatic assay that has proven to be a reliable indicator of phosphorus limitation in algae is phosphatase activity (PA). The phosphatase enzyme hydrolyzes phosphate ester bonds, thereby releasing orthophosphate (PO_4) from organic phosphorus compounds. Thus, increased PA results in more inorganic P becoming available to microorganisms in the environment. As inorganic phosphorus concentrations decline in aquatic ecosystems, PA generally increases (Healey 1973, Wetzel 1981, Currie *et al.* 1986). Thus, PA has been used to infer P limitation for aquatic microflora (Healey and Hendzel 1979, Burkholder and Wetzel 1990). Based upon their results from algal culture studies, Healey and Hendzel (1979) suggested that phosphatase levels above 0.003 mmol mg chlorophyll a^{-1} h^{-1} indicate moderate P deficiency, and those above 0.005 mmol mg chlorophyll a^{-1} h^{-1} indicate severe P deficiency.

B. Measurement of P Uptake Rates

The relationship between the nutrient concentration in the water and the rate at which nutrients are taken up by algae can be described by a hyperbolic function (Fig. 9.1). The Michaelis–Menten equation for enzyme kinetics is often used to describe this function,

$$V = V_m(S/K_s + S), \tag{9.1}$$

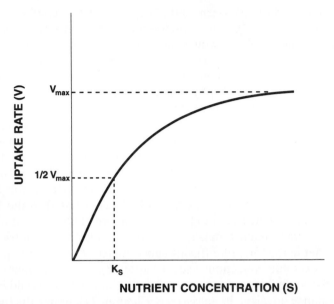

FIGURE 9.1 Relationship between nutrient concentration (*S*) and nutrient uptake rate (*V*). V_{max} represents the maximum nutrient uptake rate; K_s, the half-saturation constant, or the nutrient concentration at which the uptake rate is one-half of the maximum uptake rate.

where *V* represents nutrient uptake rate; V_m, maximum nutrient uptake rate; *S*, concentration of the nutrient; and K_s, the half-saturation constant (or nutrient concentration at which nutrient uptake is one-half the maximal uptake rate). From a biological perspective, there are two critical considerations in Fig. 9.1: (1) Nutrient uptake rates become saturated as nutrient concentration increases. Empirical studies have shown that saturation of phosphorus uptake can occur at very low concentrations in both benthic algal communities (<1 μg/liter; Bothwell 1989) and whole streams (<10 μg/liter; Mulholland *et al.* 1990). Thus, investigations examining P uptake in benthic algae must consider the possibility that saturation will influence uptake kinetics even at relatively low concentrations. (2) The constant K_s provides a useful index of a cell's affinity for a nutrient: a lower K_s suggests a greater affinity for the nutrient, which can confer a competitive advantage when the nutrient is present at low concentrations. Generally, taxa that have low K_s values have a competitive advantage at low nutrient concentrations. However, K_s values appear to be fixed, and do not appear to vary much under different environmental conditions. Rather, the physiological response most often exhibited by nutrient-limited algae, when exposed to elevated nutrient concentrations, is an increase in short-term nutrient uptake rates. This is because they can increase V_m, but not change K_s (Darley 1982, Lohman and Priscu 1992). Of course, over the long-term, elevated nutrient levels may result in altered algal community structure dominated by species with greater V_m than species in communities growing under low nutrient concentrations.

It is also important to distinguish between nutrient-limited *uptake* rates (above) and nutrient-limited *growth* rates. The relationship between nutrient concentration and algal growth can be modeled using either the Monod model or the Droop model. The Monod model relates algal growth to the external concentration of nutrients in the water, whereas the Droop model relates algal growth to internal (cellular) concentration of nutrients. The implications and details behind these models are beyond the scope of this chapter, but for additional details see Droop (1974), Rhee (1978), and Kilham and Hecky (1988).

C. Measurement of P Turnover Rates

This portion of the chapter is designed to examine phosphorus turnover rates in benthic algae. Phosphorus turnover provides an index of internal cycling in the algal community. Once an algal cell takes up phosphorus from the external medium, the P can be incorporated into structural elements, maintained in a labile pool, or excreted from the cell. Cells that are phosphorus-limited may be less likely to release the phosphorus they have taken up (back to the external medium) than cells that are phosphorus-replete (but see Cembella *et al.* 1984 and Borchardt *et al.* 1994). Thus, the phosphorus turnover rate in algae (i.e., loss of P from algal cell relative to total algal P) may be lower in P-limited cells than in P-saturated cells, assuming that both the P-limited and the P-replete cells have similar metabolic activities and are exposed to similar grazing pressures. One way to measure P turnover in algae is to label the cells with a P radioisotope (e.g., ^{32}P or ^{33}P) in the laboratory, place the algae back into the environment, and then measure the amount of radioactive phosphorus present in the cells over time. This gives an apparent P turnover rate, as turnover is being estimated from the entire periphyton matrix and not from individual cells.

D. Overview of Chapter

This chapter examines phosphorus limitation and uptake in benthic algae collected from a relatively low-phosphorus stream and a relatively high-phosphorus stream. In theory, the benthic algae growing in the low- and high-P streams should have adapted to the different ambient conditions. Specifically, if the algae in the low-P stream are P-limited, they should have greater C:P ratios than algae collected from the high-P stream, all else being equal. In addition, algae in the low-P stream should have greater phosphatase activities and lower K_s and V_m values than algae from the high-P stream. If *P* turnover rate is measured, the algae from the low-P stream should have lower apparent P turnover rates (greater retention) than algae from the high-P stream.

II. GENERAL DESIGN

This chapter describes the methodology to measure phosphorus limitation, uptake, and turnover in periphyton. Although valuable information will be gleaned from any of these exercises in isolation, we recommend combining them when possible to gain a broader understanding of phosphorus-related processes in streams.

A. Site Selection

Both a relatively high-phosphorus and a relatively low-phosphorus stream are needed for this exercise. If differences in algal response are to be detected, it is critical that the algae be exposed to ecologically meaningful differences in nutrient concentration. We recommend using an undisturbed stream (if available) for the "low-phosphorus" system, where there are no obvious impacts (e.g., point source inputs, lack of riparian zone, animals in streams). For the "high-phosphorus" system, use streams receiving either agricultural or sewage runoff or clarifying tanks at sewage treatment facilities (Davis *et al.* 1990). If all streams in the region have low levels of P, then it may be possible to enrich a stream with P for a sustained period of time (e.g., 8 weeks) to create high-P conditions (e.g., Steinman 1994). This can be done through the use of nutrient diffusing substrata (see Chapter 29) or solute additions (see Chapter 8). If all streams in the region have high levels of P, then we recommend that the two streams with the greatest difference in P concentrations be used in this exercise. Regardless of which streams are selected, collect algae from sites in the two streams that are generally comparable in terms of other environmental conditions (e.g., irradiance level, current velocity, temperature) if at all possible.

B. Limitation: Phosphatase Activity

This exercise consists of two parts: assay of phosphatase activity, followed by measurement of chlorophyll *a* (see Chapter 14). The phosphatase activity is normalized per unit chlorophyll to ensure that activities are not simply a function of how much (or how little) active biomass is present. Phosphatase acts on a variety of organic phosphorus compounds. In this exercise, we use a commercially available compound, *para*-nitrophenyl phosphate (*p*-NPP). When PO_4 is hydrolyzed from it, *p*-nitrophenol (*p*-NP) is formed, which can be measured spectrophotometrically.

C. Limitation: Chemical Composition (C:P Ratio in Algal Tissue)

This exercise consists of three components: measurement of algal ash-free dry mass (AFDM) (and conversion to C), acid digestion of combusted matter to dissolve phosphorus in ashed algal tissue (Solórzano and Sharp 1980), and then measurement of inorganic phosphorus in oxidized algal material according to standard methods. Ideally, the carbon concentration in algae would be measured with an elemental analyzer. However, this instrument is not always available. Consequently, in this exercise we present an alternative method to estimate C, based on measurements of AFDM (see Chapter 14), which is easy to perform but less accurate then elemental analysis.

D. Net Uptake: Stable Phosphorus

This procedure involves the measurement of net loss of soluble reactive phosphorus (SRP) from the water in which the algae are growing. The exercise consists of three components: sampling of water during the incubation, measuring SRP in water samples, and measurement of algal AFDM. Water samples are removed at the start of the incubation, and thereafter at 30 and 60 min, and analyzed for phosphorus according to *Standard Methods* (APHA *et al.* 1992). Large changes in the biomass: water volume ratio resulting from sampling during the incubation period should be avoided by minimizing sample volumes or sample times. Consequently, if the water volume in incubation chambers is low, the 30-min samples can be omitted. We recommend that volumetric change be limited to <10% of initial volume during sampling. Ideally, chambers attached to pumps that could recirculate water during the incubation would be available (Fig. 9.2), as water velocity will influence the uptake rate of phosphorus in benthic stream algae (Whitford and Schumacher 1964). However, if chambers and pumps are not available, the exercise can still be completed by using large (2-liter) glass chambers and stir plates. An open petri dish is glued to the bottom of the chamber, into which is placed a stir bar. Then coarse-meshed screening (e.g., chicken wire) is placed over the petri dish, thereby creating a shelf onto which are placed the substrata with attached algae. The rotation speed of the stir bar is varied until it matches approximately the current velocity in the sampled streams.

The use of stable elements to measure uptake rates is dependent on the initial, ambient nutrient concentration being great enough that it still will be possible to measure the remaining nutrient at the end of the incubation period. For example, if P concentrations are low at the beginning of the incubation, they may be below detection limits at the end. Another problem with this measurement is that if nutrient regeneration rates from

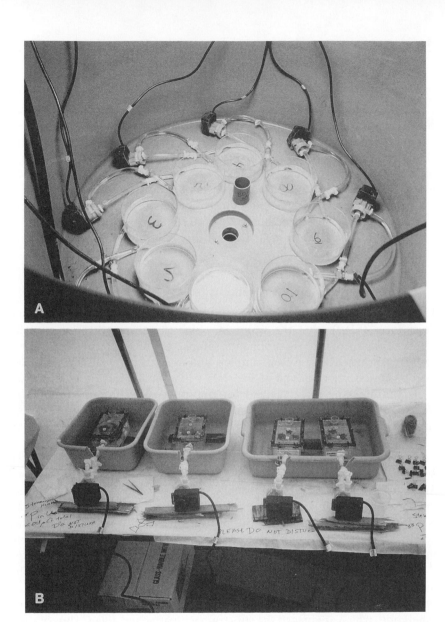

FIGURE 9.2 Examples of incubation chambers that have been used for P-uptake studies. (A) 2-Liter glass chambers fitted with adapters to accept tubing attached to submersible pumps. Pumps circulate water around the chambers. Placing clamps on tubing line can reduce flow rate, if so desired. (B) 1-Liter plexiglass chambers with detachable lids. Lids attach to main body of chamber with wing nuts; gaskets provide a leakproof seal. Chambers are attached to submersible pumps. Note the large port (far end with lip) in lid, which allows an oxygen meter to be placed directly in the chamber to measure metabolism. There are also two small ports, which allow for injection of radioisotope into the chamber.

algae are similar to nutrient uptake rates (i.e., the community is at steady state with respect to nutrient dynamics), then no net uptake will be measured. An alternative approach is to add nutrients to stream water. This elevates nutrient concentrations above ambient levels to ensure that concentrations at the end of the incubation will still be high enough to be measured and temporarily increases nutrient uptake rates above rates of nutrient regeneration. However, this approach measures only nutrient uptake potential at the higher concentration and will be an overestimate of ambient net uptake rate (see Mulholland *et al.* 1990). In this exercise, we provide instructions for measuring uptake rates at ambient nutrient concentrations (i.e., without enrichment).

E. Total Uptake: Radioactive Phosphorus

This procedure involves measuring the loss of $^{33}PO_4$ added to the water in which the algae are growing. Two advantages of measuring nutrient uptake with radioisotopes, as opposed to stable nutrients, are the ability to measure uptake at low, ambient nutrient concentrations and the ability to measure total nutrient uptake rates.[1] Uptake rates will be calculated in this exercise from algae growing in high-P and low-P streams.

This exercise should use algae growing on small artificial substrata (e.g., unglazed ceramic cylinders (Steinman *et al.* 1991a) or tiles placed in the streams for a period long enough to acquire an algal community similar to that of natural substrata). The use of small artificial substrata allows AFDM to be measured directly on the substratum without scraping, thereby minimizing contact with radioactive phosphorus in the algae. In addition, if the turnover option is to be completed (see below), the phosphorus in the algae on these substrata can be extracted with relative ease.

F. Turnover

This exercise involves measuring the rate at which radiolabeled P, incorporated into algal biomass, is lost from the algal assemblage over time.[2]

---

[1]Extreme caution must be exercised when using radioisotopes. We recommend the use of ^{33}P, instead of ^{32}P, because of its lower energy, although it is more expensive. Even with the relatively low maximum energy of ^{33}P (0.248 MeV), the small amount of radioactivity used (0.5 mCi/liter), and the short half-life of the isotope (25 days), all handling of the isotope must be done with extreme care.

[2]Extreme caution needs to be exercised when using radioisotopes. See the prior cautionary note. In addition, because this exercise requires placing radiolabeled algae back into the natural environment, we recommend that users consult with the local health physicist at their institution regarding restrictions or other potential concerns about this protocol. If this option is not viable, it is also possible to place the algae into static tanks (e.g., aquaria) filled with water of high and low phosphorus concentration, in order to evaluate the influence of P concentration on turnover.

Ideally, this procedure will be piggy-backed on the prior exercise, that of measuring phosphorus uptake rates using [33]P. The exercise consists of four parts: radiolabeling of algae, oxidation of labeled algae, extraction of P from ash, and measuring radioactivity in subsamples of extract.

After the ^{33}P uptake exercise is completed, substrata are returned to the high- and low-P streams if possible (if placing radioactive samples in natural streams is not feasible, they may be placed into aquaria containing either high or low concentrations of phosphorus). Four substrata are sampled from each stream on four different days. The algae on the substrata are oxidized, and ^{33}P is extracted from the ash. A subsample of this extract is diluted, placed into scintillation cocktail, and assayed for radioactivity using liquid scintillation spectroscopy.

Phosphorus turnover for each stream is calculated as the first-order rate constant of the decline in ^{33}P activity over time (slope of relationship between ln ^{33}P in algae vs time). A mean activity is calculated on each date from the four substrata collected and used in the regression with time. For the purposes of this exercise, we recommend normalizing ^{33}P content to unit area of substratum, as opposed to biomass. This calculation assumes relatively similar biomass levels among substrata or that sufficient samples are collected on each date to take into account the natural variability in biomass in the system. If ^{33}P content is expressed per unit biomass, it becomes necessary to introduce a growth-correction factor to account for any net growth during the period of the experiment (because the amount of radioactivity per unit biomass in the sample will decline due to the accrual of new, nonlabeled, biomass). Also, it is critical that, if the extracted ^{33}P samples are counted on different days over the period of the turnover experiment, they be corrected for radioactive decay; because of the short half-life of [33]P (25 days), some of the decline in ^{33}P content in algae will be the result of radioactive decay. Alternatively, all of the ^{33}P extract samples for the entire turnover study can be assayed on the scintillation counter at the same time at the end of the study, thereby obviating the need to correct for decay.

III. SPECIFIC EXERCISES

A. Exercise 1: Phosphatase Activity and Chlorophyll *a*

Preparation Protocol

1. At least 1 month and preferably 3–6 months, prior to the exercise, place approximately 100 3 × 3-cm (the small size allows substrata to be placed directly into sample jar for pH analysis without having to scrape algae off

algae off surface) unglazed ceramic tiles or ceramic cylinders (Steinman *et al*. 1991a; Du-Co Ceramics Co., Saxonburg, PA) in selected pool and riffle habitats in the stream to be sampled. If tiles are used, and were purchased attached to each other in sheets (as opposed to individual tiles), place the entire sheet in the stream, which minimizes the likelihood of tiles being lost if high discharge occurs. Analyze four tiles per stream (eight altogether). Alternatively, small rocks can be used but they must be small enough to fit in the incubation jars and be submersed in a small volume of water.

2. Label two tupperware containers (ca. 30 × 30 cm = 900 cm^2) by stream name or type (high-P, low-P).

3. Label 18 wide-mouth glass incubation jars (30 ml or large enough to contain the substratum) by stream type and purpose (high-P or low-P stream; PA or chlorophyll *a*): one jar for each of the 4 tiles and control per stream type for PA (10 total), 4 for each stream type for chlorophyll *a* extraction (8 total).

4. Prepare 150 m*M* *p*-NPP solution (add 2.78 g of *para*-nitrophenyl phosphate to 50 ml of double-distilled water).

Field Collection Protocol

1. Collect tiles and filter water (filter about 500 ml of stream water into a 1-liter plastic bottle using a hand pump and a Whatman GFF or Gelman type A/E glass-fiber filter) from each stream.

2. Fill two tupperware containers (one labeled high-P and the other labeled low-P) with the appropriate stream water, and place 10 tiles (2 extra, in case of loss) from each stream inside the container. Attach the lids, squeezing out excess water (keeping the containers filled with water minimizes tile movement), and place the containers in a cooler to be transported back to the laboratory.

Laboratory Protocol

1. Using a 10-ml automatic pipet, pipet 20 ml of filtered stream water (use more water if needed to completely submerse substratum) into each of the 10 incubation jars labeled for PA (5 jars for each stream type). In the laboratory, separate the sheet of tiles into individual tiles (ignore any glue that may remain attached to individual tiles following separation), and place one tile into each incubation jar. Leave one jar per stream without a tile (control).

2. Using a 1-ml pipet (set to 0.4 ml), pipet 0.4 ml of the *p*-NPP solution into the water in each of the incubation jars (or proportionately more if water volume was >20 ml), cap the jar, and gently mix. Incubate the jars at room temperature for 30 min, gently mixing the jars every 3–5 min.

3. After 30 min, filter the water in each jar by removing the water in a 25-ml plastic syringe and filtering it through a 0.45-μm pore size syringe filter (e.g., Syrfil-MF, Costar Corp., Cambridge, MA) and collecting the 10-ml filtrate in a labeled glass scintillation vial.

4. Remove the tile from each jar, rinse it by immersing it into un-amended stream water, and place it in a small plastic jar or centrifuge tube containing a known volume of 90% acetone that is sufficient to cover the substratum for extraction of chlorophyll. Follow the procedures in Chapter 14 for chlorophyll analysis.

5. Add 0.05 ml of 1 N NaOH to each vial containing the 10 ml of filtrate from each incubation jar to bring the pH up to ≥ 10 (for maximum color development of nitrophenol). Measure the absorbance of each filtrate at 410 nm against distilled water using a dual-beam spectrophotometer and 1-cm-pathlength cuvettes.

Data Analysis

1. Phosphatase activity (in mmol per 1/2 h) is calculated from the absorbance of the NPP solution as

$$PA = (Abs_{sample} - Abs_{blank}) \times 58 \times Volume_{(inc)}, \qquad (9.2)$$

where Abs_{sample} is the absorbance reading of sample at 410 nm; Abs_{blank}, the absorbance reading of blank (filtered stream water only, to correct for which each algal sample is incubated (in liters). (If 20 ml is used (as described in this exercise), this value will be 0.02.) The value 58 is the specific absorbance (at pH>10) of nitrophenol, which is the hydrolysis product of NPP. Use Table 9.1 for data entry and calculations.

TABLE 9.1
Sample Data Sheet for Determination of Phosphatase

					Stream:	
Sample	Absorbance (410 nm)	Net abs. (sample blank)	Volume (liters)	Phosphatase activity (mmol/h)	Chlorophyll *a* (mg)	Phosphatase per unit Chl (mmol mg^{-1} h^{-1})
Blank						
1						
2						
.						
.						
.						
n						

2. The phosphatase activity thus calculated is then divided by the amount of chlorophyll *a* determined for each sample, to obtain chlorophyll-normalized phosphatase activity (with units of mmol mg chlorophyll a^{-1} 1/2 h^{-1}). If phosphatase levels are very low, the incubation period can be extended to 1 h, and the values are reported per hour. Alternatively, phosphatase activity could be normalized by tile surface area to obtain area-specific PA (units of mmol cm^{-2} 1/2 h^{-1}).

B. Exercise 2: Chemical Composition

Preparation Protocol

1. Label two tupperware containers (30 × 30 cm) per stream by stream name or type (high-P, low-P).

Field Collection Protocol

1. Collect rocks from each stream. Alternatively, ceramic tiles that have been incubated in the streams for at least 3–6 months may be used.

2. Fill the two tupperware containers (one labeled high-P and the other labeled low-P) with stream water and place three small rocks from each stream inside the container. Attach the lids, squeezing out excess water (keeping the containers filled with water minimizes rock movement) and place the containers in a cooler to be transported back to the laboratory.

Laboratory Protocol

1. Follow the general procedures outlined in Chapter 14 for determination of AFDM, including the following modifications. After the algae are brushed off each rock, add the slurry (make sure the volume is less than 10 ml) to the bottom of an acid-washed 10-ml tared, glass beaker. Cover the top of the beaker with aluminum foil and etch the sample number onto the foil with a pointed object (do not write it on the beaker because it will be burned off when combusted and potentially contribute to dry mass). Dry the beaker to constant weight at 105°C (ca. 24 h). Remove the beakers from the drying oven and transfer them to desiccators until weighing.

2. After the beakers have been weighed, place them in muffle furnaces at 500°C for 1 h (make certain the ovens are at 500°C before timing the 1 h), remove, and allow them to cool to room temperature in a desiccator, and reweigh.

3. Using a 5-ml pipet, add 5 ml of 2 *N* HCl to the beaker, label the beaker with the sample number, and replace the aluminum foil with parafilm over the beaker to prevent evaporation. Digestion of ashed material should

last at least 24 h. Place beakers in the laboratory hood during digestion period.

4. After digestion, transfer contents of each beaker to a 500-ml volumetric cylinder. Rinse the beaker with distilled water and pour rinse water into the volumetric cylinder as well. Bring the total volume in the volumetric cylinder to 500 ml by adding distilled water (this will result in a leachate of 0.02 N HCl).

5. Pour each 100-ml sample into a plastic bottle, label accordingly, and analyze using standard methods for analysis of phosphorus in water (APHA et al. 1992).[3]

Data Analysis

1. Calculate the amount of carbon in the sample by multiplying the AFDM by 0.53. (Carbon content is estimated by assuming that 53% of AFDM is composed of carbon (Wetzel 1983). Although this value may vary slightly among algal groups and environmental conditions, the variance is low ($\pm 5\%$) compared to that of other cellular constituents). Use Table 9.2 for data entry and calculations.

2. Calculate the concentration of phosphorus in each sample by comparing its absorbance against a standard curve developed from the standards analyzed. The total amount of P (in mg) is then calculated by multiplying the P concentration by 0.5 (because the total volume of diluted leachate is 0.5 liter).

3. The C:P ratio is calculated by dividing the total C by the total P in each sample (converted to the same mass units) and then multiplying by 2.58 (to convert to a molar basis). Compare the ratio to the Redfield ratio (106:1) and analyze the differences between the high-P and low-P streams.

C. Exercise 3: Net Nutrient Uptake—Stable Phosphorus

Preparation Protocol

1. At least 1 month, and preferably 3–6 months, prior to the exercise, place approximately 100 3 × 3-cm unglazed ceramic tiles or ceramic cylinders (Steinman et al. 1991a) in selected high-P and low-P streams to be sampled. If tiles are used, they should be preashed to remove attached glue, which otherwise would be included in the AFDM measurement.

2. Label two tupperware containers (30 × 30 cm) by stream name or type (high-P, low-P).

3. Label six acid-washed 50-ml collection bottles per team according

[3]Standards for P analysis must be made in 0.02 N HCl to be comparable to that of samples.

TABLE 9.2

Sample Data Sheet for Determination of Chemical Composition
(Italicized Letters in Formulae Refer to Column)

	Stream:					
	A Beaker + dried material on filter	B Beaker + ashed material on filter	C AFDM = $A - B$	D C (mg) (AFDM $\times$ 0.53)	E P (mg) from digestion and SRP analysis	F Molar C:P (D/E) $\times$ 2.58
Sample						
1						
2						
3						
.						
.						
.						
.						
n						

to treatment (high-P vs low-P), team number, and time (initial and 30 and 60 min).

Field Collection Protocol

1. Collect tiles and filter water (filter 1 liter of stream water into a 1-liter plastic bottle using a hand pump and a Whatman GFF or Gelman Type A/E glass-fiber filter) from each stream.

2. Place 10 tiles into a labeled tupperware container (high-P or low-P) per team, which is filled with stream water. Attach the lids, squeezing out excess water (keeping the containers filled with water minimizes tile movement) and place the containers in a cooler to be transported back to the laboratory.

Laboratory Protocol

1. Transfer the 1 liter of filtered stream water and tiles into each stirred or recirculated incubation chamber (the number of tiles placed in the chamber is dependent on the amount of biomass attached to the substratum; a general rule of thumb would be to use at least 10 tiles if biomass is low and 5 to 10 tiles if it is high).

2. Using a 10-ml automatic pipet, remove 30 ml of streamwater from each chamber and transfer to the sample bottle labeled "initial." Filter the 30-ml water samples through a 0.45-μm pore size syringe filter (e.g., Syrfil-MF, Costar Corp., Cambridge, MA). Start either the pumps or the stir bar in the chamber.

3. Remove 30 ml of streamwater at 30 and 60 min after the start of the incubation, and transfer the water to the appropriately labeled bottle. Filter the samples as in step 2. If the water samples are not going to be analyzed for soluble reactive phosphorus within a few hours, place the bottles in the freezer until they can be analyzed for SRP levels (APHA *et al.* 1992; see below).

4. After 60 min, remove the tiles from the chamber. Place the tiles in an appropriately labeled aluminum weigh boat (etch the bottom of the boat with a sharp edge to designate the sample) and dry the tiles to constant weight at 105°C (ca. 24 h). Remove the weigh boats from the drying oven and transfer them to desiccators until weighing.

5. After the weigh boats have been weighed, place them in a muffle furnace at 500°C for at least 1 h (make certain the ovens are at 500°C before timing the 1 h), remove, allow them to cool to room temperature in a desiccator, and reweigh. Calculate AFDM as the difference between the dry mass and the combusted mass.

6. Analyze water samples for soluble reactive phosphorus (see below).

SRP Analysis

1. Make up appropriate reagents:

   a. H_2SO_4 solution: 140 ml concentrated sulfuric acid is added to 900 ml of double-distilled water.
   b. Ammonium molybdate solution: Dissolve 15 g of ammonium molybdate in 500 ml of double-distilled water (store in darkness).
   c. Ascorbic acid solution: Dissolve 2.7 g of ascorbic acid in 50 ml of double-distilled water. Make immediately before using.
   d. Antimony potassium tartrate solution: Dissolve 0.34 g of antimony potassium tartrate in 250 ml double-distilled water.
   e. Mixed reagent: Combine 25 ml of sulfuric acid solution, 10 ml of ammonium molybdate solution, 5 ml of antimony potassium tartrate solution, and 5 ml of ascorbic acid solution. Use within 6 h of preparation.
   f. Phosphorus standards: (1) stock solution, dissolve 0.5623 g of K_2HPO_4 in 1 liter of double-distilled water; (2) 1 μg P/liter solution, add 10 ml of stock solution to 1 liter of double-distilled

water; (3) 10 μg P/liter standard, add 2.5 ml of 1 μg P/liter solution to 250 ml double-distilled water.

2. Add 3.0 ml of mixed reagent to 30 ml of standard and all samples and mix thoroughly.

3. Wait for at least 20 min, but not longer than 1 h, and measure absorbance of solution at 885 nm against distilled water on a spectrophotometer using 10-cm-pathlength cuvettes.

4. Calculate SRP concentration (μg/liter) by dividing sample absorbance by the absorbance of the 10 μg/liter standard (should have a value of approximately 0.060–0.070) and multiply by 10.

Data Analysis

1. Plot the SRP concentration vs time to determine whether or not the relationship appears to be linear. Calculate the net P uptake rate using the formula

$$V = ([C_o - C_f] \times L)/t, \qquad (9.3)$$

where V is net uptake rate (mg P/h); C_o, initial SRP concentration; C_f, final SRP concentration; L, incubation volume (in liters); and t, time period of incubation (h). The net P uptake rate should then be normalized to either total biomass in the incubation (e.g., AFDM or chlorophyll a) or total substratum surface area. Use Table 9.3 for data entry and calculations.

D. Exercise 4: Phosphorus Radiotracer Method

(See cautionary notes on use of radioisotopes.)

Preparation Protocol

1. At least 1 month, and preferably 3–6 months, prior to the exercise, place small unglazed ceramic tiles or ceramic cylinders (Steinman *et al.* 1991a) in selected high-P and low-P streams to be sampled. If tiles are used, they should be preashed to remove attached glue, which otherwise would be included in the AFDM measurement.

2. Label tupperware containers (30 × 30 cm) by stream name or type (high-P, low-P).

3. Each team should have six 25-ml scintillation vials, each containing 15 ml of ecolume scintillation cocktail, labeled according to treatment (high-P or low-P), team number, and time (background and 10, 20, 30, 45, and 60 min).

TABLE 9.3
Sample Data Sheet for Determination of Net P Uptake

Stream:	
Time (min)	SRP concentration (mg/liter)
0	
10	
20	
30	
45	
60	

Calculated uptake rate:

Total AFDM or chlorophyll a in sample:

Uptake per unit AFDM (mg P mg AFDM^{-1} min^{-1}):

Uptake per unit chlorophyll a (mg P mg chlorophyll a^{-1} min^{-1}):

Field Collection Protocol

1. Collect tiles and filter water (filter 1 liter of stream water into a 1-liter plastic bottle using a hand pump and a Whatman GFF or Gelman Type A/E glass-fiber filter) from each stream.

2. Place 10 to 16 (the latter if turnover is to be measured) tiles into a labeled tupperware container (high-P or low-P), which are filled with stream water. Attach the lids, squeezing out excess water (keeping the containers filled with water minimizes tile movement) and place containers in a cooler to be transported back to the laboratory.

Laboratory Protocol

1. Transfer the 1 liter of filtered stream water and the tiles into each incubation chamber.[4]

2. Transfer approximately 50 ml of the filtered stream water to a 60-ml acid-washed plastic bottle, which will be analyzed for SRP concentration.

---

[4]If there are obvious signs of seston in the chamber water, it will be necessary to filter the subsamples before they are added to the scintillation vials (to remove radioactively labeled particulate material). This can be done by removing approximately 3 ml of water from each chamber with a 15-ml syringe, filtering the water through a 0.45-μm pore size syringe filter (eg., Syrfil-MF, Costar Corp., Cambridge, MA) into a small beaker, and then pipetting 1 ml of this filtrate into the scintillation vial.

3. Remove 1 ml of water from each chamber just prior to the ^{33}P injection. Transfer this water to the appropriately labeled scintillation vial (background) and mix thoroughly.

4. Inject 0.5 mCi of carrier-free [^{33}P]O$_4$ with a micropipette into each chamber. The micropipette tip will be extremely radioactive, so it should be removed immediately from the pipet after use and discarded in the radioactive waste bin.

5. Remove 1 ml of water from each chamber at 10, 20, 30, 45, and 60 min after the start of the incubation. Transfer the water to the appropriately labeled scintillation vial and mix thoroughly.[5]

6. After the 60-min sample is collected, carefully remove the tiles from the chamber using forceps or tongs. These tiles will then be processed for AFDM measurement, if turnover is not to be measured. For AFDM measurement, follow the procedures outlined below (step 7), making sure to avoid touching the radioactive material (keeping the tiles inside the beaker at all times minimizes this risk). If P turnover is to be measured, it is recommended that the tiles remain in the radiolabeled chamber water for an additional 5 h (6 h total) to allow a greater amount of ^{33}P incorporation by the algae. Radiolabeled tiles are then transported back out to the streams or to aquaria if turnover is to be measured. Double-bag, seal, label, and store the radioactive waste until the radioactivity decays to background levels. Wear gloves, safety glasses, and lab coats at all times when handling radioactive samples (consult the local health physicist at your institution for guidance and specific regulations associated with your site). Store the radioactive water from the incubation in sealed and labeled carboys until the radioactivity decays to background levels. It is usually necessary to store the water at least 10 half-lives before disposal (the half-life of ^{33}P is 25.4 days).

7. Finish measuring AFDM according to the methods outlined in Exercise 2 above, with the important modification of *not* brushing the algae off the substratum. Simply weigh the substratum with attached algae, and calculate AFDM as the difference in mass before and after combustion.

8. Count each scintillation vial for 10 min on a liquid scintillation counter (the counting efficiency for ^{33}P is generally >90%, and because the sample matrix is the same for all samples, no correction for counting efficiency is needed). No decay correction is needed if all samples are counted within a few hours of each other.

9. Soluble reactive phosphorus concentration of the initial stream water will be measured according to the methods outlined in Exercise C above.

---

[5]The first sample is not taken until 10 min to allow complete mixing of radioisotope within the chamber.

Data Analysis

1. Total phosphorus uptake rate is measured using the first-order rate coefficient of radiotracer depletion in the water (k), the concentration of SRP in the streamwater, and the water volume during the incubation (Steinman *et al.* 1991b). This procedure consists of three steps:

    a. Calculate k by regressing the ln-normalized scintillation count data (minus the background value determined from the sample collected just prior to ^{33}P injection) against time. Use Table 9.4 for data entry and calculations (also see Fig. 9.3).

    b. Total P uptake rate is then estimated by multiplying k by the SRP concentration and by the water volume in the incubation chamber. Based on the data from the Sample Data Sheet and depicted in Fig. 9.3, k (−0.228) is multiplied by 6.2 (SRP concentration and 1.0 (liters of water in chamber). This rate is in units of μg P/h.

    c. Total P uptake rate should then be normalized to the biomass in the chamber (μg P mg AFDM^{-1} h^{-1}) or by surface area of

TABLE 9.4
Sample Data Sheet for Determination of Total P Uptake (Radiotracer)

Stream: Laboratory stream with grazing snails			
Time (min)	Counts per minute (CPM ^{33}P)	CPM-background	ln(CPM-background)
0 (background)	49.4		
10	2009.8	1960.4	7.581
20	1939.1	1889.7	7.544
30	1841.7	1792.3	7.491
45	1761.8	1712.4	7.446
60	1669.4	1620	7.390

Calculated uptake rate constant (k): −0.228/h

Streamwater SRP concentration (μg/liter): 6.2 μg P/liter

Calculated total P uptake rate (μg P/h): 1.4136 μg P/h

AFDM (mg) or chlorophyll *a* (mg) or surface area (cm^2) in sample:

AFDM = 70.4 mg; surface area = 160 cm^2

Uptake per unit AFDM (μg P mg AFDM^{-1} h^{-1}): 0.0201

Uptake per unit chlorophyll *a* (μg P mg chlorophyll *a*$^{-1}$ h^{-1}):

Uptake per unit surface area (μg P cm^{-2} h^{-1}): 0.8835

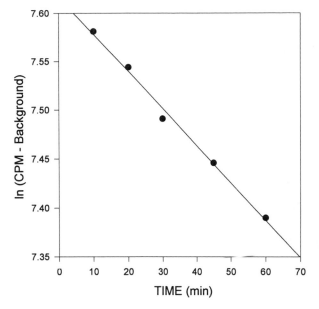

FIGURE 9.3 Radioactivity of ^{33}P in water incubated with periphyton exposed to grazing snails. Background activity of ^{33}P has been subtracted from measured activity and the data ln-normalized. Actual data used to generate the figure are based on real experiments and are included in the total P uptake sample data sheet.

substrata in the chamber (μg P cm^{-2}h^{-1}). The total uptake rate is divided either by the total AFDM in the chamber (e.g., 70.4 μg based on the Sample Data Sheet) or total substrata surface area in chamber (e.g., 160 cm^2 based on the Sample Data Sheet) to obtain a normalized uptake rate.

E. Exercise 5: Phosphorus Turnover

Preparation Protocol

1. Cover each of the 50-ml glass beakers with a square of aluminum foil and gently etch into the foil the following information: stream type (high-P or low-P), sampling date (Day 0, 2, 5, and 10), and replicate (a–d).

Field Placement and Collection (Option 1) This option is to be followed if permission is obtained by the appropriate authorities to place radiolabeled material in the chosen streams. If permission cannot be obtained, follow Option 2 (below).

1. Following the radioisotopic uptake study (Exercise 4), carefully remove the tiles from each chamber using forceps or tongs, place them into tupperware containers, and transport them to the high-P and low-P streams.

2. Place tiles in locations that have similar current velocities, irradiance levels, temperatures, and grazer densities in both streams, if possible. Remove four tiles from each stream at 1 h (the 1-h incubation allows "Day 0" samples to be rinsed by streamwater prior to sampling in order to wash off adsorbed residual ^{33}P) and again at 2, 5, and 10 days and place them in appropriately labeled, tared 50-ml glass beakers. One additional unlabeled tile also should be processed from each sticam; any radioactivity associated with the algae on this tile will be subtracted off all other counts, as it represents the naturally occurring background radioactivity. The laboratory procedures associated with this option follow Option 2 (below).

Aquarium Placement and Collection (Option 2)   For this option, radiolabeled tiles are placed in two aquaria, one with a high-P and one with a low-P concentration (actual concentrations should mimic those of the natural streams that otherwise would have been used). Ideally, the water in each aquarium should be changed daily to minimize released ^{33}P from being taken up again. However, this option is time-consuming and generates a considerable volume of contaminated waste. If water is not changed, it should be recognized that the calculated turnover rates will represent underestimates of true turnover.

1. Following the radioisotopic uptake study (Exercise 4), carefully remove the tiles from each chamber using forceps or tongs, place them into tupperware containers, and transport them to the high-P and low-P aquaria. Aquaria should be exposed to similar light and temperature regimes and be fitted with a means of circulating or moving the water (e.g., aeration or mixing).

2. Remove four tiles from each stream at 1 h (the 1-h incubation allows "Day 0" samples to be rinsed by streamwater prior to sampling in order to wash off adsorbed residual ^{33}P) and again at 2, 5, and 10 days and place them in appropriately labeled, tared 50-ml glass beakers. One additional unlabeled tile also should be processed from each aquarium; any radioactivity associated with the algae on this tile will be subtracted off all other counts, as it represents the naturally occurring background radioactivity.

Laboratory Protocol (Applicable for both Options 1 and 2)

1. On the day of collection (Days 0, 2, 5, and 10), place the beakers in a drying oven and dry the tiles to constant weight at 105°C. Because it generally requires at least 1 day for the biomass to reach constant weight, the remainder of the AFDM determination will be done the next day.

2. On the days after collection (Days 1, 3, 6, and 11), weigh dried tiles, combust them for a minimum of 1 h at a full 500°C. Remove tiles, allow them to cool to room temperature in a desiccator, and reweigh.

3. Using a 5-ml pipet, add 10 ml of 2 N HCl to the beaker (make sure this is enough to cover all the periphyton), label the beaker with the appropriate sample designation, and place parafilm over the beaker. Leaching of ashed material should last at least 24 h. Place beakers in the laboratory hood until the next lab period.

4. On Day 2 after each collection (Days 2, 4, 7, and 12), add 10 ml of distilled water to each beaker (to reduce acidity to 1 N), swirl the beaker gently to mix thoroughly, and pipet 1 ml of the diluted leachate to a scintillation vial containing 15 ml of Ecolume scintillation cocktail.

5. Count each sample on a liquid scintillation counter for 10 min. We recommend that all samples from the turnover experiment be counted during the same run (within several hours of each other) at experiment's end to preclude the need to apply a radioactive decay correction factor. The counts are used to determine turnover ate.

Data Analysis

1. Phosphorus turnover rate is computed by linear regression of $\ln[^{33}P]$ counts in algae versus time in stream (in days). Phosphorus turnover rate is therefore expressed as a first-order turnover rate constant (day^{-1}). The background radioactivity associated with unlabeled tiles should be subtracted from each sample count prior to performing the ln transformation and the regression, and the regression should be based on a mean value derived from the four substrata sampled on each day. Use Table 9.5 for data entry and calculations.

TABLE 9.5
Sample Data Sheet for Determination of Turnover

Stream:					
		ln (CPM)			
Replicate	Background	Day 0 − background	Day 2 − background	Day 5 − background	Day 10 − background
A					
B					
C					
D					

2. It should be emphasized that this determination of P turnover rate may not be an accurate physiological index of total algal P turnover rate because not all phosphorus pools within the algal cells will have reached isotopic equilibrium during the 6 h of ^{33}P exposure during the uptake part of the experiment. However, the approach described should provide a reasonable basis for comparing turnover rates in the more rapidly cycling P pools between different streams.

IV. QUESTIONS

A. Limitation: Phosphatase Activity

1. Was phosphatase activity greater in the low phosphorus stream, as hypothesized? If not, what might explain this result?

2. What other factors besides phosphorus concentration and biomass might influence the PA activity in the two streams?

3. Why is it important to normalize the PA data to an index of biomass?

4. Phosphatase is an inducible enzyme. That is, it is synthesized upon metabolic demand, as opposed to a constitutive enzyme, which is always present. What advantage is there in maintaining phosphatase as an inducible enzyme?

B. Limitation: Chemical Composition

1. C:P ratios substantially greater than 106:1 (on a molar basis) suggest phosphorus deficiency in algae. However, some algal species have greater carbon demands than others because of more carbon-based compounds in their cell walls. How would this type of demand influence the interpretation of the C:P ratio?

2. Many algal species exhibit "luxury uptake" of phosphorus, whereby they take up excessive amounts of phosphorus when it is available (e.g., during high-P episodes) and then store the P intracellularly (in polyphosphate bodies). How would luxury uptake of P influence the ratio?

C. Net Uptake: Stable Phosphorus

1. Were the net uptake rates similar or different between the two streams? If they were different, what might account for this difference?

2. Sometimes, no net uptake is measured during an incubation (i.e., the amount of phosphorus measured at the start of the experiment is the same as that at the end of the experiment). Assuming that the algae are biologically active and actively taking up phosphorus, what might account for this?

D. Total Uptake: Radiolabeled Phosphorus

1. Were the total uptake rates similar or different between the two streams? If they were different, what might account for this difference?

2. By keeping the incubation time short in this exercise, you minimize the possibility that any radioactive phosphorus that was taken up could be released within the incubation period (i.e., minimize the possibility of recycling). Thus, the radioactive phosphorus removed from the water is assumed to represent the total uptake rate. How does this differ from net uptake rate (i.e., which measure should be greater)? Why?

E. Turnover

1. Were the P turnover rates similar or different between the two streams? If they were different, what might account for this difference?

2. If the ^{33}P was allowed to come to complete isotopic equilibrium within the algae during the uptake part of the exercise, would you expect measured P turnover rates to be greater or lower than those measured? Why?

3. How might the thickness of the periphyton matrix influence turnover rates? What about grazing activity?

V. MATERIALS AND SUPPLIES

Letters in parentheses indicate in which Exercise (1, 2, 3, 4, or 5) the item is used.[6]

Field Materials

1-Liter plastic bottles (1, 3, 4)
Cooler (3, 4)
Hand pumps with GFF (or equivalent) filters (1, 3, 4)
Holder to transport beakers to and from field (5)
Tupperware containers to accommodate tiles or rocks (1, 2, 3, 4, 5)
Unglazed ceramic tiles (e.g., tiles measuring 3 × 3 cm or ceramic
    cylinders) (1, 3, 4, 5)

---

[6]Any use of isotope requires specific laboratory protocols. These protocols are available from the Safety and Health Department at the institution or laboratory licensed for isotope use. These protocols must be followed carefully. It is essential that gloves (we recommend double gloving, using vinyl gloves directly over the hands and disposable gloves over the vinyl ones), lab coats, and safety glasses be worn at all times.

Laboratory Materials

1 *N* NaOH (1)
2 *N* HCl (2, 5)
10-ml glass beakers (2, 4)
25-ml plastic syringes with syringe holders; 0.45-μm pore size (1, 3, 4)
25-ml scintillation vials (1, 4, 5)
50-ml glass beakers (5)
50-ml plastic bottles (3)
60-ml plastic bottles (4)
100 ml plastic bottles (2)
100-ml volumetric cylinders (2)
150 m*M para*-nitrophenyl phosphate solution (add 2.78 g of *para*-nitrophenyl phosphate to 50 ml of double-distilled water) (1)
90% Acetone (90 parts acetone with 10 parts saturated magnesium carbonate solution)
Aluminum foil (2, 5)
Aluminum weigh boats (3, 4)
Carrier-free ^{33}P isotope (0.5 mCi/chamber; order from New England Nuclear) (4, 5)
Coarse-bristled toothbrushes (2)
Ecolume Scintillation Cocktail (ICN Scientific, Costa Mesa, CA) (4, 5)
Large pans or trays (2)
Parafilm (2, 5)
Plastic jars or centrifuge tubes (1)
Reagants for SRP analysis (2, 4)
Wide mouth glass incubation jars (30 ml or larger for larger substrata) (1)

Laboratory Equipment

Analytical balance (2, 3, 4, 5)
Automatic pipets (1 ml, 10 ml) (1, 2, 3, 4, 5)
Cuvettes (1- and 10-cm pathlength) (1, 2, 3, 4)
Desiccators (2, 3)
Drying oven (2, 3, 4, 5)
Liquid scintillation counter (4, 5)
Muffle furnace (2, 3, 4, 5)
Recirculating chambers (with either pumps or stirrers to circulate water) (3, 4)
Spectrophotometer (narrow bandwidth: 0.5 to 2.0 nm) (1, 2, 3, 4)

REFERENCES

APHA, AWWA, and WEF. 1992. *Standard Methods for the Examination of Water and Wastewater,* 18th ed. American Public Health Association, Washington, DC.

Borchardt, M. A., J. P. Hoffmann, and P. W. Cook. 1994. Phosphorus uptake kinetics of *Spirogyra fluviatilis* (Charophyceae) in flowing water. *Journal of Phycology* **30:**403–412.

Bothwell, M. L. 1989. Phosphorus-limited growth dynamics of lotic periphytic diatom communities: Areal biomass and cellular growth rate responses. *Canadian Journal of Fisheries and Aquatic Sciences* **46:**1293–1301.

Burkholder, J. M., and R. G. Wetzel. 1990. Epiphytic alkaline phosphatase on natural and artificial plants in an oligotrophic lake: Re-evaluation of the role of macrophytes as a phosphorus source for epiphytes. *Limnology and Oceanography* **35:**736–747.

Cembella, A. D., N. J. Antia, and P. J. Harrison. 1984. The utilization of inorganic and organic phosphorus compounds as nutrients by eukaryotic microalgae: a multidisciplinary perspective: Part 1. *Critical Reviews in Microbiology* **10:**317–391.

Currie, D. J., E. Bentzen, and J. Kalff. 1986. Does algal-bacterial phosphorus partitioning vary among lakes? A comparative study of orthophosphate uptake and alkaline phosphatase activity in fresh water. *Canadian Journal of Fisheries and Aquatic Sciences* **43:**311–318.

Darley, W. M. 1982. *Algal Biology: A Physiological Approach.* Blackwell Scientific, Oxford, UK.

Davis, L. S., J. P. Hoffman, and P. W. Cook. 1990. Seasonal succession of algal periphyton from a wastewater treatment facility. *Journal of Phycology* **26:**611–617.

Droop, M. R. 1974. The nutrient status of algal cells in continuous culture. *Journal of the Marine Biological Association of the United Kingdom* **54:**825–855.

Elwood, J. W., J. D. Newbold, A. F. Trimble, and R. W. Stark. 1981. The limiting role of phosphorus in a woodland stream ecosystem: Effects of P enrichment on leaf decomposition and primary producers. *Ecology* **62:**146–158.

Grimm, N. B., and S. G. Fisher. 1986. Nitrogen limitation in a Sonoran Desert stream. *Journal of the North American Benthological Society* **5:**2–15.

Healey, F. P. 1973. Inorganic nutrient uptake and deficiency in algae. *Critical Reviews in Microbiology* **3:**69–113.

Healey, F. P., and L. L. Hendzel. 1979. Fluorometric measurement of alkaline phosphatase activity in algae. *Freshwater Biology* **9:**429–439.

Hecky, R. E., and P. Kilham. 1988. Nutrient limitation of phytoplankton in freshwater and marine environments: A review of recent evidence on the effects of enrichment. *Limnology and Oceanography* **33:**796–822.

Hecky, R. E., P. Campbell, and L. L. Hendzel. 1993. The stoichiometry of carbon, nitrogen, and phosphorus in particulate matter of lakes and oceans. *Limnology and Oceanography* **38:**709–724.

Hill, W. R., and A. W. Knight. 1988. Nutrient and light limitation of algae in two northern California streams. *Journal of Phycology* **24:**125–132.

Kilham, P., and R. E. Hecky. 1988. Comparative ecology of marine and freshwater phytoplankton. *Limnology and Oceanography* **33:**776–795.

Lock, M. A., R. R. Wallace, J. W. Costerton, R. M. Ventullo, and S. E. Charlton. 1984. River epilithon: Toward a structural–functional model. *Oikos* **42:** 10–22.

Lohman, K., and J. C. Priscu. 1992. Physiological indicators of nutrient deficiency in *Cladophora* (Chlorophyta) in the Clark Fork of the Columbia River, Montana. *Journal of Phycology* **28:**443–448.

Mulholland, P. J., A. D. Steinman, and J. W. Elwood. 1990. Measurement of phosphorus uptake length in streams: Comparison of radiotracer and stable PO_4 releases. *Canadian Journal of Fisheries and Aquatic Sciences* **47:**2351–2357.

Peterson, B. J., J. E. Hobbie, T. L. Corliss, and D. Kriet. 1983. A continuous flow periphyton bioassay: Tests of nutrient limitation in a tundra stream. *Limnology and Oceanography* **28:**582–595.

Pringle, C. M., P. Paaby-Hansen, P. D. Vaux, and C. R. Goldman. 1986. In situ assays of periphyton growth in a lowland Costa Rica stream. *Hydrobiologia* **134:**207–213.

Raven, P. H., R. F. Evert, and H. Curtis. 1981. *Biology of Plants,* 3rd ed. Worth, New York, NY.

Redfield, A. C. 1958. The biological control of chemical factors in the environment. *American Scientist* **46:**205–221.

Rhee, G. Y. 1978. Effects of N:P atomic ratios and nitrate limitation on algal growth, cell composition, and nitrate uptake. *Limnology and Oceanography* **23:**10–24.

Schindler, D. W. 1977. The evolution of phosphorus limitation in lakes. *Science* **195:**260–262.

Solórzano, L., and J. H. Sharp. 1980. Determination of total and dissolved phosphorus and particulate phosphorus in natural waters. *Limnology and Oceanography* **25:**574–578.

Steinman, A. D. 1994. The influence of phosphorus enrichment on lotic bryophytes. *Freshwater Biology* **31:**53–63.

Steinman, A. D., P. J. Mulholland, A. V. Palumbo, T. F. Flum, and D. L. DeAngelis. 1991a. Resilience of lotic ecosystems to a light-elimination disturbance. *Ecology* **72:**1299–1313.

Steinman, A. D., P. J. Mulholland, and D. B. Kirschtel. 1991b. Interactive effects of nutrient reduction and herbivory on biomass, taxonomic structure, and P uptake in lotic periphyton communities. *Canadian Journal of Fisheries and Aquatic Sciences* **48:**1951–1959.

Stockner, J. G., and K. R. Shortreed. 1978. Enhancement of autotrophic production by nutrient addition in a coastal rainforest stream on Vancouver Island. *Journal of the Fisheries Research Board of Canada* **35:**28–34.

Wetzel, R. G. 1983. *Limnology,* 2nd ed. Saunders College Publishing, Philadelphia, PA.

Wetzel, R. G. 1981. Long-term dissolved and particulate alkaline phosphatase activity in a hardwater lake in relation to lake stability and phosphorus enrichments. *Verhandlungen der Internationalen Vereinigung für Theoretische und Angewandte Limnologie* **21:**369–381.

Whitford, L. A., and G. J. Schumacher. 1964. Effect of a current on respiration and mineral uptake in *Spirogyra* and *Oedogonium*. *Ecology* **45:**168–170.

CHAPTER 10

Transport and Storage of FPOM

J. BRUCE WALLACE[*] AND JACK W. GRUBAUGH[†]

*Institute of Ecology
University of Georgia,
†Department of Biology
University of Memphis*

I. INTRODUCTION

Fine particulate organic matter (FPOM) includes particles in the size range of >0.45 μm to <1000 μm (1.0 mm) that are either suspended in the water column or deposited within lotic habitats. Size fractions of FPOM can be further divided into the categories of medium-large (250–1000 μm), small (100–250 μm), fine (45–100 μm), very fine (25–45 μm), and ultrafine (0.45–25 μm). Suspended fine particulate material is referred to as *seston*, and includes all living (e.g., bacteria, algae, protozoans, invertebrates, etc.) and nonliving material (amorphous organic matter, detritus, as well as suspended inorganic sediment) within the 0.45-μm to 1-mm size range. Seston can originate from many sources, including the breakdown of larger particles by physical forces, animal consumption, microbial processes, flocculation of dissolved substances, and terrestrial inputs (Wotton 1984, 1990). Transported loads of seston vary greatly among lotic systems from micrograms in some small streams to metric tons in larger streams and rivers (see Chapter 7). Seston functions as an important food resource for many filter-feeding invertebrates (Wallace and Merritt 1980, Benke *et al.* 1984), as well as for some vertebrates, such as paddlefish (*Polyodon spathula*), in

Methods in Stream Ecology   191

large rivers. In some situations, such as below the outflow of dams or lake outlets, filter-feeding populations can remove large portions of transported seston from the water column within a few kilometers (Maciolek and Tunzi 1968, Voshell and Parker 1985). The downstream transport of seston is also important to the theme of conceptualizing streams as longitudinally linked systems (Vannote *et al.* 1980, Minshall *et al.* 1985) and the concept of material spiralling in stream ecosystems (Newbold *et al.* 1982). Therefore, seston is important to many ecosystem processes as it represents a major pathway of organic matter transport and export, and is thus an important consideration in ecosystem organic matter budgets (e.g., Fisher and Likens 1973, Cummins *et al.* 1983).

FPOM occurs not only in the water column as seston, but is also found deposited in lotic habitats as fine benthic organic matter (FBOM). FBOM standing crops are rarely adequately assessed in stream research. Sometimes FBOM is ignored completely, or measurements are done in conjunction with benthic sampling for macroinvertebrates with a relatively large mesh size (e.g., 250 μm) that underestimates the total stored FBOM. For example, Minshall *et al.* (1982) found that standing crops of benthic organic matter may be underestimated by as much as 65% when sampling devices with 250-μm meshes are used. Additionally, standing crops of organic matter may vary greatly between erosional (e.g., riffles and outcrops) and depositional areas (e.g., pools) of streams. Debris dams, for example, often are sites of high FBOM storage (Bilby 1980, Smock *et al.* 1989). FBOM and associated microbes serve as an important resource for animals adapted for deposit feeding (collector–gatherers; see Chapter 21), which includes a wide assortment of invertebrates as well as some collector-gathering fishes ("rough" fishes). Many deposit-feeding animals have low assimilation efficiencies, and the ingestion and reingestion of FBOM and associated microbes may occur many times in longitudinally linked systems. Unfortunately, only a few studies have attempted to quantify the turnover of FPOM; Fisher and Gray (1983) estimated that fine-particle feeders ingested over four times their weight per day, and the entire standing crop of FPOM in Sycamore Creek, Arizona, was ingested and egested every 2 to 3 days. FBOM storage varies greatly within heterogeneous stream environments. In small headwater streams, the highest standing crops of FBOM are usually associated with pools and woody debris dams (Bilby and Likens 1980, Huryn and Wallace 1987, Smock *et al.* 1989). In large river systems, slack-water habitats such as sloughs and backwaters are repositories for large amounts of FBOM; during high flow conditions floodplains adjacent to large rivers can serve as both source and sink of seston and FBOM (Grubaugh and Anderson 1989).

A number of approaches have been used to estimate FPOM quality and will only be mentioned here. Organic : inorganic matter ratio is simply an estimate of the relative amount of organic and inorganic matter in seston and can be easily determined from procedures described below for seston sampling (see Exercise 1). This ratio often varies greatly for different size fractions of seston, with smaller size classes having a greater proportion of inorganic material (ash) than larger size fractions. Some studies (e.g., Angradi 1993a) have examined the organic constituents of seston such as chlorophyll *a* (see Chapter 15) while others have examined other organic material and microbial activity such as respiration (Peters *et al.* 1989; also see Chapter 26 for examples of respirometry techniques). Edwards (1987) evaluated the importance of bacteria in seston and in the growth of filter-feeding black fly larvae (Edwards and Meyer 1987). Carlough and Meyer (1991) found sestonic protozoans to be an important component of seston in a low-gradient, blackwater river. Voshell and Parker (1985) used microscopy to examine directly the frequency and type of particles in various size categories. The amounts in each category (e.g., animal, diatoms, other algae, vascular plant, and amorphous detritus) are estimated by the areal standard-unit method used in phytoplankton studies as described by Welch (1948). Wallace *et al.* (1987) used a microscope and digitizer interfaced with a computer for similar analyses; however, these latter methods are not appropriate for bacteria and protozoans. More recently, other "higher-tech" methods have been employed to study dynamics and origin of seston. For example, Cushing *et al.* (1993) used radioactively tagged particles to study movement and deposition of seston. Angradi (1993b) used stable carbon and nitrogen isotope analysis to study origin and movement of seston.

In the following exercises, we will describe seston sampling procedures for streams and rivers of various sizes and describe techniques to assess seston concentration, size distribution, and instantaneous estimates of total seston export. Next we will consider sampling techniques for FBOM in streams and emphasize the relative importance of depositional and erosional habitats in assessment of FBOM standing crops. The final exercise examines direct linkages between sestonic FPOM and filter-feeding biota. The specific objectives of these exercises are to: (1) introduce the reader to the importance and magnitude of FPOM transport in streams, (2) demonstrate techniques for collecting and analyzing seston and FBOM, (3) demonstrate the importance of hydrologic events in seston transport compared to base flow conditions, (4) compare the relative importance of erosional and depositional habitats in FBOM storage in streams, and (5) illustrate direct consumption of suspended particles by filter-feeding larvae of black flies (Diptera: Simuliidae). The reader should gain an appreciation for the methods involved in assessing FPOM transport, storage, and use in streams.

II. GENERAL DESIGN

A. Seston

Instantaneous seston concentrations (e.g., mg/liter) can be easily measured by filtering known volumes of water through preashed and preweighed glass-fiber filters. This simple approach can be used to compare seston concentrations during base flow and short-term hydrologic events (e.g., storms), or for comparing seston concentrations among streams of various sizes. Percentage ash in such measurements has been related to long-term watershed disturbance (Webster and Golladay 1984). In some cases, seston particle size has been shown to vary with stream size, as smaller headwater streams draining forested areas have larger median particle sizes than larger rivers downstream (Wallace *et al.* 1982). However, with few exceptions, the majority of the particles transported by most streams during baseflow conditions are <50 μm in diameter (Sedell *et al.* 1978, Naiman and Sedell, 1979a,b, Wallace *et al.* 1982).

A two-part sampling approach generally is adequate for sampling seston in small rivers and streams. The first part consists of collecting a 20- to 30-liter grab sample for measuring concentrations of finer seston particles (i.e., <250 μm). Under most conditions, seston particle-size distributions are strongly skewed toward smaller size fractions; therefore, the second part of sampling uses a collection net to filter a large quantity of water to obtain reasonable concentration estimates for larger seston particle sizes (i.e., >250 μm). One of the best devices for this purpose is a Miller plankton tow net fitted with a 250-μm mesh collecting net and flowmeter (Fig. 10.1).

Particle size separation requires a wet filtration system consisting of a series of stacked sieves of Nitex or bolting-cloth netting of various sizes. Sieves can be constructed with short (4–5 cm) sections of PVC pipe with netting glued over one end and joined with connectors to form a stackable

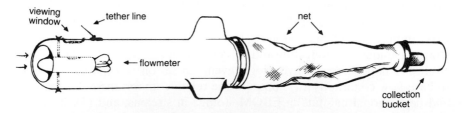

FIGURE 10.1 A Miller-type tow-net, equipped with a flowmeter to record velocity of water filtered. The flowmeter is used to calculate distance of water filtered over the time interval the net is deployed, (see seston concentration protocols and Table 10.1). A Plexiglas viewing window (optional) is ideal for viewing the dial on the current meter.

series of sieves. More elaborate wet filtration systems are constructed of threaded, stainless steel tubes fitted with stainless steel bolting-cloth filters of various dimensions. The individual filters with Teflon gaskets are inserted between threaded sections of each tube to form a series of stackable sieves with a large funnel at the top of the apparatus (Fig. 10.2). A water sample is poured into the funnel and through the sieves under vacuum. Seston particle sizes are thus separated by sieve sizes and water passing through the smallest filter can be retained for the ultrafine fraction.

Large rivers (greater than seventh order) present considerable difficulties for seston sampling. Most of these rivers are nonwadeable and sampling can be conducted only from bridges or boats. Furthermore, as a result

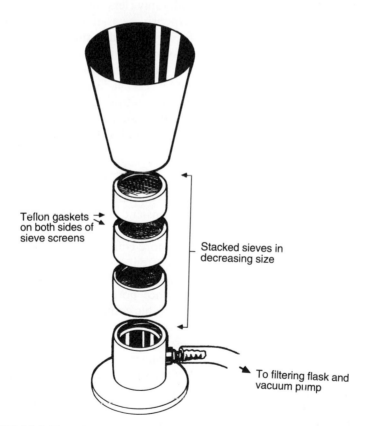

Teflon gaskets
on both sides of
sieve screens

Stacked sieves in
decreasing size

To filtering flask and
vacuum pump

FIGURE 10.2 Example of a wet filtration system using a series of stacked sieves in order of decreasing size. *Note.* Vacuum line is connected to a large filtering flask to retain filtrate for ultrafine seston and to avoid pumping water into the vacuum pump (see particle size separation protocols).

of differential settling rates and lower current velocities near the water/ substratum interface, seston concentrations and particle-size distributions can vary greatly with depth as well as with sampling location relative to the *thalweg,* or middle of the main channel. Adequate sampling of suspended material in large rivers requires depth-integrated, velocity-weighted samples taken at multiple depths along a transect. A variety of devices have been designed to collect integrated and weighted samples of total suspended sediments and these lend themselves well to seston sampling in large rivers (e.g., Grubaugh and Anderson 1989; see also Chapter 7). Drawbacks to their general use are that these samplers are expensive and require a trained operator. The reader should consult Guy and Norman (1970) for a discussion of such devices and sampling designs for use in large rivers.

Point samplers, although considerably less accurate than depth-integrating devices, are much less expensive, easier to operate, and can also be used in large rivers. The protocol provided in this exercise for large river seston collection employs point-sampling techniques. Examples of point samplers include Kemmerer or Van Dorn bottles fitted with *line depressors* or fluked weights to facilitate a vertical descent of the sampler in high-velocity rivers. To estimate seston concentration using point samplers, samples need to be collected at several locations along a given vertical gradient, with the depth and number of samples dependent on the total depth of the water column.

B. Fine Benthic Organic Matter

Laboratory protocols similar to those employed for seston can be used to determine stored FBOM concentrations and particle size distributions, if samples are resuspended in a known volume of water prior to analysis. Sampling procedures, however, can be more complicated for FBOM storage, as stream characteristics such as current velocity, substrate particle size, and the presence or absence of retention devices must also be taken into consideration. These factors influence both the physical storage of FBOM as well as the structure of the benthic community (e.g., Huryn and Wallace 1987) and hence affect transport and use of FPOM in lotic systems. To assess these influences, it is necessary to measure standing stocks of FBOM among stream habitats with differing morphological characteristics. For the purposes of this chapter, FBOM will be quantified in three size fractions: 0.45 to 250 μm, 250 to 500 μm, and 500 μm to 1.0 mm. Particle sizes that are greater than 1.0 mm constitute coarse particulate organic matter (CPOM) and are addressed in Chapter 11.

C. Linkages

Collector–gatherer and collector–filterer organisms use FBOM and seston, respectively, as food resources (see Chapter 21). One collector organism is the larval black fly, which uses cephalic fans to filter and remove small suspended particles (<300 μm diameter) from the water column. Black flies and other collector–filterers are important in energy transformations in streams and are examples of direct linkages between seston and the biota. Larval black flies also make good study organisms since they are often very abundant in many lotic habitats and have rapid (<1 h) gut passage times (e.g., Wotton 1978, 1980).

We will expose larval black flies to a dense concentration of trackable FPOM (powdered charcoal) for a brief time, which will produce a distinct band in the larval guts. Representative larvae (preferably of different instars or sizes) will be collected and preserved at 10-, 20-, and 30-min intervals following particle release. In the laboratory, larvae from each time interval will be separated by size class (body length) and their guts dissected. Using a dissecting microscope equipped with an ocular micrometer, we will make two measurements: the distance from the posterior end of the head to the band of charcoal, and the distance from the posterior end of the head to the tip of the abdomen. The ratio of these two (ratio W; Wotton 1978) gives a measure of gut passage time based on the distance the band has moved.

D. Site Selection

Seston   Exercise 1 is designed for both small streams and rivers; however, we recommend that whenever possible this exercise be restricted to lotic reaches that are safely wadeable. Application of this protocol to larger rivers requires working from either bridges or boats, both of which carry inherent risks. If a bridge site is used, the bridge should be close enough to the water surface, and water depth should be shallow enough that sampling equipment can reach the river bottom. The bridge should have sufficiently wide shoulders and limited automotive traffic to facilitate safety; sampling crews are strongly urged to exercise extreme caution and to wear "blaze-orange" garments to enhance their visibility. Boat sampling requires that attention be given not only to sampling equipment and procedures, but also to boat maintenance, proper safety equipment, and safe boating practices. Because of inherent dangers of sampling during high-flow conditions, we recommend that boat sampling be conducted only at low- to mean-flow stage conditions. Seston sampling during storms can be dangerous even in small streams and adequate precaution is again recommended.

Management of the sampling site is as important as site selection; disturbance of upstream substratum should be avoided when collecting water for seston analysis. Even minor disturbances such as wading across a stream can dislodge sufficient amounts of FPOM to greatly increase seston concentrations well above baseflow conditions. Be especially careful when working with a field team consisting of several individuals; the person or persons actually collecting seston samples should be upstream of other team members. Finally, the volume of water filtered or collected with the seston is crucial information in seston sampling. Care should be taken in both the field and laboratory when determining and recording volume measurements.

Fine Benthic Organic Matter   For Exercise 2, logistical considerations are again of primary concern and we recommend its execution in relatively shallow streams (<0.75 m depth) that are readily wadeable. Furthermore, site selection should be focused toward stream reaches with clearly hetero-geneous channel features consisting of zones of contrasting current veloci-ties, such as pools and other clearly depositional reaches versus cobble riffles or bedrock outcrops. If the site selected is in a nonwadeable, medium-sized stream or river that is too deep for devices described in the FBOM Exercise, SCUBA gear in combination with elaborate sampling devices such as dome samplers can be used to sample benthos and FBOM (Gale and Thompson 1975, Platts *et al.* 1983). Such devices are expensive and will require trained operators. In backwaters or other slow current habitats, an Eckman dredge can be used, but this device often has a high loss of FBOM. In habitats with slow to moderate currents (e.g., channel borders and sloughs) a Ponar or petite Ponar dredge can be used; in moderate to fast currents (e.g., main channel) Peterson dredges are preferred. None of these dredges will function properly if the site selected has coarse substrata.

Linkages   Exercise 3 is most easily performed in small streams less than 0.25 m in depth. The site should have a reasonably abundant black fly population; shallow outflow streams from lakes and small reservoirs are ideal locations because black fly larvae often form dense aggregations at such sites. It is also important to select areas where the charcoal slurry can be released immediately upstream of the black flies so that the slurry plume flows directly over the larvae.

Sampling site management is again important; use caution to minimize disturbing black fly aggregations prior to, during, and following release of the charcoal slurry. When collecting larvae following charcoal release, stand to one side of the area covered by the charcoal plume and remove larvae by carefully reaching into the area covered by the plume.

III. SPECIFIC EXERCISES

A. Exercise 1: Seston Concentration

Protocol for Seston Sampling in Streams and Small Rivers

1. For obtaining a *carboy* sample (steps 1–3): cover the opening of a clean, 20- to 30-liter carboy with a 250-μm mesh sieve or bolting cloth.

2. Using another carboy, bucket, or other vessel, collect a grab sample of stream water. Care should be taken not to disturb the substratum and collect resuspended benthic FPOM with the sample. Pour the water through the mesh and into the carboy until filled. This sample will be used to estimate FPOM concentrations of <250-μm particle size. If the sample is to be used for particle size analysis and seston concentration likely is low (e.g., during winter sampling or in streams with little allochthonous inputs), it is advisable to fill a second or even third carboy for sample processing.

3. Label the carboy(s) by sampling site and sample number and transport to the laboratory for processing. If the sample is to be used for particle size analysis, laboratory filtration should be completed within a few hours following collection of field samples.

4. For the *Miller net* sample (steps 4–11): record initial reading of flowmeter in the Miller net prior to putting sampler into the stream. Record start time of sampling (optional).

5. Suspend the Miller net in the water column with the front opening completely submersed. Use a tether line on the front of the sampler to secure the sampler in place (e.g., tied to a bridge rail, overhanging limb, or held by the operator if stream velocity is not prohibitive).

6. Sampling time will vary depending on the amount of suspended material. Generally, 10–30 min is an adequate sampling period, but more time might be needed if seston concentrations are low and little or no material is readily visible in the collection net at the end of sampling. Conversely, less time is needed if seston loads are heavy. If larger materials are present in the water column (i.e., leaves or sticks), check to make sure the opening of the Miller net or flowmeter is not obstructed.

7. Upon completion of sampling, record final flowmeter reading and (optionally) stop time of sampling.

8. Wash material from the collection net into a 1-mm mesh sieve nested over a 250-μm mesh sieve. This separates out CPOM which would otherwise result in an overestimate of FPOM during sample processing.

9. Wash material retained on the 250-μm sieve into a sample bag or sample jar. This collection will be used to estimate FPOM concentrations of the particle size class >250 μm. Label bag or jar by sampling site and sample number, and transport to the laboratory for processing. If the sample is to be used for particle size analysis, laboratory filtration should be completed within a few hours following collection of field samples.

10. To estimate the volume of water filtered through the Miller net, the following information is needed: radius (r in m) of front opening and distance (d in m) of water filtered as measured by the flowmeter.

11. The volume (V in liters) filtered is calculated by the equation

$$V = r^2 d \times 1000. \qquad (10.1)$$

12. If elapsed sampling time (t) was recorded in seconds, a measure of velocity (v in m/s) can also be made:

$$v = d \div t. \tag{10.2}$$

Although velocity is not crucial to estimations of seston concentration, it is an additional and easily calculated physical parameter by this method. A data sheet for recording and calculating water volume and velocity information using Miller-type tow nets is presented (see Table 10.1).

Protocol for Seston Sampling in Large Rivers

1. The protocol described is for bridge sampling. Measure river width below the bridge with a tape measure.

2. Divide the river width measurement by 11 and use the result to determine 10 equidistant points across the river. Using an erasable marking pen, clearly mark and number these 10 points on the bridge railing. These will be the locations of the sampling verticals.

3. Measure the distance from bridge rail to the water surface and from bridge rail to river bottom at each vertical; the difference between these distances is total depth at each vertical.

4. The number and depth of collections for point samples at each vertical is determined from water depth of individual verticals as follows:

Water depth	Sampling depth (measured from surface)
≤1 m	60% of water depth
1–3 m	20 and 80% of water depth
≥3 m	20, 60, and 80% of water depth

5. Lower Kemmerer or Van Dorn bottle to appropriate depth, close, and retrieve sampler. Filter samples through a 1-mm mesh sieve or bolting cloth into milk jugs to remove CPOM. Mark each jug as to collection site, vertical, and sample depth.

6. Transport samples to the laboratory to determine seston concentrations (see "Standard Processing Protocols" below).

7. To determine mean seston concentration, first calculate mean concentration for each vertical and then calculate the mean of all 10 verticals.

Standard Processing Protocols

1. Set up a microfiltration unit consisting of a filter holder, base, and funnel that can accommodate 47- or 50-mm diameter filters. The microfiltration unit is seated on a 2- to 4-liter capacity filtering flask connected

TABLE 10.1
Field Collection of Seston Using Miller-type Tow Nets

Date _____ Observers _____ Sampling location _____

Stream and site _____ Stream stage condition _____

Net mesh size _____ μm   (A) Diameter of net opening _____ m   (B) Radius of net opening (A ÷ 2) _____ cm   (C) pi = 3.1416

(D) Flowmeter conversion to meters _____   (E) Volume conversion (B^2 × C * 1000) _____

Sample and filter no.	(F) Initial flowmeter reading	(G) Final flowmeter reading	(H) Elapsed time (sec)	(I) Elapsed flow (G − F)	(J) Water filtered (m) (I × D)	(K) Volume filtered (L) (J × E)	(L) Velocity (m sec^{-1}) (J ÷ H)

with vacuum tubing to a vacuum pump. Filters are 47- or 50-mm-diameter glass-fiber filters (GFFs), preashed, and preweighed.[1]

2. For carboy and jug samples, vigorously shake carboy or jug to resuspend seston. Pour a 1- to 4-liter aliquot of the sample into a graduated cylinder. Record the volume of sample used.

3. Pour the aliquot into the microfiltration funnel and draw down onto a GFF under vacuum. Volume required will vary depending upon seston concentration in the aliquot. In general, volume should be sufficient to produce a clearly visible layer of seston on the GFF.

4. Rinse the microfiltration funnel with distilled/deionized, prefiltered water to ensure that seston particles are not adhering to the funnel. Remove GFF from the microfiltration unit with blunt forceps and return to its labeled aluminum square.

5. Repeat steps 2, 3, and 4 until three replicate aliquots have been filtered for each sample.

6. For Miller net samples, thoroughly wash seston out of the sample bag or sample jar and into the microfiltration funnel using distilled/deionized, prefiltered water and draw down onto a GFF under vacuum. Rinse microfiltration funnel and remove GFF as above.

7. To determine dry mass, seston samples and GFFs should be oven-dried (50°C for 24 h), desiccated (24 h), and weighed on an analytical balance.

8. To determine ash mass, dry-weighed seston samples and GFFs should be ashed in a muffle furnace (500°C for 0.5 to 1 h), rewetted with distilled/deionized water to restore waters of hydration (Weber 1973), oven-dried (50°C for 24 h), desiccated (24 h), and weighed on an analytical balance.

9. Masses obtained provide measures of ash-free dry mass (AFDM) or organic seston (dry mass − ash mass) and inorganic seston (ash mass). Masses from >250-μm and <250-μm fractions need to be mathematically combined if samples were collected using Protocol 1 (above). Seston concentrations may be reported directly as milligrams or grams of seston per sample volume. However, it is preferable to standardize units to either mg/liter or g/m^3.

Particle-Size Separation Protocols

1. Set up the wet filtration system consisting of a funnel, a series of stacked sieves in decreasing size order, a base to attach the sieve stack to

---

[1]Filters should be free of binder such as Gelman type A/E or equivalent. Labeled squares of aluminum foil are useful for maintaining individual preashed and weighed GFFs.

a filtering flask, and an electric vacuum pump connected to the flask with vacuum tubing (Fig. 10.2). A large capacity ($\geq$4 liters) vacuum flask should be connected between the stacked sieves and the vacuum. For carboy samples, the largest sieve size should be 250 μm, which should be the smallest sieve size for Miller net samples.

2. For carboy samples (steps 2–8), vigorously shake the carboy to resuspend seston. Slowly pour the sample from the carboy into the funnel of the filtration system. The volume of water required will vary depending upon seston concentration in the sample. In general, it will take the entire volume of the carboy, but somewhat less if seston loads are high. Under conditions of low seston concentrations, several carboys may be needed to obtain adequate samples.

3. Filtration will require turning off the vacuum periodically to empty the filtering flask to avoid pulling water into the vacuum pump. Carefully disconnect the stack and base from the flask and *record the volume of water* in the flask prior to discarding. Make sure at least 3 liters of filtrate is retained to measure ultrafine seston. Reassemble the system and continue filtration.

4. When filtration is complete, compute and record the total volume of water filtered through the system. Disassemble the wet filtration system, arranging sieves such that size fractions are clearly denoted.

5. Set up a microfiltration unit as described in Step 1 of "Standard Processing Protocols."

6. Starting with the largest sieve (i.e., 250 μm), wash retained material into the funnel of the microfiltration unit with distilled/deionized, prefiltered water and draw down onto a preashed, preweighed GFF under vacuum. Rinse funnel to ensure that seston particles are not adhering. Remove GFF from the microfiltration unit with blunt forceps and return to its labeled aluminum square.

7. Repeat Step 6 for the next smaller sieve size, carefully recording sieve sizes and corresponding GFF identification numbers. Continue until material from all sieves has been drawn down onto separate GFFs.

8. Filter reserved ultrafine seston filtrate onto a GFF. Volume required will vary depending upon seston concentration in the aliquot. In general, volume should be sufficient to produce a clearly visible layer of seston covering on the GFF. Record the volume of filtrate used.

9. Process GFFs and seston samples as described in steps 7 and 8 of *Standard Processing Protocols*.

10. For Miller net samples (steps 10–15), resuspend sampled material in distilled/deionized, prefiltered water and pour into the funnel of the wet filtration system. Carefully wash out sample bag or jar into funnel to ensure that all material is recovered.

11. Draw down material into the stacked sieve column under vacuum while rinsing funnel with distilled/deionized, prefiltered water to ensure that seston particles are not adhering.

12. When filtration is complete, disassemble the wet filtration system, arranging sieves such that size fractions are clearly denoted.

13. Set up a microfiltration unit as described in step 1 of "Standard Processing Protocols."

14. Starting with the largest sieve, wash retained material into the funnel of the microfiltration unit with distilled/deionized, prefiltered water and draw down onto a preashed, preweighed GFF under vacuum. Rinse microfiltration funnel and remove GFF as above.

15. Repeat step 13 for the next smaller sieve size, carefully recording sieve sizes and corresponding GFF identification numbers. Continue until material from all sieves has been drawn down onto separate GFFs.

16. Process GFFs and seston samples as described in steps 7 and 8 of *Standard Processing Protocols*.

17. Seston concentrations can now be determined as milligrams per sample volume for individual particle size categories. Since water volume filtered to obtain samples differs between carboy and Miller net samples, concentrations must be converted to a standard unit (e.g., mg of seston/liter) prior to comparison.

Exercise 1—Option a: Seston Export

1. Export, or total transport of seston, requires knowledge of stream discharge at the time of seston sampling (see Chapter 3 for methods of determining discharge). Estimates of total export are made by weighting seston concentration (mass per unit volume) by discharge (volume per unit time) to determine export or total transport (mass per unit time). Provided that you have the necessary data, this is easily accomplished by multiplying total seston concentration (mg AFDM/liter) $\times$ 1000 = mg AFDM/m^3. The product is multiplied by discharge (m^3/s) to estimate milligrams of FPOM exported per second.

2. One advantage to sampling large rivers is that these systems are routinely gaged and information such as mean daily discharge (in feet^3/s or m^3/s) is easily obtainable from United States Geological Survey's *Water Resources* data book published annually for each state (usually found in government publications sections of most libraries). Daily loads of transported seston (seston export) for large rivers can then be estimated by adjusting for units of measure and multiplying seston concentration by mean daily discharge.

3. Other methods of estimating export include the use of rating curves (see Chapter 7, Cummins *et al.* 1983, Webster *et al.* 1990), thus incorporating

some aspect of discharge to estimate POM concentrations. However, discharge and POM concentrations are generally poorly related (Bilby and Likens 1979, Gurtz *et al.* 1980, Cuffney and Wallace 1988). These studies indicate that infrequent sampling and poor rating curves are not good predictors of POM export.

4. Another method for continuous export measurements involves Coshocton proportional samplers, which are suitable only for small streams and require more elaborate instrumentation (Cuffney and Wallace 1988, Wallace *et al.* 1991).

Exercise 1—Option b: Seston Sampling during Storms

1. In small streams with quickly fluctuating ("flashy") discharge, the bulk of the total suspended material is carried during the rising hydrograph of storms (e.g., Gurtz *et al.* 1980, Webster *et al.* 1990, Wallace *et al.* 1991). Sampling these events can be difficult due to their unpredictable timing and short duration. Although seston sampling for particle-size analysis can be conducted under such conditions, it is extremely labor intensive and for our purposes we will examine only total seston concentrations under conditions of baseflow and rising and falling hydrographs. As storms are largely unpredictable, this will require access to a stream located near your laboratory that can be readily sampled. Small, gaged streams are ideal for this purpose. If none are available, see Chapter 3 for stream-gaging methods.

2. In some cases a meter stick anchored vertically to an area where the cross-sectional profile can be measured will suffice as a gage. Record the water height on the meter stick with each sample taken during the storm. Use standard processing procedures to calculate total dry mass, ash, and AFDM concentrations for these samples. You may wish to repeat these measurements over a several-day period if no storms occur. You should have a series of 10 to 15 clean bottles with caps (1- to 2-liter capacity) for sampling as storms approach as well as a supply of preashed and preweighed GFFs. Start your storm sampling sequence prior to the first rainfall if possible. Clearly record the time and water height for each subsequent sample as stream turbidity increases on the rising hydrograph during the storm. Brief and intense summer thundershowers are ideal for this purpose; however, severe electrical storms can be dangerous and be sure not to seek shelter under tall trees between sampling intervals. Although timing is tricky for such storms, you should attempt to sample over a period that provides a series of samples taken over both the rising and falling hydrographs. In the laboratory, process each sample separately, clearly labeling each filter from the sequential samples. Determine total dry mass, ash, and AFDM (mg/liter) for each sample in the storm sequence.

B. Exercise 2: Fine Benthic Organic Matter

Protocols for Field Collection of FBOM

1. Prepare a substantial amount of filtered streamwater: pour stream-water into a carboy or other large, clean vessel through a 250-μm mesh sieve.

2. Select sites that are characteristic of either depositional or erosional stream habitats (see Site Selection above). Place a graduated barrel (or large bucket) and paddle in close proximity to the sampling site.

3. With minimal disturbance to the substratum, force the sampling corer into the substratum. The core should be $\leq$22 cm diameter made of steel or PVC pipe. For cobble–riffle and bedrock–outcrop habitats, wrap a cloth towel around the outside base of the corer once it is in place to form an effective seal with the substratum.

4. Remove material from within the corer with either a plastic cup or hand-powered diaphragm pump (the latter works more efficiently on hard-bottomed substrata). Pass removed material through nested 1-mm and 250-μm mesh sieves that are positioned over the graduated barrel to retain water passing through the smaller mesh.

5. In riffle areas, cobbles inside the corer should be thoroughly brushed and disturbed while pumping; bedrock substratum also should be thoroughly brushed.

6. Once water has been removed from inside the coring device and the substratum cleaned of fine particles, thoroughly wash the sieves with filtered stream water, retaining material passing through the bottom sieve in the graduated barrel. Measure and record water volume in the barrel.

7. For the <250 μm fraction, stir water in the barrel thoroughly with the paddle and remove a subsample of the agitated water (0.2 to 1 liters depending on the concentration of particles). Store subsamples in either separate bottles or large, self-closing plastic bags. Three replicate subsamples (stirring before each subsample) are desirable for each sample.

8. For the >250-μm fractions, discard material retained on the 1-mm mesh sieve, which is the CPOM fraction of the sample. Wash material retained on the 250-μm mesh sieve with filtered stream water into a suitable container (e.g., large plastic bag or wide-mouth bottle), and clearly label this container and the subsamples.

9. Repeat the sampling procedure for all targeted habitats (i.e., erosional and depositional areas).

FBOM Processing Protocols

1. For the >250-μm fractions (steps 1–5), wash contents of the sample container with tap water into a large pail and resuspend in water.

2. Pour the resuspended material through nested 500-μm and 250-μm

mesh sieves. Allow time to drain samples thoroughly and transfer material to separate, labeled paper bags.

3. Oven-dry material and bags at 50°C to constant weight (24 h to several days, depending on sample size). Place bags in a desiccator for 24 h.

4. Remove material from bags and weigh on a top-loading balance to determine dry mass.

5. Ash material at 500°C (small, heavy-gauged, aluminum baking pans work well for this purpose), and reweigh to obtain AFDM for the 250- to 500-μm and 500-μm to 1.0-mm size fractions.

6. For the <250-μm fraction (steps 6–11), set up a microfiltration unit as described in step 1 of "Standard Processing Protocols."

7. Individually pour each of the three replicate subsamples into separate 1-liter graduated cylinders and record the subsample volumes.

8. Pour the first subsample into the funnel of the filtration unit. Wash any material clinging to the subsample bag or graduated cylinder into the funnel with distilled/deionized, prefiltered water. Draw material down onto a GFF, washing sides of funnel with distilled/deionized, prefiltered water. Remove filter with blunt forceps and return to aluminum square.

9. Repeat steps 7 and 8 for remaining replicates.

10. Dry, weigh, ash, and reweigh FBOM samples and GFFs following steps 7 and 8 of Standard Processing Protocols.

11. AFDM of the 0.45-μm to 250-μm size fraction is estimated as the means of the following quantity calculated for each of the three subsamples:

$$\text{AFDM} = (\text{barrel volume} \div \text{subsample volume}) \times \text{subsample AFDM.}$$
$$(10.3)$$

12. FBOM quantity is normally expressed as g AFDM/m^2 of stream bottom. This requires you to know the area of your sampling device (in cm^2). Use the following equation for FBOM standing stocks estimated for each size fraction to express your results:

$$\text{g AFDM/m}^2 = (\text{mg AFDM} \div 1000)$$
$$\times (10{,}000 \div \text{cm}^2 \text{ of area sampled}). \quad (10.4)$$

The g AFDM/m^2 for each size fraction are summed to obtain total FBOM standing crop (in AFDM) in your sample.

C. Exercise 3: Linkages of Sestonic FPOM to the Biota

Field Release and Larval Collection

1. In the field, thoroughly mix charcoal with streamwater in one or two large pails until no more charcoal remains on the surface of the water,

to form a dense slurry of suspended charcoal (some continuous stirring even during release may be required to ensure suspension).

2. Position members of the team on either side of the black fly aggregation, being careful to minimize disturbance.

3. At a location 1–2 m upstream of the black fly aggregation, slowly pour the slurry back and forth across a 0.5- to 1.0-m width of stream, ensuring that the water passing over the larval aggregation is darkly stained with charcoal particles. Pour slowly to ensure that the contents of the pail are not released as a massive instantaneous dosage. (A beaker can be used for removing the slurry from the bucket and releasing the mixture in the stream).

4. The release should take 1 to 2 min. Record the starting and ending time of the slurry release. Note the width of the slurry passing over the aggregation and the lateral boundaries of the slurry (flagging attached to wire stakes may be useful for this purpose) and keep larval collections within boundaries.

6. Collect larvae at 10-, 20-, and 30-min intervals following slurry release.

7. Use collecting forceps to pick larvae and place in a vial prelabeled with the appropriate time interval and half-filled with 70% ethanol. Collectors should strive to sample a range of larval sizes at each period. Sampling should continue for about 1 min after each 10-min interval. Following each collection period, check all vials to ensure that the time interval is correctly indicated.

Laboratory Analysis

1. Separate vials into specific time intervals and work with larvae from only one interval at a time to avoid confusion.

2. Starting with larvae collected 10 min after charcoal release, use a dissecting microscope fitted with an ocular micrometer to divide black flies into size classes to the nearest 0.5 mm. Keep size classes separate.

3. For each size class, use the ocular micrometer to measure the distance from the posterior end of head to the tip of the abdomen on each larva. Record data as Distance x.

4. Using the point of the jeweler's forceps, carefully split open the larval integument from below the head to the tip of the abdomen.

5. With two pairs of jewler's forceps, gently tease the gut out of the body cavity, keeping the head attached to the gut.

6. Measure with the ocular micrometer the distance from the posterior end of the head to the charcoal band in the gut. Record data as Distance y.

7. Repeat this procedure until all larvae from each size class and all three time intervals have been measured.

8. Upon completion of measurements, you should have recorded the following information for each larva collected:

    a. Collection interval (10, 20, or 30 min);

    b. Larval length (mm);

    c. Distance y in mm (posterior end of head to charcoal band);

    d. Distance x in mm (posterior end of head to tip of abdomen);

    e. Ratio W (Distance y ÷ Distance x).

    f. Plot the ratio of W (y-axis) against the larval length (x-axis) for each larva examined at the 10-min interval. Repeat this process for each larva measured for the 20-min interval, as well as for those from the 30-min interval. You should be able to regress the values for the ratio of W and larval length for each time interval.

IV. QUESTIONS

A. Seston

1. What is the total organic seston (in mg AFDM/liter) concentration in your stream?

2. Based on your measurements of individual size classes, what sizes are the most abundant in terms of total organic seston in transport?

3. Suppose you repeat these measurements in smaller headwater streams or a larger downstream river. Would you expect the same results? Why or why not?

4. How does seston concentration vary with stream depth? with distance from the thalweg?

5. Does seston quality (in terms of organic:inorganic ratio) change with distance from the thalweg? If so, can you hypothesize as to why this change occurs?

6. If seston concentrations are available for two rivers or sites, or the same river in different seasons, compare estimates of seston export between rivers, sites, or seasons. How do they compare and can you suggest any mechanisms to account for differences?

7. (Exercise 1—Option a) Convert seston concentration data into an estimate of total seston export. What source did you use for discharge data? Can you predict seasonal patterns of seston export for your system based on what information you have on discharge and seston concentrations?

8. (Exercise 1—Option b) How did the dry mass, AFDM, and ash concentration change over the rising and falling hydrograph of the storm?

If you are working in a gaged stream it will be useful to plot each sample concentration against discharge at the time the sample was collected. If not you can plot each sample against water depth measured on the meter stick as a rough estimate of relative discharge for each sample.

9. (Exercise 1—Option b) At what stage of the storm sampling sequence was maximum and minimum seston concentrations reached? How do you explain your results?

10. (Exercise 1—Option b) Based on your sampling results during the storm, what problems do you see with calculating organic matter export for stream ecosystems? How does this influence organic matter budgets for a given stream reach?

B. Fine Benthic Organic Matter

1. How do FBOM particle-size distributions and total FBOM standing crops compare between erosional and depositional habitats?

2. Hypothesize as to the specific physical characteristics in each habitat which account for differences in FBOM standing crops.

3. Given differences in FBOM particle-size distribution and standing crops, what are your hypotheses concerning the relative functional structure of the benthic macroinvertebrate community in each habitat? (see Chapters 16 and 21 for information concerning benthic community functional structure).

C. Linkages of Sestonic FPOM to the Biota

1. Do black fly larvae display any tendency to select food particles based on type of food available? Give reasons for your answer.

2. Black fly larvae have been described as feeding nonselectively on particles <300 μm in diameter. Based on your analyses of seston particle sizes, what significance do you attach to this observation with respect to particle size availability in lotic habitats?

3. For a specific time interval, i.e., 10, 20, or 30 min following charcoal exposure, is there any difference in gut passage times for larvae of different size classes? If so, what differences did you detect? What does the ratio of W versus larval length illustrate about gut passage times?

4. What is your best estimate of gut passage time for black fly larvae of different size classes? Did you notice any difference in charcoal bands after 30 min? How do you account for differences after longer time intervals (see Wotton 1980)?

5. What do you see as the "ecological role" microfiltering collectors such as black flies play in stream ecosystems? Explain your answer.

V. MATERIALS AND SUPPLIES

Letters in parentheses indicate in which Exercise (1, 2, or 3) the item is used.

Aluminum squares (1, 2), Approximately 60 mm side length and numbered to facilitate filter identification.

Balance (1, 2), analytical.

Balance (2), top-loading.

Bags, paper (2).

Bags, plastic (1, 2), self-closing (e.g., Ziploc or Whirl-Pak).

Bottles (2), wide-mouth, capped, 1- to 2-liter capacity.

Buckets, 10- to 15-liter capacity (3).

Charcoal, fine-powdered (3). Optional: fine powdered fluorescent pigments can be substituted for charcoal. These are more expensive than powdered charcoal but easier to locate in the gut, especially for black fly larvae with heavily pigmented integuments. They also glow when exposed to a black light source, such as a mineral light used by geologists. One source of such pigments is Radiant Color, 2800 Radiant Avenue, Richmond, CA 94804. Type P-1600 (average particle size = 5 μm) manufactured by Radiant Color have the added advantage that specimens can be mounted on glass slides without interference from the many solvents used in mounting.

Cover (2), Hand-held or stove-pipe with inside diameter 22.6 cm or greater made of steel or PVC pipe that can be forced into the substratum.

Cup, plastic (2), for sampling depositional areas.

Desiccator (1, 2), with $CaSO_4$ desiccant.

Filters (1, 2), 47- or 50-mm glass-fiber filters without binder (e.g., Gelman type A/E, Whatman GFF, or equivalent). Prior to use, filters are ashed in a muffle furnace (500°C for 0.5 to 1 h), rewetted with distilled/deionized water to restore waters of hydration, oven-dried (50°C for 24 h), desiccated (24 h), and preweighed on an analytical balance. Store glass-fiber filters on labeled aluminum squares in a desiccator.

Flags (3), (optional) attached to wire stakes.

Forceps, jewelers (3), blunt (1, 2)

Furnace, muffle (1, 2).

Graduated container (2), large pail or vinyl trash can marked for the volume of water at various depths.

Graduated cylinders, 1-liter capacity (1, 2).

Jugs, approximately 1-gallon capacity (2). Milk jugs are good for this
   purpose as they are inexpensive and many are needed.
Microfiltration unit (1, 2), includes:
   47- or 50-mm filter base, holder, and funnel
   Filtering flask, 2 to 4 liter capacity
   Vacuum pump
   Vacuum tubing
   Blunt forceps.
Marker, permanent ink (1, 2, 3).
Microscope, dissecting, binocular (3), fitted with an ocular micrometer.
Notebook, field (1, 2, 3), waterproof pages.
Oven, drying (1, 2).
Paddle, canoe (2, 3).
Pump (2), hand-held, diaphragm-type, for sampling erosional areas.
Sampler, point (1), includes:
   Kemmerer or Van Dorn sampling bottle
   Weighted messenger
   Tether line marked on 0.5-m increments
   Line depressor.
Sampler, Miller-type tow-net (1), includes:
   Sampler body with slightly tapered front (reduction fitting)
   Collecting net, 250 μm
   Catch bucket
   Flowmeter
   Tether line.
Sieves, standard testing (1, 2). Nestable, with mesh sizes of 250 μm,
   500 μm, and 1.0 mm.
Stopwatch (1, 3).
Tape, measuring (1), 10 to 50 m marked in 0.5-m increments.
Wash bottles (1, 2).
Wet filtration unit (2), includes:
   Top funnel
   Nestable sieves of decreasing mesh sizes (examples: 500, 250, 100,
      50, and 25 μm)
   Filtering flask, 2- to 4-liter capacity
   Teflon gaskets
   Vacuum pump
   Vacuum tubing.
Vials, 1 dram with stoppers (3). Vials should be half-filled with 70%
   ethanol and prelabeled to indicate 10-, 20-, and 30-min collec-
   tion intervals.
Water, distilled/deionized and prefiltered through glass-fiber filters (1, 2).

REFERENCES

Angradi, T. R. 1993a. Chlorophyll content of seston in a regulated Rocky Mountain River, Idaho, USA. *Hydrobiologia* **259:**39–46.

Angradi, T. R. 1993b. Stable carbon and nitrogen isotope analysis of seston in a regulated Rocky Mountain River, USA. *Regulated Rivers: Research and Management* **8:**251–270.

Benke, A. C., T. C. Van Arsdall Jr., D. M. Gillespie, and F. K. Parrish, 1984. Invertebrate productivity in a subtropical blackwater river. The importance of habitat and life history. *Ecological Monographs* **54:**25–63.

Bilby, R. E. 1980. Role of organic debris dams in regulating the export of dissolved and particulate matter from a forested watershed. *Ecology* **62:**1234–1243.

Bilby, R. E., and G. E. Likens. 1979. Effects of hydrologic fluctuations on the transport of fine particulate organic carbon in a small stream. *Limnology and Oceanography* **24:**69–75.

Bilby, R. E., and G. E. Likens. 1980. Importance of organic debris dams in the structure and function of stream ecosystems. *Ecology* **61:**1107–1113.

Carlough, L. A. and J. L. Meyer. 1991. Bacterivory by sestonic protists in a southeastern blackwater river. *Limnology and Oceanography* **36:**873–883.

Cuffney, T. F. and J. B. Wallace. 1988. Particulate organic matter export from three headwater streams: Discrete versus continuous measurements. *Canadian Journal of Fisheries and Aquatic Sciences* **45:**2010–2016.

Cummins, K. W., J. R. Sedell, F. J. Swanson, G. W. Minshall, S. G. Fisher, C. E. Cushing, R. C. Petersen, and R. L. Vannote. 1983. Organic matter budgets for stream ecosystems: Problems in their evaluation. Pages 299–353 *in* J. R. Barnes and G. W. Minshall (Eds.) *Stream Ecology: Application and Testing of General Ecological Theory*. Plenum, New York, NY.

Cushing, C. E., G. W. Minshall, and J. D. Newbold. 1993. Transport dynamics of fine particulate organic matter in two Idaho streams. *Limnology and Oceanography* **38:**1101–1115.

Edwards, R. T. 1987. Seasonal bacterial biomass dynamics in the seston of two southeastern blackwater rivers. *Limnology and Oceanography* **32:**221–234.

Edwards, R. T. and J. L. Meyer. 1987. Bacteria as a food source for black fly larvae in a blackwater river. *Journal of the North American Benthological Society* **6:**241–250.

Fisher, S. G., and L. J. Gray. 1983. Secondary production and organic matter processing by collector macroinvertebrates in a desert stream. *Ecology* **64:**1217–1224.

Fisher, S. G., and G. E. Likens. 1973. Energy flow in Bear Brook, New Hampshire: an integrative approach to stream ecosystem metabolism. *Ecological Monographs* **43:**421–439.

Gale, W. F., and J. D. Thompson. 1975. A suction sampler for quantitatively sampling benthos on rocky substrates in rivers. *Transactions of the American Fisheries Society* **104:**398–405.

Grubaugh, J. W. and R. V. Anderson. 1989. Upper Mississippi River: Seasonal and floodplain forest influences on organic matter transport. *Hydrobiologia* **174:**235–244.

Gurtz, M. E., J. R. Webster, and J. B. Wallace. 1980. Seston dynamics in southern Appalachian streams: effects of clear-cutting. *Canadian Journal of Fisheries and Aquatic Sciences* **37:**624–631.

Guy, H. P., and V. H. Norman. 1970. Field methods for measurement of fluvial sediment. *in Techniques of Water-Resources Investigations of the United States Geological Survey,* Book 3, Chapter C2. United States Geological Survey, Washington, DC.

Huryn, A. D., and J. B. Wallace. 1987. Local geomorphology as a determinant of macrofaunal production in a mountain stream. *Ecology* **68:**1932–1942.

Maciolek, J. A., and M. G. Tunzi. 1968. Microseston dynamics in a simple Sierra Nevada lake-stream ecosystem. *Ecology* **49:**60–75.

Minshall, G. W., J. T. Brock, and T. W. LaPoint. 1982. Characterization and dynamics of benthic organic matter and invertebrate functional feeding group relationships in the Upper Salmon River, Idaho (U.S.A.). *International Revue Gesamten Hydrobiologie* **67:**793–820.

Minshall, G. W., K. W. Cummins, R. C. Petersèn, C. E. Cushing, D. A. Bruns, J. R. Sedell, and R. L. Vannote. 1985. Developments in stream ecosystem theory. *Canadian Journal of Fisheries and Aquatic Science* **42:**1045–1055.

Naiman, R. J., and J. R. Sedell. 1979a. Characterization of particulate organic matter transported by some Cascade Mountain streams. *Journal of the Fisheries Research Board of Canada* **36:**17–31.

Naiman, R. J., and J. R. Sedell. 1979b. Benthic organic matter as a function of stream order in Oregon. *Archive für Hydrobiologie* **97:**404–422.

Newbold, J. D., P. S. Mulholland, J. W. Elwood, and R. J. O'Neill. 1982. Organic carbon spiralling in stream ecosystems. *Oikos* **38:**266–272.

Peters, G. T., E. F. Benfield, and J. R. Webster. 1989. Chemical composition and microbial activity of seston in a southern Appalachian headwater stream. *Journal of the North American Benthological Society* **8:**74–94.

Platts, W. S., W. F. Megahan, and G. W. Minshall. 1983. *Methods for Evaluating Stream, Riparian, and Biotic Conditions.* General Technical Report INT-138. United States Department of Agriculture, Forest Service, Intermountain Forest and Range Experiment Station, Ogden, UT.

Sedell, J. R., R. J. Naiman, K. W. Cummins, G. W. Minshall, and R. L. Vannote. 1978. Transport of particulate organic matter in streams as a function of physical processes. *Verhandlungen der Internationalen Vereinigung für Theoretische und Angewandte Limnologie* **20:**1366–1375.

Smock, L. A., G. M. Metzler, and J. E. Gladden. 1989. Role of debris dams in the structure and functioning of low-gradient headwater streams. *Ecology* **70:**764–775.

Vannote, R. L., G. W. Minshall, K. W. Cummins, J. R. Sedell, and C. E. Cushing. 1980. The river continuum concept. *Canadian Journal of Fisheries and Aquatic Sciences* **37:**130–137.

Voshell, J. R., and C. R. Parker. 1985. Quantity and quality of seston in an impounded and a free-flowing river in Virginia, U.S.A. *Hydrobiologia* **122:**271–288.

Wallace, J. B., A. C. Benke, A. H. Lingle, and K. Parsons. 1987. Trophic pathways of macroinvertebrate primary consumers in subtropical blackwater streams. *Archiv für Hydrobiologie Supplement 74,* **4:**423–451.

Wallace, J. B., T. F. Cuffney, J. R. Webster, G. J. Lugthart, K. Chung, and B. S. Goldowitz. 1991. Export of fine organic particles from headwater streams: effects of season, extreme discharges, and invertebrate manipulations. *Limnology and Oceanography* **36:**670–682.

Wallace, J. B., and R. W. Merritt. 1980. Filter-feeding ecology of aquatic insects. *Annual Review of Entomology* **25:**103–32.

Wallace, J. B., D. H. Ross, and J. L. Meyer. 1982. Seston and dissolved organic carbon dynamics in a southern Appalachian stream. *Ecology* **63:**824–838.

Weber, C. I. (Ed.). 1973. *Biological Field and Laboratory Methods for Measuring the Quality of Surface Water and Effluents.* United States Environmental Protection Agency, EPA 640/4-73-001, Cincinnati, OH.

Webster, J. R., and S. W. Golladay. 1984. Seston transport in streams at Coweeta Hydrologic Laboratory, North Carolina, U.S.A. *Verhandlungen der Internationalen Vereinigung für Theoretische und Angewandte Limnologie* **22:**1911–1919.

Webster, J. R., S. W. Golladay, E. F. Benfield, D. J. D'Angelo, and G. T. Peters. 1990. Effects of forest disturbance on particulate organic matter budgets of small streams. *Journal of the North American Benthological Society* **9:**120–140.

Welch, P. S. 1948. *Limnological Methods.* Blakiston Co., Philadelphia, PA.

Wotton, R. S. 1978. The feeding-rate of *Metacnephia tredecimatum* larvae in a Swedish lake-outlet. *Oikos* **30:**121–125.

Wotton, R. S. 1980. Coprophagy as an economic feeding tactic in blackfly larvae. *Oikos* **34:**282–286.

Wotton, R. S. 1984. The importance of identifying the origin of microfine particles in aquatic systems. *Oikos* **43:**217–221.

Wotton, R. S. (Ed.). 1990. *Biology of Particles in Aquatic Systems.* CRC Press, Boca Raton, FL.

CHAPTER 11

Transport and Retention of CPOM

GARY A. LAMBERTI[*]
AND STANLEY V. GREGORY[†]

*Department of Biological Sciences
University of Notre Dame
†Department of Fisheries and Wildlife
Oregon State University

I. INTRODUCTION

Coarse particulate organic matter, or CPOM, in streams is defined as any organic particle larger than 1 mm in size (Cummins 1974). CPOM can be divided further into wood and nonwoody material (Cummins and Klug 1979). Wood includes all size classes from branches to entire trees that fall into stream channels. Accumulations of wood across stream channels are known as *debris dams,* which perform important ecological functions (Bilby and Likens 1980). The nonwoody component includes materials donated by riparian vegetation (e.g., leaves, needles, fruits, flowers, seeds, frass) and materials produced within the stream (e.g., fragmented aquatic plants, dead aquatic animals). Smaller materials, both fine particulate organic matter (1 mm > FPOM > 0.45 μm) and dissolved organic matter (DOM < 0.45 μm), are considered in Chapters 7–10.

Allochthonous CPOM is a major energetic resource for stream ecosystems (Cummins 1974). CPOM provides a large proportion of the fixed carbon in small streams of both deciduous and coniferous forests and is important in larger streams (Cummins *et al.* 1983). CPOM that enters streams is *transported* downstream by the unidirectional flow of lotic ecosys-

217

tems (Vannote *et al.* 1980). Trapping of this material is essential for the subsequent microbial colonization that normally precedes consumption by shredding macroinvertebrates (Cummins and Klug 1979). The process of deposition and trapping, termed *retention*, provides the critical link between input and the long-term storage and processing of CPOM.

The retentive capacity of streams for CPOM is a function of hydrologic, substrate-related, and riparian features (Speaker *et al.* 1984). High *roughness* of the channel (e.g., large substrate particle size, abundant woody debris), combined with certain hydraulic conditions (e.g., presence of back-waters, interstitial flow), tends to increase the CPOM trapping efficiency of stream reaches. Debris dams are particularly important retention structures (Bilby 1981, Smock *et al.* 1989). Young *et al.* (1978) noted that the probability that a particle in transport will be retained is a function of the "active" entrainment efficiency of that particle size by a channel obstacle (e.g., rock, log, root, etc.) and the density of those obstacles within the channel. Particles also will be retained "passively" when current velocity is less than the velocity required to keep the particle moving in the water column or along the streambed (Jones and Smock 1991). This relationship can be expressed as a probability function,

$$P(R) = f(E, N, V), \tag{11.1}$$

where R represents retention; E, entrainment efficiency by channel obstacles; N, obstacle density in the channel; and V, critical velocity required to transport a particle. If an organic particle is retained, it subsequently will either decompose, be consumed, or, if flow conditions change, be dislodged and transported downstream (Speaker *et al.* 1984).

In this chapter, we describe a field method for assessing quantitatively the CPOM retention efficiency of a specific stream reach. The method is most easily applied to small streams (orders 1–4), but can be adapted for larger streams and rivers. The approach is intended not only to measure retention but to relate retention to hydraulics, streambed roughness, channel geomorphology, and riparian zone structure. The specific objectives are to (1) introduce the concept and importance of retention; (2) demonstrate how to measure retention, analyze data, and calculate indices of retention; and (3) illustrate the utility of retention measurements for assessing stream channel condition. This chapter focuses on short-term trapping of CPOM and does not consider long-term storage and processing (see Chapters 27 and 28).

II. GENERAL DESIGN

In practice, lotic retention can be viewed as the difference between the number of particles in transport at a given point in the stream and the number still in transport at some distance downstream (Speaker *et al.* 1984). Retention is most easily measured by releasing known numbers of readily distinguishable particles into the channel. To compare different stream reaches within a study, the experimental approach must be standardized for type and number of particles released, length of experimental reach, and duration of the retention measurement. Many types of CPOM have been released into streams, including leaves (Speaker *et al.* 1984, Ehrman and Lamberti 1992), plastic strips (Bilby and Likens 1980, Speaker *et al.* 1988), wood dowels (Ehrman and Lamberti 1992), and fish carcasses (Cederholm *et al.* 1989). In general, we believe that it is preferable to release natural (decomposable) materials into streams because particle retrieval is nearly always less than 100% and because analogs (e.g., plastic strips) may not behave the same as natural materials. In this exercise, we will demonstrate retention of leaves and small wood, but other materials significant to the specific stream can be substituted. For example, fruits are significant CPOM inputs in many tropical streams.

A. Site Selection

The selection of a stream in which to conduct this exercise may be influenced by logistical considerations. In general, wadeable, third- to fourth-order streams are ideal. Very small streams at low flow have low transport, and the method described in this chapter is difficult (and can be dangerous) to conduct in large rivers. In general, however, this method can be scaled to a wide variety of stream sizes. Within the study stream, at least two reaches with contrasting channel features should be selected by the research coordinator. Ideally, one reach would have a relatively simple channel (straight, low roughness, limited hydraulic diversity, sparse wood). The other reach should have a complex channel (sinuous, high roughness, diverse hydraulic conditions, abundant wood).

Length of the experimental reach should be scaled to stream size, with length increasing with stream order. As a rule of thumb, start with a stream length that is 10 times the wetted channel width. For example, 50 m may be an appropriate length for a second-order stream, 100 m for a third-order stream, and 200 m for a fourth-order stream. Streams of the same size in different settings will have specific retention characteristics. If possible, use a pilot study to adjust reach length such that retention is not less than 10% or greater than 90% of released particles.

B. General Procedures

CPOM Releases   Leaves are the major form of nonwoody CPOM input in many streams and their retention is an important ecological process. In retention experiments, released leaves must be distinguishable from leaves found naturally in the channel and should be easy to obtain and manipulate. We have found that, for North American streams, abscised leaves of the exotic ginkgo tree (*Ginkgo biloba*) meet these requirements. The bright yellow leaves are tough, their size approximates that of many leaf-types of riparian vegetation, and the leaves are easily identified in the channel. *Ginkgo* trees have been planted worldwide as ornamentals, which usually are male trees because female trees drop pungent fruits in the autumn. Other species of leaves can be substituted depending on their availability and the composition of local riparian vegetation. Released wood similarly must be easy to manipulate, distinguishable from natural wood in the channel, and of a realistic size. We have found that these requirements are met by wood dowels, which can be obtained at hardware stores in a range of diameters and lengths. Dowels, however, have a simpler shape than tree branches and will be a conservative estimator of wood retention. Alternatively, fallen branches can be collected and marked with fluorescent paint to distinguish them from natural wood in the channel. Keep in mind that it is more difficult to standardize branches among releases than dowels.

Data Analysis   Physical data from the stream channel should be analyzed according to the level of measurements performed (see Chapters 2–4). At a minimum, the following parameters of each reach should be calculated: gradient, discharge, cross-sectional area, planar wetted area, and volume of large woody debris. Retention data for leaves and small wood should be fit to a negative exponential decay model, from which various indices of retention (e.g., the retention coefficient, $-k$; average particle travel distance, $1/k$) can be calculated. If dye releases are conducted, various hydraulic parameters (e.g., discharge, nominal transport time) can be calculated from a plot of dye concentration over time at a downstream sampling site.

Optional Exercises   Several optional activities involving additional time or facilities are presented in this chapter. For example, different levels of physical measurement of the channel can be performed (see Chapters 2–4). An inventory of leaf and dowel entrapment in the channel can be performed after the release to describe further the pattern of retention and to quantify entrapment by specific benthic features. CPOM releases can be conducted over longer periods of time, or at different seasons and discharges, to develop relationships between retention and stream temporal dynamics (e.g., Jones and Smock 1991). CPOM retention often is correlated with *hydraulic retention* (i.e., the retention of water within a reach). Hydraulic retention and discharge can be evaluated by releasing a tracer, such as fluorescent dye, into the channel and measuring water movement and dilution through the reach. Discharge calculated from tracer releases and more conventional approaches can be compared (see Chapter 3). The use of tracers that are more conservative than dyes (e.g., chloride) is described in Chapter 8.

III. SPECIFIC EXERCISES

A. Exercise 1: Transport and Retention

Laboratory Preparation

1. In the autumn, collect and air-dry abscised "exotic" leaves, such as *G. biloba* or other identifiable species. Leaves can be spread on screens, netting, or even on the floor for drying. Leaves can be stored dry in garbage bags for a considerable length of time. Alternatively, you can use fresh-fallen leaves if the releases will be performed soon after collection (within days).

2. Count out batches of 1000 leaves to generate 2000 leaves per release in a third-order stream. Smaller or larger streams may require fewer or more leaves, respectively.[1]

3. The day before the release, soak leaves overnight in buckets of water to impart neutral buoyancy during transport. Drain water before departing to the field.

4. Obtain 60 wood dowels, each approximately 1.5 cm in diameter and 1 m in length. (*Note.* Other dowel sizes, or wood chips, can be used to test retention of variously sized CPOM.) Alternatively, natural sticks can be collected on site from the riparian zone before the release. These sticks can be marked with a spot of fast-drying spray paint to distinguish them.

Field Physical Measurements

1. Measure and flag the appropriate length (e.g., 100 m for a third-order stream) of at least two stream reaches differing in channel complexity, or in other relevant features. Stretch a meter tape along the bank over the length of the reach, with 0 m at the downstream end.

---

[1]Weighing of leaves is not a suitable substitute for enumeration. Leaf fragmentation and variation in size makes estimates of leaf number from mass unreliable.

2. Measure major channel features at a level of intensity appropriate to the research objectives. We recommend working in a research team of three (two making measurements and one recording data). Minimally, measurements should include gradient, channel cross section, average width, depth, sinuosity, and substrate composition. Repeat for each reach.

3. Measure the length (L) and average diameter (D) of all wood contacting the channel and larger than a minimum size (e.g., 1 m L × 10 cm D). Note if the wood is part of a debris dam (i.e., wood accumulation blocking some portion of stream flow).

4. If this exercise is being used for class demonstration, prior to the releases briefly discuss channel and riparian features. Predict retention for each reach (e.g., percentage of leaves that will be retained).

CPOM Releases

1. Position several researchers at the downstream end of the reach. Release the leaf batch (e.g., 2000 leaves) at the upstream end of the reach, by dispersing leaves over the entire width of the stream channel.

2. Collect nonretained leaves at the downstream end of the reach. Either of two approaches can be used to collect leaves. A beach seine can be stretched across the width of the channel, with the bottom lead-line anchored, without gaps, to the streambed with rocks (*Note*. In sand-bottom streams, tent stakes can be used). The top of the seine should be held out of the water by attaching it to a taut rope tied to trees on both banks. Alternatively, researchers can line up across the channel and collect leaves in transport with hand-held dip nets (e.g., D-frame or delta nets). The seine method is more efficient, especially if the number of researchers is low. The individual netting approach results in greater involvement of researchers in actual leaf collection, but some leaves may be missed.

3. Continue collecting leaves for a period of time specified by the coordinator, usually at least 15 min and up to 1 h, or when leaf transport ceases. Release interval should be consistent for all reaches. Count collected (i.e., nonretained) leaves.

4. Release 50 dowels or sticks into the stream channel and hand-collect nonretained wood at the downstream end of the reach. Count nonretained wood pieces. Retrieve retained dowels at the end of the exercise.

Data Analysis

1. Calculate reach physical parameters, such as gradient, planar surface, cross-sectional area, average depth, current velocity, hydraulic radius, sinuosity, and discharge (see Chapters 2–4). These fundamental physical parameters can be related empirically or theoretically to observed retention values.

2. Determine the density and total volume of woody debris in each reach, assuming that a cylinder approximates the geometry of a log:

$$\text{volume} = \pi L D^2/4. \tag{11.2}$$

3. Fit the leaf and stick retention data to a negative exponential decay model of the form

$$P_d = P_0 e^{-kd}, \tag{11.3}$$

where P_0 is the number of particles released into the reach and P_d the number of particles still in transport at some downstream distance d from the release point. Calculate the slope $-k$ (the instantaneous retention rate) and its reciprocal $1/k$ (the average distance traveled by a particle before it is retained). If particles are not inventoried after the release, then the model will be based on two data points, P_0 and P_d.

4. The optional inventory data can be used to refine the exponential model and produce a more accurate estimate of $-k$, or to fit retention data to an alternate regression model (e.g., linear, power) more appropriate for the specific reach. The reinventory most likely will not turn up all of the retained leaves; therefore, it is necessary to normalize reinventory data to a percentage of total leaves found. Graph the particle transport data for each release, using distance downstream from the release point as the x-axis and percentage of particles still in transport as the y-axis (Figs. 11.1A, 11.1B). Plot also the percentage of leaves or dowels retained by specific channel structures in each reach (Fig. 11.1C).

5. The optional dye release can be used to calculate several hydraulic parameters, among which we will discuss discharge and transport time. Discharge (Q, in liters/s) can be calculated from dye dilution using the equation

$$Q = VC_u \bigg/ \int (C_d - C_b)\, dt, \tag{11.4}$$

where V is the volume of dye released in liters; C_u, the concentration of released dye in μg/liter; C_d, instream dye concentration at time t in μg/liter; and C_b, background fluorescence in μg/liter. In general, C_b effectively will be zero and V will equal 1.0 or a very small number compared to stream discharge, and thus can be ignored. The denominator can be calculated by first plotting the measured dye concentration (in μg/liter) on the y-axis against time (in seconds) on the x-axis (see Fig. 11.1D); then integrate

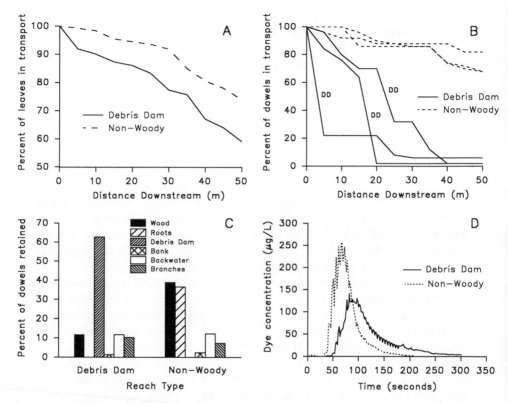

FIGURE 11.1 Examples of plots for displaying retention of particulate matter and water by two types of stream reaches. Retention of (A) leaves and (B) dowels in 50-m reaches with and without debris dams (DD) in an Indiana stream (particles were reinventoried within channel). Each leaf plot is an average of three releases; each dowel plot is a separate release. (C) Retention of dowels on different channel structures as revealed by the reinventory. (D) Slug release of fluorescein dye through stream reaches with and without debris dams; water samples were taken 50 m downstream of the release point. Note the longer hydraulic retention in the reach with debris dams (redrawn from Ehrman and Lamberti 1992).

the area under the dye concentration curve using computer digitation, numerical, or graphical methods (Gordon *et al.* 1992). Divide Q by 1000 to convert to m³/s. Nominal transport time (NTT; Triska *et al.* 1989) is an appropriate measure of hydraulic retention as indicated by a dye slug release. NTT is calculated as the time interval required for 50% of the dye to pass out of the reach. Integration of the concentration curve, starting at the origin and proceeding until 50% of the total area is found, will yield the NTT. NTT generally increases with reach complexity and the presence of certain channel features, such as large pools or significant interstitial flow.

B. Exercise 2: Importance of Different Retention Structures (Optional)

1. Conduct CPOM release as in Exercise 1.

2. Inventory the location, number, and retention structure for retained leaves and wood. This is best accomplished by dividing the reach into longitudinal increments of 5 m using the bankside meter tape.

3. Researchers should move up the channel as a single line of observers, perpendicular to flow.

4. Leaves are located and counted within each increment, noting also the retention structure (e.g., rock, wood, bank, etc.; see Table 11.1). Released wood can be inventoried simultaneously.

5. Describe the longitudinal pattern of retention and identify important retention structures within the channel.

TABLE 11.1
Sample Data Sheet for Inventory of Retained CPOM Particles

Stream:		POM type:		Date:
Location:		Total		Team:
Reach:		released:		Notes:
Length:		Total		
Duration:		captured:		
		Total		
		retained:		

Location	Unit	No. of particles retained on structure					
Meter mark	Riffle or pool	Rocks	Roots	Backwater	Bank	Wood	Debris dam
0–5							
5–10							
10–15							
.							
.							
.							
95–100							

C. Exercise 3: Long-Term Retention and Transport (Optional)

1. Conduct CPOM release as in Exercise 1, but with one or more of these suggested modifications.

2. Release dowels over a time span of several weeks or months, depending on the stream and research objectives.

3. Inventory the location of dowels in the channel, but leave them in place and reinventory after varying periods of time. Different sizes of wood also can be released. Year-classes of wood can be marked differently, permitting year-to-year evaluation of transport. Additional releases of leaves and wood can be conducted in different seasons or at different discharges to describe more precisely the temporal dynamics of retention.

D. Exercise 4: Dye Release (Optional)

1. Carefully and accurately weigh several batches of fluorescent dye (e.g., 1.0 g of fluorescein powder or rhodamine–WT liquid) into scintillation vials. Number a set of empty scintillation vials from 1 to 100.

2. Qualitatively estimate the amount of dye to be released (1.0 g is appropriate for about 0.25 m^3/s discharge; about a third-order stream). Thoroughly dissolve the dye in a small volume of water (e.g., 1 liter). Release the dye slurry at the upstream end of the reach into a constricted, turbulent zone, if available, to ensure rapid mixing of the dye with the streamwater. In slower moving water, dispense the dye evenly across the stream channel. Station the researchers at the downstream end of the reach with the numbered scintillation vials, a stopwatch, and a notebook.

3. At the downstream end of the reach, the dye concentration curve must be measured accurately by taking water samples as the plume passes through the reach. Commence sampling of water from the *thalweg* (main thread of flow) in the numbered scintillation vials immediately after the release. Sampling frequency and duration will depend on transport time related to stream size, reach length, and channel geomorphology. We recommend that water samples be drawn every 5 s as the dye plume passes through the downstream end of the reach. The interval between samples can be lengthened for the trailing edge of the plume. Continue sampling even after visible dye has passed from the reach and until the coordinator indicates to stop (e.g., 5–10 min in a third-order stream). Record elapsed time with each numbered water sample (see Table 11.2).

4. In the laboratory, calibrate a fluorometer with a standard concentration series of the released dye (within the expected dilution range, such as 0.1, 1, 10, and 100 μg/liter). Measure and record dye concentration in each water sample using the fluorometer. Dilute samples if dye concentration exceeds your calibration curve.

TABLE 11.2
Sample Data Sheet for Conducting Hydraulic Retention Study

Stream:	Dye:	Date:
Location:	Concentration:	Team:
Reach:	Volume:	Notes:
Length:		

Elapsed time (min : sec)	Vial no.
0:0	1
0:5	2
0:10	3
.	.
.	.
.	.
5:00	n

IV. QUESTIONS

1. To what features do you attribute the differences in retention of leaves and wood between the two study reaches? What were the most important retention structures in the two reaches? Were they the same for leaves and wood?

2. Were more leaves retained in pools or riffles? Why? What are the mechanisms responsible for retention in these two types of bedforms?

3. Did the exponential model adequately describe the POM retention patterns? What exactly do the parameters of this model describe? Are there more appropriate models?

4. What physical features influenced hydraulic retention? Did the measurement of discharge with dye correspond to that determined from the area–velocity technique? What are the limitations of the dye slug release approach?

5. How do you think stream size (order) would affect retention of POM and water? Speculate about retention efficiency in smaller or larger streams than the one you studied. How might discharge and season affect retention in the *same* stream?

6. In light of your findings, discuss the implications of stream and riparian management practices that tend to reduce the amount of wood loading to streams, to simplify stream channels, or to modify the hydrograph.

V. MATERIALS AND SUPPLIES

Field Materials

Buckets with leaves
Current velocity meter (optional)
Dip (D-frame) nets (1 per investigator)
Field notebook with data sheets
Flagging
Log calipers (if available)
Meter sticks
Metric tapes (100, 50, 10 m)
Scintillation vials (100 plastic; numbered)
Seine with lead line
Stadia rod and clinometer or hand level (for gradient analysis)
Stopwatch

Laboratory Materials

Buckets (four 20-liter (5-gallon), for soaking leaves)
Dried or fresh-fallen leaves (e.g., 5000 abscised *Ginkgo biloba* leaves)
Fluorescent dye (fluorescein powder or rhodamine–WT liquid)
Garbage bags (to store leaf batches)
Scintillation vials (to hold dye)
Wood dowels (60 dowels ca. 1 m L × 1.5 cm D)

Laboratory Equipment for Optional Dye Release

Electronic balance (±0.01 g)
Fluorometer with filters for specific fluorescent dye
Microcomputer and digitizer

REFERENCES

Bilby, R. E. 1981. Role of organic debris dams in regulating the export of dissolved and particulate matter from a forested watershed. *Ecology* **62:**1234–1243.

Bilby, R. E., and G. E. Likens. 1980. Importance of organic debris dams in the structure and function of stream ecosystems. *Ecology* **61:**1107–1113.

Cederholm, C. J., D. B. Houston, D. L. Cole, and W. J. Scarlett. 1989. Fate of coho salmon (*Oncorhynchus kisutch*) carcasses in spawning streams. *Canadian Journal of Fisheries and Aquatic Sciences* **46:**1347–1355.

Cummins, K. W. 1974. Structure and function of stream ecosystems. *BioScience* **24:**631–641.

Cummins, K. W., and M. J. Klug. 1979. Feeding ecology of stream invertebrates. *Annual Review of Ecology and Systematics* **10:**147–172.

Cummins, K. W., J. R. Sedell, F. J. Swanson, G. W. Minshall, S. G. Fisher, C. E. Cushing, R. C. Petersen, and R. L. Vannote. 1983. Organic matter budgets for stream ecosystems. Pages 299–353 *in* J. R. Barnes and G. W. Minshall (Eds.) *Stream Ecology: Application and Testing of General Ecological Theory.* Plenum, New York, NY.

Ehrman, T. P., and G. A. Lamberti. 1992. Hydraulic and particulate matter retention in a 3rd-order Indiana stream. *Journal of the North American Benthological Society* **11:**341–349.

Gordon, N. D., T. A. MacMahon, and B. L. Finlayson. 1992. *Stream Hydrology. An Introduction for Ecologists.* Wiley, Chichester, UK.

Jones, J. B., and L. A. Smock. 1991. Transport and retention of particulate organic matter in two low-gradient headwater streams. *Journal of the North American Benthological Society* **10:**115–126.

Smock, L. A., G. M. Metzler, and J. E. Gladden. 1989. Role of debris dams in the structure and function of low-gradient headwater streams. *Ecology* **70:**764–775.

Speaker, R. W., K. J. Luchessa, J. F. Franklin, and S. V. Gregory. 1988. The use of plastic strips to measure leaf retention by riparian vegetation in a coastal Oregon stream. *American Midland Naturalist* **120:**22–31.

Speaker, R. W., K. Moore, and S. V. Gregory. 1984. Analysis of the process of retention of organic matter in stream ecosystems. *Verhandlungen der Internationalen Vereinigung für Theoretische und Angewandte Limnologie* **22:**1835–1841.

Triska, F. J., V. C. Kennedy, R. J. Avanzino, G. W. Zellweger, and K. E. Bencala. 1989. Retention and transport of nutrients in a third-order stream: channel processes. *Ecology* **70:**1877–1892.

Vannote, R. L., G. W. Minshall, K. W. Cummins, J. R. Sedell, and C. E. Cushing. 1980. The river continuum concept. *Canadian Journal of Fisheries and Aquatic Sciences* **37:**130–137.

Young, S. A., W. P. Kovalak, and K. A. Del Signore. 1978. Distances travelled by autumn-shed leaves introduced into a woodland stream. *American Midland Naturalist* **100:**217–222.

SECTION C

Stream Biota

CHAPTER 12

Heterotrophic Microorganisms

AMELIA K. WARD
AND MARK D. JOHNSON

Department of Biological Sciences
University of Alabama

I. INTRODUCTION

Heterotrophic microorganisms in streams include bacteria, protists, and fungi, which are important components of the microbial communities associated with submerged surfaces such as rocks, leaves, and wood as well as with interstitial water of benthic sediments and in the overlying water (e.g., Aumen *et al.* 1983, Bott *et al.* 1984, Findlay *et al.* 1993, Lock 1993). These same habitats contain microscopic animals (meiofauna), which are discussed in Chapter 15. The focus of this chapter will be on heterotrophic bacteria and protists, which function primarily as decomposers of dissolved (DOM) and particulate organic matter (POM), but are also consumed by higher trophic levels (Meyer 1990).

One important role of benthic bacterial communities is in the assimilation of dissolved materials from the overlying water. Both biotic and abiotic retention of DOM (e.g., leachate from leaves) on benthic surfaces is well documented in small stream ecosystems (Cummins *et al.* 1972, Lock and Hynes 1976, Dahm 1981, McDowell 1985). The ecological importance of these processes is that they result in the transfer of organic carbon associated with DOM from the overlying stream water to surfaces, where it can then be partially or wholly metabolized by benthic, heterotrophic microbial communities. Therefore, benthic microbial communities function to retain and transform DOM, which is an important source of organic matter and energy.

The removal of DOM from overlying water to benthic habitats by stream microorganisms has primarily been documented by measuring the disappearance of DOM. These studies have revealed that some sources of DOM are removed more rapidly than others, suggesting differences in quality of DOM (Lock and Hynes 1975, Lush and Hynes 1978). However, direct measurement of metabolic responses of heterotrophic microorganisms in stream ecosystems to different sources of DOM has been much less studied. *Quality of organic matter* is an imprecise term, but relates to the ability of organisms to use it as a food source. High-quality DOM should be rapidly assimilated and quickly metabolized by heterotrophic organisms, whereas low-quality DOM should be more slowly utilized (e.g., Hedin 1990).

Increase in bacterial metabolism with increasing temperature is a commonly observed, but not universal, phenomenon in aquatic ecosystems (e.g., Bott *et al.* 1985, White *et al.* 1991, McKnight *et al.* 1993). Heterotrophic microbial production can respond to both temperature and sources of DOM (Hoch and Kirchman 1993). Shiah and Ducklow (1994) found that estuarine bacterial growth rates responded positively and most consistently to increases in temperature. They also noted that bacterial growth was more likely to be substrate limited during summer than winter. Seasonal changes in temperature as well as DOM concentration and quality are common occurrences in lotic ecosystems. Therefore, these environmental factors are important to consider in understanding bacterial metabolism in streams.

In addition to bacterial heterotrophs, particle-consuming or *phagotrophic* protists are also found in stream and river habitats as part of microbial food webs (Baldock *et al.* 1983). These single-celled, eukaryotic organisms can be important consumers of bacteria in lotic ecosystems (e.g., Carlough and Meyer 1990). Even some protists traditionally thought of as algae are *mixotrophic* and can be important bacterivores in microbial food webs (Sanders and Porter 1988, Sanders 1991a). Therefore, measuring abundances of protists is useful not only in providing a more thorough understanding of total microbial heterotrophy, but also in interpreting community and production dynamics of bacteria. Although found to some extent in almost all flowing water habitats, abundances of phagotrophic protists are typically greater in sediments, attached to surfaces, and in organically enriched sites, than in the water column (Baldock and Sleigh 1988, Bott and Kaplan 1989).

A common method of measuring microbial community metabolism is by following changes in oxygen concentrations (Bott *et al.* 1978, Naiman

1983). More recently, other techniques have been developed for investigating stream bacterial community activity, including the use of redox dyes to estimate numbers of respiring, heterotrophic bacteria (Bott and Kaplan 1985, Johnson and Ward 1993) and radioisotope techniques for estimating bacterial productivity (e.g., Findlay *et al.* 1984, Stock and Ward 1989). The use of settling chambers or fluorescence microscopy (e.g., Carlough and Meyer 1991) to enumerate preserved flagellates and ciliates works well for planktonic communities if the particulate material does not obscure protist numbers. However, quantification of protists in benthic habitats is more easily accomplished using a live count technique (e.g., Gasol 1993).

In this chapter we describe methods for the enumeration and measurement of metabolism of heterotrophic microorganisms in streams. The experiments are designed to test effects of common, environmental factors, such as temperature and source of organic substrates, that can control the abundance and metabolism of heterotrophic microorganisms. Several options that can be adapted to different laboratory situations and availability of instruments are presented. The methods range from the least complex in terms of required supplies and instrumentation (e.g., oxygen methods) to more sophisticated radioisotope techniques. The specific objectives are to (1) introduce procedures for quantifying numbers and activity of heterotrophic microorganisms in streams, (2) present experiments that address the effects of temperature and source of organic matter on bacterial communities, and (3) provide protocols for data organization and analysis.

II. GENERAL DESIGN

The following exercises are designed in such a manner that measurements of microbial abundances and metabolic activity can be compared and contrasted among different stream habitats. Alternatively, treatment effects of *dissolved organic carbon* (DOC) amendments and/or temperature can be evaluated in an experimental design. The exercises are organized into four sections addressing different methods of determining possible treatment effects: (A) changes in oxygen concentration as a measure of microbial respiration, (B) protist abundances, (C) bacterial abundances and biomass, and (D) heterotrophic bacterial productivity.

A method for generating leachate from natural sources for DOC amendments is described here and can be used for any of the exercises below that incorporate this treatment. Notes are included in the exercise procedures to indicate when these treatments can be used. Similar notes are made with appropriate exercises for temperature effects. For best results and overall interpretation, one or more of these exercises could be performed in conjunction with exercises from other chapters, particularly those describing organism abundances/metabolism (e.g., Chapters 15, 25) and nutrient uptake (e.g., Chapters 9, 29).

Stream microbial communities are typically exposed to a mixture of dissolved organic compounds generated from plant decomposition processes rather than high concentrations of a simple sugar or a single carbohydrate. Therefore, the use of natural sources of DOC provides a better estimation of lotic microbial response to dissolved organic matter than does glucose or some other simple carbohydrate. Generation of leachates for the DOC amendment and DOC quality treatments should use material collected from the same location that stream water and benthic substrata are obtained. Aquatic macrophytes, algae, or riparian plant leaves can be leached easily to provide a source of DOC for the experiments in this exercise. DOC concentrations can be measured by use of a total organic carbon analyzer.

Protocol for Leachate (DOC) Generation[1]:

1. Collect the chosen plant type(s). For a comparison of DOC qualities, select for example: (a) coniferous needles vs. deciduous leaves, (b) dead and dry vs. fresh leaves, or (c) a specific species of plant vs. another. Plant material can be dried by spreading it on tables or nets and air-drying for several days, depending on the water content of the plant.

2. Weigh an equal amount of each plant type and add to an accurately measured volume of deionized (DI) water (e.g., 500 ml) in a large beaker or Erlenmeyer flask. Less than 100 g of dried plant material is usually sufficient. (A range of concentrations of DOC can be generated by diluting the original stock leachate. The important point here is that if a carbon analyzer is not available, qualitative or relative quantification is all that will be possible.)

3. Place a magnetic stir bar in the beaker with the plant material and DI water, cover with aluminum foil, and set on a magnetic stirrer. Adjust speed on the magnetic stirrer so that the liquid moves in a slow, swirling

---

[1]Leachate material is extremely susceptible to microbial contamination. Take as many precautions as possible to retard bacterial/fungal contamination before the leachate is used in experiments. These precautions should include preparing leachate material just prior to use (or as close in time as possible). If storage is necessary, store in clean, preferably sterile, stoppered containers in a refrigerator. The leachate material can be filter-sterilized, but probably should not be autoclaved since that process can cause precipitation of organic matter or other physical/chemical changes in the leachate. If a lyophilizer is available, leachate material can be freeze–dried, stored dry in sealed containers, and reconstituted with clean, deionized water prior to use.

motion. If possible, place in a cold room (e.g., 4–10°C) or cold growth chamber overnight.

4. After leaching, the liquid in the beaker should appear light to dark brown in color. Remove large particulate material from the liquid by pouring the contents of the beaker through clean cheesecloth, Nitex netting, or a stack of sieves (e.g., 1-cm, 1-mm, and 43-μm mesh sizes). Once the large particles have been removed, filter the remaining material through glass fiber or polycarbonate filters, preferably 0.5 μm or less pore size. Place filtrate in a sterile, stoppered bottle.

III. SPECIFIC EXERCISES

A. Exercise 1: Changes in Oxygen Concentration as a Measure of Microbial Respiration (Fig. 12.1)

Aerobic microorganisms consume oxygen when metabolically active. The metabolic response of heterotrophic microbial communities to treatment effects of DOM source or temperature can be measured by (1) measuring oxygen concentration in stream water, (2) placing stream water and samples of microbial communities in a sealed, dark container of known volume, (3) applying the treatment (e.g., increased temperature or different DOM sources), (4) incubating the container for a known period of time, and (5) measuring oxygen concentration in the overlying water in the containers at the end of the incubation period. Respiration of the sample can be estimated by calculating the difference in oxygen concentration at the beginning and end of the incubation period. The Winkler titration is an inexpensive, easy, and accurate method for measuring dissolved oxygen in water (see Chapter 5). An alternative procedure testing these (following) hypotheses using oxygen electrodes and water recirculating in chambers could be used if such equipment is available (e.g., Naiman 1983, Bott *et al.* 1985).

Hypothesis: *Increased temperature and/or DOC concentration will result in increased community respiration and thus lowered oxygen concentrations.*

Protocol for Preparation of Oxygen Laboratory

1. Assemble an appropriate number of 60-ml bottles (with sintered necks and beveled stoppers) for dissolved oxygen (DO) measurement, 300-ml bottles (with sintered necks and beveled stoppers) for biological oxygen demand (BOD) determination, and small aluminum foil sheets. The BOD bottles and top of stoppers should be wrapped in black, plastic electrical tape while the bottles are dry and the stopper–bottle junction wrapped in

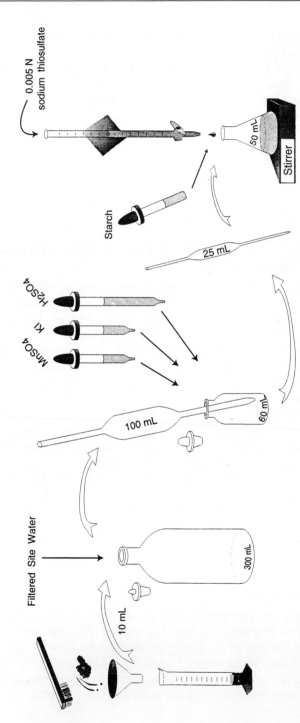

FIGURE 12.1 Procedures for use of changes in oxygen concentration as a measure of sediment metabolic response.

aluminum foil at the beginning of incubation. At a minimum, duplicate bottles for each treatment should be used. Three or more replicates for each treatment would, of course, be preferable. Label all bottles appropriately.

2. Assemble Winkler titration equipment, including burette, volumetric pipets, reagents, and indicator solutions (see detailed equipment and supplies list in Chapter 5 and in Section V of this chapter).

3. Organize bottles, field reagents, and pipets into a box suitable for transport into the field.[2]

Protocol for Oxygen Field Activity

1. Collect streamwater in 60-ml DO bottles.[3] Fill each bottle completely, and then gently place a beveled stopper in each one, making sure no air bubbles are included. Remove the stopper and add 0.75 ml each of $MnSO_4$ and alkaline KI. Replace stopper; then invert bottle several times to mix reagents.

2. Collect 10-cm^3 samples of sediment. In order to replicate the amount of sediment used in each bottle, either scoop sediment with spatula into a plastic, graduated cylinder, or use a 10-ml plastic syringe with cut-off tip to collect a measured volume of sediments. Put each sediment sample into a separate 300-ml BOD dark bottle (using a squeeze bottle of streamwater to rinse graduated cylinder, if used). Fill bottles with stream water.

3a. Treatment effect: The effect of added DOC can be evaluated by removing 10 ml of water from each of the treatment BOD bottles and replacing with 10 ml of leachate (see leachate generation procedure in Section II). All DOC-amendment and no-amendment bottles should be placed back in the stream for a measured incubation period (several hours) at streamwater temperatures. If field incubations are inconvenient, all bottles can be brought back to the laboratory and incubated at the appropriate temperature.

3b. Optional treatment effect: The effect of different temperatures on oxygen consumption can be evaluated by transporting streamwater and stream sediment back to the laboratory and dispensing to 300-ml BOD bottles as described in step 2. Bottles can then be incubated at different temperatures in water baths and/or growth chambers.

4. At the end of the incubation period, use a 100-ml volumetric pipet to carefully fill a labeled 60-ml DO bottle from each incubated 300-ml treatment bottle, allowing the excess to overflow. Field reagents should be

---

[2]CAUTION: Field reagents include caustic chemicals—handle with care.

[3]Note that the oxygen determination method described here is a "micro-Winkler" technique modified from the technique described in detail in Chapter 5. You may find it useful to consult Chapter 5 as a reference for standard titration technique.

added to these subsamples as described in step 1 above. Remember to record accurately the initial time (time when bottles are stoppered and placed in the stream or water baths) and ending time (time when incubation bottles are sampled and reagents are added).

Protocol for Oxygen Laboratory Activity[4]

1. Just prior to titration, add 0.5 ml concentrated H_2SO_4 to each 60-ml DO bottle, replace stopper, and shake well.

2. Transfer a 25-ml subsample from each 60-ml DO bottle into a 50-ml flask using a volumetric pipet with suction bulb. Add a magnetic stir bar to the flask and place the flask on a magnetic stirrer with a white background.

3. Titrate to pale yellow with 0.005 M sodium thiosulfate from a finely graduated 10- or 25-ml burette.

4. Add one or two drops of starch glycerin solution to the flask. The subsample should turn dark blue. Titrate until the dark blue color of the 25-ml subsample completely clears. A light blue color should return in 10 to 15 s if the sample was not overtitrated.

5. Read the volume of titrant used while holding a piece of white paper or aluminum foil at an angle behind the burette to highlight the meniscus.

6. Optional: if you wish to increase resolution in the experiment, follow steps 1 through 9 on 300-ml bottles of incubated streamwater (without sediments) in order to factor out respiratory oxygen consumption of the plankton from benthic metabolism.

Oxygen Calculations

1. Determine oxygen concentration (O_c, in mg/liter) in the sample using the equation

$$O_c = \frac{V_t \times M_{ts} \times 8000}{V_{ss} \times (B_{DO} - R/B_{DO})}, \tag{12.1}$$

where V_t represents titrant volume (ml) of sodium thiosulfate used to titrate the 25-ml subsample; M_{ts}, molarity of thiosulfate; V_{ss}, subsample volume (ml); B_{DO}, BOD bottle volume (ml); and R, volume of reagents (mL).

2. If the volumes and concentrations are the same as described in the above procedure, then the equation reduces to[5]:

---

[4]Modifications of this basic oxygen protocol for situations involving high iron, DOC, or suspended solid content are discussed in Wetzel and Likens (1991).

[5]For increased accuracy, each bottle used in the experiment should be measured separately and the specific volumes of each bottle used in Eq. (12.1), rather than the shorter Eq. (12.2).

$$\text{mg } O_2/L = \text{(titrant volume) (1.64)}. \qquad (12.2)$$

3. Subtract the oxygen concentration (mg O_2/L) of the incubated treatment from initial streamwater oxygen concentration to obtain oxygen consumption. (Correction for streamwater respiration values from optional step 6 above should be done at this point in the protocol as well.)

4. If 10 cm³ of sediment was added to a 300-ml BOD bottle, then oxygen consumption/cm³ of sediment can be determined by multiplying the oxygen consumption value from step 3 by 29 (300 ml − 10 ml sediment volume = 290 ml; 290 ml streamwater/10 ml sediment = 29 ml streamwater/cm³ sediment).

5. Assuming simple respiratory processes, use the following formula ($C_6H_{12}O_6 + 6O_2 = 6CO_2 + 6H_2O$) and your oxygen consumption data from the above to determine the amount of carbon metabolized in respiration. As six oxygen molecules (mw 192) are consumed per molecule of glucose respired (with mw of carbon = 72), multiply the milligrams of oxygen consumed as determined above by 0.375 to obtain amount of carbon used.

6. Divide the value obtained in step 5 by the number of hours of incubation. You should now have the amount of carbon respired per cubic centimeter (or gram) of sediment per hour in your sample (mg C cm^{-3} h^{-1}).

B. Exercise 2: Protist Abundances

Free-living, phagotrophic protists (protozoa) that are likely to be encountered in lotic ecosystems can be divided into three groups based upon general morphology: flagellates, ciliates, and sarcodynes. These organisms are morphologically and functionally diverse (e.g., Sleigh 1989), but some characteristics that are useful in distinguishing one group from another are given below as well as information on typical abundances and some representative illustrations.

Flagellates are a polyphyletic group of microorganisms that use one to several undulipodia[6] for locomotion (Fig. 12.2a). Nanoflagellates range in size from 2 to 20 μm and most are capable of bacterivory, even photosynthetic forms (Sanders 1991a, b). Some flagellates can have a significant impact as grazers of algae, while others function osmotrophically (Sanders 1991b). Heterotrophic microflagellates, ranging in size from 20 to 200 μm, are most likely to be found in waters high in organic matter, such as just downstream of sewage plant effluents (Sladecek 1973). Streamwater commonly contains from 10 to several 1000 flagellates/ml, depending upon

[6]As flagella and cilia are indistinguishable ultrastructurally the term Undulipodia has been suggested for all membrane-covered, cell motility organelles with a 9 + 2 microtubule ultrastructure (Margulis *et al.* 1989).

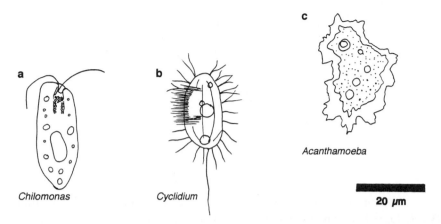

FIGURE 12.2 Examples of typical free-living, phagotrophic protist types: (a) flagellates, (b) ciliates, and (c) sarcodynes.

the season and water quality. Lotic sediments often contain an order of magnitude greater abundance per volume than the overlying water. Small flagellates can be in the same size range as large, motile bacteria. However, close observation with a phase contrast microscope will often reveal the undulipodia. Bacterial flagellar bundles are typically not visible with light microscopy at magnifications of 400× or less.

   Ciliates are a monophyletic group of single-celled protists, united by nuclear ultrastructure and the use of many undulipodia for motion at some stage in their life cycle (Fig. 12.2b). Some also use cilia for feeding. Ciliates feed on bacteria, detritus, algae, flagellates, and other protists depending upon species and environmental conditions. Some also contain endosymbiotic algae. Ciliates typically range in size from approximately 10 to several 100 μm in length. Sandy sediments often harbor long, thin, worm-like forms. Attached ciliates such as *Vorticella* can be found in abundance on stones (e.g., Harmsworth *et al.* 1992) and submerged plant surfaces (e.g., Baldock *et al.* 1983). Depending upon the time of year and the conditions in the stream, ciliate abundances can range from near 0 to several 100 ciliates per milliliter in streamwater and up to several 1000/cm³ in benthic habitats. Although ciliates are typically less abundant than nanoflagellates, their large size can make them dominant in terms of protist biomass.

   Sarcodynes comprise a polyphyletic assemblage of organisms including various amoeboid forms, heliozoa, and others (Fig. 12.2c). Although not commonly seen in flowing water and hard to detect visually in decaying debris and particulate detritus, they do occur in high numbers in some benthic habitats, such as organic-rich sediments. In addition to feeding on bacteria, detritus, and algae, some prey upon flagellates and ciliates. Some sarcodynes contain chloroplasts or endosymbiotic algae. Amoeboid forms range in size from <5 μm to >1 mm in diameter. Typically, once samples to be counted are placed on the microscope, they need to be left undisturbed by motion or vibration for a few minutes before amoeboid motion will resume. *Tests* (rigid shells around organisms that are less close-fitting than a cell wall) of large forms such as *Difflugia* are often seen in sediments, but should not be counted unless the cytoplasm of the cell can be seen, because the tests are often empty. Small sarcodynes are often insinuated within particulate detritus and visually undetectable. Culture methods are used to determine more accurately their abundances (see Page 1988).

Protist taxonomy is a fascinating field of investigation, but determination of species identity is not necessary for this exercise. However, identification of the organisms does provide better resolution than quantifying on the basis of the general categories described above. It is also helpful to know the variety of protist morphologies possible in order to improve the search image for specific groups. In order to facilitate recognition of protist groups, line drawings of sample organisms (Fig. 12.3) demonstrate size ranges of protists likely to be encountered in stream samples.[7]

Protist Experimental Approach Hypotheses and experiments described in previous sections using temperature or DOC treatments can also be applied to stream samples containing protists. Abundances of protists can then be reported under the different treatment levels and compared to controls. Alternatively, protist communities can be described for different stream habitats, such as overlying water, organic sediments, sand, rock surfaces, etc., and compared with abundances of other groups of organisms such as bacteria, macroinvertebrates, meiofauna, and/or periphyton (see Section C for bacterial abundances and Chapters 13–17). Also, sampling before and after major leaf inputs into a stream, or above and below a pollution source, can provide insightful comparisons.

Counts of small protists typically require fluorescence microscopy of aldehyde fixed samples concentrated upon stained, membrane filters (Sherr *et al.* 1993). However, it is difficult to quantify protists in samples of sediment or periphyton scrapings in this manner. The amount of dilution necessary to reduce the concentration of obscuring particles would make enumeration of several whole filters necessary for an accurate count in many cases. As human vision is better at detecting motion than differentiating among

---

[7]The text descriptions, drawings, and light-microscope, color photographs in Patterson and Hedley (1992) are very useful for initial recognition and identification and make it possible to identify readily many common, heterotrophic protists to genus.

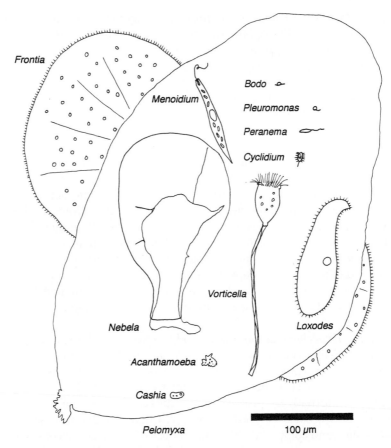

FIGURE 12.3 Examples of phagotrophic protist sizes as illustrated by different protist genera.

particle morphology, counting live samples not only gives more realistic numbers of small nanoflagellates and ciliates than preserved sediment samples, it is also easier! The following technique was modified from that described in Gasol (1993). Diluted samples of sediment are scanned with 100× phase contrast microscopy either on slides with raised coverslips or in Palmer–Maloney counting chambers.

Protist Field Sampling

1. Hold a 10-ml plastic syringe (with the tip cut off) just above the surface of an area of loose, fine sediment. Insert the barrel housing of the syringe into the sediments a couple of centimeters while holding the plunger

still. Holding the syringe in place, slide a stopper or flat surface under the open end of the syringe. A piece cut from the lid of a plastic container works well.

2. Optional treatment effect: If treatments such as DOC amendments or temperature are to be used, samples should be transferred to appropriate incubation vessels (e.g., large test tubes or small bottles with cotton plugs, not glass stoppers) at this point. Refer to incubation procedures in the oxygen section. The incubation period can be extended to accommodate generation times of protists in order to see treatment effects (e.g., to several days in the stream or laboratory). Keep in mind that the longer the incubation period, the more artifacts may result from enclosure effects. Do not preserve samples at the end of the incubation. Sample incubation vessel and proceed as described below.

3. Place contents into a small jar or container for transport, or use an additional plunger to plug open end of syringe. Keep samples in a cooler on crushed ice until just prior to counting.

Protist Laboratory

1. Pipet 0.9 or 1.9 ml of filtered streamwater, depending on sediment type and abundances of organisms, into a small container such as an Eppendorf tube. Use a serial dilution initially if necessary to determine an optimum dilution factor.[8]

2. Pipet a 0.1-ml subsample of sediment sample using an automatic pipet (e.g., 1-ml Rainin style with a tip that has been trimmed to provide a slightly larger opening) into the container with streamwater.

3. Before sediment settles, pipet a 0.05-ml subsample of diluted sediment using an automatic pipet and trimmed pipet tip into the center of a Palmer–Maloney counting chamber.[9]

---

[8]If sediment sample is from below-surface, organic mud, it is likely to be anaerobic. Oxygen shock will result from mixing aerobic streamwater with the sediment. Using supernatant from an ultracentrifuged bulk sediment sample is likely to work better. You can also use water that has just been flushed with another gas such as nitrogen, Ringers solution, or boiled water that was placed in a tightly sealed jar while cooling.

[9]*Helpful hint.* Palmer–Maloney counting cells, which are slides with small chambers for receiving liquids, are ideal for this procedure. However, if they are not available, substitutes can be constructed using slides, coverslips, and a little Vaseline. Place four small blobs of Vaseline on the slide with the tip of a toothpick or dissecting probe in such a manner that the grease will support the four corners of a coverslip. Pipette the 0.05-ml subsample onto the slide in the center of the grease supports. Gently place the coverslip on top so that the underside of the coverslip contacts the sample. Water depth on the slide under the cover slip needs to be sufficiently thin so that the depth of field at 100× phase contrast allows detection of nanoflagellate motion throughout the thickness of the sample, but thick enough not to crush large, delicate ciliates and sarcodynes (0.3–0.4 mm is about right).

4. Lower a large coverslip so that the spread-out droplet does not touch the sides of the Palmer–Maloney chamber. In this way organisms do not become obscured from view by contact with the walls of the chamber.

5. Place the chamber on a microscope stage under low power (100× total magnification), and adjust phase contrast to provide an image of bright organisms against a darker background.

6. If the working distance on the microscope allows it, carefully change the objective to a higher power objective (e.g., 400× total magnification) to inspect the morphology of the small, motile microorganisms. Large motile bacteria can fall within the same size range as small nanoflagellates. Characteristic shapes, motile behavior, and whether or not flagella can be seen distinguishes between these kingdoms of organisms. For example, bacterial flagella cannot be seen at 400× magnification, so if the organisms you observe at this magnification have visible undulipodia, they are eukaryotic.

7. Scanning relatively slowly, count the number of flagellates and ciliates from one side of the chamber to the other. Move the slide over to the edge of the field of view, and scan the next "strip" of the diluted subsample (Fig. 12.4). Continue in this manner until the entire subsample has been viewed and counted. After this initial count has been made, check any testate amoebae encountered for pseudopodial activity. These sarcodynes typically take several minutes to reemerge after disturbance.

8. Some large ciliates can move extremely quickly and from one counting strip to another, increasing the probability of counting one individual several times. To avoid this possibility, count large ciliates using a dissecting scope at 25 to 50×, a magnification at which the entire sample droplet can be seen in one field, and the large organisms in it can be counted quickly and with less error.

9. For increased accuracy, count as many subsamples per treatment as time allows. Record your data on the "protist abundance calculation worksheet" found on Table 12.1.

10. Multiply number of organisms counted by dilution factor and subsample volume (e.g., if you counted a 0.05-ml subsample of a 1 : 10 dilution, multiply the number of organisms counted by 200 to derive the number of organisms per cm^3 of sediment).

C. Exercise 3: Bacterial Abundances and Biomass

The application of fluorescent dyes to samples from aquatic environments has become the most accepted technique for enumerating bacteria in recent years (Fry 1990, Kemp et al. 1993). The basis of this technique is that the dyes bind with specific cellular components that fluoresce when exposed to light of a particular wavelength. The fluorescing bacterial cells can be seen more easily under the microscope than unstained cells. Other

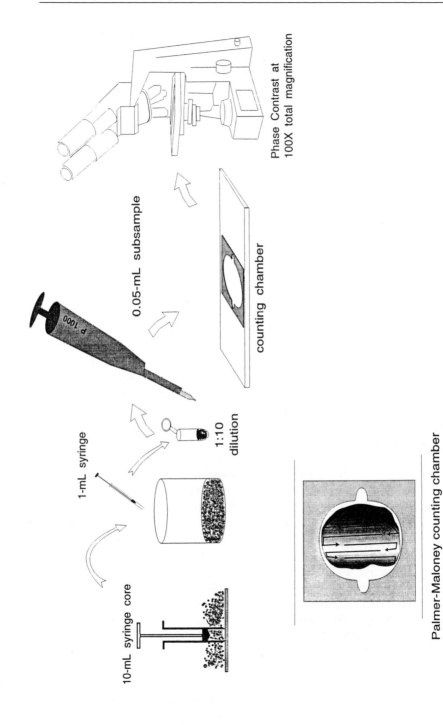

10-mL syringe core

1-mL syringe

1:10 dilution

0.05-mL subsample

counting chamber

Phase Contrast at 100X total magnification

Palmer-Maloney counting chamber subsample viewing path

FIGURE 12.4 Steps in protocol for counting live protists.

TABLE 12.1
Protist Abundance Calculation Worksheet

	(a)	×	(b)	×	(c)	=	(d)
Type of organism	No. of organisms counted		1/Subsample size		Dilution factor		Totals (no./ml)
Nanoflagellates 2–20 μm							
Heterotrophic microflagellates 20–200 μm							
Ciliates							
Amoeboid forms							

techniques which involve culturing bacteria on agar plates typically underestimate bacterial abundances greatly because relatively few types of bacteria found in natural environments culture easily (Herbert 1990). Acridine orange (AO) and 4',6'-diamidino-2-phenylindole (DAPI) are two of the most common fluorescent stains used in direct counts. AO fluoresces green or red. AO red can obscure autofluorescing chlorophyll, and AO can also bind with nonbacterial, particulate material. DAPI fluoresces blue or yellow. Fluorescence of some clay/silt minerals is similar to the DAPI yellow. The stain that works best for a given set of circumstances must be determined a priori by the individual investigator. Additional details regarding the use of these and other fluorescent stains can be found in Fry (1990) and Kemp *et al.* (1993).

Direct counts of bacteria using a fluorescent stain give higher counts than the use of culture plate techniques, but do not differentiate between metabolically active and dormant, senescent or dead bacteria. A metabolic indicator such as INT–Formazan distinguishes between respiring and nonrespiring bacteria and can be used in conjunction with fluorescent dyes to assess metabolically active versus nonactive bacteria (e.g., Bott and Kaplan 1985, Johnson and Ward 1993). The number of respiring bacteria in native environments, including streams, is usually quite low, often less than 30% of total, fluorescently stained cells (Boon 1991, Marxsen 1988). The reasons for this high percentage of inactive bacteria and the environmental cues that may stimulate bacteria to change from dormant to metabolically active states are not well understood in lotic or other aquatic environments. Steps for the use of INT are included in the protocol below as an option. More information on the details of the procedure can be found in Johnson and Ward (1993) and the references therein.

Typical bacterial abundances in lotic plankton range from approximately 0.08 to 55.5 $\times$ 10^6 cells/ml with highest abundances associated with agriculturally enriched and blackwater riverine ecosystems (Barcina *et al.* 1991, Carlough and Meyer 1991, Vaques *et al.* 1993, and Johnson 1995). Bacterial abundances on surfaces in streams generally range from 8×10^{10} to 4.4×10^{12} cells/m^2 (Haack and McFeters 1982, Stock and Ward 1989). Benthic bacteria in streams have been found to have a higher percentage of metabolically active cells than stream bacterioplankton (Johnson and Ward 1993).

In addition to assessing bacterial numbers in this exercise, an estimation of biomass from measurement of bacterial biovolumes is also possible. While different stains can result in slightly different measurements of the same bacteria, in general the volume of a bacterium can be determined by using the formula for a sphere, a cylinder, or an oblate spheroid. In some cases, a formula for a straight-sided rod with hemispherical ends (volume = (diameter^2) (π/4) (length $-$ diameter/3)) may be more appropriate. For other bacterial morphotypes, different formulas may be used to increase precision. The volume is multiplied by a conversion factor of 310 fg C/μm^3 to obtain the biomass (Fry 1990).

Hypothesis: *An increased concentration of labile organic carbon will result in increased bacterial abundances, biomass, and percent of bacteria respiring.*[10]

Protocol for Bacteria Laboratory Option   The number of treatments and replicates you use will depend upon your experimental design. We describe here a nonreplicated design to facilitate clarity of description.

1. Set up the following temperature treatments: (A) Shaker table in cold chamber at 5–10°C and (B) shaker table at room temperature of 20–25°C.

2. Make DOC leachate as described above.

3. Place 250 ml streamwater in each of four 500-ml flasks.

4. As described in the microscopy section below, count and measure the bacteria from each flask at the beginning of the experiment.

---

[10]Elegant experimental designs can be developed with replications of multiple temperatures and regression analysis or a two-way ANOVA factorial design testing the interactive effects of temperature and DOC concentrations.

4a. For the INT-formazan option, incubate 1 ml of 0.2% INT–chloride and 10 ml of streamwater in foil wrapped test tubes for at least 30 min. Terminate incubation with an equal volume of phosphate-buffered 5% formalin. Quantify percentage bacteria respiring using protocol below.

5. Add 20 ml leaf leachate to each of two flasks.

6. Place one flask with DOC added into each water bath.

7. Place one flask without DOC added into each water bath.

8. Incubate for 1 to 7 days.

9. Subsample and mix 1 : 1 with phosphate-buffered 5% formalin.

Protocol for Bacteria Field Option (Leachate Treatment Only)

1. Make DOC leachate as described in Section II.

2. Anchor six plastic wide-mouth jars (e.g., 500 ml) *in situ* with opening above stream surface.

3. Place gravel or cobble in chamber and cover with 250 ml stream-water.

4. At start of experiment, scrape sample rocks and count and measure bacteria per protocol below.

5. Add 20 ml DOC leachate to three jars, and leave three jars without the amendment (20 ml additional DI or streamwater can be added to these treatments instead).

6. Incubate for 1 to 7 days.

7. Sample cobble surfaces by scraping with a toothbursh and preserve 1 : 1 with phosphate-buffered 5% formalin.

7a. For quantification of percentage bacteria respiring, add INT–chloride to incubation containers, or subsamples from the containers, a half-hour (or more) prior to subsampling for fixation.

8. Count and measure bacteria from treatment containers.

Bacteria Quantification Protocol (Sample Preparation) (Fig. 12.5)   To the extent possible, work in a very dimly lit room when working with fluorescent stains. Bright light can induce distracting background fluorescence in the samples.

1. Make up 1 mg/ml stock solution of DAPI in a small vial. Wrap vial in aluminum foil to protect from light when not in use.

2a. Scrape a rock with a toothbrush.

2b. Measure the rock by covering with aluminum foil, one layer thick on all surfaces scraped. Weigh foil and calculate area based upon weight of foil of known area.

3. Dilute sample 1 : 200 (more or less depending upon bacterial abundance in the sample) in 0.2 μm filtered streamwater or DI water.

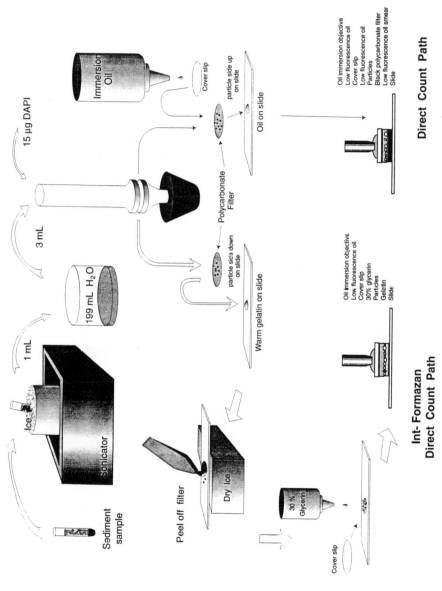

FIGURE 12.5 Steps in enumerating bacteria only (right flowpath) and enumerating percentage of respiring bacteria by INT–formazan procedures (left flowpath).

4. To remove bacteria from particle surfaces, sonicate 5-ml subsamples three times (1 min each sonication) in test tubes packed in ice (allow 30 s cooling time between sonication periods) in a typical bath sonicator. Alternative: Use a probe sonicator to deliver a total of 4 W-min of energy in short periods at the lowest power setting. Allow cooling times between delivery periods (e.g., two periods of 30 s at a 4-W setting.)

5. Allow sonicated material to settle for 30 min. (Initially, settled material can be checked using the same general protocol below to determine the percent removal of bacteria.)

6. Dampen the filter tower frit with a drop of sterile, DI water. Place a 0.45- to 3.0-μm pore size cellulose acetate, backing filter on frit. Place prestained (black) 0.2-μm pore size polycarbonate filter,[11] shiny side up, on top of the damp backing filter. Clamp filter tower onto filter/frit stack. (Black filter not needed for INT option.)

7. Place 2 ml supernatant from the settled, sonicated sample into the reservoir of the filter tower. (Add additional filtered DI water to filter tower if less than 2 ml of preserved sample is to be used. Total volume should be at least 1 ml/cm^2 of filter surface area to provide an even distribution of bacteria.)

8. Add ~1–2 μg DAPI/ml in tower for plankton samples, or ~5 μg DAPI/ml for low dilution samples of sediment or epiphyton samples.

9. Wait 10 minutes.

10. Turn on the pump. Use a pressure of less than 150 mm Hg.

11. Smear a small drop of immersion oil on slide.

12. When filtration is complete, transfer the filter, particle side up, to a slide.

13. Place a small drop of immersion oil on top of the filter. Place a No. 1 coverslip on top of the filter. Press down on coverslip with pencil eraser to flatten filter and squeeze out any excess oil. Gently remove any excess oil from around the coverslip with a Kim-Wipe to prevent getting oil on the microscope stage.

INT Option (substitute for steps 11–13 above)

   a. Place a small drop of 1% gelatin on a warm slide.
   b. Invert damp polycarbonate filter from filter tower while still slightly damp onto liquid gelatin, particle side down. Leave an edge of the filter hanging over the edge of the slide.

---

[11]*Helpful hint.* Polycarbonate filters can be purchased or stained in the lab. To stain your filters, dissolve Sudan black B (Merck) in absolute ethanol. Dilute to 50% with particle free DI water. Final concentration should be ~1 : 15,000. Soak filters for 1 day. Rinse thoroughly with particle-free DI water (Zimmerman *et al.* 1978).

   c. Place slide, filter side up, onto a piece of dry ice, or a supercooled metal bar (e.g., presoaked in liquid nitrogen).

   d. Remove slide from metal or dry ice. Use forceps to remove filter from frozen gelatin. (The gelatin should now contain the particulates from the filter.)

   e. Mount a coverslip on top of the gelatin with 30% glycerol.

Bacteria Quantification Protocol (Enumeration and Measurement)

1. For abundance, count the bacteria in at least 10 randomly selected grid fields with an epifluorescence microscope at 1000× total magnification using a UV filter set (365 nm excitation). An average bacterial density of 30 bacterial cells per grid is optimal (Kirchman 1993). More information can be obtained by quantifying relative abundances of different morphotypes of bacteria (e.g., rods, cocci, spirals).

2. For biomass calculations, you need to estimate the size of the bacteria in your samples. Sometimes an increase in the average size of bacteria will result from the treatment rather than an increase in bacterial numbers. Measure at least 10 to 20 randomly selected bacteria of each common morphotype in your sample, using an ocular micrometer or image analysis system (if the analysis system can distinguish between bacteria and other particles).

INT Option: You need to view each field with both fluorescence microscopy and bright field microscopy. This is facilitated by covering the white light source with a flat surface (e.g., piece of cardboard or aluminum foil) while counting and measuring bacteria inside the graticule (grid). Then remove the white-light barrier to identify any reddish-brown formazan crystals. In practice, you need to switch back and forth between bright field and fluorescence microscopy a number of times per grid.

Bacteria Quantification (Calculations)

1. To calculate abundance:

   a. Determine correction factor for your microscope by calculating the number of grids per sample area on the filter. Measure area of filter covered by sample (¥) using your particular filter apparatus. Measure area of your grid reticule at 1000× (β) using an ocular micrometer. Divide ¥ by β.

   b. Multiply average number of bacteria in a grid by the correction factor.

   c. If 10 ml of preserved sample (i.e., 5 ml sample and 5 ml formalin)

were placed in the tower, then divide the product in step b by 5.
If a different subsample size was used in the tower, this needs to
be taken into account.
d. Multiply by the dilution.
e. Divide by the surface area of the sample.

INT Option: Divide formazan containing bacteria by the total number
to determine percentage of bacteria respiring.

2. To calculate biomass use the "bacteria biomass calculation worksheet" found on Table 12.2 and follow these steps:

a. Divide total measured bacterial biovolume by number of bacteria
measured to determine average size.
b. Multiply number of bacteria/cm^2 by the average bacterial size in
μm^3. This gives you the biovolume of the bacteria.
c. Multiply by 0.31 pg to estimate biomass.

D. Heterotrophic Bacterial Productivity

The methods for estimating heterotrophic bacterial productivity in
aquatic ecosystems have been developed more recently than those for
measuring productivity of other organisms such as algae, aquatic vascular
plants, and macroanimals. Early methods focused on the measurement of
nucleotide ([^{3}H]thymidine) incorporation into bacterial DNA during DNA
synthesis as an estimate of the rate of bacterial biomass production (e.g.,
Fuhrman and Azam 1982). These techniques were initially developed for
ocean/estuarine plankton systems, but have been adapted to lotic ecosystems, including benthic habitats (Findlay *et al.* 1984, Stock and Ward 1989,
Hudson *et al.* 1990, Kaplan *et al.* 1992). More recently, the use of amino
acid ([^{3}H]leucine) incorporation into bacterial protein has also gained ac-
ceptance as an estimate of bacterial productivity (Kirchman *et al.* 1985,
Wetzel and Likens 1991). [^{3}H]leucine was chosen over other amino acids
because the leucine content of bacterial protein remained more constant
than other amino acids. The results from the bacterial protein production
(BPP) method have shown good correspondence with results from
[^{3}H]thymidine studies (Kirchman and Hoch 1988), and the technique is a
more direct and less complicated procedure than the [^{3}H]thymidine method
for estimating bacterial productivity.

The development of a technique for estimating heterotrophic bacterial
productivity has been valuable to the field of aquatic ecology for several
reasons. It provides a unit of measurement that can be compared with

TABLE 12.2
Bacteria Biomass Calculation Worksheet

Bacteria type	(a) No. bacteria counted	×	(b) Microscope correction factor	/	(c) No. of fields	×	(d) Milliliters of sample/2	=	(e) No. of bacteria/ml	×	(f) Biovolume per cell	=	(g) Biomass (pg C/ml)
Cocci													
Rods													
Spirals-vibrio													
Total													

estimates of productivity for organisms at other trophic levels (i.e., C per volume or area per time). Therefore, both the quantitative importance of bacterial productivity in aquatic ecosystems and the roles of bacteria in microbial food webs can now be more thoroughly evaluated (e.g., Findlay *et al.* 1986, Simon and Azam 1989). Despite much progress in this area in recent years, radioisotope methods for estimating bacterial productivity in aquatic ecosystems are not without technical and interpretative problems. Recent reviews (e.g., Robarts and Zohary 1993) and journal articles (e.g., Jørgensen 1992a, b) discuss methods and/or problems associated with non-specific labeling, isotope dilution, and calculation of conversion factors to arrive at units of growth or productivity.

In this exercise, a method for measuring [³H]leucine incorporation into bacterial protein is outlined for estimating bacterial productivity. These procedures were adapted from the literature cited above and from isotope dilution experiments in planktonic and benthic freshwater habitats (Thomaz and Wetzel 1995, Ward unpublished data). Methods are described for both planktonic and benthic habitats. For each habitat, triplicate killed controls and triplicate samples are used. Samples are collected, transferred to large test tubes, and incubated *in situ* for short time intervals (e.g., 30 min). The incubation is terminated by adding formalin, and samples are taken back to the laboratory for further processing.

Treatments consisting of DOC amendments and temperature can easily be incorporated into the exercise. DOC amendments should be made to samples and killed controls prior to addition of the [³H]leucine. Preincubation with DOC can vary from one-half hour to several hours or longer before the addition of [³H]leucine. Temperature effects can be evaluated by transporting samples to the laboratory before addition of the [³H]leucine and incubating samples in temperature controlled waterbaths or growth chambers.

The amount of leucine added in this type of experiment must saturate bacterial uptake kinetics without causing increased metabolism because of increased nutrient availability. Freshwater, planktonic bacterial uptake of leucine typically saturates at a concentration of about 100 nM (Jørgensen 1992a, b). In order to conserve expensive radiolabeled leucine, a mixture of both radiolabeled and nonradiolabeled leucine is used to accomplish this saturation. Sediments and epilithic microhabitats typically have higher saturation requirements than planktonic habitats. We suggest a leucine concentration of 300 nM in the benthic habitat protocol below. The actual concentrations needed to saturate leucine uptake can vary from one system to another.

Hypothesis: *An increased concentration of labile organic carbon will result in increased bacterial productivity.*

General Preparation for Production Laboratory

1. Assemble and label necessary supplies (see supplies list; Section V). Autoclave test tubes, volumetric flasks to dilute radiolabeled and nonradiolabeled leucine, and pipet tips.

2. Arrange field supplies in an appropriate field kit (see Section V).

Production Procedure (Laboratory Preparation)[12]

1a. (Plankton samples) Make 10 μM leucine (nonradioactive) solution by dissolving 0.0013 grams L-leucine in 1 liter of sterile, DI, high-purity water in a volumetric flask.

1b. (Benthic samples) Make 30 μM leucine (nonradioactive) solution by dissolving 0.0039 grams L-leucine in 1 liter of sterile, DI, high-purity water in a volumetric flask.

2. Make [^{3}H]leucine solution by diluting stock solution from manufacturer so that 50 μl of the final, diluted, radiolabeled solution added to a 10-ml sample yields a 10 nM [^{3}H]leucine solution. For example, add 1.0 ml of a [^{3}H]leucine solution of specific activity of 143 Ci/mmol (5 mCi/ml) to 16.5 ml of sterile DI water of high purity.

3. Add 2.5 ml of the [^{3}H]leucine solution from step 2 to 5.0 ml of the unlabeled leucine solution from step 1. Transfer to a sterile, capped container that can be transported to the field site in the field kit.

Production Procedure (Field) Option: If temperature effects are to be investigated, collect samples and transport immediately to the laboratory for the following procedures:

1. Pipet 0.4 ml 100% formalin into killed control tubes.

2a. (Plankton) Pipet 10.0 ml of unfiltered site water into all test tubes (killed controls and samples) with screw tops.

2b. (Benthos) Pipet 10.0 ml of filter-sterilized site water into all test tubes (killed controls and samples), add the benthic sample to the tubes, (e.g., rock chips, portions of macrophytes or leaves, or a measured volume of sediment), and attach screw tops to tubes.

3. Equilibrate all test tubes at *in situ* temperature for 30 min.

4. Unscrew tops.

5. Pipet 150 μl radiolabeled/nonradiolabeled leucine mix solution to all tubes. Record time accurately (digital watch) on data form.[13] It requires about 5 s to pipet the leucine solution into each tube.

---

[12]Use sterile procedures as much as possible (in laminar flow hood, if available).

[13]One investigator can monitor a digital watch, indicate when to pipet, and note the time on the data sheet while another pipets.

6. Replace tops and swirl tubes gently to mix (or vortex in laboratory); place back into sampling habitat or in laboratory incubator for 30 min.

7. At end of incubation, add 0.4 ml 100% formalin to each *sample* tube and mix. Record time accurately on form. Transport to laboratory as soon as possible.

Production Procedure (Laboratory)

1. (Benthic sample only) Sonicate all tubes for 5 min in a bath sonicator.

2. Add 3.25 ml of 15% TCA (trichloroacetic acid) to all tubes and vortex carefully.

3. Place tubes in a 95°C water bath for 30 min.

4. Remove from water bath and cool at room temperature for approximately 20 min.

5a. (Plankton) Filter the entire sample onto a 25-mm-diameter, polycarbonate filter of 0.2 μm pore size. Use a filtration pressure of no more than 150 mm Hg.

5b. (Benthos) Filter appropriate aliquots depending upon sample type (e.g., 0.5 ml for sediments).

6. After filtration of the sample, but while the filter is still on the filtration frit, rinse each filter at least seven times with 5-ml aliquots of DI water to remove as much abiotically adsorbed radiolabel as possible.

7. Remove each filter from the filtration apparatus and place in a labeled scintillation vial with 10 ml of an appropriate scintillation cocktail (e.g., AquaSol-2). Samples should be radioassayed by liquid scintillation to yield DPM/vial.

Production Calculations

1. Subtract average DPM of killed control tubes from average DPM of sample tubes.

2a. Use the following formula to calculate mol leucine$_{inc}$ liter^{-1} h^{-1} for planktonic samples:

$$\frac{(dpm_{sample} - dpm_{killed})(ml/liter)(min/h)(\text{formalin addition factor, 1.03})}{(dpm/Ci)(\text{specific activity in Ci/mmol})(\text{vol filtered, ml})(\text{incubation time, in min})(mmol/mol)}, \quad (12.3)$$

where dpm/Ci $= 2.2 \times 10^{12}$, and mmol/mol $= 1000$.

2b. Use the following formula to calculate mol leucine$_{inc}$ cm^{-2} hr^{-1} for benthic and surface samples:

$$\frac{(dpm_{sample} - dpm_{killed})(min/h)(sample + TCA, 13.25\ ml)}{(dpm/Ci)(specific\ activity\ in\ Ci/mmol)(surface,\ cm^2)} \cdot \frac{(formalin\ factor,\ 1.03)}{(incubation\ time,\ in\ min)(mmol/mol)(aliquot)}, \quad (12.4)$$

where aliquot = ml subsample filtered of 13.25 ml in the incubation tube.

3. In addition to having DPM measurements, the other factors used in the equations below need to be considered. Determine BPP by multiplying the moles of exogenous leucine incorporated by 100/mol% of leucine in protein (7.3), and by the gram molecular weight of leucine (131.2). This value is multiplied by an intracellular isotope dilution (ID) of leucine if one is known. A typical ID factor of 2 (Wetzel and Likens 1991) is based on an intracellular isotope dilution of 50%, determined from studies of marine bacterioplankton (Simon and Azam 1989). However, Jørgensen (1992a) concluded that freshwater bacteria have more variable isotope dilutions than marine bacteria, ranging from 30 to 90%, which would result in ID factors of 3.3 to 1.1. We include a factor of 2 in the calculations below as representative of a reasonable middle value. External dilution (ED) is entered into the calculations to account for the ratio of labeled to unlabeled leucine. ED is composed of two parts: the dilution caused by the ambient amount of leucine in the water and the dilution caused by the amount of unlabeled leucine added with the [^{3}H]leucine. In the planktonic procedure described above a 10 nM [^{3}H]leucine concentration and a 100 nM leucine concentration are added to ambient concentrations in the streamwater, resulting in a dilution factor of 11. In the benthic procedure described above, a 10 nM [^{3}H]leucine concentration and a 300 nM leucine concentration are added to ambient concentrations in the streamwater, resulting in a dilution factor of 31. The amount of leucine normally found in freshwater systems is quite low, normally less than 5 nM and often less than 1 nM. If the ambient concentration is known, this factor can also be added. We do not include the naturally occurring, ambient, leucine ED in the calculations below.[14] The complete formula is

$$BPP(g) = (mol\ leucine_{inc})(100/mol\%\ leucine)(leucine\ mw)(ID)(ED), \quad (12.5)$$

[14]The formulas here are derived from Simon and Azam (1989). See that reference for further details and rationale for the method.

or simplified,

$$BPP(g) = (mol\ leucine_{inc})(100/7.3)(131.2)(ID)(ED), \qquad (12.6)$$

or further simplified,

$$BPP(g) = (mol\ leucine_{inc})(1797)(2)(ED). \qquad (12.7)$$

5. Convert BPP to bacterial carbon production by multiplying BPP by 0.86.

6. The resulting value is bacterial carbon production per unit volume or area per hour.

IV. QUESTIONS

A. Changes in Oxygen as a Measure of Metabolic Response

1. How do you think the lack of water current in the bottles could have affected your measurements of oxygen changes? How could stream flow be more accurately simulated and incorporated into the procedure of this exercise? How do you think enclosures affect experimental results?

2. If you used light bottles, instead of, or in addition to, dark bottles, what additional information would you have gained in this procedure?

3. If you had removed metazoa from your samples, e.g., by passing the sample through a 63 or 100 μm mesh, how might that have affected your results?

4. Do you think the *quality* of leachate generated by the procedure described in this chapter differs substantially from dissolved organic carbon measurable in the stream water? If so, how do you think it is different and why?

5. What problems arise from the assumption of simple carbohydrate being the form of carbon used by the organisms for their respiration?

B. Protist Abundances and Biomass

6. In assessing total microbial response to an experimental treatment such as DOC amendments, why could measuring only bacterial abundances or community respiration give an inaccurate evaluation?

7. If bacterial abundance was similar between a high-DOC and a low-DOC treatment, what would a generally higher abundance of heterotrophic nanoflagellates and ciliates in the high-DOC treatment suggest?

8. In the live-count procedure described for protist quantification, how do you think accuracy would be affected by counting more slowly or more quickly? Why?

C. Bacterial Abundances and Biomass

9. Why do many types of bacteria found in natural environments generally not culture well on standard bacterial media (e.g., peptone, beef extract, general nutrient agar)?

10. Although it is generally agreed that direct counts are more accurate than plate counts for enumeration of bacteria from natural habitats, how could direct counts overestimate bacterial abundances?

11. If one group of investigators count and measure the bacteria in the DOC amendment treatment, and another group of investigators count and measure the bacteria in the "control" treatment, what are possible reasons for variation between treatments?

12. Would you expect bacterial abundances and/or biomass to correlate positively with DOC concentrations? Why or why not?

13. Would you expect bacterial abundances and/or biomass to correlate positively or negatively with protist abundances/biomass? Why?

14. What could be some possible reasons for the low percentage of metabolically active (respiring) bacteria in many natural environments as compared to laboratory cultures?

D. Bacterial Production

15. Typically, starved bacteria divide repeatedly without growing. What effect would this have on interpreting results from the [^{3}H]leucine incorporation into protein method and the [^{3}H]thymidine incorporation into DNA?

16. In assessing overall ecological importance of bacteria, what are the advantages of knowing rates of bacterial production in addition to abundance and/or biomass?

17. Some anaerobic bacteria do not take up appreciable amounts of dissolved thymidine or leucine. If your sample substrata contain anaerobic microzones, how would this affect your interpretation of the results?

18. In the experiments described in this exercise, leucine concentrations are increased to a value that theoretically is in the saturating range for leucine uptake by native bacterial communities. The assumption is that measurable, natural leucine concentrations are low (e.g., 5 nM or less). How would substantially higher, natural concentrations of leucine affect the calculations?

19. Would you expect that heterotrophic bacterial productivity would always be positively correlated with bacterial abundances and/or biomass? Why or why not?

V. MATERIALS AND SUPPLIES

Changes in Oxygen as a Measure of Metabolic Response

10- or 25-ml Burette
25-ml Volumetric pipets
50-ml Erlenmeyer flasks
60-ml DO bottles
100-ml Pipets
300-ml Dark BOD bottles
Beakers
Concentrated sulfuric acid
In each of two dark-glass dropper bottles:

    1. 400 g $MnSO_4 \cdot 2H_2O$ in 1 liter distilled water
    2. 960 ml of [500 g NaOH and 150 g KI in 1 liter H_2O] plus 10 g NaN_3 in 40 ml distilled water

Starch–glycerin solution of Van Landingham (1960)
    3 g powdered soluble starch in 100 ml of glycerin; heat to 190°C
Stir bar
Suction bulbs
Titrant: 50 ml of 0.1 M $Na_2S_2O_3$ in 950 ml distilled water

Protist Abundances and Biomass

10-ml Plastic syringe with tip cut off
1000-μl Pipet (Rainin type)
A flat piece of plastic ($\sim$10 cm^2)
Eppendorf type 1.5- to 2-ml snap-top tubes
Palmer–Maloney counting chamber (available from Wildco)
Phase contrast microscope with stage manipulator
Slides, coverslips, Vaseline, and a toothpick
Small jars ($\sim$40 ml)
Trimmed pipet tips

Bacterial Abundances and Biomass

0.2-μm Polycarbonate filters
0.45-μm Cellulose acetate filters
500-ml Flasks
DAPI (4',6'-diamidino-2-phenylindole)
Epifluorescence microscope equipped with: (1) oil immersion lens
    $\geq$1000$\times$ total magnification, (2) UV light source and filter set, (3)
    and ocular counting grid and measuring reticule
Filter tower and flask or filter manifold
Jars ($\sim$40 ml)
Kim-Wipes
Low-fluorescence immersion oil
Microscope slides and coverslips
Phosphate-buffered formalin
    For a 0.1 *M* phosphate buffered 5% formalin:
        57.7 ml 1 *M* K_2HPO_4 at pH 7.0
        42.3 ml 1 *M* KH_2PO_4 at pH 7.0
        Dilute to 1 liter with filtered DI water
        Remove 50 ml, add 50 ml formaldehyde
Pipets
Toothbrush
Vacuum pump
For INT option:
    2-(*p*-Iodophenyl)-3-(*p*-nitrophenyl)-5-phenyl   tetrazolium   chloride (Sigma)
    30% Glycerol
    Dry ice or metal bar and liquid nitrogen

Heterotrophic Bacterial Production

[^{3}H]leucine ($\sim$150 Ci/mmol)
0.2-μm Polycarbonate filters
15% Trichloroacetic acid (TCA)
Aquasol-2
Autopipetter(s) and tips
Disposable 0.2-μm filter unit (field)
Filter tower(s) or manifold
Formalin
Hand pump (field)
Large (65 ml) screw-top test tubes

Liquid scintillation counter
Scintillation vials
Sonicator
Test tube rack(s)
Vacuum pump
Vortex
Water bath and thermometer

REFERENCES

Aumen, N. G., P. J. Bottomley, G. M. Ward, and S. V. Gregory. 1983. Microbial decomposition of wood in streams: Distribution of microflora and factors affecting ^{14}C-lignocellulose mineralization. *Applied and Environmental Microbiology* **46:**1409–1416.

Baldock, B. M., J. H. Baker, and M. A. Sleigh. 1983. Abundance and productivity of protozoa in chalk streams. *Holarctic Ecology* **6:**238–246.

Baldock, B. M., and M. A. Sleigh. 1988. The ecology of benthic protozoa in rivers: seasonal variation in numerical abundance in fine sediments. *Archiv für Hydrobiologie* **111:**409–421.

Barcina, I., B. Ayo, A. Muela, L. Egea, and J. Iriberri. 1991. Predation rates of flagellate and ciliated protozoa on bacterioplankton in a river. *FEMS Microbiology Ecology* **85:**141–150.

Boon, P. I. 1991. Bacterial assemblages in rivers and billabongs of Southeastern Australia. *Microbial Ecology* **22:**27–52.

Bott, T. L., J. T. Brock, C. E. Cushing, S. V. Gregory, D. King, and R. C. Petersen. 1978. A comparison of methods for measuring primary productivity and community respiration in streams. *Hydrobiologia* **60:**3–12.

Bott, T. L., J. T. Brock, C. S. Dunn, R. J. Naiman, R. W. Ovink, and R. C. Petersen. 1985. Benthic community metabolism in four temperate stream systems: An inter-biome comparison and evaluation of the river continuum concept. *Hydrobiologia* **123:**3–45.

Bott, T. L., and L. A. Kaplan. 1985. Bacterial biomass, metabolic state, and activity in stream sediments: Relation to environmental variables and multiple assay comparisons. *Applied and Environmental Microbiology* **50:**508–522.

Bott, T. L., and L. A. Kaplan. 1989. Densities of benthic protozoa and nematodes in a Piedmont stream. *Journal of the North American Benthological Society* **8:**187–196.

Bott, T. L., L. A. Kaplan, and F. T. Kuserk. 1984. Benthic biomass supported by streamwater dissolved organic matter. *Microbial Ecology* **10:**335–344.

Carlough, L. A., and J. L. Meyer. 1990. Rates of protozoan bacterivory in three habitats of a southeastern blackwater river. *Journal of the North American Benthological Society* **9:**45–53.

Carlough, L. A., and J. L. Meyer. 1991. Bacterivory by sestonic protists in a southeastern blackwater river. *Limnology and Oceanography* **36:**873–883.

Cummins, K. W., M. J. Klug, R. G. Wetzel, R. C. Petersen, K. F. Suberkropp, B. A. Manny, J. C. Wuycheck, and F. O. Howard. 1972. Organic enrichment with leaf leachate in experimental lotic ecosystems. *BioScience* **22:**719–722.

Dahm, C. N. 1981. Pathways and mechanisms for removal of dissolved organic carbon from leaf leachate in streams. *Canadian Journal of Fisheries and Aquatic Sciences* **38:**68–76.

Findlay, S. G., L. A. Carlough, M. T. Crocker, H. K. Gill, J. L. Meyer, and P. J. Smith. 1986. Bacterial growth on macrophyte leachate and fate of bacterial production. *Limnology and Oceanography* **31:**1335–1341.

Findlay, S. G., J. L. Meyer, and R. T. Edwards. 1984. Measuring bacterial production via rate of incorporation of (H^3)thymidine into DNA. *Journal of Microbiological Methods* **2:**57–72.

Findlay, S. G., D. Strayer, C. Goumbala, and K. Gould. 1993. Metabolism of streamwater dissolved organic carbon in the shallow hyporheic zone. *Limnology and Oceanography* **38:**1493–1499.

Fry, J. C. 1990. Direct methods and biomass estimation. Pages 41–85 *in* R. Grigorova and J. R. Norris (Eds.) *Methods in Microbiology,* Vol. 22. Academic Press, London, UK.

Fuhrman, J. A., and F. Azam. 1982. Thymidine incorporation as a measure of heterotrophic bacterioplankton production in marine surface water: Evaluation and field results. *Marine Biology* **66:**109–120.

Gasol, J. M. 1993. Benthic flagellates and ciliates in fine freshwater sediments: calibration of a live counting procedure and estimation of their abundances. *Microbial Ecology* **25:**247–262.

Haack, T. K., and G. A. McFeters. 1982. Microbial dynamics of an epilithic community in a high alpine stream. *Applied and Environmental Microbiology* **43:**702–707.

Harmsworth, G. C., M. A. Sleigh, and J. H. Baker. 1992. The abundance of different peritrich ciliates on stone surfaces in contrasting lowland streams throughout the year. *Journal of Protozoology* **39:**58–65.

Hedin, L. O. 1990. Factors controlling sediment community respiration in woodland stream ecosystems. *Oikos* **57:**94–105.

Herbert, R. A. 1990. Methods for enumerating microorganisms and determining biomass in natural environments. Pages 1–39 *in* R. Grigorova and J. R. Norris (Eds.) *Methods in Microbiology,* Vol. 22. Academic Press, London, UK.

Hoch, M. P., and D. L. Kirchman. 1993. Seasonal and inter-annual variability in bacterial production and biomass in a temperate estuary. *Marine Ecology Progress Series* **98:**283–295.

Hudson, J. J., J. C. Roff, and B. K. Burnison. 1990. Measuring epilithic bacterial production in streams. *Canadian Journal of Fisheries and Aquatic Sciences* **47:**1813–1820.

Johnson, M. D. 1995. *Ecosystem Dynamics of Bacteria and Protists in a Freshwater Wetland.* Ph.D. dissertation, University of Alabama, Tuscaloosa, AL.

Johnson, M. D., and A. K. Ward. 1993. A comparison of INT-formazan methods for determining bacterial activity in stream ecosystems. *Journal of the North American Benthological Society* **12:**168–173.

Jørgensen, N. O. G. 1992a. Incorporation of ^{3}H-leucine and ^{3}H-valine into protein of freshwater bacteria: Uptake kinetics and intracellular isotope dilution. *Applied and Environmental Microbiology* **58:**3638–3646.

Jørgensen, N. O. G. 1992b. Incorporation of ^{3}H-leucine and ^{3}H-valine into protein of freshwater bacteria: Field applications. *Applied and Environmental Microbiology* **58:**3647–3653.

Kaplan, L. A., T. L. Bott, and J. K. Bielicki. 1992. Assessment of [^{3}H]thymidine incorporation into DNA as a method to determine bacterial productivity in stream bed sediments. *Applied and Environmental Microbiology* **58:**3614–3621.

Kemp, P. F., B. F. Sherr, E. B. Sherr, and J. J. Cole (Eds.). 1993. *Handbook of Methods in Aquatic Microbial Ecology*. Lewis Publishers, Boca Raton, FL.

Kirchman, D. L. 1993. Statistical analysis of direct counts of microbial abundance. Pages 117–119 *in* P. F. Kemp, B. F. Sherr, E. B. Sherr, and J. J. Cole (Eds.) *Handbook of Methods in Aquatic Microbial Ecology*. Lewis Publishers, Boca Raton, FL.

Kirchman, D. L., and M. P. Hoch. 1988. Bacterial production in the Delaware Bay estuary estimated from thymidine and leucine incorporation rates. *Marine Ecology Progress Series* **45:**169–178.

Kirchman, D. L., E. K'Nees, and R. Hodson. 1985. Leucine incorporation and its potential as a measure of protein synthesis by bacteria in natural aquatic systems. *Applied and Environmental Microbiology* **49:**599–607.

Lock, M. A. 1993. Attached microbial communities in rivers. Pages 113–138 *in* T. E. Ford (Ed.) *Aquatic Microbiology: An Ecological Approach*. Blackwell Scientific, Oxford, UK.

Lock, M. A., and H. B. N. Hynes. 1975. The disappearance of four leaf leachates in a hard and soft water stream in South Western Ontario, Canada. *International Revue gestamen Hydrobiologie* **60:**847–855.

Lock, M. A., and H. B. N. Hynes. 1976. The fate of "dissolved" organic carbon derived from autumn-shed maple leaves (Acer saccharum) in a temperate hardwater stream. *Limnology and Oceanography* **21:**436–443.

Lush, D. L., and H. B. N. Hynes. 1978. The uptake of dissolved organic matter by a small spring stream. *Hydrobiologia* **60:**271–275.

Margulis, L., J. Corliss, M. Melkonian, and D. Chapman (Eds.). 1989. *Handbook of the Protoctista. The Structure, Cultivation, Habitat and Life Histories of the Eukaryotic Microorganisms and Their Descendants Exclusive of Animals, Plants, and Fungi*. Jones and Bartlett, Boston, MA.

Marxsen, J. 1988. Investigations into the number of respiring bacteria in groundwater from sandy and gravelly deposits. *Microbial Ecology* **16:**65–72.

McDowell, W. H. 1985. Kinetics and mechanisms of dissolved organic carbon retention in a headwater stream. *Biogeochemistry* **1:**329–352.

McKnight, D. M., R. L. Smith, R. A. Harnish, C. L. Miller, and K. E. Bencala. 1993. Seasonal relationships between planktonic microorganisms and dissolved organic material in an alpine stream. *Biogeochemistry* **21:**39–59.

Meyer, J. L. 1990. A blackwater perspective on riverine ecosystems. *BioScience* **40**:643–651.

Naiman, R. J. 1983. The annual pattern and spatial distribution of aquatic oxygen metabolism in boreal forest watersheds. *Ecological Monographs* **53**:73–94.

Page, F. C. 1988. *A New Key to Freshwater and Soil Gymnamoebae.* Freshwater Biological Association, Cumbria, UK.

Patterson, D. J., and S. Hedley. 1992. *Free-Living Freshwater Protozoa: A Color Guide.* Wolfe Publishing and CRC Press, Boca Raton, FL.

Robarts, R. D., and T. Zohary. 1993. Fact or fiction—Bacterial growth rates and production as determined by methyl-^{3}H-thymidine? Pages 371–425 *in* J. G. Jones (Ed.) *Advances in Microbial Ecology,* Vol. 13. Plenum, New York, NY.

Sanders, R. W. 1991a. Mixotrophic protists in marine and freshwater ecosystems. *Journal of Protozoology* **38**:76–81.

Sanders, R. W. 1991b. Trophic strategies among heterotrophic flagellates. Pages 21–38 *in* D. J. Patterson and J. Larsen (Eds.) *The Biology of the Free-Living Heterotrophic Flagellates.* Systematics Association Special Vol. 45. Clarendon Press, Oxford, UK.

Sanders, R. W., and K. G. Porter. 1988. Phagotrophic phytoflagellates. Pages 167–192 *in* K. C. Marshall (Ed.) *Advances in Microbial Ecology,* Vol. 10. Plenum, New York, NY.

Sherr, E. B., D. A. Caron, and B. F. Sherr. 1993. Staining of heterotrophic protists for visualization via epifluorescent microscopy. Pages 213–227 *in* P. F. Kemp, B. F. Sherr, E. B. Sherr, and J. J. Cole (Eds.) *Handbook of Methods in Aquatic Microbial Ecology.* Lewis Publishers, Boca Raton, FL.

Shiah, F. K., and H. W. Ducklow. 1994. Temperature and substrate regulation of bacterial abundance, production and specific growth rate in Chesapeake Bay, USA. *Marine Ecology Progress Series* **103**:297–308.

Simon, M., and F. Azam. 1989. Protein content and protein synthesis rates of planktonic marine bacteria. *Marine Ecology Progress Series* **51**:201–213.

Sladecek, V. 1973. System of water quality from the biological point of view. *Archiv für Hydrobiologie Beihefte Ergebnisse der Limnologie* **7**:1–218.

Sleigh, M. 1989. *Protozoa and Other Protists.* Hodder & Stoughton, London, UK.

Stock, M. S., and A. K. Ward. 1989. Establishment of a bedrock epilithic community in a small stream: microbial (algal and bacterial) metabolism and physical structure. *Canadian Journal of Fisheries and Aquatic Sciences* **46**:1874–1883.

Thomaz, S. M., and R. G. Wetzel. 1995. [^{3}H]Leucine incorporation methodology to estimate epiphytic bacterial biomass production. *Microbial Ecology* **29**:63–70.

Van Landingham, J. W. 1960. A note on a stabilized starch indicator for use in iodometric and iodimetric determinations. *Limnology and Oceanography* **5**:343–345.

Vaques, D., M. L. Pace, S. G. Findlay, and D. Lints. 1993. Fate of bacterial production in a heterotrophic ecosystem: Grazing by protists and metazoans in the Hudson estuary. *Marine Ecology Progress Series* **89**:155–163.

Wetzel, R. G., and G. E. Likens. 1991. *Limnological Analyses,* 2nd ed. Springer-Verlag, New York, NY.

White, P. A., J. Kalff, J. B. Rasmussen, and J. M. Gasol. 1991. The effect of temperature and algal biomass on bacterial production and specific growth rate in freshwater and marine habitats. *Microbial Ecology* **212:**99–118.

Zimmermann, R., Iturriaga, R., and J. Becker-Birck. 1978. Simultaneous determination of the total number of aquatic bacteria and the number thereof involved in respiration. *Applied and Environmental Microbiology* **36:**926–935.

CHAPTER 13

Benthic Stream Algae: Distribution and Structure

REX L. LOWE AND GINA D. LALIBERTE

Department of Biological Sciences
Bowling Green State University

I. INTRODUCTION

The algae are an ubiquitous group of photosynthetic organisms responsible for the majority of photosynthesis in most sunlit streams. Benthic algae are dominant members of the periphyton community and live on submerged substrata in the photic zone of most aquatic ecosystems including both marine environments and freshwater.

Although algae are usually studied as a group of organisms that are closely related at a functional level, they are very diverse on an evolutionary level and have been assigned to several different kingdoms based on morphological, chemical, and ecological parameters. Some algae are closely related to bacteria (blue–green algae) while others are more similar to animals (dinoflagellates and chrysophytes). Only one group of algae are true plants in the evolutionary sense (green algae), but, because algae possess chlorophyll *a* and liberate oxygen during photosynthesis, they have traditionally been studied by botanists. This chapter discusses the collection and identification of benthic algae from lotic habitats and the investigation of microhabitat fidelity among the algae.

An alga (singular) or algae (plural) can be defined as thalloid organisms, bearing chlorophyll *a* and lacking multicellular gametangia, and their colorless relatives. This definition excludes higher plants (macrophytes) such as *Potamogeton, Lemna,* and other flowering aquatic plants that have well developed tissue and organ systems. The definition also excludes aquatic bryophytes (mosses and liverworts), which have multicellular gametangia, and aquatic fungi, which have no chlorophyll. Since the group of organisms classified as algae is ecologically closely related but evolutionarily distantly related, algae are a taxonomically "unnatural" group of organisms. Thus, some algae such as the blue–green algae are also bacteria (Cyanobacteria). Some algae such as dinoflagellates are also Protozoa and most algae are also considered to be Protista. This lack of taxonomic cohesiveness is often frustrating to aquatic researchers who prefer clean unambiguous systems of classification. Still, algae share many physiological, morphological, and ecological features that make them a logical group of aquatic organisms to investigate as a community.

A. Taxonomic Classification

Algae can be classified in several different ways. We focus here on two different classification schemes: taxonomic classification and ecological classification. Algae belong to at least 10 different taxonomic divisions; divisions in plant taxonomy are equivalent to phyla in zoological classification. Classification is based on four major considerations: (1) pigmentation (kinds and quantities), (2) internal storage products (chemistry and structure), (3) cell wall (chemistry and structure), and (4) flagellation (number and type). All of the above taxonomic parameters are considered to be evolutionarily conservative and thus are good tools for recognizing divisions of taxonomically related algae. Taxonomic differences among the five major divisions of algae common in periphyton communities of freshwater streams are presented in Table 13.1.

Diatoms (Bacillariophyta) are probably the most widespread and abundant of all divisions of benthic algae. They are recognizable in the field where they form buff to brown to gold colored films on submerged objects. Under microscopic examination, the two cell walls of a diatom (called valves) fit together like the halves of a petri dish to form the frustule, or siliceous cell wall. The valves are penetrated by many pits and pores (punctae) that are symmetrically arranged in rows (striae). Many diatoms possess a slit through the cell wall (raphe) that allows them to be motile (Fig. 13.1 in Appendix 13.1). Diatoms are classified to genus primarily on the basis of symmetry and cell ornamentation.

Green algae (Chlorophyta) in the periphyton are usually filamentous with subspherical to cylindrical cells attached end to end. In the field, green algae have a color similar to green terrestrial plants. Filaments of green algae may be branched or unbranched. Occasionally, nonfilamentous cells

TABLE 13.1

Patterns of Pigment Content, Cell Wall Chemistry, Storage Chemistry, and Flagellation among the Divisions of Algae Most Commonly Encountered in Freshwater Periphyton

Division	Pigmentation	Cell wall	Storage products	Flagella
Bacillariophyta (diatoms)	Chlorophylls a and c but with carotenoid pigments dominant. Cells usually gold to brown in color	Mostly SiO_2 composed of two overlapping halves	Oil and leucosin	Absent vegetatively
Chlorophyta (green algae)	Chlorophylls a and b dominant	Cellulose and pectin	Plant starch	Usually 2–4 of equal length when present
Cyanophyta (blue–green algae)	Chlorophyll a and phycobilins, blue green to olive green in color	Peptidoglycan, gram negative	Glycogen-like	Absent
Chrysophyta (yellow green algae)	Chlorophylls a and c yellow green in color	Pectin and cellulose	Oil and leucosin	Absent vegetatively
Rhodophyta (red algae)	Chlorophyll a and phycoerythrin, olive green to maroon in color	Mannans and xylans (slimy)	Glycogen-like	Absent

or colonies of green algae also may be present in periphyton communities but are rarely dominant (Fig. 13.2 in Appendix 13.1).

Blue-green algae (Cyanophyta) are also called Cyanobacteria because of their prokaryotic nature. They may appear as mats of olive green to blue–green to brown growth and often have a characteristic musty odor. Filamentous forms are most common in benthic habitats.

Yellow-green algae (Chrysophyta) are represented by only a few genera in the periphyton (*Vaucheria, Hydrurus,* and *Tribonema*). All three of these genera are filamentous and may become locally abundant.

Red algae (Rhodophyta) are most abundant in oceans but several filamentous genera occur in freshwater. They usually appear olive green to maroon in color.

B. Ecological Classification

Algae can colonize almost any submerged substratum and can be classified according to the microhabitat that they occupy. *Epilithon* is the name given to benthic algae growing on rocks, *epiphyton* grows on plants (including filamentous algae), *epidendron* grows on wood, *epipelon* grows on fine sediment, *epipsammon* grows on sand, and *epizoon* grows on aquatic animals. Finally, algae that are only loosely associated with the substrata, such as cloudy masses of *Spirogyra,* are called *metaphyton*. It should be noted that substrata availability and abundance is a function of other stream variables such as current velocity, catchment geology, and the nature of terrestrial riparian vegetation. Within each microhabitat present in a stream, there is a considerable amount of algal microhabitat specificity as specialists have evolved to occupy specific microhabitats. These microhabitat specialists are often the same across similar streams within a region. For example, epipelic microhabitats are usually dominated by highly motile diatoms capable of moving over and between fine particles of sediment. On the other hand, epilithon is usually dominated by firmly attached diatoms or green filamentous algae.

C. Physiognomy

Algal community physiognomy addresses the physical or architectural structure of the community. Benthic algae can develop a complex physical structure similar to that of a terrestrial forest (Hoagland *et al.* 1982) except on a much smaller scale (Fig. 13.3 in Appendix 13.1). Different algal growth forms include: (1) nonmotile and prostrate, (2) attached by a mucilaginous pad at one end, (3) attached to the end of long mucilaginous stalks, or (4) filamentous. Some species are motile and travel throughout the structured community. Benthic algal communities on different substrata develop different physiognomies, which can affect interspecific interactions among the algae. Physiognomy also can impact the nature of interactions between benthic algae and stream invertebrates (Steinman *et al.* 1992).

D. Roles of Benthic Algae in Stream Communities

Benthic algae play several roles of fundamental importance to stream ecosystems. As organisms at the base of the food web, they are at the interface of the physical–chemical environment and the biological community. Photosynthesis by benthic algae provides oxygen for aerobic organisms in the ecosystem and the fixed carbon provides food for algivores. In many stream habitats, the contribution of organic carbon to the food web from algal photosynthesis is considerable (Lamberti 1996, Steinman 1995). Benthic algae may enter the food web through direct consumption from the substrata by benthic invertebrates such as snails or insects (Barnese *et al.* 1990, Steinman *et al.* 1987) or through capture of drifting benthic algae by filter-feeders that strain the water column (Hauer and Stanford 1981, Barnese and Lowe 1992).

A second role of benthic algae that stems from their position at the interface of abiotic and biotic stream components is their utility as water quality indicators. Benthic algae possess many attributes that make them ideal organisms to employ in water quality monitoring investigations. Because benthic algae are sessile they cannot swim away from potential pollutants. They must either tolerate their surrounding abiotic environment or die. Benthic algal communities are usually species-rich and each species, of course, has its own set of environmental tolerances and preferences (Lowe 1974, Beaver 1981). Thus, the entire assemblage represents an information-rich system for environmental monitoring. The short life cycles of most benthic algal species result in a rapid response to shifts in environmental conditions. Extant benthic algal communities are typically very representative of current environmental conditions. Identification is not exceedingly difficult. Taxonomy of benthic algae is usually based on cell or thallus morphology easily discernible through the light microscope and excellent taxonomic keys exist for identification of benthic algae in most parts of the world.

It is the objective of this exercise to introduce stream benthic algae to the investigator who has little previous experience with algae. You will learn how to recognize periphyton growth in the field in a variety of microhabitats. You will also learn how to identify dominant algal taxa under the microscope. And finally, you will learn how to estimate algal density on a substratum and calculate community parameters based on two measures of algal abundance: cells/unit area and biovolume/unit area.

II. GENERAL DESIGN

A. Site selection

As mentioned above, benthic algae grow on submerged substrata that receive ample sunlight. In large, relatively clear rivers this may include substrata several meters deep. For the exercises in this chapter, however, we will focus on wadeable stream sections. Try to select a stream reach that has a variety of substratum types to maximize the types of benthic algae that you will see. Ideally, the stream reach would include examples of epilithic, epidendric, epipelic, epiphytic, epipsammic, and (if you're fortunate) epizoic habitat. Stream reaches including all of these substratum types are unusual, but streams generally contain from one to four microhabitats.

B. Field Sampling

Epilithic habitats are usually found in sections of the stream that experience relatively fast current. Thus, epilithic algae are often tightly attached to the substratum. This necessitates scraping the rock substratum with a knife, scalpel, or similar tool. If the water current is extremely swift, the stone should be removed from the stream before scraping so that scraped algae are not washed away. If the stone is too large to remove from the stream, the benthic algae can be scraped from the stone under water and captured in a small plankton net as algae are washed downstream.

Epidendric habitats may take the form of woody debris from riparian vegetation or from submerged woody tissue of living vegetation, such as alder or willow trees, or from bank-exposed roots. The best method for collecting this community is similar to methods for collecting epilithon.

Epipelic algae are often motile and only loosely associated with the substrata. Since epipelic habitats occur in areas of little or no current where fine sediment can accumulate, one need not worry about the benthic algae washing away while collecting. Epipelon is best collected with a turkey baster or with a pipet and rubber bulb. Care must be taken not to penetrate the sediments too deeply. Most of the benthic algae will be on top of the sediment or within the first millimeter. Collection of deeper layers will obscure live cells during microscopic analysis. This community can be collected quantitatively by defining the area on the sediment to be collected. A rubber hose washer or O-ring works nicely to isolate the sediment surface area for quantitative collections.

Epiphytic algae are usually tightly attached to their plant hosts (filamentous algae or aquatic macrophytes). A small portion of the plant host should be placed into a bottle with streamwater, but leaving air space in the top of the bottle. The bottle should then be shaken vigorously to remove the epiphytes from the plant host. The host can then be removed from the bottle, wrung out, and discarded. This procedure will leave some tightly attached epiphytes remaining on the plant host. These can be observed directly by mounting a portion of the plant epidermis on a microscope slide and studying the epiphytes while they are still attached to the host.

As with epiphyton, in sampling *epipsammic* algae a small quantity of the sand substrate should be agitated in a bottle containing water. The sand will quickly fall to the bottom of the bottle after agitation and the suspended algae can be decanted into another container. As with plant epidermis, sand grains are usually transparent enough to be observed directly under the microscope. This technique allows the investigator to study the microdistribution of algae on individual sand grains.

Epizoic algae are most likely to be collected from larger animals with a rigid covering such as snails, clams, and turtles. Algae should be scraped from these organisms as if they were epilithic habitats. Smaller animals such as midges (Pringle 1985) and caddis flies (Bergey and Resh 1994) can also host unique epizoic algal floras. These habitats can be investigated by carefully collecting the host and observing sections of its case directly under the microscope.

C. Preservation and Labeling

Samples should be transferred into plastic vials or bottles which are labeled externally with a waterproof pen. It is a good idea to place a small label into the collection vial as well in case samples are later transferred to a different container. The label should include information about the date, stream, microhabitat, and surface area sampled for quantitative samples. Both formaldehyde and glutaraldehyde at a final concentration of 3 to 5% will preserve samples well for later examination. In addition to labeling samples directly, a field notebook should also be annotated with details about each collection identified by collection number. A workable system for numbering samples is the month-day-year-collection number method. For example, sample 12-28-95-7 would be the seventh sample collected on December 28th of 1995. Information entered into the notebook should include field observations such as water temperature, pH, depth, substratum type, color of growth collected, and any other information that seems pertinent to the collection.

D. Laboratory Processing

Cleaning and Mounting of Diatoms[1]   The algal samples can be examined directly in a wet mount with a compound microscope but in order to

---

[1]This procedure should be performed in a fume hood or out of doors. Hydrogen peroxide is a strong oxidant! The investigator should wear eye protection, a laboratory coat, and protective gloves.

identify diatoms to genus they must first be "cleaned" (after Van Der Werff 1955). This process involves oxidizing the organic matter in the sample so that just the silica cell walls of diatoms remain. The empty valves and frustules are then mounted in a mounting medium of high refractive index such as Naphrax. A detailed step-by-step procedure for cleaning and mounting diatoms is presented here.

1. Place a small amount of the sample to be cleaned in a 1000-ml beaker. Five or 10 ml of sample is usually adequate.

2. Add about 80 ml of 30% H_2O_2 and allow the mixture to stand for 24 h. If necessary you can proceed to step 3 without waiting 24 h, but oxidation of organic matter may be incomplete.

3. Add a microspatula of $K_2Cr_2O_7$ to the mixture under a fume hood. This will initiate a violent exothermic reaction. At the completion of the reaction, 5 to 10 min, the solution will change from purple to a golden color.

4. Transfer the solution to a tall 200-ml beaker and add distilled water until the beaker is full.

5. Allow the mixture to stand for a minimum of 4 h as the cleaned diatoms settle to the bottom. Decant the mixture, carefully removing and discarding the liquid[2] but being careful not to disturb the diatom–cell sediment on the bottom of the beaker. About 30 ml of the mixture should remain in the beaker. Refill the beaker with distilled water.

6. Repeat step 5 until the mixture is colorless.

7. Pipet a portion of the 30-ml concentrate onto an alcohol-cleaned cover glass and air dry.

8. Place a drop of mounting medium on a clean microscope slide and invert the cover glass on the drop.

9. Heat the slide on the "high" setting on a laboratory hot plate for about 30 s or until the bubbling of the medium slows.

10. Remove the slide and by applying gentle pressure with forceps force air bubbles from beneath the cover glass.

11. Allow the slide to cool and label with collection number.

Semipermanent Mounts of Soft Algae   Permanent slides of "soft" algae, as nondiatom algae are often called, also can be made quite easily and inexpensively but it is not necessary for soft algal identification. An easy and inexpensive technique for preparing semipermanent slides of soft algae is the modification by Stevenson (1984) of Taft's (1978) glucose mounts, presented here.

---

[2]Be certain to follow institution guidelines for waste disposal.

1. Make the following Stock Solutions:

   a. 4% Formaldehyde
   b. 100% Taft's glucose: (7 parts 4% formaldehyde solution + 3 parts light Karo corn syrup)
   c. 10% Taft's glucose: (9 parts H_2O + 1 part 100% Taft's glucose solution)

2. Preserve material in approximately 2% glutaraldehyde fixative.[3]

3. Place material with fixative onto a coverslip. Add 10% Taft's glucose solution. (material + fixative: 10% Taft's approximately 1:1.)

4. Allow material to dry to tackiness.

5. Place a drop of 100% Taft's glucose onto a microscope slide. Invert the dried material on coverslip onto slide.

6. Lightly press coverslip to evenly spread the mountant under coverslip.

7. Allow the material to harden. If the glucose solution pulls away from the edge of coverslip add additional 100% Taft's glucose solution.

8. Ring and seal with fingernail polish along all four edges of coverslip.

Collections of algae from the field should be mounted in a wet mount on a microscope slide, covered with a cover glass and examined on a compound microscope. It is best to first examine the slide with relatively low magnification (about 100×) to get comfortable with the range of taxa and types of morphologies present. After a few minutes, switch to higher magnification (430–450×) and continue to examine the collection. When you are comfortable with the microscope and have some good representative specimens, begin to key out some of the most common taxa. A simplified key to genera of stream algae is presented in Appendix 13.1. This key does not include all of the genera found worldwide but instead focuses on dominant genera of rivers in North America. As you work through the key, the first few times take notes on which dichotomy you choose. You may have to backtrack if you go the wrong way in the key. If you have chosen a rare or uncommon alga to key out, it may not be in this key and you will reach a dead end. Several more comprehensive keys are available for more in-depth investigations into riverine algae (Appendix 13.2).

Benthic algae can be enumerated in a quantitative fashion that allows the estimation of cells of taxa per cm^2 of substrata. Field colleciton techniques are the same as detailed above except that the surface area sampled is measured and recorded. This is most easily done on flat surfaces of stones

---

[3]Glutaraldehyde is a strong fixative with little noticeable odor; use it carefully and keep it away from your own cells.

where a section (2 to 10 cm^2) can be isolated from the rest of the stone with duct tape.

In the laboratory, the sample volume is measured and a subsample of the suspension is pipetted into a Palmer–Maloney nanoplankton counting chamber.[4] This is a device that is 0.4 mm deep and holds exactly 0.1 ml of suspension. The volume of sample contained in a single microscope field can be calculated by the formula for the volume of a cylinder,

$$V_f = \pi r^2 d, \qquad (13.1)$$

where V_f represents volume of one microscope field in the counting chamber, r, radius of the microscope field; and d, depth of the counting chamber (0.4 mm).

The number of fields examined in the counting chamber to quantify benthic algae is a function of the density of algae in the sample but it is customary to examine enough fields to enumerate 300–500 algal cells. The density of any given alga is calculated by the formula

$$D_i = N_i V_s / V_c / A, \qquad (13.2)$$

where D_i is density of cells of the ith taxon; N_i, number of cells of the ith taxon counted in the counting chamber; V_s, total volume of the sample; V_c, volume of the sample counted [(fields counted)(V_f)]; and A, area in cm^2 of substratum sampled.

The total density of benthic algae on the substratum can be calculated by summing D_i for all taxa encountered, 1 through n. One can also report algal densities on the basis of biovolume/cm^2. This may be a more appropriate measure of the success of an algal population than cell number since a single cell of one species may be over 1000 times larger than the cell of a second species. Biovolumes can be calculated simply by measuring cells with a calibrated ocular micrometer and using appropriate geometric formulae to calculate their volumes. Most cells can be viewed as cylinders, cones, spheres, or elongated cubes for purposes of biovolume estimates (Wetzel and Likens 1991). Several cells (5–10) from each population should be measured to obtain an average biovolume for each algal taxon of interest. The density of an alga based on biovolume is the product of its biovolume per cell and its cellular density.

---

[4]Palmer–Maloney counting chambers are relatively expensive and it is possible to make your own chamber by pounding a clean hole through a plastic coverslip with a cork borer and cementing the coverslip to a microscope slide. The depth of the well inside the cored cover slip can be calculated with precision calipers.

III. SPECIFIC EXERCISES

A. Exercise 1: Investigation of Algal Microhabitats

As discussed above, algae occupy a variety of microhabitats within stream systems. In this exercise microhabitats will be sampled qualitatively to examine the role of habitat variables in the distribution of algal taxa.

1. Select a stream reach that provides a variety of substratum types and a range of physical variables such as light quantity and current speed. A variety of current speeds can be obtained, for example, by sampling different sides of a boulder. Different light intensities can be obtained by sampling inside or outside shady patches resulting from stream-side vegetation. Select five to eight different microhabitats from which to collect benthic algae. Collect and preserve benthic algal samples in the field while taking detailed notes concerning each microhabitat. Remember, this is a microbial community and microhabitats are often quite small. Each collection should be limited to a few cm^2.

2. Upon returning to the laboratory examine the collections from each microhabitat to make certain that you have collected a healthy benthic algal community. If specimens are mostly dead or if the sample contains too much debris to view algae clearly, discard the sample. Bad collections can be avoided by closely examining the microhabitats and collecting carefully. If time permits, make permanent diatom mounts and semipermanent glucose mounts from each collection.

3. Identify the five most numerically abundant algal genera in each sample and record them on the Benthic Algal Survey Sheet (Table 13.2).

B. Exercise 2: Analyses of Algal Density: Cellular vs Volumetric Analyses

It is often useful to determine the absolute abundance of algal taxa on a specific substratum. In this exercise, epilithic benthic algal communities from two contrasting current speeds will be compared quantitatively.

1. In the field, select pool and riffle habitats that are representative of the stream. Carefully remove three stones from the pool and three stones from the riffle for quantitative benthic algal analysis. Try to select flat-topped stones whose surfaces lie parallel to the stream bed. On the stream bank, isolate 4 cm^2 of each of the stone surfaces by overlapping four strips of duct tape at right angles leaving a 4-cm^2 opening in the center. Brush and scrape benthic algae from each isolated surface into a collecting basin using a razor blade and a firm-bristle toothbrush. Flush the sample into a bottle and preserve it. This procedure will result in three replicate samples from each pool and riffle habitat.

TABLE 13.2
Sample Data Sheet: Investigations of Algal Microhabitats

Stream _____ Date _____ Location _____ Investigators _____

Microhabitat type	Notes about microhabitat: current?, color of growth?, sunny or shaded?	Dominant algal genera
1		1a. _____
		1b. _____
		1c. _____
		1d. _____
		1e. _____
2		2a. _____
		2b. _____
		2c. _____
		2d. _____
		2e. _____
3		3a. _____
		3b. _____
		3c. _____
		3d. _____
		3e. _____
4		4a. _____
		4b. _____
		4c. _____
		4d. _____
		4e. _____
5		5a. _____
		5b. _____
		5c. _____
		5d. _____
		5e. _____

2. In the laboratory bring all samples to an equal volume (between 20 and 50 ml). Transfer a subsample of the sample into a counting chamber and examine enough fields to identify 300 algae to genera. A bench sheet (Table 13.3) is provided for data entry. Process each sample in this manner, with the results of individual enumerations averaged for each habitat.

3. Calculate the number of cells/cm^2 of the five most abundant algal taxa. Determine the average biovolume/cell of the abundant taxa and calculate biovolume/cm^2 of each of the same taxa. Compare differences in domi-

TABLE 13.3
Analyses of Algal Density; Cellular vs Volumetric Analyses

Stream _____ Date _____ Location _____ Investigators _____

Microhabitat type	Dominant algal taxa	Cells/cm^2	Biovolume/cm^2
Riffle, stone 1	R1a. _____	_____	_____
	R1b. _____	_____	_____
	R1c. _____	_____	_____
	R1d. _____	_____	_____
	R1e. _____	_____	_____
Riffle, stone 2	R2a. _____	_____	_____
	R2b. _____	_____	_____
	R2c. _____	_____	_____
	R2d. _____	_____	_____
	R2e. _____	_____	_____
Riffle, stone 3	R3a. _____	_____	_____
	R3b. _____	_____	_____
	R3c. _____	_____	_____
	R3d. _____	_____	_____
	R3e. _____	_____	_____
Pool, stone 1	P1a. _____	_____	_____
	P1b. _____	_____	_____
	P1c. _____	_____	_____
	P1d. _____	_____	_____
	P1e. _____	_____	_____
Pool, stone 2	P2a. _____	_____	_____
	P2b. _____	_____	_____
	P2c. _____	_____	_____
	P2d. _____	_____	_____
	P2e. _____	_____	_____
Pool, stone 3	P3a. _____	_____	_____
	P3b. _____	_____	_____
	P3c. _____	_____	_____
	P3d. _____	_____	_____
	P3e. _____	_____	_____

nant taxa between habitats based on cell number. Compare differences in dominant taxa between habitats based on cell biovolume.

IV. QUESTIONS

1. How does current speed affect the availability of substratum types available for benthic algal colonization?

2. Which microhabitat has the most filamentous green algae? Why?

3. Where are motile (raphe-bearing) diatoms most abundant? Why would motility have an adaptive advantage in this microhabitat? Where would motility be a disadvantage?

4. Do different divisions of algae seem to do better in well illuminated or shaded microhabitats? Why?

5. How do you think grazers affect the patterns of algal distribution that you found?

6. Do different algal genera dominate pools and riffles?

7. Are algal densities of pools different than densities of riffles? Why?

8. What is the relationship between the size and the numerical abundance of an algal taxon?

9. What do you feel is a better measure of the success of an algal taxon, cells/cm^2 or biovolume/cm^2? Why?

V. MATERIALS AND SUPPLIES

Field

10 Collection bottles (10- to 50-ml, plastic preferable)
Knife scalpel, or double-edged razor blades
Plankton net
Toothbrush
Turkey baster or medicine dropper
Waterproof field notebook and pencil

Laboratory

30% Hydrogen peroxide
Compound light microscope
Corn syrup
Formaldehyde
Glass beakers (1000 and 250 ml tall)
Glutaraldehyde

Granular potassium dichromate
Hot plate
Nail polish
Naphrax mounting medium
Slides and cover glasses

APPENDIX 13.1: ILLUSTRATED KEY TO THE MOST COMMON LOTIC ALGAL GENERA

1a. Pigments in cells localized in chloroplasts and not diffused throughout the cell; pigments usually some shade of green or gold, 4
1b. Pigments diffuse in the cells and not localized in chloroplasts; cells often very small and olive green to blue–green to brownish in color (blue–green algae), 2

2a. Filamentous, 3
2b. Not filamentous, *Chroococcus, Microcystis, Aphanothece, Merismopedia,* or closely related genus (Fig. 4)

3a. Filaments composed of cells that all look alike, *Oscillatoria, Phormidium, Schizothrix, Lyngbya,* or related genera (Fig. 5)
3b. Filaments with some cells that look different (heterocysts), *Anabaena, Nostoc, Tolypothrix, Calothrix,* or related genera (Fig. 6)

4a. Chloroplasts grass green, alga may be filamentous, single-celled or colonial, 5
4b. Chloroplasts not grass green but yellow green, olive green, gold, brown, or pink; may be filamentous, single-celled, or colonial, 23

5a. Alga filamentous, 6
5b. Alga unicellular or colonial (unusual in lotic periphyton and not considered in this key; see Prescott (1962, 1964) for more details).

6a. Large macroscopic "plant-like" alga often several decimeters long; branches whorled at nodes, *Chara* or *Nitella* (Fig. 7)
6b. Thallus usually much smaller and lacking whorled branches, 7

7a. A true filament or pseudofilament; multicellular linear arrangement, branched or unbranched, 8
7b. A long branching unicellular tube without cross walls (coenocytic); cell wall may be occasionally constricted, *Vaucheria* (Fig. 8)

8a. Filament branched, 9
8b. Filament not branched, 16

9a. Growing on the back of a snapping turtle!; branching only near the base, *Basicladia* (Fig. 9)
9b. Not growing on the back of a snapping turtle, 10

10a. Short prostrate filament growing epiphytically on a larger filament; bristle-like setae arising from some of the cells, *Aphanochaete* (Fig. 10)

10b. Not as above, 11

11a. Filament in tough mucilage, often forming macroscopic hemispherical green growths difficult to mash in wet mount; tips of filaments tapered to points or hair-like bristles, *Chaetophora* (Fig. 11)

11b. Not in a tough mucilage, 12

12a. Some cells bearing bristles, setae, or filaments tapering to fine points, 13

12b. No setae present; ends of filaments rounded, 15

13a. Setae with large bulbous bases, *Bulbochaete* (Fig. 12)

13b. Filaments without bulbed setae, 14

14a. All cells of the filament about the same diameter, highly branched, *Stigeo-clonium* (Fig. 13)

14b. Central axis of large cylindrical cells with cells of branches much smaller in diameter, *Draparnaldia* (Fig. 14)

15a. Sparsely branched cells many times longer than broad, *Rhizoclonium* (Fig. 15)

15b. Profusely branched and bushy; often covered with epiphytes, *Cladophora* (Fig. 16)

16a. Cells slightly or grossly constricted in the middle, 17

16b. Cells of filament not constricted, 18

17a. Constriction very slight; cylindrical cell often with broad sheath, *Hyalotheca* (Fig. 17)

17b. Constriction deeper or cells not cylindrical, *Desmidium, Spondylosium,* or *Bambusina* (Figs. 18–20)

18a. H pieces present (filaments fragment in the center of cells rather than between cells), *Tribonema* or *Microspora* (Figs. 21,22)

18b. H pieces absent, 19

19a. Chloroplast parietal, with a large water-filled vacuole in the center of the cell, 20

19b. Chloroplast axial; a plate or star-shaped, 22

20a. Chloroplast a parietal spiral, *Spirogyra* (Fig. 23)

20b. Chloroplast not a parietal spiral, 21

21a. Chloroplast a parietal reticulum, cells sometimes slightly club-shaped, *Oedo-gonium* (Fig. 24)

21b. Chloroplast a parietal bracelet, *Ulothrix* (Fig. 25)

22a. Chloroplast an axial plate, *Mougeotia* (Fig. 26)

22b. Two axial star-shaped chloroplasts per cell, *Zygnema* (Fig. 27)

23a. Alga filamentous; chloroplasts pink, maroon, or sometimes olive green–red alga, *Audouinella* or *Batrachospermum* (Figs. 28,29)

23b. Alga filamentous, colonial or single-celled. Chloroplast yellow green, gold, or brown, probably a diatom, 24

24a. Alga single-celled or colonial, 26
24b. Alga filamentous, cells arranged end to end, 25

25a. Cells box-like and touching to form filaments or ribbons, *Fragilaria, Aulacoseira,* or *Melosira* (Figs. 30–32).
25b. Cells more loosely arranged; alga usually in mountain streams, *Hydrurus* (Fig. 33)

26a. Cells usually in a colony or mucilaginous tube, 27
26b. Cells usually single or in groups at the end of stalks, 29

27a. Colony in the form of a zig-zag chain, *Tabellaria* or *Diatoma* (Figs. 34,35)
27b. Colony otherwise; arranged in a tube or fan-shaped, 28

28a. Colony "fan-shaped" or sometimes forming a tight circle of cells; common in cold water, *Meridion* (Fig. 36)
28b. Colony not fan-shaped; cells aggregated within a tube of mucilage, *Cymbella, Nitzschia,* or *Navicula* (Figs. 37–39)

29a. Cells attached on one end by a stalk or pad; may occur singly or in groups, 30
29b. Cells not attached at one end, 32

30a. Cells wedge-shaped or club-shaped (one end wider than the other) *Gomphonema, Gomphoneis, Rhoicosphenia,* or *Meridion* (Figs. 40–43)
30b. Cells rectangular or bent in girdle view (both ends of equal width), 31

31a. Cell rectangular in girdle view; often long and narrow, *Synedra* (Fig. 44)
31b. Cell a bent rectangle in girdle view; occurring most often in turbulent water, *Achnanthidium* (Fig. 45)

At this point it is recommended that cleaned and mounted specimens be observed. The remainder of the key beginning at couplet 32 assumes that you will observe frustular details from permanently mounted specimens employing an oil immersion objective at 1000X magnification.

32a. Cells circular in outline, *Cyclotella* or *Stephanodiscus* (Figs. 46,47)
32b. Cells not circular in outline, 33

33a. Cells symmetric to both apical and transapical axes, 38
33b. Cells asymmetric to one or both axes, 34

34a. Cells sigmoid (S-shaped), *Gyrosigma* or *Nitzschia* (Figs. 48,49)
34b. Cells not sigmoid; either lunate or club-shaped, 35

35a. Cells club-shaped, 37
35b. Cells lunate, 36

36a. Cells with a clear axial area containing the raphe, *Cymbella, Amphora,* or *Reimeria* (Figs. 50–52).

36b. Cells lacking a clear axial area containing the raphe, *Eunotia, Hannaea, Epithemia,* or *Rhopalodia* (Figs. 53–56)

37a. Cells with a raphe on one or both valves, *Rhoicosphenia, Gomphonema,* or *Gomphoneis* (Figs. 40–42)

37b. Cells lacking a raphe, *Meridion* or *Martyana* (Figs. 43,57)

38a. Cells with a raphe in the center of both valves. Many genera fit this description including the common stream genera *Navicula, Stauroneis, Neidium, Sellaphora, Pinnularia, Craticula, Anomoeoneis, Brachysira,* and *Diploneis* (Figs. 58–66)

38b. Cells with a raphe lacking on one or both valves or with raphe along the margin of the cell, 39

39a. Raphe present on one valve only; cells often tightly attached to the substratum, *Cocconeis* and *Achnanthidium* (Figs. 67,68)

39b. Raphe lacking or if present difficult to see along the margin of the valve, 40

40a. Raphe lacking-*Tabellaria, Diatoma* or *Synedra* (Figs. 69–71)

40b. Raphe present along the margin of the cell and difficult to see, *Surirella, Nitzschia,* and *Cymatopleura* (Figs. 72–74)

APPENDIX 13.1

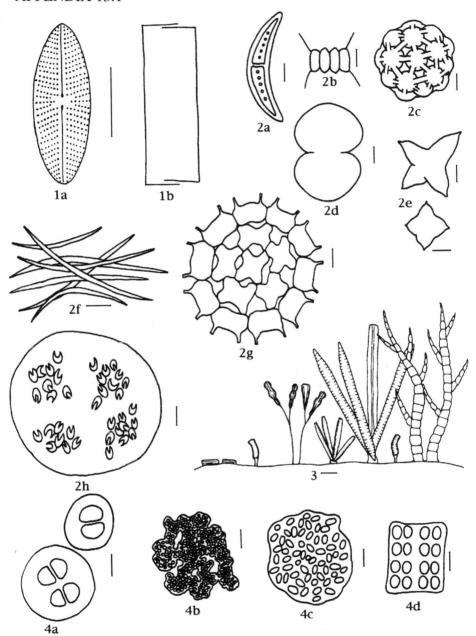

APPENDIX 13.1—*continued*

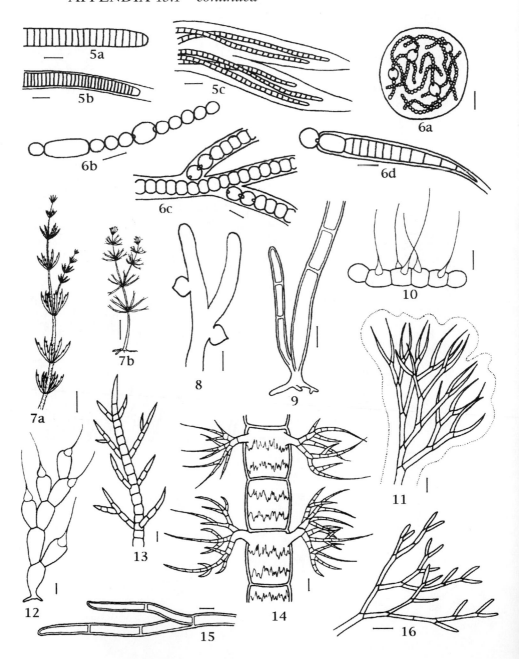

APPENDIX 13.1—*continued*

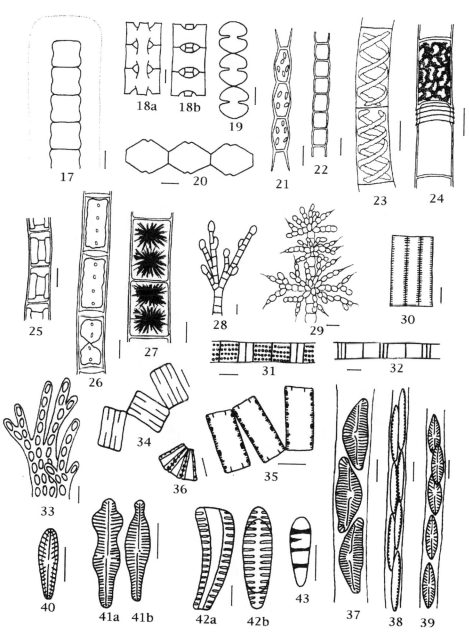

APPENDIX 13.1—*continued*

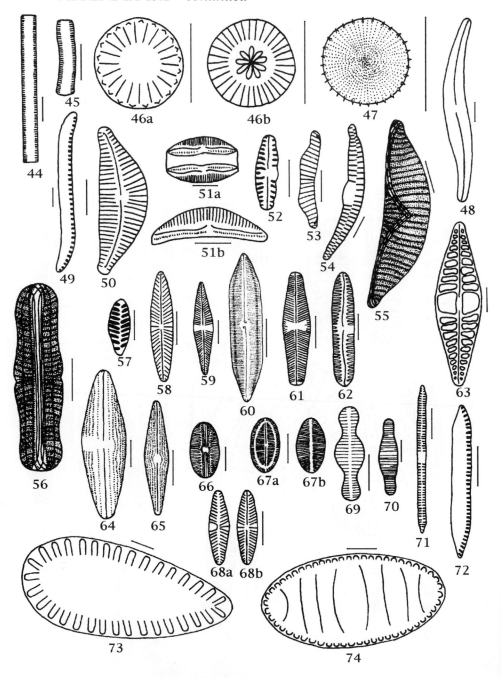

APPENDIX 13.1 LEGENDS: FIGURES OF BENTHIC ALGAL GENERA (ALL SCALE BARS = 10 μm UNLESS LABELED OTHERWISE)

Figure 13.1 Diatom frustule in valve (a) and girdle (b) views. Figure 13.2 Nonfilamentous green algae usually uncommon in stream periphyton: (a) *Closterium* (scale bars = 100 μm), (b) *Scenedesmus*, (c) *Coelastrum*, (d) *Cosmarium*, (e) *Tetraëdron*, (f) *Ankistrodesmus*, (g) *Pediastrum*, (h) *Kirchneriella*. Figure 13.3 Diagram of periphyton physiognomy with diatoms (left) and a filamentous green alga (extreme right). Figure 13.4 Coccoid blue–green algae. (a) *Chroococcus*, (b) *Microcystis*, (c) *Aphanothece*, (d) *Merismopedia*. Figure 13.5 Filamentous blue green algae lacking specialized cells: (a) *Oscillatoria*, (b) *Lyngbya*, (c) *Schizothrix*. Figure 13.6 Filamentous blue green algae with specialized cells: (a) *Nostoc*, (b) *Anabaena*, (c) *Tolypothrix*, (d) *Calothrix*. Figure 13.7 Macroscopic "plant-like" green algae. (a) *Chara*, (b) *Nitella*. Scale bar = 1 cm. Figure 13.8 *Vaucheria*, a siphonaceous green Chrysophyte. Scale bar = 100 μm. Figure 13.9 *Basicladia*. Figure 13.10 *Aphanochaete*. Figure 13.11 *Chaetophora*. Figure 13.12 *Bulbochaete*. Figure 13.13 *Stigeoclonium*. Figure 13.14 *Draparnaldia*. Figure 13.15 *Rhizoclonium*. Figure 13.16 *Cladophora*. Figure 13.17 *Hyalotheca*. Figure 13.18 *Desmidium* from front (a) and back (b) views. Figure 13.19 *Spondylosium*. Figure 13.20 *Bambusina*. Figure 13.21 *Tribonema*. Figure 13.22 *Microspora*. Figure 13.23 *Spirogyra*. Figure 13.24 *Oedogonium*, details of parietal reticulate chloroplast illustrated in upper cell; cells separated by apical caps. Figure 13.25 *Ulothrix*. Figure 13.26 *Mougeotia*. Figure 13.27 *Zygnema*. Figure 13.28 *Audouinella*. Figure 13.29 *Batrachospermum*. Figure 13.30 *Fragilaria*. Figure 13.31 *Aulacoseira*. Figure 13.32 *Melosira*. Figure 13.33 *Hydrurus*. Figure 13.34 *Tabellaria* colony. Figure 13.35 *Diatoma* colony. Figure 13.36 *Meridion* colony. Figure 13.37 *Cymbella* colony. Figure 13.38 *Nitzschia* colony. Figure 13.39 *Navicula* colony. Figure 13.40 *Gomphoneis*. Figure 13.41 *Gomphonema:* (a) *G. acuminatum*, (b) *G. parvulum*. Figure 13.42 *Rhoicosphenia:* (a) girdle view, (b) valve view. Figure 13.43 *Meridion*, valve view. Figure 13.44 *Synedra*, girdle view. Figure 13.45 *Achnanthidium*, girdle view. Figure 13.46 *Cyclotella:* (a) *C. meneghiniana*, (b) *C. stelligera*. Figure 13.47 *Stephanodiscus*. Figure 13.48 *Gyrosigma*. Figure 13.49 *Nitzschia*. Figure 13.50 *Cymbella*. Figure 13.51 *Amphora:* (a) girdle view, (b) valve view. Figure 13.52 *Reimeria*. Figure 13.53 *Eunotia*. Figure 13.54 *Hannaea*. Figure 13.55 *Epithemia*. Figure 13.56 *Rhopalodia*. Figure 13.57 *Martyana*. Figure 13.58 *Navicula*. Figure 13.59 *Stauroneis*. Figure 13.60 *Neidium*. Figure 13.61 *Sellaphora*. Figure 13.62 *Pinnularia*. Figure 13.63 *Craticula*. Figure 13.64 *Anomoeoneis*. Figure 13.65 *Brachysira*. Figure 13.66 *Diploneis*. Figure 13.67 *Cocconeis:* (a) raphe valve, (b) rapheless valve. Figure 13.68 *Achnanthidium:* (a) rapheless valve, (b) raphe valve. Figure 13.69 *Tabellaria*. Figure 13.70 *Diatoma*. Figure 13.71 *Synedra*. Figure 13.72 *Nitzschia*. Figure 13.73 *Surirella*. Figure 13.74 *Cymatopleura*.

APPENDIX 13.2: DETAILED TAXONOMIC REFERENCES FOR IDENTIFICATION OF BENTHIC STREAM ALGAE

Bourrelly, P. 1966. *Les Algues D'eau Douce*, Vol. 1, Les Algues Vertes. N. Boubée & Cie, Paris, France.

Bourrelly, P. 1968. *Les Algues D'eau Douce*, Vol. 2, Les Algues jaunes et brunes. N. Boubée & Cie, Paris, France.

Bourrelly, P. 1970. *Les Algues D'eau Douce,* Vol. 2, Les Algues bleues et rouges. N. Boubée & Cie, Paris, France.

Dillard, G. E. 1989. Freshwater algae of the Southeastern United States. 1. Chlorophyceae: Volvocales, Tetrasporales, and Chlorococcales. *Bibliotheca Phycologica* **81**:1–278.

Dillard, G. E. 1989. Freshwater algae of the Southeastern United States. 2. Chlorophyceae: Ulothrichales, Microsporales, Cylindrocapsales, Sphaeropleales, Chaetophorales, Cladophorales, Schizogoniales, Siphonales, and Oedogoniales. *Bibliotheca Phycologica* **83**:1–248.

Dillard, G. E. 1990. Freshwater algae of the Southeastern United States. 3. Chlorophyceae: Zygnematales: Zygnemataceae, Mesotaeniaceae, and Desmidiaceae. *Bibliotheca Phycologica* **85**:1–276.

Germain H. 1981. *Flore des Diatomées.* Diatomophcyées eaux douces et saumâtres du Massif Armoricain et des contrées voisines d'Europe occidentale. Boubée, Paris, France.

Hustedt, F. 1930. Bacillariophyta (Diatomeae). Pages 1–466 *in* A. Pascher (Ed.) *Die Süsswasser-Flora Mitteleuropas,* Heft 10. Gustav Fisher, Jena, Germany.

Patrick, R., and C. W. Reimer. 1966. *The Diatoms of the United States Exclusive of Alaska and Hawaii,* Vol. 1, Monograph 13. Academy of Natural Science, Philadelphia, PA.

Patrick, R., and C. W. Reimer, 1975. *The Diatoms of the United States Exclusive of Alaska and Hawaii,* Vol. 2, Part 1, Monograph 13. Academy of Natural Science, Philadelphia, PA.

Prescott, G. W. 1962. *Algae of the Western Great Lakes Area.* Brown, Dubuque, IA.

Prescott, G. W. 1964. *How to Know the Freshwater Algae.* Brown, Dubuque, IA.

Prescott, G. W., H. T. Croasdale, and W. C. Vinyard. 1972. *Desmidiales. I. Saccodermae, Mesotaeniaceae.* North American Flora Series II (6).

Prescott, G. W., H. T. Croasdale, and W. C. Vinyard. 1975. *A Synopsis of North American Desmids. II. Desmidiaceae Placodermae,* Section 1. University of Nebraska Press, Lincoln, NE.

Prescott, G. W., H. T. Croasdale, and W. C. Vinyard. 1977. *A Synopsis of North American Desmids. II. Desmidiaceae Placodermae,* Section 2. University of Nebraska Press, Lincoln, NE.

Prescott, G. W., H. T. Croasdale, W. C. Vinyard, and C. E. de M. Bicudo. 1981. *A Synopsis of North American Desmids. II. Desmidiaceae Placodermae,* Section 3. University of Nebraska Press, Lincoln, NE.

REFERENCES

Barnese, L. E., R. L. Lowe, and R. D. Hunter. 1990. Comparative grazing efficiency of six species of sympatric snails in Douglas Lake, Michigan. *Journal of the North American Benthological Society* **9**:35–44.

Barnese, L. E., and R. L. Lowe. 1992. Effects of substrate, light and benthic invertebrates on algal drift in small streams. *Journal of the North American Benthological Society* **11**:49–59.

Beaver, J. 1981. *Apparent Ecological Characteristics of Some Common Freshwater Diatoms.* Ontario Ministry of the Environment, Technical Support Section, Don Mills, Ontario, Canada.

Bergey, E. A., and V. H. Resh. 1994. Effects of burrowing by a stream caddisfly on case-associated algae. *Journal of the North American Benthological Society* **13:**379–390.

Hauer, F. R. and J. A. Stanford. 1981. Larval specialization and phenotypic variation in *Arctopsyche grandis* (Trichoptera: Hydropsychidae). *Ecology* **62:**645–653.

Hoagland, K. D., S. C. Roemer, and J. R. Rosowski. 1982. Colonization and community structure of two periphyton assemblages with emphasis on the diatoms (Bacillariophyceae). *American Journal of Botany* **62:**188–213.

Lamberti, G. A. 1996. The role of periphyton in benthic food webs. Pages 533–572 *in* R. J. Stevenson, M. L. Bothwell, and R. L. Lowe (Eds.) *Algal Ecology: Freshwater Benthic Ecosystems.* Academic Press, San Diego, CA.

Lowe, R. L. 1974. *Environmental Requirements and Pollution Tolerance of Freshwater Diatoms.* Report EPA-670/4-74-007. United States Environmental Protection Agency, Environmental Monitoring and Support Laboratory, Cincinnati, OH.

Pringle, C. M. 1985. Effects of chironomid (Insecta: Diptera) tube-building activities on stream diatom communities. *Journal of Phycology* **21:**185–194.

Steinman, A. D. 1996. Effects of grazers on freshwater benthic algae. Pages 341–373 *in* R. J. Stevenson, M. L. Bothwell and R. L. Lowe (Eds.) *Algal Ecology: Freshwater Benthic Ecosystems.* Academic Press, San Diego, CA.

Steinman, A. D., C. D. McIntire, S. V. Gregory, G. A. Lamberti, and L. R. Ashkenas. 1987. Effects of herbivore type and density on taxonomic structure and physiognomy of algal assemblages in laboratory streams. *Journal of the North American Benthological Society* **6:**175–188.

Steinman, A. D., P. J. Mulholland, and W. R. Hill. 1992. Functional responses associated with growth form in stream algae. *Journal of the North American Benthological Society* **11:**229–243.

Stevenson, R. J. 1984. Procedures for mounting algae in syrup medium. *Transactions of the American Microscopical Society* **103:**320–321.

Taft, C. E. 1978. A mounting medium for freshwater plankton. *Transactions of the American Microscopical Society* **97:**263–264.

Van Der Werff, A. 1955. A new method of concentrating and cleaning diatoms and other organisms. *Verhandlungen der Internationalen Vereinigung für Theoretische und Angewandte Limnologie* **12:**276–277.

Wetzel, R. G., and G. E. Likens. 1991. *Limnological Analyses,* 2nd ed. Springer-Verlag, New York, NY.

CHAPTER 14

Biomass and Pigments of Benthic Algae

ALAN D. STEINMAN* AND GARY A. LAMBERTI[†]

*Department of Ecosystem Restoration
South Florida Water Management District
†Department of Biological Sciences
University of Notre Dame

I. INTRODUCTION

Determination of biomass is one of the most fundamental measurements made in ecology. Although it is a static index, it provides information on the relative importance of either a taxonomic unit within a community (which has implications for competitive interactions) or of a community within an ecosystem (which has implications for trophic level interactions). The measurement of benthic algal biomass is complicated by the fact that benthic algae usually exist as part of a complex assemblage variously referred to as periphyton, aufwuchs, or biofilm. This assemblage consists of algae, bacteria, fungi, and microzoans that are held within a polysaccharide matrix secreted by the microorganisms themselves (Lock *et al.*, 1984). Consequently, assessment of algal biomass alone is confounded by the presence of other organic constituents in the benthic assemblage. Numerous techniques have been developed for measuring benthic algal biomass, some of which attempt to account for the presence of nonalgal components, and others that treat the assemblage as a whole. In this chapter, our objectives are threefold: (1) provide a context for the study of benthic algal biomass, (2) discuss in detail two of the more commonly used approaches to measure benthic algal biomass, and (3) describe a field exercise to examine the influence of stream channel morphology on algal biomass, whereby the two approaches can be employed and compared with each other to assess their individual performance.

Two specific questions will be addressed in this chapter: (1) Does benthic algal biomass differ between pools and riffles in streams? (2) Do different approaches for measuring benthic algal biomass produce consistent results? In addition, optional methods allow researchers to determine if benthic algal biomass differs between natural and artificial substrata and whether algal biomass is correlated with specific environmental parameters. Because the drying and extraction times involved in the measurement of biomass and pigments are relatively long, we have designed the laboratory portion of the chapter mindful of the need for two separate sessions.

Although the energy base for stream ecosystems often derives from allochthonous sources (Hynes 1975, Vannote *et al.* 1980), autotrophic production can contribute a substantial fraction of fixed carbon to many streams (Naiman 1976, Minshall 1978, Cushing and Wolf 1984). Indeed, even in streams commonly considered heterotrophic, algal biomass may play a critical role as a food resource for herbivores (Gregory 1980, Mayer and Likens 1987, Steinman 1992). Low algal biomass does not necessarily mean that the algae present are unimportant. The food quality of algae often is much higher than that of detritus, thus representing a more nutritious food resource for invertebrates (Cummins and Klug 1979). Modeling (McIntire 1973) and empirical (Gregory 1980, Sumner and McIntire 1982) studies have both revealed that lotic algal assemblages can support grazer biomasses approximately 10–20 times greater than their own because of high algal turnover rates. Although algal biomass does provide an index of how much food is potentially available to herbivores in a system at a point in time, it provides no information on how large a herbivore population can be supported by that biomass.

Not all algal biomass is consumed by herbivores. Calculations by Lamberti *et al.* (1989) on the fate of algal biomass grown in laboratory streams indicated that when algae are grown under high light conditions, the majority of material is exported from the system and not consumed by snails. Thus, sloughed algal biomass also can serve as a detrital source or be collected by filtering heterotrophs downstream.

Short-term changes in algal biomass can be used as an index of productivity in a system (e.g., Bothwell 1988). However, one must be careful to control for immigration and emigration processes, as well as losses from grazing, which could confound the biomass accrual rate. In general, this calculation must be done with young assemblages (this reduces the chance of major emigration events) and under controlled conditions if the production rate is to have rigor.

Algal biomass levels can change rapidly in streams because of disturbance events (Steinman and McIntire 1990) or rapid growth responses to changing environmental conditions (Sheath *et al.* 1986). Consequently, drawing conclusions about stream conditions based on isolated sampling events of algal biomass can be misleading. Understanding the relationship between algal biomass and the stream ecosystem is best achieved through a systematic sampling regime, where biomass is sampled from the same general location(s) on a weekly or biweekly basis throughout an entire year (or longer, if feasible). In this way, both seasonal factors and chance events can be accounted for, and a more representative picture of algal biomass dynamics will emerge. Often, environmental factors are measured simultaneously with biomass, and correlations will be calculated between biomass and factors such as inorganic nutrient concentration, current velocity, temperature, irradiance, or grazer density (Burkholder and Sheath 1985). Based on the correlation results, hypotheses can then be generated regarding what factor(s) control algal biomass within a stream or at a site, which then can be tested using controlled experiments.

A. Different Approaches to Measure Biomass (General Overviews)

Ash-Free Dry Mass   This gravimetric approach involves drying the collected samples to a constant weight, oxidizing them in a muffle furnace, and reweighing the oxidized samples. The loss in weight upon oxidation is referred to as ash-free dry mass (AFDM). The method has the advantages of requiring only basic laboratory instrumentation and being relatively non-labor-intensive. However, it does not allow the investigator to distinguish algal material from other organic material (e.g., fungi, bacteria) in the sample, nor does it account for the physiological state of the organic material (i.e., it may be senescent). Further, drying by heat may volatilize certain organic compounds and carbonates, leading to an underestimation of true AFDM. If one is interested in estimating algal biomass alone, this method may prove unsatisfactory, especially if there is a large fraction of nonalgal organic material in the sample.

Pigment Analysis   This approach involves extracting pigments from the collected samples and measuring each pigment's absorbance of light at a series of wavelengths. Because different pigments absorb light maximally at different wavelengths, it is possible to estimate the concentrations of different pigments from an individual alga or from an assemblage. Chlorophyll *a* is the most abundant pigment in plants (although not always in algae), and consequently its absorbance is measured most frequently. However, it is also possible to measure the absorbance of chlorophylls *b* and *c*, other accessory pigments, and pheopigments. The main advantages of pigment analysis are: (1) its relative simplicity, (2) its ability to differentiate algal biomass from other organic constituents in the assemblage, and (3) its ability to provide information on algal community structure if an entire suite of pigments is analyzed. Its main disadvantages include: (1) the need for either a spectrophotometer or fluorometer, (2) the amount of chlorophyll in an algal cell can change depending on ambient conditions, and (3) extract concentration can change depending on solvent type and extraction method.

Biovolume Measurement   Measurement of individual algal cell volume provides a direct, but labor-intensive, method for estimating benthic algal biomass. The methodology involves examination of cells under a light microscope, measuring cell dimensions using an ocular micrometer, and calculating cell volumes for each species by applying appropriate geometric formulae to each form (e.g., Wetzel and Likens 1991). This measurement can be included with the measurement of algal community composition, thereby providing information on both algal taxonomic structure and biomass simultaneously. One of the main advantages of this combined approach is its ability to assess the biotic impacts of toxic chemicals or other environmental stressors, which may restructure the community but have no net impact on biomass (Schindler 1987). However, biovolume measurement suffers from the difficulty in distinguishing between live and dead cells when cell counts are being made using a microscope. Usually, only those cells with intact plastids are considered "live." The protocol associated with this procedure is covered in Chapter 13.

B. Overview of Chapter

This chapter will compare algal accumulation on substrata in pools and riffles. Pools and riffles have distinctly different geomorphic and hydrologic characteristics, which influence the character of their biotic components (see also Chapters 1 and 2 for additional information on geomorphology). Hydraulically, pools have a water surface slope of near zero, lack turbulent flow, and can have substantial depth. Riffles have a surface slope of about 1%, contain zones of turbulent, high-velocity flow, and are shallow. The morphology of these features is maintained by the highest flows; sediments are eroded from pools and deposited in riffles during floods. Conversely, pools are depositional and riffles are erosional during low-flow conditions (Morisawa 1968).

The greater velocities and turbulence in riffles results in greater exchange of water (thereby replenishing dissolved nutrients and gases) and thinner boundary layers over substrata than in pools. However, greater current velocities in riffles can result in higher shear stress, which can remove benthic algae from substrata. Although local environmental conditions (e.g., grazer density, differences in canopy) also will influence benthic algal biomass, in this chapter we focus on the hydraulic differences between pools and riffles. As you evaluate the pool and riffle habitats, consider what factors are most different between the two and how those factors might influence algal biomass. Predict where you think algal biomass will be greatest before starting this exercise.

II. GENERAL DESIGN

A. Overall Exercise Design

This exercise is designed to examine potential differences in algal biomass between pools and riffles and to examine potential differences in methods of measuring benthic algal biomass (AFDM vs pigment analysis). Researchers will sample a paired pool/riffle habitat, collect artificial substrata from the pool and riffle, and analyze AFDM and chlorophyll a from each substratum type. A very simple measurement will be made by each team in each pool and riffle sampled: water velocity will be estimated by timing how long it takes a float to travel a known distance. See also Chapters 3 and 4 for additional information on hydrology and hydraulics in streams.

B. Site Selection

The stream selected must have a reasonable number of discrete riffle/pool reaches. In general, wadeable second- or third-order streams are ideal for the exercise. Each team of researchers (two to four individuals) will be assigned one riffle/pool reach, so the total number of riffle/pool reaches needed for the exercise will depend on the number of research teams.

C. Analytical Procedures (Overview)

Substratum Type   Natural substrata vary in size, texture, and origin, thereby introducing an element of variation into the sampling process. As an alternative, investigators sometimes use artificial substrata of known size and texture to reduce substratum variability. Historically, glass slides were used as artificial substrata (e.g., Ivlev 1933, Patrick *et al.* 1954), but some studies have shown that algal communities grown on glass slides do not accurately represent natural communities (Brown 1976, Tuchman and Blinn 1979). As a consequence, other substratum types have been employed. Both Tuchman and Stevenson (1980) and Lamberti and Resh (1985) reported that clay tiles produced more reproducible results than sterilized rocks. In this chapter, benthic algae will be sampled from artificial (clay tile) substrata. In an optional exercise, comparisons of biomass can be made between natural and artificial substrata.

Collection of Algae   The type of device used to remove algae from substrata for biomass determination will depend on the location, size, and texture of the substratum being sampled. Many different types of devices have been designed for collecting benthic algae (see reviews by Sladeckova 1962 and Aloi 1990). In general, if the substratum is too large or too deep to move, then a sampling device similar to, or modified from, the one suggested by Douglas (1958) can be utilized. Recent advances have resulted in a Plexiglass chamber or tube, which is placed over the portion of the substratum that is to be sampled. A neoprene gasket at the base of chamber improves the seal between the chamber and the substratum, preventing leakage of scraped material. A scalpel or brush is placed inside the chamber and used to remove attached algae from the substratum. A syringe is then either placed inside or attached to the chamber and the scraped slurry is sucked out (e.g., Loeb 1981).

In situations where the substratum can be removed from the stream, there is no concern of scraped material being washed downstream. In this case, the entire exposed surface or a specific area of the substratum can be scraped or brushed; in the case of pigments, the entire substratum (if small enough) can be placed in solvent for pigment extraction. For substrata with substantial algal biomass, it is best to remove most material first with a razor blade. This should be done over a small pan or photographic tray, so that none of the loose material is lost. After scraping, the remaining material can be removed by vigorous brushing with a hard-bristled toothbrush. It is important to clean the brush thoroughly before it is used again on a separate substratum. For substrata without much algal biomass, removal can begin with the brush. However, even vigorous brushing is not 100% effective in removing algal biomass (Cattaneo and Roberge 1991). Once a slurry is collected it should be either filtered in the field or placed on ice in the dark until transported back to the laboratory for filtration and processing (see Exercise A below for details on filtration).

Ash-Free Dry Mass   After filtration, the preweighed filter is placed into a numbered (to keep track of each sample) aluminum weigh boat or porcelain crucible and dried to constant weight at 105°C (usually for 24 h). The dried material is weighed to the nearest 0.1 mg, oxidized at 500°C, and reweighed. Some researchers recommend rewetting the oxidized material with reagent-grade water and redrying to constant weight at 105°C. This will reintroduce the water of hydration in clay, which is not lost at 105°C, but is volatilized at 500°C (Nelson and Scott 1962). It has been our experience that this loss is extremely low (<1% of AFDM) and can be ignored for most purposes, but if clay is very abundant in the stream, it may be worthwhile to conduct some pilot studies to determine whether rewetting is necessary. Dry mass (DM) is calculated as the weight of the dried material plus filter (usually in mg) minus the original filter weight, divided by the area of the sampled substratum (usually in cm^2). AFDM is calculated as the weight of the DM minus the residual ash, divided by the area of the sampled substratum (see Data Analysis section for details).

Chlorophyll and Degradation Products There are three different methods for measuring chlorophyll *a* in benthic algae: spectrophotometry, fluorometry, and high-performance liquid chromatography (HPLC). In this chapter, we will describe only the spectrophotometric method, which requires instrumentation that is generally more available than fluorometers or HPLC. Fluorometry is more sensitive than spectrophotometry, requires less material, and can be used for *in vivo* measurements (Lorenzen 1966), which makes it ideal for pigment analysis of phytoplankton but usually unnecessary for benthic algae. Details of the fluorometric method can be found in *Standard Methods* (APHA *et al.* 1992). HPLC is a very sensitive method that can measure a wide spectrum of accessory pigments as well as the chlorophyll degradation products. However, the instrumentation is expensive and it is not standard equipment in many ecology laboratories.

The two most common degradation products of chlorophyll *a* are pheophorbide *a* and pheophytin *a*. All chlorophylls contain a central magnesium ion, which is bonded to four nitrogen atoms in a ring structure (Fig. 14.1). The Mg^{2+} is lost if chlorophyll is exposed to an acidic environment. Pheophytins are created when the magnesium is lost from the structure. Because the above degradation products have been reported to contribute up to 60% of the measured chlorophyll *a* content in freshwater (Marker *et al.* 1980) and can absorb light in the same region of the spectrum as does chlorophyll *a*, their presence can interfere with the estimation of chlorophyll *a* concentration. Thus, pigment analysis requires chlorophyll *a* absorbance to be measured both prior to, and following, acidification, in order to correct for pheophytin that may have been present.

After filtration of the sample, the algae can either be extracted immediately or frozen (in the dark) for analysis the following week. Physical disruption of the assemblage with a tissue grinder or sonicator is often recommended to facilitate extraction, but Axler and Owen (1994) found no statistically significant difference in chlorophyll *a* content of phytoplankton extracted with and without grinding. However, assemblages dominated by

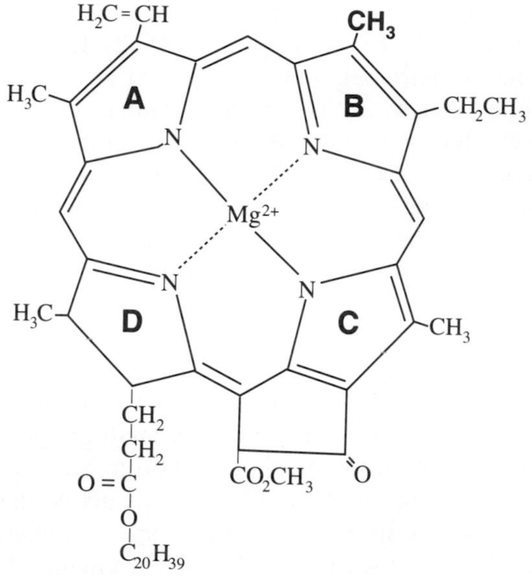

FIGURE 14.1 Structure of chlorophyll *a*, showing the central core of magnesium held in a porphyrin ring. Chlorophyll *b* is identical to chlorophyll *a* except for the substitution of a –CHO group in place of the –CH$_3$ (bold) group in ring B.

cyanobacteria may require grinding (Marker *et al.* 1980). For the purposes of this exercise, we do not recommend grinding unless cyanobacteria are the dominant algal class present in the sample; the potential disadvantages of filters sticking to the bottom of grinding tubes and potential loss of sample during processing argue against grinding under most circumstances (Axler and Owen 1994). The extract is allowed to steep in solvent for a minimum of 2 h and centrifuged gently (if needed) to separate the supernatant (containing chlorophyll) from the pellet (disrupted organic material). The absorbance of a known volume of supernatant is read on a spectrophotometer, acidified with 0.1 *N* HCl (see exact volumes below), and reread on the spectrophotometer (see Section III for details).

D. Optional Exercises

Two optional exercises are suggested if time and interest permit. The first option involves comparing biomass quantity on natural vs artificial substrata. Researchers collect both artificial and natural substrata in each pool/riffle habitat and analyze both AFDM and chlorophyll *a* on both substratum types. *T* tests are conducted at the end to examine if differences exist between artificial and natural substrata for AFDM and chlorophyll in pools and in riffles. The second option is to sample the environmental conditions in the pool and riffle habitat (e.g., incident light level, grazer densities) and correlate these data with the biomass data. It is important to remember that a significant correlation does not mean cause and effect, but provides a basis for generating hypotheses on possible mechanisms controlling benthic algal biomass, which could then be tested with experiments.

E. Data Analysis

The AFDM and pigment data will be normalized per unit area sampled. The following t tests will be performed:

1. AFDM in pools vs AFDM in riffles;
2. Chlorophyll a in pools vs chlorophyll a in riffles.

In addition, an autotrophic index (AI) will be calculated. This index (AFDM/chlorophyll a) provides information on the trophic status or relative viability of the periphyton community. If large amounts of nonliving organic material are present, the numerator becomes inflated, and the ratio exceeds the normal range of 50–200 (APHA *et al.* 1992).

III. SPECIFIC EXERCISES

A. Exercise 1: Pool vs Riffle Comparison Using Artificial Substrata

Preparation Protocol

1. At least 1 month, and preferably several months, prior to the exercise, place unglazed ceramic tiles in selected pool and riffle habitats in the stream to be sampled. If the tiles were purchased in attached sheets, as opposed to separate units, place the entire sheet in the stream, which minimizes the likelihood of the tiles being removed by floods. Assume that each team will analyze eight tiles per pool (four for AFDM, four for chlorophyll a) and eight tiles per riffle (again, four for AFDM and four for chlorophyll a). The total number of tiles sampled may vary depending on the number of researchers per team; we assume that each team member will sample four tiles altogether: one tile for AFDM and one tile for chlorophyll a in the pool and riffle. *Note.* If the tiles are placed in the stream as separate units (i.e., not in sheets), set out twice the anticipated number of tiles to be used in the exercise (storm events may result in the loss of tiles during the colonization period).

2. Label tupperware containers (ca. 900 cm^2) by pool and riffle number, which correspond to sites where tiles are located.

3. Provide each team with vials (20-ml scintillation vials work well if available and researchers are conservative with sample volume) labeled according to pool or riffle, AFDM or chlorophyll, and tile number.

4. Weigh at least 100 precombusted glass-fiber filters (Whatman GF/F; 0.45-μm pore size) to the nearest 0.1 mg and place in appropriately numbered containers (aluminum weigh boats with a number etched on their bottom with a sharp instrument work well).

Protocol for Field Collection and Hydraulic Measurements

1. Collect tiles from pool and riffle habitats.

2. If samples are to be processed in the laboratory, fill two tupperware containers (one labeled pool and the other labeled riffle) with streamwater. Place 10 tiles (two extra, to accommodate mishandling or unanticipated problems) from each habitat inside the container. Attach the lids, squeezing out excess water (keeping the containers filled minimizes tile movement), and place the containers in coolers to be transported back to the laboratory.

3. Estimate water velocity in the pool and riffle. Use protocols for determination of water velocity found in Chapter 3.

Laboratory Protocol: Procedure 1

1. In the laboratory, separate the sheet of tiles into individual tiles (ignore any glue that may remain attached to individual tiles following separation).

2. Using a hard-bristled toothbrush, brush off the algae on the tile, collecting the brushed material into a pan (or tray). Use a squirt bottle filled with distilled water to periodically wash the tile and toothbrush. Be conservative in the use of the squirt bottle. Carefully pour the removed material into appropriately labeled vials.

3. Filter each slurry onto a glass fiber filter (precombusted and preweighed in the case of biomass) using a standard filtration apparatus and no more than 15 p.s.i. of vacuum (to avoid rupturing of cells).

4a. (AFDM) Put the filter back into the weigh boat and place inside a drying oven set at 105°C (APHA *et al.* 1992). Generally, it will take at least 24 h for the biomass to reach constant weight. Consequently, the remainder of the AFDM determination will be done in the second laboratory session. To facilitate finishing the exercise, we recommend that a researcher or team leader remove the weigh boats from the drying oven during the week and transfer them to desiccators.

4b. (Chlorophyll *a*) After filtration, place the filter in a small container or centrifuge tube containing a known volume (just enough to cover filter) of 90% buffered acetone (90 parts acetone and 10 parts saturated magnesium carbonate solution).[1] Steep samples at least 2 h, and preferably 24 h, at 4°C (or on ice) in the dark.[2]

Laboratory Protocol: Procedure for AFDM

1. If the filters have not been removed from the drying ovens during the week, remove them and place them in desiccators at the start of the lab. Keep filters in the desiccator until they cool to room temperature.

2. Weigh the filter (use forceps to remove filter from weigh boat) on an analytical balance (to the nearest 0.1 mg).

3. After weighing the filter, return it to the aluminum weigh boat and oxidize the material at 500°C for 1 h. Keep in mind that it may take an hour or two (depending on size and style) to bring the muffle furnace temperature to a constant 500°C.

4. Remove filters with oxidized material from the muffle furnace and allow them to cool in the desiccator.

5. Weigh the filter (use forceps to remove filter from weigh boat) on an analytical balance (to the nearest 0.1 mg).

Laboratory Protocol: Procedure for Chlorophyll a

1. Room lights should be dimmed during chlorophyll measurements to avoid changes in absorbance values. Remove the chlorophyll extract from the freezer. Centrifuge the sample if grinding was employed.

2. Transfer 3 ml of extract to a 1-cm cuvette and read optical density (OD) at 750 and 664 nm. (The absorption at 750 nm is substracted from the reading at 664 nm to correct for the presence of turbidity and colored materials (Wetzel and Likens 1991).

3. Acidify the extract in cuvette with 0.1 ml of 0.1 N HCl. Gently agitate the acidified extract (lightly "flick" the cuvette with finger), wait 90 s, and read OD at 750 and at 665 nm.

4. Rinse the cuvette with 90% acetone and shake dry prior to measurement of the next sample.

B. Exercise 2: Analysis of Pigments on Natural vs Artificial Substrata (Optional)

This exercise allows the researcher to compare the effectiveness of ceramic tiles in representing natural substrata (cf. Lamberti and Resh 1985). It requires sterilization of cobble prior to the exercise.

---

[1]Other solvents may be used to extract chlorophyll, but the equations given in this chapter are based on the use of acetone. Always wear gloves, lab coat, and eye shields when working with acetone or other solvents.

[2]Chlorophyll is subject to photodegradation. Extraction of chlorophyll must be done in the dark.

Preparation Protocol

1. At least 1 month, and preferably several months prior to the exercise, collect natural substrata (preferably small cobble with relatively flat surfaces) from the stream. Scrub the substrata and either autoclave or immerse in acetone for 24 h to remove and kill all attached organic material.

2. Place the sterilized rocks and unglazed ceramic tiles adjacent to each other in the pools and riffles of the stream.

Field Collection

1. Collect four rocks and four tiles from each pool and riffle and place in tupperware containers as outlined above. Instead of scraping the substrata, place them directly into separate wide-mouth jars containing sufficient 90% acetone to cover the exposed surface of the substratum (approximately 30 ml). When adding the substrata to the solvent, make certain that the upper surface of the substrata (the exposed surface in the stream) is placed face down in the jar (to ensure immersion of algae) and add them slowly, to minimize splashing and loss of solvent.

Laboratory Procedures

1. Following removal of the algae from the rock substrata, estimate rock surface area. We recommend use of the "aluminum foil method." If the rocks have been placed in solvent to extract pigments, allow them to dry completely in a hood (place them on a labeled paper towel that corresponds to the sample number associated with the rock). After the rocks are dry, wrap each rock completely, avoiding overlap, in aluminum foil. Trim excess foil with scissors. Remove the foil wrap and weigh on a balance. Also, weigh an unfolded square of known area. The following equation is used to calculate the unknown rock area (A_r):

$$A_r = \left(\frac{A_k}{W_k}\right) \cdot W_{rf}, \qquad (14.1)$$

where A_k represents known area; W_k, known weight; and W_{rf}, weight of rock foil. To estimate "colonized" surface area, A_r typically is divided by two (assume that only the top half of the rock is covered by periphyton).

2. Follow laboratory procedures in Exercise 1 (above) to measure chlorophyll *a*.

C. Exercise 3: Correlating Biomass with Environmental Variables and Relating Biomass to Algal Taxonomic Structure (Optional)

1. At each site where algae are collected, measure irradiance as described in Chapter 5.

2. At each site where algae are collected, measure inorganic nitrogen and soluble reactive phosphorus as described in Chapters 8 and 9, or in *Standard Methods* (APHA *et al.* 1992).

3. At each site where algae are collected, measure densities of grazing invertebrates as described in Chapters 16 and 19.

4. Coordinate algal biomass collection with measurement of algal community structure (Chapter 13) to determine which taxa account for the majority of biomass in each community.

D. Data Analysis

AFDM[3]

1. Calculate the dry mass (DM, in mg/cm^2) and AFDM (in mg/cm^2) of the biomass on each tile (or rock),

$$DM = \frac{(W_a - W_f)}{A_{t/r}}, \tag{14.2}$$

where W_a is dried algae on filter (mg); W_f, filter weight (mg); and $A_{t/r}$, area of tile or rock (cm^2) (see Table 14.1); and

$$AFDM = \frac{(W_a - W_f - W_{ash})}{A_{t/r}}, \tag{14.3}$$

where W_{ash} is material on filter (mg) after ashing.

Pigments

1. Calculate the chlorophyll *a* and pheophytin concentrations of the periphyton biomass on each tile or rock (see Table 14.2);

$$\text{Chlorophyll } a \ (\mu g/cm^2) = 26.7 \ (E_{664b} - E_{665a})$$
$$\times V_{ext}/\text{area of tile } (cm^2) \times L \tag{14.4}$$

---

[3]Some researchers report AFDM as g/m^2 instead of mg/cm^2; to convert from mg/cm^2 to g/m^2, multiply by 10.

TABLE 14.1
Sample Data Sheet for Ash-Free Dry Mass

Stream: Date:

Location: Investigators:

Pool/ riffle no.	Rep no.	Weigh boat #	Filter wt (mg)	DM + filter (mg)	DM (mg) = $B - A$	Ashed material + filter (mg)	AFDM (mg) = $B - D$
			A	*B*	*C*	*D*	*E*
P-1	a						
"	b						
"	c						
"	d						
R-1	a						
.	.						
.	.						
R-4	d						

TABLE 14.2
Sample Data Sheet for Pigments

Stream: Date:

Location: Investigators:

Pool/ riffle no.	Rep no.	Preacidification		Postacidification		Chl *a* (μg/cm^2)	Phaeo- phytin (μg/cm^2)
		750 nm	664 nm	750 nm	665 nm		
P-1	a						
"	b						
"	c						
"	d						
R-1	a						
.	.						
.	.						
R-4	d						

and

$$\text{Pheophytin } (\mu g/cm^2) = 26.7\,(1.7E_{665a} - E_{664b}) \\ \times V_{ext}/\text{area of tile } (cm^2) \times L, \qquad (14.5)$$

where

$E_{664b} =$
  [{Absorbance of sample at 664 nm − Absorbance of blank at 664 nm}
  − {Absorbance of sample at 750 nm − Absorbance of blank at 750 nm}]
  before acidification;

$E_{665a} =$
  [{Absorbance of sample at 665 nm − Absorbance of blank at 665 nm}
  − {Absorbance of sample at 750 nm − Absorbance of blank at 750 nm}]
  after acidification;

V_{ext} represents volume of 90% acetone used in the extraction (ml);

L, length of path light through cuvette (cm);

26.7 = absorbance correction (derived from absorbance coefficient for chlorophyll a at 664 nm [11.0] × correction for acidification [2.43]);

1.7 = maximum ratio of $E_{664b} : E_{665a}$ in the absence of pheopigments.

Autotrophic Index (AI)[4]

$$\text{AI} = \text{AFDM } (mg/cm^2)/\text{chlorophyll } a \ (mg/cm^2).$$

Conduct the Following Statistical Comparisons (Using T Tests; See Also Chapter 31)

1. Algal biomass and pigments in riffles versus pools.
2. Algal pigments on natural versus artificial substrata.

IV. QUESTIONS

1. Assuming that different biomass levels were measured between pools and riffles, what environmental factors might have accounted for this difference?

---

[4]Remember to convert chlorophyll a from $\mu g/cm^2$ to mg/cm^2 (divide by 1000) before calculating this index.

2. Were both AFDM and chlorophyll *a* greater in one channel unit type than the other (i.e., did they show a similar pattern in the riffles vs pools)? If not, what might have accounted for the variation?

3. Do you think the different biomass levels between pools and riffles stay constant in streams of different order? Would you expect greater differences in a first-order or a fourth-order stream? Why?

4. If you were able to conduct this study over an entire year, how might you expect algal AFDM or chlorophyll *a* to change over the four seasons in your stream? What about in a desert stream vs a deciduous woodland stream?

5. Some streams with very low periphyton abundance support very large populations of grazing macroinvertebrates. How is this possible?

6. How might algal biomass differences between pools and riffles change during flood periods? During extended low flow periods?

7. If comparisons of substratum type (i.e., natural vs artificial) were made, which type gave the most consistent (i.e., lowest variability) pigment results?

V. MATERIALS AND SUPPLIES

Field Materials

Materials and equipment necessary to measure current velocity (see
   Chapter 3), irradiance (see Chapter 5), inorganic nutrients (Chapters 8 and 9), and grazer density (Chapters 16 and 19) if this optional
   exercise is chosen
Tupperware containers (one per site; large enough to accommodate
   10 tiles, labeled)
Unglazed ceramic tiles (e.g., 100 tiles measuring 5 × 5 or 10 × 10 cm)

Laboratory Materials

0.1 N Hydrochloric acid
90% Acetone (90 parts acetone with 10 parts saturated magnesium
   carbonate solution)
Aluminum foil
Aluminum weigh boats or porcelain crucibles
Coarse-bristled toothbrushes
Kim Wipes
Pipets (5 ml and Pasteur)
Saturated magnesium carbonate solution (1.0 g finely powdered $MgCO_3$
   added to 100 ml distilled water)

Scintillation vials (100 20-ml volume, labeled)
Squirt bottles
Whatman GF/F filters, or equivalent (100)

Laboratory Equipment

Analytical balance (sensitivity of 0.1 mg)
Cuvettes (1 cm path length)
Desiccator
Drying oven
Filtration apparatus with vacuum pump and solvent-resistant filter assembly
Forceps
Muffle furnace
Tongs (for handling hot crucibles)
Spectrophotometer (narrow bandwidth: 0.5 to 2.0 nm)

REFERENCES

Aloi, J. E. 1990. A critical review of recent freshwater periphyton field methods. *Canadian Journal of Fisheries and Aquatic Sciences* **47:**656–670.

APHA, AWWA, and WEF. 1992. *Standard Methods for the Examination of Water and Wastewater*, 18th ed. American Public Health Association, Washington, DC.

Axler, R. P., and C. J. Owen. 1994. Measuring chlorophyll and phaeophytin: whom should you believe? *Lake and Reservoir Management* **8:**143–151.

Bothwell, M. L. 1988. Growth rate responses of lotic periphyton diatoms to experimental phosphorus enrichment: The influence of temperature and light. *Canadian Journal of Fisheries and Aquatic Sciences* **45:**261–270.

Brown, H. D. 1976. A comparison of attached algal communities of a natural and an artificial substratum. *Journal of Phycology* **12:**301–306.

Burkholder, J. M., and R. G. Sheath. 1985. Characteristics of softwater streams in Rhode Island. I. A comparative analysis of physical and chemical variables. *Hydrobiologia* **128:**97–108.

Cattaneo, A., and G. Roberge, 1991. Efficiency of a brush sampler to measure periphyton in streams and lakes. *Canadian Journal of Fisheries and Aquatic Sciences* **48:**1877–1881.

Cummins, K. W., and M. J. Klug. 1979. Feeding ecology of stream invertebrates. *Annual Review of Ecology and Systematics* **10:**147–172.

Cushing, C. E., and E. G. Wolf. 1984. Primary production in Rattlesnake Springs, a cold desert spring-stream. *Hydrobiologia* **114:**229–236.

Douglas, B. 1958. The ecology of the attached diatoms and other algae in a small stony stream. *Journal of Ecology* **46:**295–322.

Gregory, S. V. 1980. *Effects of Light, Nutrients, and Grazing on Periphyton Communities in Streams.* Ph.D. dissertation. Oregon State University, Corvallis, OR.

Hynes, H. B. N. 1975. The stream and its valley. *Verhandlungen der Internationalen Vereinigung für Theoretische und Angewandte Limnologie* **19:**1–15.

Ivlev, V. S. 1933. Ein Versuch zur experimentellen Erforschung der Ökologie der Wasserbiocoenosen. *Archiv für Hydrobiologie* **25:**177–191.

Lamberti, G. A., L. R. Ashkenas, S. V. Gregory, A. D. Steinman, and C. D. McIntire. 1989. Productive capacity of periphyton as a determinant of plant-herbivore interactions in streams. *Ecology* **70:**1840–1856.

Lamberti, G. A., and V. H. Resh. 1985. Comparability of introduced tiles and natural substrata for sampling lotic bacteria, algae, and macroinvertebrtaes. *Freshwater Biology* **15:**21–30.

Lock, M. A., R. R. Wallace, J. W. Costerton, R. M. Ventullo, and S. E. Charlton. 1984. River epilithon: Toward a structural–functional model. *Oikos* **42:**10–22.

Loeb, S. 1981. An in situ method for measuring the primary productivity and standing crop of the epilithic periphyton community in lentic systems. *Limnology and Oceanography* **26:**394–399.

Lorenzen, C. J. 1966. A method for the continuous measurement of in vivo chlorophyll concentration. *Deep Sea Research* **13:**223–227.

Marker, A., E. Nusch, H. Rai, and B. Reimann. 1980. The measurement of photosynthetic pigments in freshwaters and standardization of methods: conclusions and recommendations. *Archiv für Hydrobiologie Beihefte Ergebnisse der Limnologie* **14:**91–106.

Mayer, M. S., and G. E. Likens. 1987. The importance of algae in a shaded headwater stream as food for an abundant caddisfly (Trichoptera). *Journal of the North American Benthological Society* **6:**262–269.

McIntire, C. D. 1973. Periphyton dynamics in laboratory streams: a simulation model and its implications. *Ecological Monographs* **43:**399–420.

Minshall, G. W. 1978. Autotrophy in stream ecosystems. *BioScience* **28:**767–771.

Morisawa, M. 1968. *Streams: Their Dynamics and Morphology.* McGraw–Hill, New York, NY.

Naiman, R. J. 1976. Primary production, standing stock, and export of organic matter in a Mohave thermal stream. *Limnology and Oceanography* **21:**60–73.

Nelson, D. J., and D. C. Scott. 1962. Role of detritus in the productivity of a rock outcrop community in a piedmont stream. *Limnology and Oceanography* **7:**396–413.

Patrick, R., M. H. Hohn, and J. H. Wallace, 1954. A new method for determining the pattern of the diatom flora. *Notulae Naturae* **259:**1–12.

Schindler, D. W. 1987. Detecting ecosystem responses to anthropogenic stress. *Canadian Journal of Fisheries and Aquatic Sciences* **44:**6–25.

Sheath, R. G., J. M. Burkholder, M. O. Morison, A. D. Steinman, and K. L. Van Alstyne. 1986. Effect of tree canopy removal by gypsy moth larvae on the macroalgal community of a Rhode Island headwater stream. *Journal of Phycology* **22:**567–570.

Sladeckova, A. 1962. Limnological investigation methods for the periphyton ("aufwuchs") community. *The Botanical Review* **28:**287–350.

Steinman, A. D., and C. D. McIntire. 1990. Recovery of lotic periphyton communities after disturbance. *Environmental Management* **14:**589–604.

Steinman, A. D. 1992. Does an increase in irradiance influence periphyton in a heavily-grazed woodland stream? *Oecologia* **91:**163–170.

Sumner, W. T., and C. D. McIntire. 1982. Grazer-periphyton interactions in laboratory streams. *Archiv für Hydrobiologie* **93:**135–157.

Tuchman, M. L., and D. W. Blinn. 1979. Comparison of attached algal communities on natural and artificial substrata along a thermal gradient. *British Phycological Journal* **14:**243–251.

Tuchman, M. L., and R. J. Stevenson. 1980. Comparison of clay tile, sterilized rock, and natural substrate diatom communities in a small stream in southeastern Michigan, USA. *Hydrobiologia* **75:**73–79.

Vannote, R. L., G. W. Minshall, K. W. Cummins, J. R. Sedell, and C. E. Cushing. 1980. The river continuum concept. *Canadian Journal of Fisheries and Aquatic Sciences* **37:**130–137.

Wetzel, R. G., and G. E. Likens. 1991. *Limnological Analyses*, 2nd ed. Springer-Verlag, New York, NY.

CHAPTER 15

Meiofauna

MARGARET A. PALMER[*] AND DAVID L. STRAYER[†]

*Department of Zoology
University of Maryland
†Institute of Ecosystem Studies*

I. INTRODUCTION

The meiofauna are defined as those benthic animals that pass through a 500-μm sieve but are retained on a 40-μm sieve (Fenchel 1978, Higgins and Thiel 1988). Although rarely studied by stream ecologists, and often thought of as a curiosity, the meiofauna dominate benthic animal communities in terms of numbers and species richness and play important roles in community and ecosystem processes. Stream meiofauna communities are usually dominated by rotifers, harpacticoid and cyclopoid copepods, young chironomids, naidid and enchytraeid oligochaetes, and nematodes (Whitman and Clark 1984, Pennak and Ward 1986, Strayer 1988, Palmer 1990) and may also contain flatworms, gastrotrichs, tardigrades, cladocerans, ostracods, mites, and the young of various insects. Local communities typically contain hundreds of species. The meiofauna often are divided into the permanent meiofauna (animals that spend their whole lives as meiofauna) and the temporary meiofauna (animals such as insects that start off as meiofauna but grow into macrofauna).

Interstitial meiofauna live between grains of sand and typically are small and worm-shaped. Many interstitial species have adhesive organs for attaching to sand grains. Burrowing meiofauna live in fine sediments and often have robust bodies for pushing aside mud and silt. Epibenthic meiofauna live on the streambed or on wood, leaves, or plants. These typically are the largest members of the meiofauna and often are good swimmers.

Like other animals, meiofauna are patchy in distribution (e.g., Rouch 1991), but factors responsible for this patchiness have not been well studied (Pennak and Ward 1986). By analogy with marine studies, both biotic and abiotic factors likely regulate the distribution and abundance of stream meiofauna. With respect to biotic factors, predation by fish and macrobenthos have been shown to structure some marine meiobenthic assemblages (Coull 1990). In addition, there is some evidence from both marine and lotic systems that interference competition might be an important determinant of species distribution (Chandler and Fleeger 1987, Van de Bund and Davids 1993).

Dissolved oxygen and water flow may be the most important abiotic factors regulating meiofaunal populations. Most stream meiofauna are obligate aerobes, and several studies have found a correlation between oxygen and meiofaunal populations (e.g., Rouch 1991, Boulton et al. 1991). Perhaps in response to gradients of dissolved oxygen, most meiofauna are found in the top few centimeters of sediment (Danielopol 1976, Coull 1988, Palmer 1990). During spates, however, the upper layers of sediment may be denuded of meiofauna as substratum is eroded and the animals are swept downstream (Marmonier and Cruezé des Châtelliers 1991, Palmer et al. 1992). Because dissolved oxygen and interstitial water flow are influenced by sediment grain size, the latter may be a good predictor of meiofaunal abundance and composition (Coull 1988, Pennak 1988, Ward and Voelz 1990). Gravel harbors an abundant and diverse meiofauna, particularly including rotifers, copepods, and tardigrades. Sands and silts are inhabited chiefly by oligochaetes, chironomids, and nematodes. Other physical factors (e.g., temperature and pH) become important regulators only at their extremes (Hummon et al. 1978, Rundle and Hildrew 1990). Meiofauna also are fairly sensitive to various pollutants, and may be useful as indicator species (Coull and Chandler 1992).

Meiofauna show marked seasonality in reproduction and abundance. In most temperate streams, meiofaunal populations reach peak abundances in late spring through early fall (up to $6,000,000/m^2$; Danielopol 1976, Hummon et al. 1978, Strayer 1988, Strayer and Bannon-O'Donnell 1988, Palmer 1990, Shiozawa 1991, Suren 1992). They may constitute >95% of the benthic animals in most streams and may be energetically important. The production of a single species of meiofaunal copepod was shown to be equal to that of the dominant macrobenthic shredder in a headwater stream (O'Doherty 1985). Poff et al. (1993) estimated that the meiofauna and a single species of macrofauna were responsible for most of the benthic respiration in Goose Creek, Virginia. Further, the activities of stream meiofauna have been shown to affect microbial communities and detrital dynamics (Perlmutter and Meyer 1991, Borchardt and Bott 1995). Finally, meiobenthos may be preyed upon by stream macrobenthos and fish (e.g., Sherberger and Wallace 1971, Benke and Wallace 1980, Brown *et al.* 1989, Armores-Serrano 1991, Rundle and Hildrew 1992), although detailed studies of the role of the meiofauna in stream food webs have not been conducted. More generally, because densities of stream meiofauna are similar to those observed for marine meiofauna, the meiofauna may be as important in streams as in marine systems. In the latter systems, meiofauna are important in energy flow (Gerlach 1971, Kennedy 1994), in linking microbes to fish and large invertebrate predators (Montagna 1984, De Morais and Bodiou 1984, Palmer 1988a, Service *et al.* 1992), in processing detritus (Findlay and Tenore 1982, Alkemade *et al.* 1992), and in determining community dynamics, especially following disturbance (Chandler and Fleeger 1983).

The objectives of this chapter are to introduce the meiofauna as a taxonomic and ecological group and to introduce the methods used in the study of stream meiofauna. First, we provide a methodological overview of how to collect and observe meiofauna for both qualitative, observational studies and for quantitative, ecological investigations. Second, we describe three exercises to introduce the dominant members of the meiofauna, to explore the important relationship between substrata characteristics and mciofauna communities, and to demonstrate field experimentation with meiofauna assemblages.

II. GENERAL DESIGN

A. Site Selection

Studies of meiofauna may be either qualitative or quantitative, depending on the research goals. When selecting a research site, the investigator may wish to first just scoop up fresh stream substrata and observe the material live. For quantitative ecological work, samples are easily collected and preserved using well defined techniques. For identification of major taxa, the researcher can use general references such as Pennak (1989), Higgins and Thiel (1988), Thorp and Covich (1991), or Giere (1993). If more detailed (i.e., species-level) identifications are needed, the bibliographies of these sources should be consulted.

The new investigator may want to work in a relatively pristine stream and at a site with good water flow and medium to coarse substrata. Finer substrata (e.g., mud) do harbor large numbers of animals, but samples must be collected only in the top layer of sediment and care must be taken not to "poison" the sample with deeper, anoxic mud prior to preservation.

Additionally, in the warm, summer months animals left unpreserved will decay in hours.

B. Collection of the Live Fauna: Qualitative Sampling

Examination of live fauna is quite instructive and a variety of data may be collected on locomotion, feeding, and sexual behavior. Additionally, for many taxa, species-level identifications cannot easily be made on preserved material (e.g., bdelloid rotifers, turbellarians, gastrotrichs).

Aliquots of substratum should be collected from the field by scooping sediment, leaves, moss, and rocks directly into buckets. The meiofauna can be concentrated by adding water to the bucket, swirling the sediment into the water to suspend the meiofauna, then pouring the water through a sieve. The composition and abundance of the fauna collected depends strongly on the mesh size of the sieve (Hummon 1981). A 125-μm sieve[1] is fine for laboratory exercises; 40-μm sieves are more appropriate for research settings although this mesh size still may result in the loss of a small fraction of the animals (Hummon 1981).

The animals retained on the sieve are then rinsed, using a wash bottle, into a second bucket with several liters of fresh streamwater. The swirl-and-decant process should be repeated five or six times to ensure that an adequate sample of the fauna has been extracted from the sediment. This swirl–decant process can be performed in the field or in the laboratory as long as fresh streamwater is used. If animals are to be examined later, it is imperative that the buckets of fresh material be kept cool. We routinely keep fauna alive for several days if we store the buckets in a cool environmental chamber or a refrigerator.

For observation of live fauna, remove the bucket from the environmental chamber, suspend the settled substratum into the water, and pour an aliquot through a sieve. This sample can be transferred to a petri dish for direct observation using a stereomicroscope. As the sample begins to reach room temperature, the animals will become more active and are easily observed. "Cool," fiber-optic light is preferred but "warm," transmitted or reflected light is acceptable.

If higher levels of magnification are desired for further examination, animals should be transferred to a drop of water on a glass slide, a small piece of hair or wax inserted as a spacer, and a coverslip placed on top. Anesthetics (e.g., 6% $MgCl_2$ = 73.2 g/liter) may be added to the petri dish sample to slow animals. This application will facilitate their transfer to

---

[1]Sieves are most economically made from Nitex mesh and large plastic jars. Cut off the end of the jar and a large circle out of the jar lid. Secure the mesh under the jar lid by screwing the lid on tightly over the mesh.

slides using pipets or Irwin loops. The latter are small, wire inoculating loops used in bacteriological work. Inexpensive, small-bore pipets ("meiofaunal-sized") may be made by drawing out Pasteur pipets. Once animals are on a slide, neosynephrine (available at any drugstore) may be used as a narcotic. It can be bled under a coverslip by putting a drop along one side of the coverslip and letting it work its way under the slip. Alternatively, most animals can be slowed down by simply placing a tissue at the edge of the coverslip and drawing out just enough of the mounting water to squeeze the animal a bit.

There are several methods for collecting and concentrating live animals. (1) Some animals such as many copepods migrate toward a light source; thus, the use of a beam of cool light at the top of a container holding animals in water attract the animals to the surface from where they can then be decanted from the top layer. (2) Bou–Rouch pumps[2] are useful in streambeds that are difficult to penetrate. The pump stand has a perforated lower hollow column into which water and associated fauna can seep and from which water is then pumped up and sieved (Bou 1974). (3) A turkey baster or similar suction device is particularly useful for the qualitative sampling of the upper layers of muds and silts that are sucked up and placed directly into a bucket. (4) Temperature gradients have been used to concentrate animals as many meiofauna move away from cold (freezing) surfaces. (5) Bubbling of air into a bucket with fresh sediment causes many animals to float on the water surface. These animals may then be collected by skimming or by using blotting paper with subsequent rinsing. These and additional techniques ar discussed by Higgins and Thiel (1988).

C. Collection of Preserved Samples: Quantitative Sampling

A vast array of sampling devices has been designed for streambeds, only a few of which we will mention here. If the substratum can be penetrated easily, the best and simplest sampling device is a corer made of PVC or clear acrylic pipe. If sampling is to be relatively shallow, then a corer made out of a cut-off 30-cc syringe works quite well. The investigator should stand downstream of the sampling site and the corer should be inserted into the sediment in an area that has not been disturbed. If the sediment is fine enough (muds), a cork can be placed in the top of the corer and

---

[2]A Bou–Rouch pump can be made for about $100, using materials available at any plumbing store. Buy a wellpoint, punch through the screening of all openings in the lowest 15 cm of the pipe, and using bathtub caulking, seal up the rest of the openings (i.e., those above 15 cm). A hand pump that pulls a large volume per stroke (e.g., a "pitcher pump" in plumbing store jargon) works quite well. Occasionally, the caulk will have to be replaced, but this apparatus is durable, adaptable, and inexpensive.

then the corer can be removed from the streambed without losing the sample. In all other types of substratum, the investigator will need to push his/her hand down under the core bottom before pulling the corer out of the streambed. Samples collected in gravel or sand substrata should be collected to a depth of at least 10 cm into the bed; in muds and silts, the coring can be shallower ($\approx$1 cm) because generally the depth of oxygen penetration is less.

For sampling deeper in sandy beds, a standpipe corer (Williams and Hynes 1974) works well. This corer allows one to collect intact samples from discrete depths in the streambed. For streambeds that simply cannot be cored, a Bou–Rouch pump (see above) may be used and has the advantage that it may be left in place between sampling dates. The disadvantage of pump sampling is that animals are often damaged and it is difficult to quantify samples; one can report numbers of animals per volume of water pumped but it is almost impossible to know from what area of the streambed this water originated. Another alternative is to use *in situ* freeze–coring devices. Bretschko (1990) has used this technique quite effectively and has minimized faunal avoidance of the sample by electroshocking the area prior to sample collection, which stuns the animals so they do not migrate from the freezing surface.

Once samples have been retrieved from the streambed, they should be transferred to a sample container and several milliliters of $MgCl_2$ anesthetic (see above) added. The sample should be stirred and left to sit for ca. 5 min, after which it should be rinsed through a sieve using fresh streamwater. The contents of the sieve should then be rinsed back into the sample container using deionized water and a wash bottle. Several milliliters of 10% buffered Rose Bengal–formalin solution should then be added to the sample. Rose Bengal (1 g/liter of 10% formalin) is a protein stain that greatly facilitates microscopic sorting (but will kill live animals). Some animals will stain in 15 min but many require 48 h for optimum staining.

D. Extraction and Identification of Animals

To facilitate microscopic identification, animals may be extracted from preserved sediment using a variety of techniques. If the substratum is coarse (sand, gravel, cobbles), the best technique is a simple swirl-and-decant procedure. Pour the contents of the preserved sample onto the appropriately sized sieve and rinse with tap water to remove the formalin. Transfer the sample from the sieve into a 1000-ml Erlenmeyer flask. Add ca. 200 ml of tap water and vigorously swirl the sample to suspend fine particles and animals into the water. Quickly pour the supernatant through the sieve. Repeat this procedure five or six times and most of the animals will be extracted, assuming that you properly relaxed the animals (see anesthetics above) prior to preservation. If this technique is being used for a quantitative study, the method must be "calibrated" for each sediment type. Using a dissecting microscope, the sorter should examine the sediment remaining in the flask after six decantations to be certain most of the animals were removed.

For fine sediments, either the entire sample must be microscopically examined (fortunately, much less sediment is collected when sampling muds) or a more complicated extraction procedure used. Many workers employ a density gradient technique. Here, the preserved sample is rinsed into a 100-ml Nalgene centrifuge tube. The tube is filled with water and the sample centrifuged at high speed for several minutes. To remove formalin and excess water, the supernatant is poured through a sieve, with the sample "pellet" left in the tube. Occasionally a few animals will be retained on the sieve, so the contents of the sieve should be rinsed into a jar to be used for your extracted fauna. About 50 ml of a colloidal silica solution (Ludox-TM distributed by DuPont and described in DeJonge and Bouwman 1971) is added to the sample pellet. The sample should be stirred thoroughly for several minutes to be certain that the entire pellet and associated animals are suspended. The sample is then centrifuged at a lower speed (500–1500 rpm) for 3–5 min. Pour the supernatant into the sieve and, using a wash bottle, carefully rinse the sides of the tube (but not the sediment pellet) into the sieve. Most of the animals will be floating in the silica supernatant. As with the decantation procedure, this process requires calibration for each sediment type. The centrifuge speed and time greatly affect the separation of animals from sediments. This technique is described in more detail and compared to other procedures in Pfannukuche and Thiel (1988).

Once the animals have been extracted from the sediment, a small amount of animal material (see Fleeger *et al.* 1988 for a discussion of subsampling) should be placed in a gridded petri dish. Small rectangular dishes (ca. 10 × 5 cm) with shallow sides (ca. 1.5 cm depth) constructed of thin Plexiglas work well and can be made inexpensively. The bottom can be "gridded" by simply scratching the surface of the plexiglas; the size of the grids should correspond to the stereoscopic field of vision. The sample, which will contain some sediment, can then be scanned (one grid square at a time) at 12–18× to locate meiofauna. Higher magnification (up to 50×) on the stereoscope is desirable as many of the rotifers are small and this higher magnification is necessary to identify them. It is wise to check the efficiency of this process periodically by having one or more individual investigators count a single sample several times.

For more exact identification, the animals should be transferred to a drop of glycerol on a glass slide and viewed with a compound microscope.

If a coverslip crushes the animals, place several very small pieces of hair under the coverslip. Hulings and Gray (1971) provide a more lengthy discussion of mounting techniques.

III. SPECIFIC EXERCISES

A. Exercise 1: Observing Living Meiofauna and Their Adaptations

The purpose of this exercise is to familiarize you with the dominant members of the meiofauna, their morphologies, locomotion, mode of reproduction, and feeding.

1. Use the methods described under Collection of Live Fauna above to collect sand, mud, and leaf or algal material from the field. Keep each of these three substrata in separate containers to compare the associated meiofauna. Place a small aliquot from each of the three samples in a separate petri dish. Examine the contents under a dissecting microscope and identify as many animals as you can to major taxonomic levels. If you do not have many animals in your sample, you can concentrate a larger volume from the bucket by swirling and decanting through a "live samples only" sieve (one that was not used for formalin samples). At a minimum, you should be able to identify the following animals: bdelloid and monogonont rotifers, chironomid larvae, cyclopoid and harpacticoid copepods (adults and nauplii), oligochaetes, turbellarians, and nematodes. Depending on the material, you may be able to find gastrotrichs, tardigrades, cladocerans, and various insect larvae. A copy of Pennak (1989), Thorp and Covich (1991), or other invertebrate text will facilitate this exercise. In all cases where animals are mounted on glass slides, you may want to first anesthetize the animal, add a drop of Protoslo prior to adding the coverslip, or carefully squeeze the animal by removing water from under the coverslip with a piece of tissue.

2. Note the general differences in taxa, body shapes, and body sizes among the habitats. The sand sample will probably contain small, slender representatives of each taxonomic group and will have fewer chironomids and oligochaetes than the mud or litter sample. Many of these sand-dwelling animals exhibit other adaptations for an interstitial existence including adhesive organs, such as the "toes" of rotifers and gastrotrichs. Find at least one of these structures by mounting an animal on a glass slide and observing it under a higher power on a compound microscope. Adhesive organs are believed to reduce the chance of displacement from the highly mobile sandy substratum.

3. Isolate several copepods from the sand sample, the mud sample, and the litter sample. Note the difference in appendages (this may require

high magnification). The sand-dwelling (interstitial) copepods, especially the harpacticoids, may have smaller legs than the copepods from the other substrata. The legs of these interstitial animals may closely adhere to the narrow or bullet-shaped body. Compaction of the legs close to a fusiform body makes it easier for the animal to "glide" among the interstices of sand. The mud-dwelling copepods will generally be larger than the interstitial forms and will have more robust bodies and stout appendages that are used to help the animal push mud apart while it burrows. The leaf or algal-dwelling copepod will have a large, often somewhat flattened cephalothorax and appendages (especially the first leg), because it spends its life clinging and swimming among structures.

4. Isolate a few mud-dwelling worms in an area of the petri dish to study their locomotion. Note that they are poor swimmers compared to the copepods. Nematodes have only longitudinal muscles and thus they make jerky, side-to-side movements; they move forward by pushing off a substratum. Oligochaetes are more adept burrowers than nematodes due to their well developed longitudinal and circular muscles. Rotifers have complex musculature and loop along by pushing and pulling, often using their toes or anterior end for adhering to a sand grain.

5. Depending on the time of year and how gentle you have been with your sample, you may be able to find individuals in various states of reproduction. Copepods are often found *in copula* within samples: the male grasps the female's urosome with his first antennae and will eventually pass a sperm sac to her. Eggs (usually 2–10) may be seen attached ventrally to her abdomen. Cladoceran eggs may be visible through the body wall. Aquatic nematodes may bear live young or lay eggs. Examine specimens under the compound microscope. You should be able to distinguish males from females by the presence/absence of copulatory spicules that are used in sperm transfer. Sexual reproduction is rarely seen in rotifers; however, you may see amictic eggs developing through the body wall. Some rotifers have separate sexes; however, most reproduction is parthenogenic. Freshwater meiobenthic oligochaetes may reproduce asexually or sexually, so you may find individuals in various states of budding, or see eggs (cocoons) attached to substrata in your sample. The immature chironomids you see are, of course, only one stage in the animal's life cycle. These larvae emerge at particular times of the year, and the adult lives a brief aerial existence before it flies back to the stream to lay eggs.

6. Meiofauna feed on bacteria, diatoms, other meiofauna, and protozoa. It is difficult to observe them feeding directly (unless you have live turbellarians), but you may find evidence of past feeding (Kennedy 1994). Most meiofauna are transparent enough to see their gut contents. Be sure to mount some chironomids on glass slides and examine the guts for traces of diatoms or other animals. Often whole rotifers can be seen in their guts.

In the digestive tract of rotifers, diatoms (or even other rotifers!) may be easily seen.

B. Exercise 2: Substratum–Meiofaunal Relationships

A number of studies have examined the spatial and temporal patterns of meiofaunal abundance and biomass. Population and community regulation is complex and involves both biotic (e.g., predation, competition) and abiotic (e.g., temperature, desiccation tolerances) components. Our purpose here will be to compare meiobenthos abundance and taxonomic composition from (1) a coarse sandy or pebble streambed or area of a stream, (2) a finer sandy area, and (3) a muddy area. We do not recommend that you sample cobble (or coarser) substrata as one of your three areas because sampling that substratum type requires different methods. Because grain size is so critical to the meiofauna, we will also compare median grain size in the areas. If a muffle furnace is available, measure percentage organic matter in the sediment as a possible indication of food resources available to the fauna (see Chapters 10 and 14). To make statistically valid comparisons between the sites requires careful sampling and replication. Since meiofauna are numerous and time-consuming to count, take small samples using 30-cc syringes as corers (cut off the pointed end of the syringe).

Field Protocol

1. Insert the core to a depth of 10 cm in the sand. Place your thumb over the end and turn the corer slightly to its side so you can slip your hand under the bottom of the core and remove it from the bed without losing any sediment. Place the sample in a prelabeled jar containing ≈ 10 ml of 6% $MgCl_2$ (to promote faunal "release" from the sand grains). Rinse the inside of the core into the sample jar using a wash bottle filled with streamwater. This small amount of streamwater will not impede faunal relaxation. After about 5 min, pour the excess water from the jar through a sieve, leaving the sediment sample in the jar. Rinse the contents of the sieve back into the sample jar. Add buffered formalin with Rose Bengal. Take five such cores from the sand area.

2. Repeat this procedure in the other areas except only insert the core to a depth of 1 cm in the mud area. Anesthetizing agents ($MgCl_2$) are not required for mud samples.

3. For evaluation of grain size and organic matter content, use a corer or a small shovel and scoop up some sediment (at least 50 cc but the exact amount is unimportant) from each of the areas into separate jars. Samples should be kept in a cool place or on ice until return to the laboratory. If these samples will not be processed within 1 day, they should be frozen.

TABLE 15.1
Sample Data Sheet for Identification and Enumeration of Meiofauna

Sample site: _____   Sorter: _____

Taxa	Notes	Core 1	Core 2	Core 3	Core 4	Core 5	Mean	%
Chironomids								
Cladocerans								
Copepods								
Gastrotrichs								
Nematodes								
Oligochaetes								
Ostracods								
Rotifers								
Tardigrades								
Turbellarians								
Total meiofauna								

If time permits, more than one sediment sample within each area (i.e., replicate samples) should be collected.

Laboratory and Data Analysis Protocols

1. In the laboratory, rinse the contents of each faunal sample separately through a 40-μm sieve and into a counting tray. Split the sample into several trayfulls such that only small amounts of sediment are in the bottom of the tray, thus making it easier to locate animals.

2. Enumerate and identify to major taxa the animals in each core using Table 15.1. Make notes on the general shape and size of individuals within each taxon. To speed the counting process, samples may be distributed among researchers.[3] If this exercise is used in a classroom setting, it is important for each person to at least qualitatively examine one sample from each of the three areas so that he/she is familiar with the fauna.

[3]A word of caution here: the efficiency with which animals are seen and properly identified differs among individual researchers. For research settings, you will want to minimize the variance in faunal abundance and composition introduced by this sorter effect. You may want to run calibration "exercises" to be certain efficiency levels are similar among sorters or you may decide to have the same person sort all of the samples.

3. After all samples have been processed, calculate the mean abundances (usually reported as No. meiofauna/10 cm^2) and 95% confidence intervals for each of the three sampling areas. Also, calculate the percentage taxonomic composition for each site (e.g., copepods made up x percentage of the total meiofauna found at the mud site).

4. Prior to statistical analysis, the data may have to be transformed (e.g., log) to ensure homogeneity of variances (Zar 1984). Using analysis of variance (ANOVA) followed by multiple contrasts (e.g., Zar 1984), compare the abundances (total meiofauna and individual taxa) between the three sites to determine if they are significantly different.

Protocol for Median Grain Size Determination   An easy way to calculate the median grain size is to wash a sediment sample through a sieve series and determine the weight fraction retained on each sieve. Sediments finer than 62 μm are analyzed using a method based on sedimentation rates in graduated cylinders. The following method is modified from Buchanan (1984).

1. Thaw the frozen sediment samples collected for measurement of grain size.

2. Stack the following sieve series on top of a bucket: 63 μm (on the bottom), 125 μm, 250 μm, and 500 μm (on the top). Place a sediment sample on the top sieve and rinse the sample carefully (the water should be passing through the sieve series into the bucket). Remove the fraction retained on each sieve in separate preweighed, labeled containers (small glass beakers or aluminum pans work well).

3. Fill the water/sediment mixture in the bucket to a known volume. Disperse the sediment by stirring for at least 1 min. Stop stirring; after exactly 1 min and 56 s,[4] withdraw a 20-ml pippet sample from the mid-depth center of the container and transfer to a preweighed, labeled container. The mass of this material after drying will represent the sediment <31 μm in the suspension.

4. Remix the suspension for 1 min. After 7 min and 44 s, withdraw another 20-ml sample and transfer to a preweighed container. The mass of this material will represent the sediment <16 μm in the suspension.

5. Since you only sampled 20 ml each of the 31-μm size range and 16-μm size range, you must adjust for the total volume in the bucket in making your calculations (i.e., multiply by appropriate factor).

---

[4]Exact timing is necessary since separation of sediments into various size classes is based on the differences in fall velocities for different sized sediment particles.

TABLE 15.2
Sample Data Sheet for Grain Size Analysis of Meiofauna Habitats

Sediment sample _____   Location _____

Particle size (μm)	Sample no.	Tare	Tare + sample	Sample	Mass (% of total)	Cumulative %[a]
16						
31						
63						
125						
250						
500						

[a]Calculate the cumulative percentage beginning with the 16-μm size category and ending at 100% when the 500-μm size category is added.

6. Dry all of the samples in an oven at 70°C until the water is driven off (overnight), cool in a desiccator, and weigh. Enter data in Table 15.2.

7. Plot the cumulative % (y-axis) versus the particle size (x-axis) and determine the median grain size (50th percentile). Repeat this process for each of the sediment samples collected from the three areas. Compare median grain size and also the general shape of the cumulative plots. This procedure is a useful way of characterizing grain size unless the sample was poorly sorted (e.g., a few large, heavy rocks and lots of sand). In this case, plots will not exhibit typical sigmoidal shapes and median grain size is not necessarily a good habitat descriptor (see Buchanan 1984 for additional ways to compare graunulometry).

8. Organic matter content can be assessed by transferring a small amount of sediment from each sample to a preweighed aluminum weigh boat (see also Chapters 10 and 14). Dry the sample + aluminum weigh boat overnight in a 70°C oven, place in desicator to cool, then reweigh and record data. Place the sample + weigh boat in a muffle furnace and ash the sample (6 h at 550°C). Let sample cool, rewet with deionized water, dry the sample overnight in the 70°C oven, cool the sample in a desiccator, and reweigh after it is cool. Calculate the percentage loss in mass (= organic matter).

C. Exercise 3: Meiofauna as Experimental Tools

Many streams are subject to intense and frequent flooding. Floods can have a dramatic impact on benthic invertebrates, causing massive mortality or increases in downstream drift (Resh *et al.* 1988, Marmonier and Cruezé des Châtelliers 1991). Even when faunal losses from the streambed are great, most benthic populations, including meiofauna, are able to recover fairly rapidly (Palmer *et al.* 1996). This recovery may result from animals that survive in the drift or in refuges within the channel or floodplain and then subsequently recolonize streambeds (Sedell *et al.* 1990). Rapid recolonization may also be facilitated if animals that survived the flood are able to reproduce rapidly or if fauna have terrestrial adults that are available to deposit eggs in the stream just after a flood (e.g., Grimm and Fisher 1989).

Meiofauna are ideal animals for studying the recolonization process because they are small, easy to enumerate, and recover rapidly from disturbances (Coull and Palmer 1984). In this exercise, the mechanisms by which meiofauna recolonize stream substrata will be examined. The goal is to compare the relative importance of faunal inputs from the water column (drift and settlement) versus inputs from the streambed (infaunal immigration through the hyporheic zone) to the colonization process. Since stream meiofauna are known to drift in large numbers (Schram *et al.* 1990, Richardson 1990, Palmer 1992), water column dispersal may exceed dispersal through infaunal immigration (Palmer 1988b, Palmer *et al.* 1992). The experiment described is for a sandy-bottom channel; however, it can be modified for use in streams with coarse substrata by using different types of colonization chambers (e.g., shallow plastic boxes instead of the deeper tubes described below).

Employing the hypothesis, *meiofaunal drift and settlement from the water column will contribute more to the colonization of defaunated sediments than infaunal immigration through the hyporheic zone*, the experiment will involve comparing faunal abundance in three types of recolonization chambers designed to allow colonization from different sources (drift, hyporheic, and drift + hyporheic) and in ambient sediments over time (as in Chandler and Fleeger 1983 and Palmer *et al.* 1992; for a similar experiment, see Boulton *et al.* 1991). The chambers should be filled with azoic substrata (see below), placed in the streambed, sampled at Time 0 (immediately after deployment), and then sampled after 1 week, 2 weeks, and 1 month. The timing may be varied depending on the season and current velocity. Our experience has shown that if the water is fairly warm (e.g., $\geq 15°C$) and flow is moderate (≥ 5 cm/s), colonization generally occurs within weeks.

There are four "treatments" (three chamber types and ambient cores) that will be compared in this experiment. One chamber type has closed sides, a closed bottom but an open top; this allows for colonization from the *drift* (see Chapter 17). The second chamber type has mesh sides, a mesh bottom and a closed top; this allows for colonization from the *hyporheic zone* (see Chapters 6 and 30). The third chamber type has mesh sides, a mesh bottom, and an open top; it will serve as a *control* chamber allowing for colonization from all sources. Each time chambers are collected, researchers take cores from the streambed to determine ambient meiofaunal abundances. At the end of the experiment, calculate the mean faunal abundance and taxonomic composition for each of the four "treatments" at each of the four times. Plots of faunal abundance versus time yield three colonization curves and one "curve" representing variation in ambient abundances during the experiment.

Preparation Protocols Prior to the Experiment

1. Collect sand by scooping material from the top 10 cm of the streambed into buckets. Spread the sand in thin layers on plastic sheets and leave to air dry (preferably in the sun) for at least 1 week. If the material has been thoroughly dried, few animals will survive, thus, you have "azoic" stream substrata for your experiments.

2. Colonization chambers can be made from Nalgene sample tubes (purchase with caps) ca. 10 cm deep and 3 cm in diameter (Fig. 15.1). For the drift colonization treatment, the tubes are unmodified. For the hyporheic colonization treatment and the control treatment, the sides of the tubes are cut away except for four thin vertical strips. This procedure leaves four open "windows" on the tube. Cover the windows with 500-μm mesh by taping the mesh on to the vertical strips and the top and bottom of the

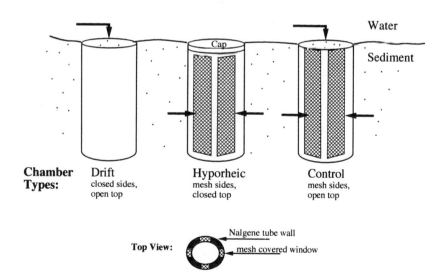

FIGURE 15.1 Schematic illustration of colonization chambers. Arrows indicate source of colonists.

tubes using nylon strapping tape. Assuming five replicates per treatment and four sampling times, you will need 20 drift chambers and 40 cut-away chambers. A few extra chambers should have both the tops removed and the bottom cut away and they will serve as corers for collection of ambient samples. After the chambers are constructed, fill each with azoic sand and cap them.

The Experiment

1. Push chambers (with caps on) into the streambed in four "blocks" of 15 chambers such that each block has 5 drift chambers, 5 hyporheic chambers, and 5 control chambers. The chamber types should be randomly assigned within each block (e.g., Fig. 15.2). The four blocks should be arranged longitudinally in the stream so that the most downstream block can be collected at time zero without disturbing upstream blocks.

2. After all four blocks of chambers have been deployed, start at the upstream block and carefully remove the caps from the drift and control chambers by progressively moving to downstream blocks. Be sure to stand downstream of the chambers and leave caps on the hyporheic colonization chambers.

3. After all caps have been removed from the drift and the control chambers, you are now at Time 0. The first sample should be collected soon after by replacing the caps on all chambers in the most downstream block and then pulling the chambers out of the streambed. Then collect

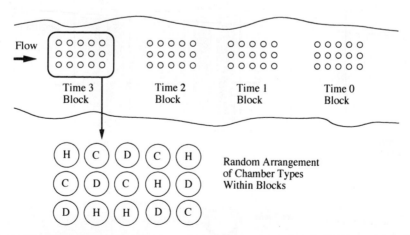

FIGURE 15.2 Schematic illustration of the arrangement of treatment blocks along a stream. One of the blocks is shown in more detail to illustrate randomization of treatments within a single block.

five ambient cores from the streambed lateral to the block being sampled. Each core and the contents of each chamber should be emptied into a separate jar that has been labeled with respect to the appropriate treatment. This sampling procedure should be repeated at the second, third, and fourth sampling times.

4. Add approximately 10 ml of 6% $MgCl_2$ solution to each jar to relax fauna. After 5 min, rinse the sample onto a 40-μm sieve using filtered streamwater. The sample should then be returned to the jar and formalin–Rose Bengal added. The samples are then brought back to the laboratory and after 48 h are ready for faunal enumeration.

Data Analysis

1. Make separate graphs for each major taxon and plot mean abundance ($\pm$sd) versus time for each treatment (e.g., Palmer *et al.* 1992). Thus there will be four lines on each graph, one line corresponding to the abundance of fauna in the open-top chambers (allowed colonization from the drift only), one line for the open-sided chambers (allowed colonization from the hyporheic zone only), one line from the open-top/open-side chambers (allowed colonization from both the drift and the streambed), and one line corresponding to ambient faunal abundances during the experiment.

2. After log-transformation of the data, perform a two-way analysis of variance (for each taxon) with multiple contrasts (Zar 1984) from the three colonization treatments (drift, hyporheic, drift + hyporheic) to determine if the abundance of fauna varied over time and among treatments. The effect of most interest in the analysis is the time $\times$ treatment interaction; a significant interaction would indicate that the colonization *rate* for the different chamber types differed significantly. Faunal abundances from the ambient cores permit comparisons to determine if colonization was complete at time 4. A one-way analysis of variance may be used to statistically compare faunal abundances at the final sampling time (i.e., at time 4) among the four groups: drift, hyporheic, drift + hyporheic, ambient.

IV. QUESTIONS

1. A sandy-bottom stream with year-round flow and an extensive, thoroughly oxygenated hyporheic zone would be expected to have a diverse meiofauna community. If the watershed was subjected to intense development resulting in large amounts of silt deposition in the stream, and flow was reduced dramatically, how would you expect the faunal composition to change?

2. Given their vast difference in modes of locomotion and reproduction, you might expect meiofauna to differ greatly in their dispersal abilities. Which taxonomic group(s) would you expect to have the greatest dispersal potential? The lowest potential? Why?

3. Some of the animals you observed reproduce asexually while others have separate sexes and internal fertilization. What are the advantages of each mode of reproduction for these fauna? If a stream is prone to unpredictable, severe disturbances (floods, droughts, etc.), which reproductive mode might prevail? Why?

4. The three samples you collected were from vastly different habitats from a meiobenthic "perspective." What features of the habitats may be most important in influencing taxonomic composition and meiofaunal abundance?

5. Did the three sites differ in abundance of total meiofauna, median grain size, and organic matter content? Which site had the highest faunal abundances? Speculate on what factors may regulate meiofaunal abundances.

6. Calculate the coefficient of variation (standard deviation/mean) of faunal abundances for each of the three sites. Were they comparable? Did one site have more poorly sorted sediments (i.e., was not dominated by a single grain size) than another? What might this tell you about the diversity of meiofauna and their spatial variability at each of the sites?

7. Did the taxonomic composition differ among the three sites? What groups dominated at each site? Can you relate this to granulometry or organic matter content (if you did not complete Protocol 1, then go back and read that section for help here)? What other factors that we did not measure might be important in influencing taxonomic composition?

8. If you found little relationship between organic matter content and faunal abundance across these sites, speculate on how refractory carbon and differences in productivity (carbon turnover) at the sites may influence your ability to detect such a relationship.

9. Were there significant differences among faunal abundances in each of the three chamber types during the experiment?

10. At the end of your experiment, was colonization complete for each or any of the chambers (i.e., by time 4 did faunal abundances in the chambers equal abundances in the ambient sediments)? Which chambers were the closest to complete colonization? Why?

11. What do your answers to questions 1 and 2 tell you about the relative importance of the various modes (drift versus hyporheic) in recolonization by meiofauna?

12. What was the function of the open-top/open-sided chambers? For each of the four sampling periods, did average abundances in the open-

top (drift) chambers + average abundances in the open-sided (hyporheic) chambers approximately equal abundances in the open-top/open-sided chambers? If not, what can you say about your design? What were some potential experimental artifacts?

13. Did different taxa have different colonization rates? Why do you think these differences may have occurred?

V. MATERIALS AND SUPPLIES

Field Materials

6% MgCl$_2$ solution (73.2 g/liter)
10% Buffered Formalin (buffering: saturate with sodium borate)
125-μm or smaller sieve
Buckets (that have not been exposed to formalin)
Corer (ca. 2 cm diameter, 10 cm long; can be made from 30-cc syringe)
Irwin loops or small-bore Pasteur pipettes
Rose Bengal stain (add 1 g of Rose Bengal powder per liter of 10% formalin)
Sample jars
Trowel or piece of plastic cut from a milk container for scooping up sediment
Wash bottles

Laboratory Materials

125-μm or smaller sieve
1000-ml Erlenmeyer flasks
Micropipets
Microscope slides and cover slips
Plastic petri dishes or other sorting trays
Protoslo, MgCl$_2$, or other anaesthetizing agent

Extra Materials Required for Exercise 2

20-ml Volumetric pipet
500, 250, 125, and 63-μm sieves
500-ml glass beakers (18)
Aluminum weigh boats

Extra Materials Required for Exercise 3

70 Nalgene tubes (3 cm diameter, 10 cm deep) with caps
500-μm Mesh netting
Nylon strapping tape
Plastic sheeting

Laboratory Equipment

Balance
Desiccator
Drying oven
Muffle furnace

REFERENCES

Alkemade, R., A. Wielemaker, and M. A. Hemminga. 1992. Stimulation of decomposition of *Spartina alterniflora* leaves by the bacterivorous marine nematode *Diplolaimelloides bruciei* (Monhysteridae). *Journal of Experimental Marine Biology and Ecology* **159**:267–278.

Amores-Serrano, R. R. 1991. *Stream-Dwelling Meiofauna: Temporal Abundance, Distribution and Utilization by Larval Fish*. Ph.D. dissertation. University of Arkansas, Little Rock, AR.

Benke, A. C., and J. B. Wallace. 1980. The trophic basis of production among net-spinning caddisflies in a southern Applachian stream. *Ecology* **61**:108–118.

Borchardt, M. A., and T. L. Bott. 1995. Meiofaunal grazing of microbial food resources in a Piedmont stream. *Journal of the North American Benthological Society* **14**:278–298.

Bou, C. 1974. Les methodes de recolte dans les eaux souterraines interstitielles. *Annales de Spéléologie* **29**:611–619.

Boulton, A. J., S. E. Stibbe, N. B. Grimm, and S. G. Fisher. 1991. Invertebrate recolonization of small patches of defaunated hyporheic sediments in a Sonoran Desert stream. *Freshwater Biology* **26**:267–277.

Bretschko, G. 1990. The effect of escape reactions on the quantitative sampling of gravel stream fauna. *Archiv für Hydrobiologie* **120**:41–49.

Brown, A. V., R. L. Limbeck, and M. D. Schram. 1989. Trophic importance of zooplankton in streams with alluvial riffle and pool geomorphometry. *Archiv für Hydrobiologie* **114**:349–367.

Buchanan, J. B. 1984. Sediment analyses. Pages 41–65 *in* N. A. Holmes and A. D. McIntyre (Eds.) *Methods for the Study of Marine Benthos*. IBP Handbook 16, 2nd ed. Blackwell Scientific, London, UK.

Chandler, G. T., and J. W. Fleeger. 1983. Meiofaunal colonization of azoic estuarine sediment in Louisiana: Mechanisms of dispersal. *Journal of Experimental Marine Biology and Ecology* **69**:175–188.

Chandler, G. T., and J. W. Fleeger. 1987. Facilitative and inhibitory interactions among estuarine meiobenthic copepods. *Ecology* **68**:1906–1919.

Coull, B. C. 1988. Ecology of the marine meiofauna. Pages 18–38 *in* R. P. Higgins and J. Thiel (Eds.) *Introduction to the Study of Meiofauna*. Smithsonian Institution Press, Washington, DC.

Coull, B. C. 1990. Are members of the meiofauna food for higher trophic levels? *Transactions of the American Microscopical Society* **109**:233–246.

Coull, B. C., and G. T. Chandler. 1992. Pollution and meiofauna: field, laboratory, and mesocosm studies. *Oceanography and Marine Biology Annual Reviews* **30:**191–271.

Coull, B. C., and M. A. Palmer. 1984. Field experimentation in meiofaunal ecology. *Hydrobiologia* **118:**1–19.

Danielopol, D. L. 1976. The distribution of fauna in the interstitial habitats of riverine sediments of the Danube and the Piesting (Austria). *International Journal of Speleology* **8:**23–51.

DeJonge, V. N., and L. A. Bouwman. 1971. A simple density separation technique for quantitative isolation of meiobenthos using the colloidal silica Ludox-TM. *Marine Biology* **42:**143–148.

De Morias, L. T., and J. Y. Bodiou. 1984. Predation on meiofauna by juvenile fish in a Western Mediterranean flatfish nursery ground. *Marine Biology* **82:**209–215.

Fenchel, T. M. 1978. The ecology of micro- and meiobenthos. *Annual Review of Ecology and Systematics* **9:**99–121.

Findlay, S. G., and K. R. Tenore. 1982. Effect of a free-living marine nematode (*Diplolaimella chitwoodi*) on detrital carbon mineralization. *Marine Ecology Progress Series* **8:**161–166.

Fleeger, J. W., D. Thistle, and H. Thiel. 1988. Sampling equipment. Pages 115–125 in R. P. Higgins and J. Thiel (Eds.) *Introduction to the Study of Meiofauna.* Smithsonian Institution Press, Washington, DC.

Gerlach, S. A. 1971. On the importance of marine meiofauna for benthos communities. *Oecologia* **6:**176–190.

Giere, O. 1993. *Meiobenthology: The Microscopic Fauna in Aquatic Sediments.* Springer-Verlag, New York, NY.

Grimm, N. B., and S. G. Fisher. 1989. Stability of periphyton and macroinvertebrates to disturbance by flash floods in a desert stream. *Journal of the North American Benthological Society* **8:**293–307.

Higgins, R. P., and J. Thiel (Eds.). 1988. *Introduction to the Study of Meiofauna.* Smithsonian Institution Press, Washington, DC.

Hulings, N. C., and J. S. Gray. 1971. *A Manual for the Study of Meiofauna.* Smithsonian Contributions to Zoology, No. 78. Smithsonian Institution Press, Washington, DC.

Hummon, W. D. 1981. Extraction by Sieving: A biased procedure in studies of stream meiobenthos. *Transactions of the American Microscopical Society* **100:**278–284.

Hummon, W. D., W. A. Evans, M. R. Hummon, F. G. Doherty, R. H. Wainberg, and W. S. Stanley. 1978. Meiofaunal abundance in sandbars of acid mine polluted, reclaimed, and unpolluted streams in southeastern Ohio. Pages 188–203 in J. H. Thorp and J. W. Gibbons (Eds.) *Energy and Environmental Stress in Aquatic Ecosystems.* DOE Symposium Series (CONF-771114), National Technical Information Service, Springfield, VA.

Kennedy, A. D. 1994. Carbon partitioning within meiobenthic nematode communities in the Exe Estuary, UK. *Marine Ecology Progress Series* **105:**71–78.

Marmonier, P., and M. Cruezé des Châtelliers. 1991. Effects of spates on interstitial assemblages of the Rhone River: Importance of spatial heterogeneity. *Hydrobiologia* **210:**243–251.

Montagna, P. A. 1984. *In situ* measurement of meiobenthic grazing rates on sediment bacteria and edaphic diatoms. *Marine Ecology Progress Series* **18:**119–130.

O'Doherty, E. C. 1985. Stream-dwelling copepods: Their life history and ecological significance. *Limnology and Oceanography* **30:**554–564.

Palmer, M. A. 1988a. Epibenthic predators and marine meiofauna: separating predation, disturbance, and hydrodynamic effects. *Ecology* **69:**1251–1259.

Palmer, M. A. 1988b. Dispersal of marine meiofauna: A review and conceptual model explaining passive transport and active emergence with implications for recruitment. *Marine Ecology Progress Series* **48:**81–91.

Palmer, M. A. 1990. Temporal and spatial dynamics of meiofauna within the hyporheic zone of Goose Creek, Virginia. *Journal of the North American Benthological Society* **9:**17–25.

Palmer, M.A. 1992. Incorporating lotic meiofauna into our understanding of faunal transport processes. *Limnology and Oceanography* **37:**329–341.

Palmer, M. A., A. E. Bely, and K. E. Berg. 1992. Response of invertebrates to lotic disturbance: a test of the hyporheic refuge hypothesis. *Oecologia* **89;**182–194.

Palmer, M. A., P. Arensberger, A. P. Martin, and D. W. Denman. 1996. Disturbance and patch-specific responses: The interactive effects of woody debris and floods on lotic invertebrates. *Oecologia* **105:**247–257.

Pennak, R. W. 1988. Ecology of the freshwater meiofauna. Pages 39–60 *in* R. P. Higgins and J. Thiel (Eds.) *Introduction to the Study of Meiofauna*. Smithsonian Institution Press, Washington, DC.

Pennak, R. W. 1989. *Freshwater Invertebrates of the United States*, 3rd ed. Wiley, New York, NY.

Pennak, R. W., and J. V. Ward. 1986. Interstitial faunal communities of the hyporheic and adjacent groundwater biotopes of a Colorado mountain stream. *Archiv für Hydrobiologie Supplement* **74:**356–396.

Perlmutter D. G., and J. L. Meyer. 1991. The impact of a stream-dwelling harpacticoid copepod upon detritally associated bacteria. *Ecology* **72:**2170–2180.

Pfannkuche, O., and H. Thiel. 1988. Sample processing. Pages 134–145 *in* R. P. Higgins and J. Thiel (Eds.) *Introduction to the Study of Meiofauna*. Smithsonian Institution Press, Washington, DC.

Poff, N. L., M. A. Palmer, P. L. Angermeier, R. L. Vadas, C. C. Hakenkamp, A. Bely, P. Arensburger, and A. P. Martin. 1993. Size structure of the metazoan community in a Piedmont stream. *Oecologia* **95:**202–209.

Resh, V. H., A. V. Brown, A. P. Covich, M. E. Gurtz, H. W. Li, G. W. Minshall, S. R. Reice, A. L. Sheldon, J. B. Wallace, and R. C. Wissmar. 1988. The role of disturbance in stream ecology. *Journal of the North American Benthological Society* **7:**433–455.

Richardson, W. B. 1990. Seasonal dynamics, benthic habitat use, and drift of zooplankton in a small stream in southern Oklahoma, U.S.A. *Canadian Journal of Zoology* **69:**748–756.

Rouch, R. 1991. Structure du peuplement des Harpacticides dans le hyporheique d'un ruisseau des Pyrenees. *Annals of Limnology* **27:**227–241.

Rundle, S. D., and A. G. Hildrew. 1990. The distribution of micro-arthropods in some southern English streams: The influence of physicochemistry. *Freshwater Biology* **23:**411–431.

Rundle, S. D., and A. G. Hildrew. 1992. Small fish and small prey in the food webs of some southern English streams. *Archiv für Hydrobiologie* **125:**25–35.

Schram, M. D., A. V. Brown, and D. C. Jackson. 1990. Diel and seasonal drift of zooplankton in a headwater stream. *American Midland Naturalist* **123:**135–143.

Sedell, J. R., G. H. Reeves, F. R. Hauer, J. A. Stanford and C. P. Hawkins. 1990. Role of refugia in recovery from disturbances—Modern fragmented and disconnected river systems. *Environmental Management* **14:**711–724.

Service, S. K., R. J. Feller, B. C. Coull, and R. Woods. 1992. Predation effect of three fish species and a shrimp on macrobenthos and meiobenthos in microcosms. *Estuarine, Coastal, and Shelf Science* **34:**277–293.

Sherberger, F. F., and J. B. Wallace. 1971. Larvae of the southeastern species of *Mollana*. *Journal of the Kansas Entomological Society* **44:**217–224.

Shiozawa, D. K. 1991. Microcrustacea from the benthos of nine Minnesota streams. *Journal of the North American Benthological Society* **10:**286–299.

Strayer, D. L. 1988. Crustaceans and mites (Acarina) from hyporheic and other underground waters in Southeastern New York. *Stygologia* **4:**192–207.

Strayer, D. L., and E. Bannon-O'Donnell. 1988. Aquatic microannelids (Oligochaeta and Aphanoneura) of underground waters of southeastern New York. *American Midland Naturalist* **119:**327–335.

Suren, A. M. 1992. Meiofaunal communities associated with bryophytes and gravels in shaded and unshaded alpine streams in New Zealand. *New Zealand Journal of Marine and Freshwater Research* **26:**115–125.

Thorp, J. H., and A. P. Covich (Eds.). 1991. *Ecology and Classification of North American Freshwater Invertebrates*. Academic Press, San Diego, CA.

Van de Bund, W. J., and C. Davids. 1993. Complex relations between freshwater macro- and meiobenthos: Interactions between *Chironomus riparius* and *Chydorus piger*. *Freshwater Biology* **29:**1–6.

Ward, J. V., and N. J. Voelz. 1990. Gradient analysis of interstitial meiofauna along a longitudinal stream profile. *Stygologia* **5:**93–99.

Whitman, R. L., and W. J. Clark. 1984. Ecological studies of the sand-dwelling community of an East Texas stream. *Freshwater Invertebrate Biology* **3:**59–79.

Williams, D. D., and H. B. N. Hynes. 1974. The occurrence of benthos deep in the substratum of a stream. *Freshwater Biology* **4:**233–256.

Zar, J. H. 1984. *Biostatistical Analysis*, 2nd ed. Prentice–Hall, Englewood Cliffs, NJ.

CHAPTER 16

Benthic Macroinvertebrates

F. RICHARD HAUER[*] AND VINCENT H. RESH[†]

*Flathead Lake Biological Station
The University of Montana
†Department of Environmental Science, Policy and Management
University of California, Berkeley*

I. INTRODUCTION

Freshwater macroinvertebrates are ubiquitous; even the most polluted or environmentally extreme lotic environments usually contain some representatives of this diverse and ecologically important group of organisms. By convention, the term *macro* invertebrates refers to invertebrate fauna retained by a 500-μm net or sieve. However, the early life stages of many macroinvertebrates pass through mesh openings that are this size. Because these early stages are important in understanding species-specific life histories, trophic relations, secondary production, and a multitude of other ecological relationships, there has been a general trend among stream ecologists to use collecting methods employing finer-meshed collecting nets (e.g., 125 to 250 μm) (Hauer and Stanford 1981, Cuffney *et al.* 1993).

In many lotic environments, the macroinvertebrate community consists of several hundred species from numerous phyla (e.g., Allan 1975, Morse *et al.* 1980, Benke *et al.* 1984, Ward and Stanford 1991) including arthropods (insects, mites, scuds, and crayfish), molluscs (snails, limpets, mussels, and clams), annelids (segmented worms), nematodes (roundworms), and platyhelminthes (flatworms). Most stream macroinvertebrate species are associated with surfaces of the channel bottom (e.g., bedrock, cobble, finer sediments) or other stable surfaces (e.g., fallen trees, snags, roots, and submerged or emergent aquatic vegetation) rather than being routinely free-swimming. Because of their propensity for bottom habitats, most stream macroinvertebrates are referred to as being *benthic*, or collectively as the macrozoobenthos (Greek: *benthos* = depth).

Partly because of their importance within the stream community as a fundamental link in the food web between organic matter resources (e.g., leaf litter, algae, detritus) and fishes, and partly because of their diversity and ubiquity, the study of macroinvertebrates has been a central part of stream ecology (Hynes 1970, Allan 1995). Earlier chapters within this book have focused on the multitude of interactive physical, chemical, and biological variables that constitute the stream ecosystem. For example, geology, climate, and other landscape features directly affect hydrologic patterns, and the movement and storage of inorganic and organic materials. Nutrients and the downstream transport of solutes are affected by channel and substratum complexity, the interactions of ground- and surface waters, and by the stream biota itself. Interactions between the stream channel, hyporheic zone, and riparian floodplains likewise are important features in the structure and function of the entire stream corridor (Gregory *et al.* 1991, Stanford and Ward 1993). These and many other factors affect the microhabitat structure of the stream and the distribution and abundance of stream macroinvertebrates.

A. Phylogeny and Adaptations

The origin of stream macroinvertebrates includes groups that are terrestrially derived (e.g., the insects) and groups that are marine in origin (e.g., molluscs and crustaceans). Of the various taxonomic groups that comprise the stream macroinvertebrate community, no group has been studied more than the aquatic insects. Not only are the aquatic insects extremely diverse both taxonomically and functionally, but they also are frequently the most abundant large organisms collected in stream benthic samples. Thirteen orders of aquatic insects occur in North America (Merritt and Cummins 1996), but only five of these are composed strictly of aquatic species (i.e., species that have at least one life-history stage that is obligatorily aquatic): the dragonflies and damselflies (Odonata), the stoneflies (Plecoptera), the mayflies (Ephemeroptera), the caddisflies (Trichoptera), and the hellgrammites (Megaloptera). Although the remaining eight orders have primarily terrestrial inhabitants, several of these orders exhibit high species richness (often thousands of species) in aquatic habitats. For example, the beetles (Coleoptera) and true flies (Diptera) each contain more aquatic species than are found among any of the completely aquatic orders (Merritt and Cummins 1996).

The phylogeny of aquatic insects is part of what makes these organisms so interesting. There is no single line of aquatic insect evolution; rather insects invaded the freshwater environment many different times and in many different ways (Resh and Solem 1996). As a result, problems of living in the stream environment, such as how to obtain oxygen or remain in a fixed position, have been solved repeatedly. Moreover, the mechanisms developed to overcome these obstacles involve a variety of different approaches and morphological adaptations. For example, some lotic species have developed structures to obtain oxygen from the atmosphere (analogous to snorkeling), others use temporary storage of an air bubble (analogous to SCUBA diving), a few species use respiratory pigments (analogous to vertebrate hemoglobin), and many species have developed tracheal gills for obtaining oxygen dissolved in the water (Eriksen *et al.* 1996). Likewise, morphological adaptations for existence in a running water environment include sclerotized projections along trailing edges of legs and body to form hydrofoils that press the organism onto the substratum, streamlining of body shape to offer reduced resistance while swimming, suckers and modified gills to provide attachment to smooth surfaces, and leg and anal hooks to provide attachment to a variety of surfaces, to name but a few (Resh and Solem 1996). The Trichoptera, Lepidoptera, and Diptera also use silk in a myriad of ways such as for attachment (e.g., free-living caddisflies and black flies), food gathering (e.g., net-spinning caddisflies), and shelter construction (e.g., midge larvae, moth larvae, and cased caddisflies).

Life history features that govern the reproduction and survival of lotic macroinvertebrates also show adaptations to specific characteristics of running water environments. Many stream environments are very dynamic (hydrologically, spatially, thermally, trophically, etc.), and macroinvertebrate life histories reflect this through tremendous diversity and adaptability (Butler 1984). For example, some species are specially adapted to ephemeral streams by having dormant egg stages that hatch as they are hydrated when flow resumes (Williams 1987). Also, closely related species that perform a similar trophic function may temporally separate growth and adult emergence within the same stream reach (Hauer and Stanford 1982a, 1986). Other life history adaptations can be seen in the seasonal timing of larval diapause (Hauer and Stanford 1982b) or pheromone release by adults for mate attraction (Resh *et al.* 1987). There is also considerable variation in the length of life cycles: some species may have several complete life cycles per year (multivoltine), two life cycles per year (bivoltine), one life cycle per year (univoltine), or may require 2 or 3 years to complete a life cycle (semivoltine). Specific life histories also may be very different across the geographic distribution of a species, where in one portion of its range a species may be univoltine and in another portion (generally colder) it is semivoltine. For example the limnephilid caddisfly *Dicosmoecus gilvipes* is univoltine in coastal streams of California and Oregon but semivoltine in mountain streams of Montana (Hauer and Stanford 1982b).

Behavioral adaptations are evident in aquatic insects as well, and these include regulatory behaviors to increase the control that an individual exerts over its own metabolic status, foraging behavior that involves the gathering and processing of food resources, or reproductive behavior that is responsible for the successful continuation of life into the next generation (Wiley and Kohler 1984). For example, *behavioral drift*, the intentional entry of benthic animals into the water column and their subsequent downstream transport, is a topic that has greatly interested stream ecologists for over two decades (Waters 1972, Müller 1974, Brittain and Eikeland 1988; see also Chapter 17) and may be essential to colonization processes, search for food, or predator avoidance.

Hydrologic processes, food resources, nutrient dynamics, riparian vegetation, and many other factors intimately affect the structure and function of stream ecosystems (Hynes 1970, 1975, Cummins 1974, Allan 1995). A fundamental characteristic of these factors is that they change along the longitudinal gradient of the stream ecosystem (Vannote *et al.* 1980); these factors may be affected by various anthropogenic influences (e.g., stream regulation; Ward and Stanford 1983). Macroinvertebrate species composition also changes between headwaters, middle reaches, and large rivers, in response to changes in the stream environment. For example, a stream reach flowing through a deciduous forest with a dense overhanging canopy may have a large number of macroinvertebrates that specialize in feeding on leaf litter, but that same stream upon entering a meadow (and thus having an open canopy) may be dominated by species that graze on periphyton. Within functionally similar groups (e.g., those that feed on similar food resources and use similar feeding mechanisms; see Merritt and Cummins 1996), species replacement along the river continuum is also very common. For example, among the net-spinning caddisflies (Hydropsychidae) numerous species may occur within a large river basin and be distributed in a very predictable manner along the longitudinal stream gradient (Hildrew and Eddington 1979, Alstad 1980, Cudney and Wallace 1980, Hauer and Stanford 1982c, Stanford *et al.* 1988). Some species will occur only in first- and second-order streams, other species will replace the headwater species in third- through fifth-order middle reaches, and still other species will occur only in larger rivers. Each species of hydropsychid spins a silk-thread catchnet that filters food particles from the flowing waters. Yet, through selection of particular habitats, food resources, and temperature regimes these species exhibit very predictable landscape-scale distributions.

People collecting stream macroinvertebrates for the first time are often amazed at both the complexity of the community and the wondrous variety of habitats in which they are found. Some species exist exclusively in very turbulent, high-velocity waters where they use sucker discs, hooks, or silk to keep themselves attached to the substratum. Other species occur in pools where stream currents are slow and their specialized body structures permit them to move across the fine sediments that accumulate in pools and backwaters. Many species can be found in leaf packs where they are surrounded by the food they eat and still others bore under the bark and through the boles of coarse woody debris (Merritt and Cummins 1996, see also Chapters 21 and 27).

In this chapter, we describe exercises to expose stream researchers to selected field and laboratory methods for the collection and study of stream macroinvertebrates. Over the past two to three decades there have been numerous detailed scholarly works dedicated to the collection and analyses of stream macroinvertebrates, such as collecting and sampling (Merritt *et al.* 1996), sampling design (Resh 1979, Norris *et al.* 1992), and statistical analyses and study design (Elliott 1977, Allan 1984, Morin 1985, Norris and Georges 1993). The purpose of this chapter is neither to synthesize or replace these detailed examinations, but rather to introduce stream ecology students or researchers that have not worked previously with macroinvertebrates to this interesting and diverse group of organisms.

Although the collecting methods we describe are most easily performed in wadeable, small to mid-sized streams, they can be adapted to larger rivers. As you work through the various exercises of this chapter be aware of the tremendous variety of habitats and the different ways in which macroinvertebrates are adapted to use resources. The specific objectives are to: (1) familiarize students and researchers with a variety of techniques for sampling stream macroinvertebrates, (2) describe how to preserve and process samples for laboratory examination, (3) introduce the concepts of abundance and diversity of stream macroinvertebrates, (4) examine large-scale distribution patterns, and (5) investigate microhabitat utilization and movement.

II. GENERAL DESIGN

A. Field Sampling

Over the past several decades, many different types of sampling devices have been invented for the systematic collection of stream macroinvertebrates, yet only a few standard sampling devices are used for most studies. Stream macroinvertebrates generally can be collected by disturbing bottom sediments (e.g., gravel, cobble) and catching organisms in a net held downstream. Most samplers are designed to delineate a certain area of stream bottom (often 0.1–0.5 m^2 or 1 foot^2). Then, in a combination of hand and/or foot action, the substratum materials are disturbed and organisms captured as they are dislodged and swept into the net by the current.[1]

The Surber sampler (Surber 1937) and Hess sampler (Hess 1941) are two standard collecting devices for stream macroinvertebrates (Fig. 16.1). Both samplers generally are small, limited to sampling stream depths <30 cm, and designed for streams with small substrata (gravel and small cobble). Another standard collecting device for stream macroinvertebrates is the kick-net, so-named because of the kicking action done in front of the net. The simplest kick-net is easy to make: two wooden dowels about 1.25 m long and 2–3 cm diameter support a 1 × 1-m square of 500-μm mesh Nitex netting (Fig. 16.1). Efficient use of the kick-net technique requires a two-person team. One person opens the kick-net above the water surface oriented perpendicular to the stream current. The bottom of the screen is lowered into the current, which forces the leading edge of the net against the substrata. The top of each stick is then held with the top of the net slightly tilted back. The area to be sampled is delineated with a 0.5-m^2 frame made of 1/4- or 3/8-inch-diameter steel re-bar placed in front of the net. The second person then picks up cobble from within the frame, one stone at a time, and carefully brushes/washes the organisms from the stone with their hands and a small brush. After the surface stones are washed in front of the net and set to the side, the sample area is then vigorously disturbed by stepping into the framed area and kicking back and forth for about 1 min. The kick-net can be modified to increase capture efficiency by adding a 1.5-m long, tube-shaped net to the center. Hauer and Stanford (1981) constructed the outer net of 250-μm mesh and the inner net of 125-μm mesh Nitex net. They found that these modifications pervented back-welling and captured the early instars of aquatic insects that are often lost using a 500-μm mesh net.

Numerous sampling devices also have been designed for collecting macroinvertebrates from other stream substratum types. Coring or dredging devices have been used to sample soft sediments such as sand or mud and frequently are necessary for sampling in large rivers or wherever soft sediments are prevalent. The D-frame net (Cuffney *et al.* 1993)

---

[1]A video (Resh *et al.* 1990) that demonstrates over 20 collecting techniques usable in a variety of stream habitats is available through the Office of Media Services, University of California, Berkeley, or by writing directly to VHR.

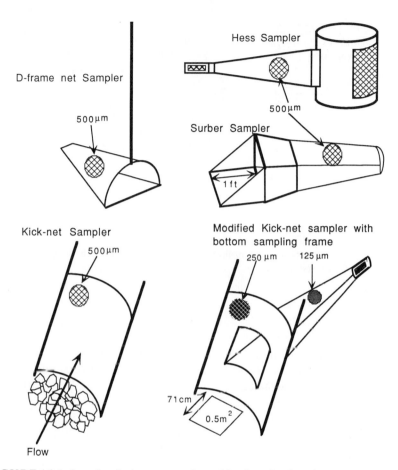

FIGURE 16.1 Sampling devices commonly used for the collection of stream macroinvertebrates.

may be used to sample gravel or cobble substrata, soft sediments, or woody snags.

After a sample is collected, the organisms are rinsed into the end of the net (or in the case of the Hess or modified kick-net sampler illustrated in Fig. 16.1, into a detachable, meshed bucket). At streamside, the macrobenthic organisms and any inorganic or dead organic material in the sample are washed into a 20-liter plastic bucket. This permits the researchers to remove all organisms that may be clinging to the inside of the sampler and add them to the sample. The contents of the bucket are then poured through a fine-meshed net or sieve ($\leq$500 μm, depending on original mesh size of

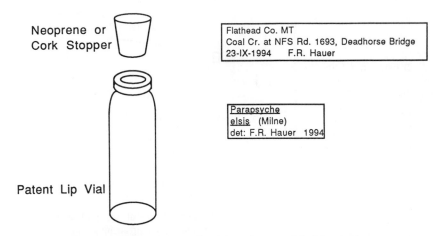

Neoprene or
Cork Stopper

Flathead Co. MT
Coal Cr. at NFS Rd. 1693, Deadhorse Bridge
23-IX-1994   F.R. Hauer

Parapsyche
elsis (Milne)
det: F.R. Hauer 1994

Patent Lip Vial

FIGURE 16.2 Illustration of patent-lip vials and proper label for the long-term storage of macroinvertebrate specimens.

sampler) to remove the excess water from the sample. Samples that are to be returned to the laboratory should be placed into a plastic jar or zip-lock bag and preserved in a 70% ethanol solution. This may be best done by adding 95% ethanol to the sample and estimating the remaining water, body fluids, and other organic matter to dilute the ethanol concentration to 70%.[2] Ideally, samples should be sorted within 24–48 h after collection to prevent specimens from deteriorating. After identification, stream macroinvertebrates that are intended for long-term storage should be curated in a glass patent-lip vial with a neoprene stopper, clean 70% ethanol, and a proper label (Fig. 16.2).

By using a Surber, Hess, or modified kick-net sampler (or some other sampling device with a defined area), quantitative samples may be collected from a known area and in a standardized fashion to obtain sample replicates. From these replicates a sample mean and variance can be calculated to estimate population size and variability. Quantitative sampling is necessary for most ecological investigations and involves a variety of decisions, including choice of sampling sites, depth of penetration of sampling into the substrata, frequency of sampling, etc. Bias (i.e., the lack of congruence between what is in a sample and what actually occurred in the sample area) can result from factors related to the characteristics of the sampler (e.g.,

[2]Traditionally, samples were preserved with 5% formalin solution or Kahle's fluid (28% ethanol, 11% formalin, 2% glacial acetic acid, and 59% water). Although formalin (or Kahle's which contains formalin) is a very effective preservative, it is hazardous to use. It is not recommended for classroom purposes.

backwashing of the sample from a clogged net), the organisms being sampled (e.g., tight attachment to substrata, movements to avoid being caught), and inconsistency among the users of the sampling devices (Resh 1979). The nonrandom distributions of most stream macroinvertebrate populations may require that large numbers of samples be collected in quantitative studies.

B. Laboratory Procedures

Sorting, which is done in the laboratory, involves the separation of the benthic macroinvertebrates from the substrata, organic matter, and other unwanted material that occurs in the sample. Sorting can be time consuming, but using sieves to separate out larger particles, dyes to stain macroinvertebrates (e.g., Rose Bengal), sugar floatation to "float off" organisms, and subsampling of very large samples can greatly reduce sorting time and effort. Large organsms can be easily seen and sorted without the aid of magnification; however, small species (e.g., microcaddis, midge larvae) and early instars of even the large species of macroinvertebrates require scrutiny under a good dissecting microscope.

After samples have been sorted, the organisms must be identified, counted, and the results analyzed. Essentially all macroinvertebrate families can be separated with some training, and generic level keys are available for most aquatic organisms; however, to ensure accuracy substantial training is required. What level of identification is required? The answer depends on the objective of the study (see discussion in Resh and McElravy 1993). We have included a simple flow-key to the more common stream macroinvertebrates (Appendix 16.1). Detailed keys to the families of North American freshwater invertebrates (Thorp and Covich 1991) and the genera of North American aquatic insects (Merritt and Cummins 1996) and noninsects (Pennak 1989) are excellent starting points for more detailed identification of stream macroinvertebrates. McCafferty (1981) has excellent illustrations of aquatic insects to the family level. Species-level keys are usually confined to a single genus (e.g., Szczytko and Stewart 1979) or region (e.g., Nimmo 1971, Baumann et al. 1977). Detailed, species-specific identification may require consultation with specialists and occasionally rearing of aquatic insects to the adult stage.

III. SPECIFIC EXERCISES

A. Exercise 1: Distributions and Habitat Relationships

If you carefully examine a reach of stream, you will discover that many populations of lotic macroinvertebrates are not distributed evenly throughout the reach. Rather, species tend to be found in particular micro-habitats. Habitat-specific distributions may be found at fairly large scales (e.g., riffles, pools, runs, woody snags, backwaters; see Chapter 2) or at very small scales of resolution (e.g., bottoms compared to tops of stones, along points of laminar flow, see Chapter 4). This exercise is designed to introduce the concepts of abundance and diversity within the stream macroinvertebrate community and how these features may differ among habitats within a single stream reach. You will be using the methods described above under Field Sampling to obtain quantitative samples from a variety of readily identifiable stream habitats.

Laboratory Preparation

1. One set of field collecting gear (listed below) should be available for every two field researchers working together as a team.

2. Select an appropriate stream segment from the study stream. This may be based on knowledge that you have acquired from other field work or specifically from work associated with earlier chapters in this book.

Field Collection

1. At streamside, identify several different habitat types along a stream length approximately equal to 10 times the width of the stream. Within this segment you may observe several different habitats.

2. Sketch a simple diagram of the stream reach that you are going to sample (see Chapter 4 for an example of method and detail).

3. If this site has been used for other exercises, refer to your notes regarding patterns of current velocity, substratum size, channel cross section, and large woody debris.

4. Delineate and note the range of microhabitats present.

5. Enter the stream and carefully look for macroinvertebrates on cobble, rock out-crops, large woody debris, or other hard surfaces. Make notes concerning your observations.

6. *Special note.* Some species of macroinvertebrates have very narrow microhabitat requirements and/or may achieve very high densities when environmental conditions are favorable. For example, look closely for black fly larvae (Diptera: Simuliidae). Black fly larvae generally have well developed "fans" on the head used for straining food particles from the stream current. Because they have narrow flow requirements, black flies often occur in very high abundance in very specialized microhabitats that have a stable substratum and smooth-laminar flows. Black fly larvae may exclude other larvae from areas around them by nipping and biting; this often results in uniform spatial distribution patterns. Look for these and other

macroinvertebrates that may occur in easily observable areas and note similarities and dissimilarities of microhabitat.

7. Take a sample from each of the different habitat types that you identified. Use a Surber, Hess, kick-net, or D-frame sampler, depending on the type(s) of habitats that are present.[3]

8. Empty the contents of the sampler into a 20-liter plastic bucket. Examine each sample for the presence of macroinvertebrates while the sample is in the bucket. At this juncture you must decide whether samples will be returned to the laboratory for detailed examination or sorted in the field. If the samples are to be preserved immediately and returned to the laboratory for further analysis go to Step 9. If samples are to be processed in the field go to the next section on field sorting.

9. Pour the sample contents from the 20-liter bucket into a sieve and let the water drain from the sample. Transfer the sample into an appropriate sized container (e.g., 500-ml jar, 1-quart zip-lock bag), place a label containing date, site, and sample number into the container (use a small piece of paper and pencil), and preserve with 95% ethanol to cover the sample completely and to reach a final concentration of 70% ethanol.[4]

Field Sorting and Identification

1. If samples are to be sorted and identified at streamside, pour the contents of the sample from the bucket into an appropriate meshed sieve (e.g., 250, 500 μm). Refloat the sample by immersing the sieve in water (filling the enamel pan used in the next step with stream water is a convenient way to do this), being careful not to allow the water to breach the upper lip of the seive and thereby lose sample contents. Distribute the sample evenly on the seive-screen and remove the seive from the water.

2. Using a spoon or butter knife, divide the sample into approximately four equal sections on the surface of the seive-screen. You have now divided your sample into 1/4 subsamples. Remove one of the 1/4 subsamples, place it into a white enamel pan, add stream water, and distribute the sample around the pan. You should be able to observe many macroinvertebrates crawling or swimming about the pan.

3. It is important to have a sufficient sample size (i.e., one consisting of several hundred individuals). If there are tens of individuals in the pan, then add additional 1/4 subsamples from the sieve, as needed. If there are

---

[3]Collection, sorting, and identification of organisms may take several hours per sample; thus, if this exercise is being used within a class setting, we recommend a careful examination of time allocation.

[4]Add a small amount of Rose Bengal to the sample to stain invertebrates and aid in the separation of organisms from debris.

thousands of individuals, then you will need to further subsample by taking the 3/4 sample remaining on the sieve, returning it to the 20-liter bucket and placing the current contents of the 1/4 subsample in the pan back into the sieve. Now go back to step 2 and further subsampling procedure. Remember that now the subsamples are 1/16 of the original sample.

4. While invertebrates are still in the enamel pan, examine the body shape of the benthic animals you see from the different microhabitats.

5. Using a forceps, remove all macroinvertebrates from the sample (or subsample) and sort them into easily recognized groups. At a minimum, you should be able to identify taxa to the phyla and order levels using the general key provided in Appendix 16.1.

6. Place each taxon in a different container.[5] Count and record the number of individuals within each taxon.

7. For detailed identification and enumeration, the sorted sample must be returned to the laboratory to be examined using a binocular dissecting microscope. Place each sorted taxon into a separate container (e.g., jar, zip-lock bag), record the date, site, and sample number on a label for each container using a piece of paper and pencil (do not use a pen), and preserve with 70% ethanol.

Laboratory Sorting, Identification, and Enumeration

1. If samples were not field sorted, use the laboratory sink and empty the contents of a sample into an appropriate meshed sieve (e.g., 250, 500 μm). Rinse the sample thoroughly with tap water, being careful not to lose any material.[6] Go through the subsample and sorting procedures described in steps 1–5 above under Field Sorting and Identification

2. Use the key provided (Appendix 16.1) to identify the most commonly occurring taxa to the family level for insects and class level for noninsects. If the organism you are identifying does not "key out" or you desire greater resolution in identifications use McCafferty (1981), Pennak (1989). Thorp and Covich (1991), and Merritt and Cummins (1996) to separate the various taxa. A binocular dissecting microscope will be needed to view the morphological structures that are used to identify the organisms.

3. *Special note.* It will not be possible for the beginning student to identify in a single laboratory exercise all the various organisms that are typically found in an unpolluted stream. It generally takes months of work to develop the skills to identify organisms to the generic level and a lifetime of work to the species level.

---

[5]Wells of a muffin tin, plastic ice-cube tray, or styrofoam egg carton work well for this purpose.
[6]This also dilutes the alcohol concentration entering the sewage treatment system.

4. Observe the diversity of species within these taxonomic groups.

5. Select one or two taxonomic groups to examine in detail. For example, you may select to look at the net-spinning caddisflies (superfamily Hydropsychoidea, especially the family Hydropsychidae) or the predaceous stoneflies in the family Perlidae.

6. Carefully sort and identify the individuals from the taxa you have chosen to study.

7. List the genera/species collected from each of the differnt habitats. If identification beyond the taxonomic level presented in Appendix 16.1 is beyond your current expertice, then genera/species within an identified taxa may be further separated as species A, B, C, etc.

Data Analyses

1. Enumerate the selected taxa from each sample collected.

2. Calculate mean density and standard deviation for each selected taxon by habitat.

3. These data for the macroinvertebrate assemblages (e.g., abundance, taxa richness) can be used to calculate various population descriptors. Calculate two common diversity indices that are relative measures of species richness and equitability; the Shannon–Wiener index and Simpson's index. The Shannon–Wiener information theory index (H') is calculated as

$$H' = -\Sigma p_i \log p_i, \qquad (16.1)$$

where p_i represents proportion of the total number of individuals in the ith species.

Simpson's index (λ) is the probability that any two individuals picked at random will be of the same species and is calculated as

$$\lambda = \Sigma p_i^2. \qquad (16.2)$$

Simpson's index is a measure of how individuals in a sample are concentrated into a few species. See Chapter 31 for biotic indices used with stream macroinvertebrates.

Special note. Sorting, identification, and enumeration of macroinvertebrate samples can be very time consuming for even the most accomplished aquatic entomologist or benthic ecologist. Merritt and Cummins (1996) provide an excellent flow diagram summarizing the general procedures for analyzing benthic samples. Depending on taxonomic complexity, abundance, and extent of analyses (e.g., enumeration, wet weights, ash-free dry mass) a single sample may take 8–10 h (or more) spread over several days to complete.

B. Exercise 2: Watershed Scale Distribution

Stream ecologists have noted that particular macroinvertebrate species often occur only within very restricted stream reaches. In some cases, this is because the habitats that particular species require only occur within certain well-defined reaches. However, in some stream systems, various habitat types occur along the entire length of the stream (e.g., riffles) yet there remain reach-specific species. This exercise is designed to illustrate the macroinvertebrate species replacement that can occur along the downstream gradient of a lotic ecosystem. In this exercise we will collect macroinvertebrates from riffle habitat at three different stream reaches along the river continuum (sensu Vannote *et al.* 1980). We will then examine the macroinvertebrate assemblages from each stream reach looking for changes in species composition with an emphasis on closely related species. We suggest that you focus your attention on the net-spinning caddisflies for this exercise, because they occur in almost all unpolluted running water systems, especially in riffles of gravel-bed streams or on stable substrata (e.g., woody snags) in sandy-bottom streams or rivers. Also, the net-spinning caddisflies are easily recognized and larval taxonomy is fairly well known for many areas of North America and Europe.

Laboratory Preparation

1. Select a fourth- to fifth-order stream network using a detailed watershed map(s) (e.g., USGS quadrangle map, scale 1:24,000).

2. Select three sampling sites along the river, with consideration for ease of access and diversity of habitats. The first site should be located in a first- to second-order stream, the second site in a third-order stream, and the third site at the most downstream end of the selected watershed, preferably fifth-order.

3. Gather field collecting supplies.

Field Collection

1. Obtain quantitative samples from riffle habitat using the technique described above under Field Sampling and in Exercise 1 (e.g., a kick-net, Surber, or Hess sampler).

2. Examine each sample for the presence of the macroinvertebrate guild of interest (e.g., Hydropsychid caddisflies, Perlid stoneflies, or Ephemerelid mayflies). Place contents of the sample into a white enamel pan; use forceps to remove the specific taxa of interest from the sample material.

Place larvae into a moderate size (500-ml) plastic jar with 70% ethanol, using a single jar for each sample. The remainder of the live sample may be preserved for additional examination or returned to the stream.

3. Make written notes of your observations of the target species. Jot down readily apparent differences in taxa present and their abundance.

Laboratory Analysis

1. After bringing samples back to the laboratory, identify target taxa to the genus or species level.

2. Use a binocular dissecting scope to observe the organisms and their key morphological structures. Merritt and Cummins (1996) provide generic-level keys; however, various regional keys for identification of species are available. (For illustrative purposes, you can use "morphospecies"; e.g., species A, B, C, etc.)

3. Observe the diversity of species within the taxonomic and functional group with which you are working.

4. Carefully sort, identify, and enumerate the individuals of each taxon from each sample.

5. Calculate the mean density and standard deviation for each species by location.

C. Exercise 3: Population Dynamics and Movement

Populations change in size over time. They increase from new births and the immigration of individuals from other areas, and they decrease from death and emigration. In this exercise we will mark members of populations of aquatic insects to observe their movements over time, as well as losses (emigration, death) or gains (immigration, births) to the population.

Water striders (Family Gerridae, and the species you probably have is *Aquarius* (=*Gerris*) *remigis*) occur on several continents and *A. remigis* likely is the most widely distributed species of aquatic insect worldwide. Individuals commonly occur in the slow-flowing margins and pools of streams (see Spence and Anderson (1994) for a detailed review of the biology of water striders). Using a handnet, catch one of these surface-dwelling creatures. Note that its "back" (i.e., the dorsum of the thorax) is where we will apply our marking tag—a dab of typewriter correction fluid ("white-out," which comes in a variety of colors and by using up to three marks on an insect and different colors, scores of individuals can be marked and followed).

Behavioral Observations

1. Working in pairs, we will describe the spatial distribution of *A. remigis* (or some other insect such as a cased limnephilid caddisfly like *Dicosmoecus*, see below).

2. Map a segment of stream ($\approx$50 m reach) following the example of Chapter 2 or 4 (or if you are working in the same location use the maps created in an earlier exercise).

3. Collect and sex (in the case of water striders) each individual (seeing two genders side by side makes this pretty clear); then mark them by using different colors (e.g., white for males, yellow for females, and blue for nymphs) and release them where you caught them.

4. Observe the behavior of individual water striders with respect to their location in the stream, and their resting, mating, searching, fighting, and feeding behaviors. Watch each individual for 10–15 min, and be sure to compare individuals of different genders (which you now can distinguish from the different colored marks on their thorax).

5. Make detailed notes on each of these behaviors and the time spent doing each behavior.

Mark and Recapture

1. A final exercise is a mark–recapture study.

2. This involves sampling with a D-frame net on 2 days, about 1 week apart.

3. Record the number of individuals originally marked on Day 1, the number collected on Day 2 that were marked and unmarked, and then calculate density (N),

$$N = \frac{M \times C}{R}, \tag{16.3}$$

where M is the No. originally marked; C, total catch on Day 2; and R, the No. of Day-2 recaptures (i.e., those originally marked on Day 1).

4. *Special note.* A second group of organisms appropriate for typewriter correction-fluid marking are the larger cased caddisflies in the family Limnephilidae (e.g., *Dicosmoecus*). In marking, remove the larva with case from the water, pat the case dry, add the mark (letting it dry for 1 min or less), and then return the caddisfly to the stream at the place collected. Another interesting exercise is to compare upstream–downstream movements of marked *Dicosmoecus* larvae that have all been released at a single point.

D. Exercise 4: Laboratory Artificial Stream Experiments

There are many experiments that can be conducted in laboratory streams (e.g., behavior, response to dynamics of flow, growth) (see Hauer 1993). Lamberti and Steinman (1993) provide many designs and applications of laboratory streams. In this exercise, we will construct several small air-lift chambers that provide the microhabitat-flow requirements needed by black flies (Simuliidae) and determine larval growthh rates in different environmental conditions (after Hauer and Benke 1987).

Set-up and Experimention

1. Construct at least four artificial stream tanks based on Fig. 16.3.
2. Using tropical fish aquarium supplies (listed below) arrange the artificial stream tanks to provide an "airlift" current when the air pump is on.
3. Water level in the artificial stream tanks should be maintained near full, but not so full that water spills over the top.
4. Obtain black fly larvae from a nearby stream and return live specimens to the laboratory in a large bucket. Collect at least 200–250 individuals in mid-size classes (3–4 mm).
5. Select a random sample of 15–20 animals of the population and preserve in 70% ethanol.
6. Distribute the remaining animals randomly among the four artificial

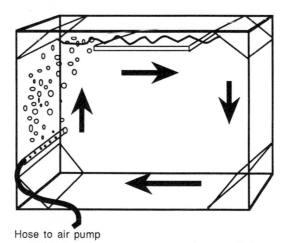

Hose to air pump

FIGURE 16.3 Illustration of an air-lift tank used for the simulation of stream conditions in black fly growth experiments. Dimensions: 10 cm W, 15 cm L, 12 cm H.

stream tanks. Maintain tanks and permit larvae to feed and grow over a 1- to 2-week period. It is necessary to add replacement streamwater to the tanks twice daily throughout the experiments to maintain water levels and natural food levels. This is best done by bringing unfiltered stream water to the laboratory in 20-liter carboy containers. You will need to calculate the amount of water added over the duration of the experiments.

7. Experimental conditions may be varied among the artificial stream tanks. For example, some tanks may be kept at cool temperatures in a refrigerator or environmental chamber while other tanks are maintained at room temperature. Likewise, some tanks may be given a supplemental food source of either cultured algae, natural seston collected from a stream, or small quantities of granular baker's yeast.

Analysis of Growth Experiments

1. Terminate the growth experiments after 10–14 days or as soon as the first individuals begin to pupate, whichever comes first.

2. Keeping larvae from each tank separate, collect and preserve animals in 70% ethanol.

3. Using a dissecting microscope fitted with an ocular micrometer, measure the total length of all larvae from each experimental stream tank and the larvae that were preserved at the start of the experiments.

4. Dry mass (DM) of each larva may be predicted from the regression

$$DM = 0.0031 \times BL^{2.64}, \qquad (16.4)$$

where BL is body length in mm.

5. Calculate daily instantaneous growth rates (g) for larvae as

$$g = \frac{\ln (DM_f DM_i)}{t}, \qquad (16.5)$$

where DM_f is the mean dry mass (in mg) of larvae at the end of the growth experiment; DM_i, the mean dry mass of the larvae at the start of the experiment; and t, the No. of days for the particular trial.

V. QUESTIONS

1. Did you observe specifc patterns of microhabitat preference? How do these relate to the morphological and behavioral adaptations described in the Introduction?

2. Were you able to see morphological differences among the species that you collected from different habitats? How did the morphology of species collected from riffles differ from species collected from pools, debris dams, and or leaf litter?

3. Consider the breadth of different habitats that you have observed in stream ecosystems. Imagine that you are standing next to a stream whose bottom and sides are concrete. In your mind or on paper consider how you would remake this concrete channel into a living stream. What structural components would you add to increase microhabitat complexity (and hence abundance of organisms)? Consider the stream bed. How would you integrate the hyporheic zone into your imagined stream? Where do factors such as riparian vegetation and nutrient sources come into play? You have now begun to think about stream restoration.

4. Did you observe distinctly different species within the larger taxonomic group that you identified. Even though you may not have been able to identify your specimens to the species level, based on your observations how many different taxa were you able to distinguish?

5. Did you observe a pattern of different species between sample sites along the longitudinal gradient of the stream within the taxonomic group that you studied in detail? What general patterns of species distributions or replacements did you observe?

6. How do water striders respond to differences in flow? On your original map, record areas of fast, medium, and slow flow (see Chapter 4) and compare to strider distribution. Were there gender, age, or other factors within the water strider experiments that appear to influence distribution? Can food resources you provide be used to alter microhabitat selection? What are some assumptions that we make about the effect of the mark on the animal when we conduct such experiments?

7. If you conducted growth experiments, what was the growth rate of larvae from each of the experimental stream tanks? Did different temperatures or different levels of food resources affect growth rate?

8. Consider instantaneous growth (g). What relationship does g have to secondary production? (see Chapter 26).

V. MATERIALS AND SUPPLIES

Field Materials

20-liter plastic bucket
≤500-μm Sieve or bagnet
95% Ethanol

Current velocity meter
D-frame net
Kick-net
Meter sticks
Stopwatch
Surber or Hess sampler (optional)
Typewriter correction fluid (various colors; for Exercise 3)

Laboratory Equipment and Materials

70% Ethanol
Appendix 16.1 and reference books mentioned in text
Artificial stream tank, air pump, tubing (Exercise 4)
Binocular dissecting microscope
Forceps
Scintillation vials or patent-lip vials with neoprene stoppers
White enamel pans for sorting

APPENDIX 16.1 A SIMPLIFIED KEY FOR THE RAPID IDENTIFICATION OF THE MOST COMMONLY OCCURRING STREAM MACROINVERTEBRATES

Noninsect taxa are described to the phyla or order level. Insect taxa are described to the family level. Many more stream macroinvertebrates occur than are presented here; this key is to only serve as a starting point for their identification. (Illustrations taken from Betten 1934, McCafferty 1981, and Thorp and Covich 1991, with permission).

APPENDIX 16.1—*continued*

KEY 1

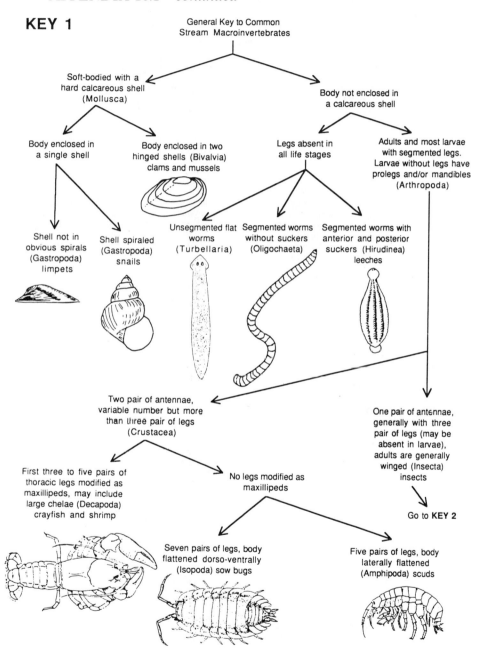

General Key to Common Stream Macroinvertebrates

Soft-bodied with a hard calcareous shell (Mollusca)

Body not enclosed in a calcareous shell

Body enclosed in a single shell

Body enclosed in two hinged shells (Bivalvia) clams and mussels

Legs absent in all life stages

Adults and most larvae with segmented legs. Larvae without legs have prolegs and/or mandibles (Arthropoda)

Shell not in obvious spirals (Gastropoda) limpets

Shell spiraled (Gastropoda) snails

Unsegmented flat worms (Turbellaria)

Segmented worms without suckers (Oligochaeta)

Segmented worms with anterior and posterior suckers (Hirudinea) leeches

Two pair of antennae, variable number but more than three pair of legs (Crustacea)

One pair of antennae, generally with three pair of legs (may be absent in larvae), adults are generally winged (Insecta) insects

First three to five pairs of thoracic legs modified as maxillipeds, may include large chelae (Decapoda) crayfish and shrimp

No legs modified as maxillipeds

Go to **KEY 2**

Seven pairs of legs, body flattened dorso-ventrally (Isopoda) sow bugs

Five pairs of legs, body laterally flattened (Amphipoda) scuds

APPENDIX 16.1—*continued*

KEY 2

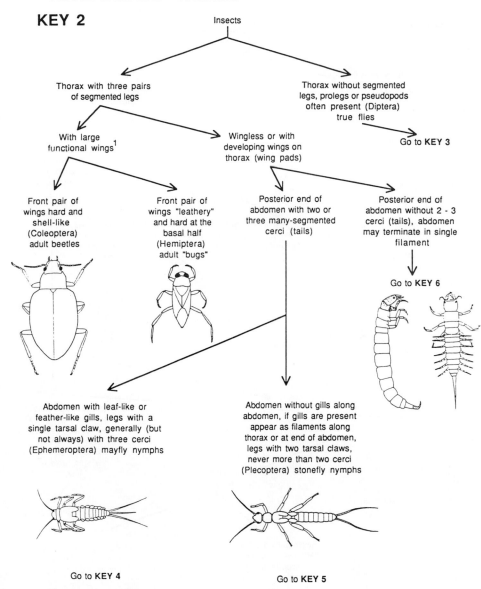

Insects

Thorax with three pairs
of segmented legs

Thorax without segmented
legs, prolegs or pseudopods
often present (Diptera)
true flies

Go to **KEY 3**

With large
functional wings[1]

Wingless or with
developing wings on
thorax (wing pads)

Front pair of
wings hard and
shell-like
(Coleoptera)
adult beetles

Front pair of
wings "leathery"
and hard at the
basal half
(Hemiptera)
adult "bugs"

Posterior end of
abdomen with two or
three many-segmented
cerci (tails)

Posterior end of
abdomen without 2 - 3
cerci (tails), abdomen
may terminate in single
filament

Go to **KEY 6**

Abdomen with leaf-like or
feather-like gills, legs with a
single tarsal claw, generally (but
not always) with three cerci
(Ephemeroptera) mayfly nymphs

Abdomen without gills along
abdomen, if gills are present
appear as filaments along
thorax or at end of abdomen,
legs with two tarsal claws,
never more than two cerci
(Plecoptera) stonefly nymphs

Go to **KEY 4**

Go to **KEY 5**

[1] NOTE: Other winged insects may be present from either
terrestrial forms or the aerial adults of aquatic larvae

APPENDIX 16.1—*continued*

KEY 3

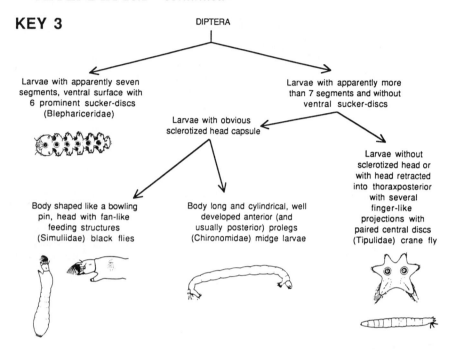

DIPTERA

Larvae with apparently seven segments, ventral surface with 6 prominent sucker-discs (Blephariceridae)

Larvae with apparently more than 7 segments and without ventral sucker-discs

Larvae with obvious sclerotized head capsule

Larvae without sclerotized head or with head retracted into thoraxposterior with several finger-like projections with paired central discs (Tipulidae) crane fly

Body shaped like a bowling pin, head with fan-like feeding structures (Simuliidae) black flies

Body long and cylindrical, well developed anterior (and usually posterior) prolegs (Chironomidae) midge larvae

APPENDIX 16.1—*continued*

KEY 4

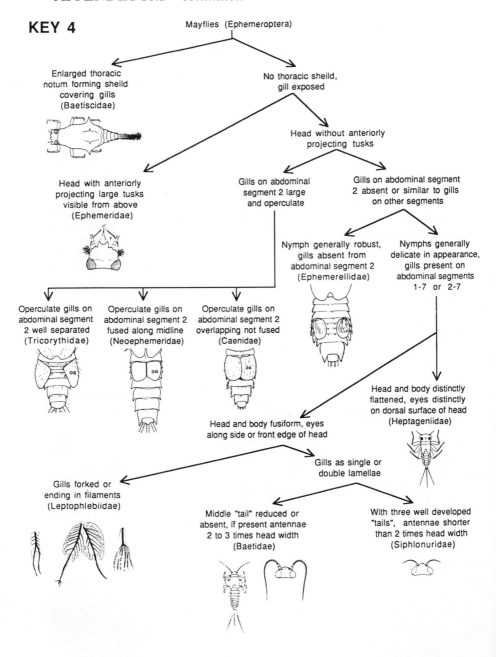

Mayflies (Ephemeroptera)

Enlarged thoracic notum forming sheild covering gills (Baetiscidae)

No thoracic sheild, gill exposed

Head without anteriorly projecting tusks

Head with anteriorly projecting large tusks visible from above (Ephemeridae)

Gills on abdominal segment 2 large and operculate

Gills on abdominal segment 2 absent or similar to gills on other segments

Nymph generally robust, gills absent from abdominal segment 2 (Ephemerellidae)

Nymphs generally delicate in appearance, gills present on abdominal segments 1-7 or 2-7

Operculate gills on abdominal segment 2 well separated (Tricorythidae)

Operculate gills on abdominal segment 2 fused along midline (Neoephemeridae)

Operculate gills on abdominal segment 2 overlapping not fused (Caenidae)

Head and body distinctly flattened, eyes distinctly on dorsal surface of head (Heptageniidae)

Head and body fusiform, eyes along side or front edge of head

Gills as single or double lamellae

Gills forked or ending in filaments (Leptophlebiidae)

Middle "tail" reduced or absent, if present antennae 2 to 3 times head width (Baetidae)

With three well developed "tails", antennae shorter than 2 times head width (Siphlonuridae)

APPENDIX 16.1—*continued*

KEY 5

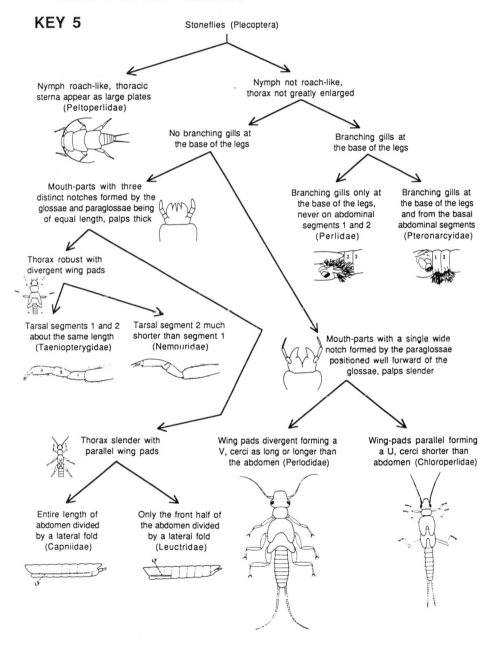

Stoneflies (Plecoptera)

Nymph roach-like, thoracic sterna appear as large plates (Peltoperlidae)

Nymph not roach-like, thorax not greatly enlarged

No branching gills at the base of the legs

Branching gills at the base of the legs

Mouth-parts with three distinct notches formed by the glossae and paraglossae being of equal length, palps thick

Branching gills only at the base of the legs, never on abdominal segments 1 and 2 (Perlidae)

Branching gills at the base of the legs and from the basal abdominal segments (Pteronarcyidae)

Thorax robust with divergent wing pads

Tarsal segments 1 and 2 about the same length (Taeniopterygidae)

Tarsal segment 2 much shorter than segment 1 (Nemouridae)

Mouth-parts with a single wide notch formed by the paraglossae positioned well forward of the glossae, palps slender

Thorax slender with parallel wing pads

Wing pads divergent forming a V, cerci as long or longer than the abdomen (Perlodidae)

Wing-pads parallel forming a U, cerci shorter than abdomen (Chloroperlidae)

Entire length of abdomen divided by a lateral fold (Capniidae)

Only the front half of the abdomen divided by a lateral fold (Leuctridae)

APPENDIX 16.1—*continued*

KEY 6

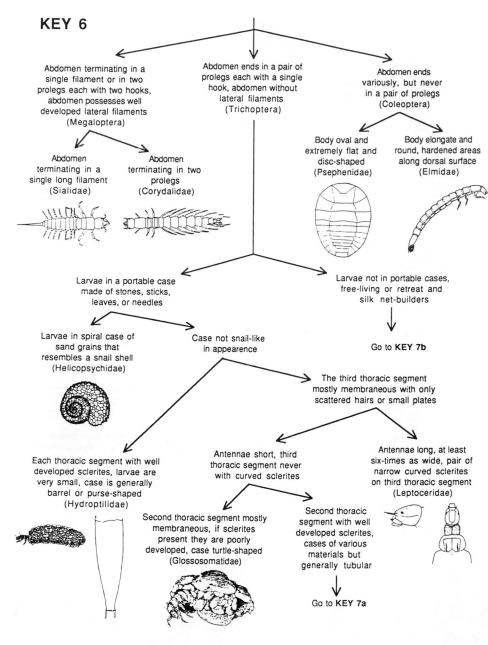

Abdomen terminating in a single filament or in two prolegs each with two hooks, abdomen possesses well developed lateral filaments (Megaloptera)

Abdomen ends in a pair of prolegs each with a single hook, abdomen without lateral filaments (Trichoptera)

Abdomen ends variously, but never in a pair of prolegs (Coleoptera)

Abdomen terminating in a single long filament (Sialidae)

Abdomen terminating in two prolegs (Corydalidae)

Body oval and extremely flat and disc-shaped (Psephenidae)

Body elongate and round, hardened areas along dorsal surface (Elmidae)

Larvae in a portable case made of stones, sticks, leaves, or needles

Larvae not in portable cases, free-living or retreat and silk net-builders

Larvae in spiral case of sand grains that resembles a snail shell (Helicopsychidae)

Case not snail-like in appearance

Go to **KEY 7b**

The third thoracic segment mostly membraneous with only scattered hairs or small plates

Each thoracic segment with well developed sclerites, larvae are very small, case is generally barrel or purse-shaped (Hydroptilidae)

Antennae short, third thoracic segment never with curved sclerites

Antennae long, at least six-times as wide, pair of narrow curved sclerites on third thoracic segment (Leptoceridae)

Second thoracic segment mostly membraneous, if sclerites present they are poorly developed, case turtle-shaped (Glossosomatidae)

Second thoracic segment with well developed sclerites, cases of various materials but generally tubular

Go to **KEY 7a**

APPENDIX 16.1—*continued*

KEY 7

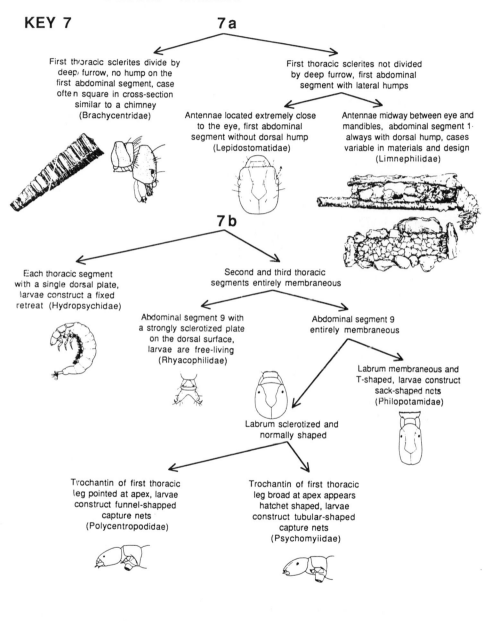

7a

First thoracic sclerites divide by deep furrow, no hump on the first abdominal segment, case often square in cross-section similar to a chimney (Brachycentridae)

First thoracic sclerites not divided by deep furrow, first abdominal segment with lateral humps

Antennae located extremely close to the eye, first abdominal segment without dorsal hump (Lepidostomatidae)

Antennae midway between eye and mandibles, abdominal segment 1 always with dorsal hump, cases variable in materials and design (Limnephilidae)

7b

Each thoracic segment with a single dorsal plate, larvae construct a fixed retreat (Hydropsychidae)

Second and third thoracic segments entirely membraneous

Abdominal segment 9 with a strongly sclerotized plate on the dorsal surface, larvae are free-living (Rhyacophilidae)

Abdominal segment 9 entirely membraneous

Labrum membraneous and T-shaped, larvae construct sack-shaped nets (Philopotamidae)

Labrum sclerotized and normally shaped

Trochantin of first thoracic leg pointed at apex, larvae construct funnel-shapped capture nets (Polycentropodidae)

Trochantin of first thoracic leg broad at apex appears hatchet shaped, larvae construct tubular-shaped capture nets (Psychomyiidae)

REFERENCES

Allan, J. D. 1975. The distributional ecology and diversity of benthic insects in Cement Creek, Colorado. *Ecology* **56:**1040–1053.

Allan, J. D. 1984. Hypothesis testing in ecological studies of aquatic insects. Pages 485–507 *in* V. H. Resh and D. M. Rosenberg (Eds.) *The Ecology of Aquatic Insects.* Praeger, New York, NY.

Allan, J. D. 1995. *Stream Ecology.* Chapman & Hall, London, UK.

Alstad, D. N. 1980. Comparative biology of the common Utah Hydropsychidae (Trichoptera). *American Midland Naturalist* **103:**167–174.

Baumann, R. W., A. R. Gaufin, and R. F. Surdick. 1977. *The Stoneflies (Plecoptera) of the Rocky Mountains.* American Entomological Society, Philadelphia, PA.

Benke, A. C., T. C. Van Arsdall, Jr., D. M. Gillespie, and F. K. Parrish. 1984. Invertebrate productivity in a subtropical blackwater river: The importance of habitat and life history. *Ecological Monographs* **54:**25–63.

Betten, C. 1934. *The Caddis Flies or Trichoptera of New York State.* New York State Museum Bulletin, The University of the State of New York, Albany, NY.

Brittain, J. E., and T. J. Eikeland. 1988. Invertebrate drift—A review. *Hydrobiologia* **166:**77–80.

Butler, M. G. 1984. Life histories of aquatic insects. Pages 24–55 *in* V. H. Resh and D. M. Rosenberg (Eds.) *The Ecology of Aquatic Insects.* Praeger, New York, NY.

Cudney, M. D., and J. B. Wallace. 1980. Life cycles, microdistribution and production dynamics of six species of net-spinning caddisflies in a large southeastern (U.S.A.) river. *Holarctic Ecology* **3:**169–182.

Cuffney, T. F., M. E. Gurtz, and M. R. Meador. 1993. *Methods for Collecting Benthic Invertebrate Samples as Part of the National Water-Quality Assessment Program.* United States Geological Survey. Open File Report 93-406. Raleigh, NC.

Cummins, K. W. 1974. Structure and function of stream ecosystems. *BioScience* **24:**631–641.

Elliott, J. M. 1977. *Some Methods for the Statistical Analysis of Samples of Benthic Invertebrates,* 2nd ed. Freshwater Biological Association Scientific Publication 25, Cumbria, UK.

Eriksen, C. H., V. H. Resh, and G. A. Lamberti. 1996. Aquatic insect respiration. Pages 29–40 *in* R. W. Merritt and K. W. Cummins (Eds.) *An Introduction to the Aquatic Insects of North America,* 3rd ed. Kendall/Hunt, Dubuque, IA.

Gregory, S. V., F. J. Swanson, W. A. McKee, and K. W. Cummins. 1991. An ecosystem perspective of riparian zones. *Bioscience* **41:**540–551.

Hauer, F. R. 1993. Artificial streams for the study of macroinvertebrate growth and bioenergetics. *Journal of the North American Benthological Society* **12:**313–384.

Hauer, F. R., and A. C. Benke. 1987. Influence of temperature and river hydrograph on blackfly growth rates in a subtropical blackwater river. *Journal of the North American Benthological Society* **6:**251–261.

Hauer, F. R., and J. A. Stanford. 1981. Larval specialization and phenotypic variation in *Arctopsyche grandis* (Trichoptera: Hydropsychidae). *Ecology* **62:**645–653.

Hauer, F. R., and J. A. Stanford. 1982a. Ecology and life histories of three net-spinning caddisfly species (Hydropsychidae: *Hydropsyche*) in the Flathead River, Montana. *Freshwater Invertebrate Biology* **1**:18–29.

Hauer, F. R., and J. A. Stanford. 1982b. Bionomics of *Dicosmoecus gilvipes* (Trichoptera: Limnephilidae) in a large western montane river. *American Midland Naturalist* **108**:81–87.

Hauer, F. R., and J. A. Stanford. 1982c. Ecological responses of hydropsychid caddisflies to stream regulation. *Canadian Journal of Fisheries and Aquatic Sciences* **39**:1235–1242.

Hauer, F. R., and J. A. Stanford. 1986. Ecology and coexistence of two species of *Brachycentrus* (Trichoptera) in a Rocky Mountain river. *Canadian Journal of Zoology* **64**:1469–1474.

Hess, A. D. 1941. New limnological sampling equipment. *Limnological Society of America Special Publication* **6**:1–5.

Hildrew, A. G., and J. M. Eddington. 1979. Factors facilitating the coexistence of hydropsychid caddis larvae (Trichoptera) in the same river system. *Journal of Animal Ecology* **48**:557–576.

Hynes, H. B. N. 1970. *The Ecology of Running Waters*. University of Toronto Press, Toronto, Canada.

Hynes, H. B. N. 1975. The stream and its valley. *Verhandlungen der Internationalen Vereinigung für Theoretische und Angewandte Limnologie* **19**:1–15.

Lamberti, G. A., and A. D. Steinman (Eds.). 1993. Research in artificial streams: Applications, uses, and abuses. *Journal of the North American Benthological Society* **12**:313–384.

McCafferty, W. P. 1981. *Aquatic Entomology: The Fisherman's and Ecologist's Illustrated Guide to Insects and Their Relatives*. Science Books International, Boston, MA.

Merritt, R. W., and K. W. Cummins (Eds.). 1996. *An Introduction to the Aquatic Insects of North America*, 3rd ed. Kendall/Hunt, Dubuque, IA.

Merritt, R. W., V. H. Resh, and K. W. Cummins. 1996. Design of aquatic insect studies: Collecting, sampling and rearing procedures. Pages 12–28. *in* R. W. Merritt and K. W. Cummins (Eds.) *An Introduction to the Aquatic Insects of North America*, 3rd ed. Kendall/Hunt, Dubuque, IA.

Morin, A. 1985. Variability of density estimates and the optimization of sampling programs for stream benthos. *Canadian Journal of Fisheries and Aquatic Sciences* **42**:1530–1540.

Morse, J. C., J. W. Chapin, D. D. Herlong, and R. S. Harvey. 1980. Aquatic insects of Upper Three Runs Creek, Savannah River Plant, South Carolina. Orders other than Diptera. *Journal of the Georgia Entomological Society* **15**:73–101.

Müller, K. 1974. Stream drift as a chronobiological phenomenon in running water ecosystems. *Annual Review of Ecology and Systematics* **5**:309–323.

Nimmo, A. P. 1971. The adult Rhyacophilidae and Limnephilidae (Trichoptera) of Alberta and eastern British Columbia and their post-glacial origin. *Quaestiones Entomologicae* **7**:3–234.

Norris, R. H., and A. Georges. 1993. Analysis and interpretation of benthic macroinvertebrate surveys. Pages 230–282 *in* D. M. Rosenberg and V. H. Resh (Eds.) *Freshwater Biomonitoring and Benthic Macroinvertebrates.* Chapman & Hall, New York, NY.

Norris, R. H., E. P. McElravy, and V. H. Resh. 1992. The sampling problem. Pages 282–306 *in* P. Calow and G. E. Petts (Eds.) *The Rivers Handbook: Hydrological and Ecological Principles*, Vol. 1. Blackwell Scientific, Oxford, UK.

Pennak, R. W. 1989. *Fresh-Water Invertebrates of the United States. Protozoa to Mollusca*, 3rd ed. Wiley, New York, NY.

Resh, V. H. 1979. Sampling variability and life history features: Basic considerations in the design of aquatic insect studies. *Journal of the Fisheries Research Board of Canada* **36:**290–311.

Resh, V. H., J. K. Jackson, and J. R. Wood. 1987. Techniques for demonstrating sex pheromones in trichoptera. Pages 161–164 *in* M. Bournaud and H. Tachet (Eds.) *5th International Symposium on Trichoptera.* Dr. W. Junk Publishers, The Hague, Netherlands.

Resh, V. H., J. W. Feminella, and E. P. McElravy. 1990. *Sampling Aquatic Insects.* 38-Minute videotape, VHS format. Office of Media Services, University of California, Berkeley, CA.

Resh, V. H., and E. P. McElravy. 1993. Contemporary quantitative approaches to biomonitoring using benthic macroinvertebrates. Pages 157–191 *in* D. M. Rosenberg and V. H. Resh (Eds.) *Freshwater Biomonitoring and Benthic Macroinvertebrates.* Chapman & Hall, New York, NY.

Resh, V. H., and J. O. Solem. 1996. Phylogenetic relationships and evolutionary adaptations of aquatic insects. Pages 98–107 *in* R. W. Merritt and K. W. Cummins (Eds.) *An Introduction to the Aquatic Insects of North America*, 3rd ed. Kendall/Hunt, Dubuque, IA.

Spence, J. R., and N. M. Anderson, 1994. Biology of water striders: Interactions between systematics and ecology. *Annual Review of Entomology* **39:**101–128.

Stanford, J. A., F. R. Hauer, and J. V. Ward. 1988. Serial discontinuity in a large river system. *Verhandlungen der Internationalen Vereinigung für Theoretische und Angewandte Limnologie* **23:**1114–1118.

Stanford, J. A., and J. V. Ward. 1993. An ecosystem perspective of alluvial rivers: connectivity and the hyporheic corridor. *Journal of the North American Benthological Society* **12:**48–60.

Surber, E. W. 1937. Rainbow trout and bottom fauna production in one mile of stream. *Transactions of the American Fisheries Society* **66:**193–202.

Szczytko, S. W., and K. W. Stewart. 1979. The genus *Isoperla* (Plecoptera) of western North America: holomorphology and systematics, and a new stonefly genus *Cascadoperla. Memoirs of the American Entomological Society* **32:**1–120.

Thorp, J. H., and A. P. Covich (Eds.). 1991. *Ecology and Classification of North American Freshwater Invertebrates.* Academic Press, San Diego, CA.

Vannote, R. L., G. W. Minshall, K. W. Cummins, J. R. Sedell, and C. E. Cushing. 1980. The river continuum concept. *Canadian Journal of Fisheries and Aquatic Sciences* **37:**130–137.

Ward, J. V., and J. A. Stanford. 1983. The serial discontinuity concept of lotic ecosystems. Pages 29–42 *in* T. D. Fontaine and S. M. Bartell (Eds.) *Dynamics of Lotic Ecosystems*. Ann Arbor Science Publishers, Ann Arbor, MI.

Ward, J. V., and J. A. Stanford. 1991. Benthic faunal patterns along the longitudinal gradient of a Rocky Mountain river system. *Verhandlungen der Internationalen Vereinigung für Theoretische und Angewandte Limnologie* **24:**3087–3094.

Waters, T. F. 1972. The drift of stream insects. *Annual Review of Entomology* **17:**253–272.

Wiley, M., and S. L. Kohler. 1984. Behavorial adaptations of aquatic insects. Pages 101–133 *in* V. H. Resh and D. M. Rosenberg (Eds.) *The Ecology of Aquatic Insects*. Praeger, New York, NY.

Williams, D. D. 1987. *The Ecology of Temporary Waters*. Timber Press, Portland, OR.

CHAPTER 17

Macroinvertebrate Movements: Drift, Colonization, and Emergence

LEONARD A. SMOCK

Department of Biology
Virginia Commonwealth University

I. INTRODUCTION

Dispersal, or the movement of individuals from one area to another, is an activity exhibited by most species. In streams, both active and passive dispersal movements are common among benthic macroinvertebrates in response to a number of factors. The continuous flow of water in lotic environments provides a convenient and energetically efficient mechanism for downstream dispersal, but it can force unwanted displacement of individuals to downstream areas and can make upstream dispersal difficult. Overland dispersal also occurs, primarily by the flight of adult insects that emerge from streams. Dispersal of aquatic macroinvertebrates also is a key process in the recolonization of disturbed areas of streams, such as areas scoured and denuded by spates.

This chapter introduces key concepts and sampling methods concerning the dispersal of stream-dwelling, benthic macroinvertebrates. Exercises 1 and 2 are concerned with *drift*, the most common form of long-distance movement by benthic fauna in streams. Exercises 3, 4, and 5 focus on the direction of movements by benthic macroinvertebrates and the mechanisms of *colonization* of disturbed sediments. The final exercise (5) examines the *emergence* of adult insects.

A. Drift

Although most macroinvertebrates that occur in streams and rivers are benthic, a net placed in the water column often will collect many individuals. These organisms are drifting, an activity whereby they enter the water column and are transported downstream by the current. It is one of the most important mechanisms for the dispersal to and colonization of downstream habitats by a wide variety of stream macroinvertebrates. Drift also is one of the most studied activities of benthic fauna and has been the subject of a number of review articles (Waters 1972, Müller 1974, Wiley and Kohler 1984, Brittain and Eikeland 1988).

An interesting aspect of drift, and one that has intrigued stream ecologists for decades, is that drift usually exhibits a distinct diel periodicity whereby the number of individuals drifting changes over a 24-h period. The majority of species drift in maximum numbers sometime during the night. The most common drift pattern is that of higher numbers drifting shortly after sunset (Fig. 17.1), although peaks in the middle of the night and just prior to sunrise also occur.

Whereas some individuals may passively enter the drift, for example by accidently being swept away by the current (Kovalek 1979), others exhibit what is known as *drift behavior,* or active entry into the water column (Wiley and Kohler 1984). Changes in ambient light intensity, although not the ultimate reason for drift behavior, serve as the trigger or phase-setting

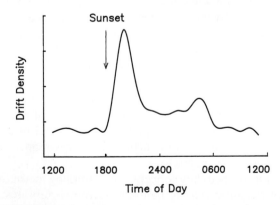

FIGURE 17.1 Typical macroinvertebrate diel drift pattern.

agent for drift. Most species have a threshold light intensity below which active drift may be initiated, as might occur at sunset.

A variety of reasons for macroinvertebrates to actively drift have been suggested. These include dispersal in search of suitable resources such as food and substratum (Otto 1976, Walton *et al.* 1977), escape from predators and competitive interactions (Corkum and Pointing 1979, Malmqvist and Sjostrom 1987), avoidance of unfavorable environmental conditions including various forms of pollution (Ciborowski *et al.* 1977, Wallace *et al.* 1989), and movements associated with life history events such as egg hatching, pupation, and emergence (Otto 1976, Krueger and Cook 1981, Ernst and Stewart 1985).

While there are many potential benefits of drifting, there are also potential costs. One of those is the increased risk of capture by fish, as macroinvertebrates are very vulnerable to fish predation while drifting. Many stream-dwelling fish, however, are visual feeders, needing to see their prey to capture them. Drifting at night, especially by large individuals that would be more readily visible than small individuals, may be an adaptation to decrease the risk of fish predation (Allan 1978, 1984).

The objectives of the two exercises on drift are to (1) introduce the phenomenon of macroinvertebrate drift in streams, (2) demonstrate the general methodology used to quantify drift, and (3) illustrate differences in the numbers and the size composition of macroinvertebrates drifting between day and night.

B. Colonization and Movement

Colonization is a process whereby organisms move to and become established in new areas or habitats or in disturbed habitats in which they previously were present (more accurately called recolonization). It occurs over broad and variable spatial and time scales (Sheldon 1984). Colonization in streams usually occurs on the sediment surface, although other substrata such as woody debris can also be colonized (Thorp *et al.* 1985). Colonization of new or denuded substratum is a common phenomenon in streams, occurring primarily as a response to sediment-scouring storms but also to other disturbances such as toxic pollutants and drying of the streambed during periods of drought.

Numerous studies have shown that macroinvertebrates can quickly colonize new or disturbed substrata, although the rate of colonization differs among species and with distance from colonizing sources (Gore 1982), time of the year (Williams 1980), and the physical characteristics of the substratum, in particular substratum particle size (Wise and Molles 1979). Particle size is an important factor in determining community structure in streams (Minshall 1984) and many species have morphological or physiological adaptations suited to their preferred substratum. These preferences help determine the likelihood of an individual colonizing and remaining on specific substrata.

Colonization of substrata requires movement of organisms from source areas to the new or disturbed substratum. Williams and Hynes (1976) noted four routes of colonization for macroinvertebrates. *Downstream movement* of organisms, primarily by drift but also from movement along the sediment, is usually the most important mechanism of movement, and hence colonization, in streams. *Upstream movements* along the sediment, possibly the result of a positive rheotaxis by macroinvertebrates, also occur. Many species exhibit upstream movements, but the distances traveled and usually low rates of movement limit their importance to the colonization process (Söderström 1987, Delucchi 1989).

Movements from the subsurface, or hyporheic zone, have also been documented in streams (Benzie 1984, Delucchi 1989, see also Chapters 6, 30). The hyporheic area of streams can have high densities of a wide variety of macroinvertebrate species and may serve as a refuge for organisms during unfavorable conditions on the surface (Sedell *et al.* 1990). It thus may be an important source of animals for colonizing surface substrata following disturbances.

Colonization by *aerial sources* is a potentially important mechanism of colonization in all streams (Gray and Fisher 1981, Benzie 1984, Cushing and Gaines 1989). Ovipositing by aerial adults is the primary mechanism of aerial colonization and is greatly affected by the time of year and the distance from source areas.

The objectives of the three protocols on colonization are to (1) introduce the concepts and mechanisms of substratum colonization in streams, (2) illustrate the process of colonization over time, (3) determine the effects of sediment particle size preferences on colonization by macroinvertebrates, and (4) demonstrate the different routes of movement and colonization in streams.

C. Adult Emergence

The sampling of aquatic insects has historically focused on their immature stages. Most species of aquatic insects, however, metamorphose into adults that emerge from the water and are then active in the terrestrial environment. While the primary activity of adults is reproduction, including mating and ovipositing, most individuals are capable of dispersal flights that can result in the laying of eggs far from the site of emergence. Indeed, a *colonization* cycle has been hypothesized, the central component of which is the possibility that the flight of females prior to ovipositing is primarily directed upstream, thereby compensating for the predominately downstream movement of the immature individuals living in the water (Müller 1982).

Collecting adults can provide important information not always obtainable from immature forms. Species-level identifications often can be made only on adult specimens because the taxonomy of the immature forms of many groups of aquatic insects is not completely known. Collecting adults also can provide information critical to the understanding of the population biology and life history of a species.

The objectives of the protocol on emergence are designed to (1) introduce the methodology for sampling emerging adults and (2) examine the differences in emergence that occur in different areas of a stream and, optionally, between day and night.

II. GENERAL DESIGN

A. Site Selection

The effective sampling of drift is best accomplished in wadeable, rocky-bottomed streams with a riffle–glide geomorphology and moderate water velocity. Drift nets become difficult to maintain or inefficient for measuring drift accurately under conditions of very high or very low water velocity. Nets can be positioned at any location within the channel, but placement in mid-stream at the downstream end of a riffle usually is most productive.

Measuring macroinvertebrate colonization and direction of movement is best performed in wadeable streams that can be easily reached on a daily basis. Security from vandalism of the traps and trays also is a consideration. Streams with a gravel to cobble substratum and moderate discharge are preferable. The traps and trays become clogged in streams with a predominately fine-grained sediment or a high suspended solids load, thus requiring considerable attention.

The tent trap suggested for sampling adult insect emergence is best used in shallow, rocky streams. If sampling in deeper water is desired, other types of emergence traps can be used (see Davies 1984).

B. General Procedures

Drift Drift is easily sampled in most streams by using drift nets set in the water for specified lengths of time (Fig. 17.2). Comparisons of the species, drift densities, and/or mean size of drifting individuals can be made between different streams or time periods (e.g., between day and night or over a 24-h cycle). Various factors must be considered when sampling drift, including the mesh size of the net, the number of nets needed for adequate

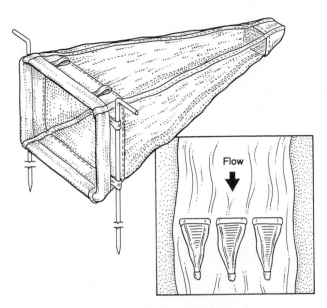

FIGURE 17.2 Drift net (modified from Merritt *et al.* 1996).

replication, sampling location, length of the sampling period, and the manner of data presentation (Brittain and Eikeland 1988).

The data often are quantified as *drift density*, which requires knowing the number of macroinvertebrates captured by the nets per volume of water passing through the nets during a sampling period. Drift density is usually best expressed as numbers of macroinvertebrates drifting per 100 m^3 of water:

$$\text{Drift Density} = \frac{(N)(100)}{(t)(W)(H)(V)(3600 \text{ s/h})},$$

where N represents No. of macroinvertebrates in a sample; t, time that the net was in the stream (h), W, net width (m); H, mean height of water column in the net mouth (m); and V, mean water velocity at the net mouth (m/s).

Colonization and Movement   Three experiments are designed to examine colonization and movements of lotic macroinvertebrates. The first experiment illustrates colonization of a substratum by macroinvertebrates over time. Trays filled with a substratum are placed into a stream at regular intervals over 2–3 weeks. All trays are collected at the end of the experiment and changes in the species composition and numbers of organisms colonizing the substratum over time are determined.

The second experiment examines the colonization by macroinvertebrates of substrata of different sizes and thus focuses on organism-substratum preferences. Trays are filled with different-sized sediment and placed in a stream for a specified time period. The species and numbers of organisms colonizing the different sediment are determined.

The third experiment examines the different routes of colonization in a stream. It follows the design of Williams and Hynes (1976) in using experimental traps placed in a stream. The traps are constructed and placed such that colonization from the upstream, downstream, and hyporheic areas (and optionally from aerial colonization) are separated and quantified over a short time period.

The colonization protocols require the use of trays or traps to hold substratum that serves as a colonization site for macroinvertebrates. A wide variety of designs for trays and traps have been used for colonization studies. Two types are suggested for the following three exercises.

Colonization trays filled with clean sediment are one of the easiest methods to measure colonization. Trays are built of wood or 1.5-cm mesh galvanized-wire screening. A solid wooden frame allows use of fine-grained sediment that would not be retained by a wire mesh. A wire mesh tray, however, is preferable because the screening allows colonization to occur from all sides rather than only from the top. Dimensions of the trays are 30 cm L × 30 cm W × 10 cm H.

Colonization traps, adapted from the design of Williams and Hynes (1976), measure the direction of colonization, and hence movement, in a stream. The traps consist of a basic wooden frame (50 cm L × 25 cm W × 25 cm H) with left and right sides covered with polyethylene plastic (Fig. 17.3). The "upstream" and "downstream" traps also have covered tops and bottoms. Stakes placed through eyebolts at the corners of each trap are driven into the sediment to secure the traps.

Downstream traps, allowing colonization only by macroinvertebrates moving downstream, are open on the upstream end and have a tapered net attached to the downstream end. The end of the net is open and suspended in the water column, thereby allowing detritus and drifting macroinvertebrates that do not colonize the substratum in the trap to pass through but inhibiting macroinvertebrates from crawling upstream into the trap. *Upstream traps*, allowing colonization only by organisms moving upstream, are open only on the downstream end. The upstream end is covered with a 250-μm mesh netting that is protected by an outer layer of 1.5-cm mesh wire screening. *Subsurface traps*, allowing colonization only vertically from the sediment, have a solid top and a bottom covered only

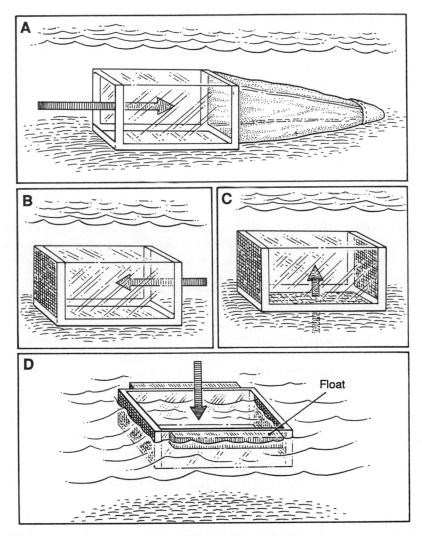

FIGURE 17.3 Colonization traps (design modified from Williams and Hynes 1976). (A) Downstream movement trap, (B) upstream movement trap, (C) upward subsurface movement trap, (D) aerial colonization trap.

with wire screening. The ends of the traps are covered with wire screening and netting.

Aerial colonization traps have a solid bottom and upstream and downstream ends covered with wire screening and netting. Polystyrene blocks are attached to each side for flotation. Guy ropes, attached to the traps and to stakes driven into the channel, hold the traps in place while allowing them to rise and fall with the water level.

A *control trap* consists of the wooden frame with a wire screening bottom and with all sides and the top open. Colonization of this trap thus can occur from all directions; it theoretically should have the highest numbers of macroinvertebrates at the end of the colonization period.

Adult Emergence   Adult insects emerging from a stream are easily collected using emergence traps. A number of different types of traps have been developed; their designs, applicability for use under different sampling conditions, and the factors affecting their performance are discussed in detail by Davies (1984). For this exercise a simple tent trap is suggested (Fig. 17.4). The trap consists of a triangle- or pyramid-shaped wooden frame enclosing an area of 0.5–1.0 m² and approximately 1 m high. The sides are covered with 500-μm Nitex netting. A sample bottle, with a funnel or cone-shaped entrance to prevent insects from returning to the net, is mounted at the apex of the trap to facilitate removal of captured adults, many of which will move to the trap's top and into the bottle. The traps are placed directly on and anchored into the streambed, thereby sampling insects emerging from a known area. Traps are left in place for 1 day, allowing capture of insects emerging during both the day and the night.

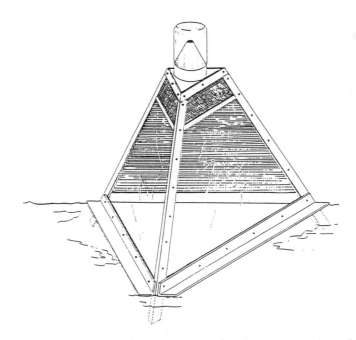

FIGURE 17.4 Mundie pyramid emergence trap (modified from Merritt *et al.* 1996).

III. SPECIFIC EXERCISES

A. Exercise 1: Quantifying Macroinvertebrate Drift

The objective of this exercise is to determine the drift density of macro-invertebrates in a stream. It can be accomplished during any time of the day and for any length of sampling period, but is most informative if samples taken during the day and night are compared.

1. Place drift nets across the stream channel (Fig. 17.2). The suggested mesh size of the nets is 200–300 μm. Nets with a larger mesh size will not retain small individuals that often are abundant in drift, resulting in inaccurate conclusions regarding the species composition and magnitude of drift. Nets with a smaller mesh size can be used, but care must be taken to ensure that they do not become clogged with seston. Nets are placed in the stream with the net face perpendicular to the direction of flow and anchored with rods driven into the substratum. Nets should be positioned at mid-depth in the water column or, if the stream is shallow, the bottoms of the nets should be 2–3 cm above the sediment to reduce the possibility of macroinvertebrates crawling into the nets. Use at least three nets positioned side-by-side (on a transect across the stream) to provide replicate measurements (Fig. 17.2).

2. Calculate the volume of water passing through each drift net by determining the area of the water column being sampled and the average water velocity at the mouth of each net. Record the width of the net mouth and then measure the depth of water entering the net at three equidistant locations across the mouth. Use a flowmeter to measure water velocity at the three locations (see Chapter 3).

3. Record the time the nets are placed in and removed from the stream. Generally strive to keep the nets in the stream for 1-h periods. This period may have to be shortened if high seston concentrations cause the net mesh to clog.

4. Remove the nets from the stream and wash the contents into a bucket partly filled with water. Use forceps to remove any macroinvertebrates that remain clinging to the inside of the nets. Wash the contents of the bucket through a sieve with a mesh size equal to or smaller than that of the net. Preserve the material from each net separately in bottles or sealable bags with 70% ethanol (final concentration). Label the samples with date of collection, net number, and time period of sampling.

5. Repeat steps 1–4 at a minimum for three consecutive 1-h periods both before and after sunset. If possible, sample over an entire 24-h period. Nets should be placed in the same location during each sampling interval.

6. In the laboratory, separate all organisms from the debris in the

TABLE 17.1
Sample Data Sheet for Macroinvertebrate Drift

Stream name _____   Date _____
Stream location _____   Investigator _____
Net mesh size _____   Net mouth width _____

	Net 1	Net 2	Net 3
Time net put in			
Time net taken out			
Total sampling time (hours)			
Water depth at net (cm)			
Point A			
Point B			
Point C			
Water velocity at net (m/s)			
Point A			
Point B			
Point C			
Number of macroinvertebrates			
Mass of macroinvertebrates (mg)			
Species composition			
Taxon A			
Taxon B			
Taxon C			
Etc.			

samples. This is best accomplished using a stereomicroscope at low power. Count the number of macroinvertebrates in each sample. Record data on Table 17.1.

7. Calculate the mean drift density of macroinvertebrates in the stream during each time interval. Construct a curve showing the change in drift density over time (e.g., Fig. 17.1). Use a *t* test or nonparametric Mann–Whitney *U* test to test the null hypothesis that there was no significant difference between day and night drift densities.

B. Exercise 2: Species and Size Composition of Drifting Macroinvertebrates

The objective of this exercise is to determine if differences exist in the species composition or mean size of macroinvertebrates drifting at different times of the day. The exercise uses the samples collected for Exercise 1.

1. To determine species composition, identify the macroinvertebrates collected in the samples for Exercise 1. Determine if there was a change in the species drifting at different times of the day.

2. Determine the size of each individual collected in the samples for Exercise 1. This can be accomplished by either of two methods.

  a. Measure all macroinvertebrates in each sample using an ocular micrometer fitted in a stereomicroscope. Calculate the mean size of the organisms in each sample.

  b. Place all macroinvertebrates collected in a sample together in an aluminum weighing pan and dry them in a drying oven for 24 h at 60°C. Weigh the pooled macroinvertebrates on an electronic balance. Calculate the mean individual dry mass of the drifting organisms (= pooled dry mass/number of individuals in the sample).

3. Construct a figure showing the mean size or mass of macroinvertebrates drifting during each time period. Use a *t* test or Mann–Whitney *U* test to test the null hypothesis that there was no significant difference between the mean size or mass of macroinvertebrates drifting during the day and that measured during night.

C. Exercise 3: Colonization Over Time

The objective of this exercise is to determine changes in the species and numbers of macroinvertebrates colonizing substrata over time.

1. Fill a minimum of 15 colonization trays with sediment of a uniform size that is similar to that of the predominant particles in the stream. Fifteen trays provide three replicate trays per time period. Additional trays can be used to increase the number of replicates per time period or to increase the number of time periods that trays are collected.

2. Place three replicate trays into the stream on each of 5 days spaced over 2–3 weeks (e.g., at 4-day intervals). Trays should be buried such that their tops are flush with the stream bed.

3. Retrieve all trays 1–2 days after the last set of trays are placed into the stream. Thus, the *simultaneous-removal* method for determining colonization is used for this exercise. Trays are placed in the stream periodically over the entire colonization period. All trays are retrieved together at the end of the study, thereby subjecting all substrata to the same potential colonizing species and to the same environmental conditions at the end of the colonization period (Shaw and Minshall 1980).

4. Wash macroinvertebrates associated with the sediment in a tray into

TABLE 17.2
Sample Data Sheet for Macroinvertebrate Colonization

Stream name _____   Date _____

Stream location _____   Investigator _____

	Tray no.			
	1	2	3	4
Day tray put in				
Day tray retrieved				
Total days in stream				
Number of macroinvertebrates				
Mass of macroinvertebrates (mg)				
Species composition				
Taxon A				
Taxon B				
Taxon C				
Etc.				

a bucket. Pass the contents of the bucket through a 250-μm mesh sieve and preserve and label the material retained by the sieve. Add a small amount of Rose Bengal to the sample to aid in the separation of organisms from debris.

5. In the laboratory, remove, identify, and count all macroinvertebrates in each sample. Record data on Table 17.2.

6. Calculate the mean number of taxa and mean number of individuals that colonized the sediment during each time period. Construct colonization curves illustrating changes in numbers of individuals and taxa over time (e.g., Fig. 17.5). Use an analysis of variance or Kruskal–Wallis test to test the null hypothesis that no significant difference existed in the number of taxa or individuals colonizing the substratum over time. Use a multiple comparison test (e.g., the Scheffe or the Tukey procedure) to determine the number of days it took until there was no significant increase in the number of taxa or individuals that had colonized the substratum.

7. *Optional.* Dry (60°C for 24 h) and weigh the macroinvertebrates to determine changes in biomass of the organisms colonizing the substratum over time. Use regression analysis to determine if there was a trend in the mean size of colonizing individuals over time (mean size is the biomass in a sample divided by the number of organisms in the sample).

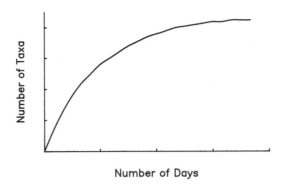

FIGURE 17.5 Sample colonization curve showing changes in number of taxa colonizing a substratum over time.

D. Exercise 4: Effect of Substratum Size on Colonization

The objective of this exercise is to determine the effect of sediment particle size on the colonization of substrata by macroinvertebrates.

1. Collect sediment and separate by sieving into four size categories. The suggested range in particle sizes is from about 0.2 to 63 mm. Potential size categories are 0.2–2, 2–16, 16–37, 37–63 mm. Wash and dry the sediment to remove all macroinvertebrates.

2. Place the sediments into a minimum of 12 colonization trays, three each for each of the size categories. Use additional trays if possible to increase replication. Bury the trays in the stream with the tops flush with the stream bed. Locate the trays such that water velocity is similar across their tops.

3. Retrieve the trays after 7–14 days. Preserve samples with ethanol, label, and add Rose Bengal.

4. In the laboratory, remove, identify, and count all macroinvertebrates in each sample.

5. Calculate the mean number of taxa and mean number (and optionally biomass) of individuals that colonized each particle size. Use an analysis of variance or Kruskal–Wallis test to test the null hypothesis that there was no significant difference in the number of taxa or individuals colonizing the different particle sizes. Use a multiple comparison test (e.g., the Scheffe or the Tukey procedure) to identify significant differences among the means.

6. *Optional.* A more extensive version of this exercise is to combine the procedures of Exercises C and D, thereby examining if significant

differences occur in the rate of colonization of different sediment particle sizes.

E. Exercise 5: Mechanisms of Colonization by Macroinvertebrates

The objective of this exercise is to determine the relative importance of different mechanisms of colonization of substrata by lotic macroinvertebrates.

1. Wash and dry sediment from the stream to remove all macroinvertebrates.
2. Fill colonization traps with the sediment to a depth of 5 cm. Use a minimum of three downstream, upstream, subsurface, and control traps; aerial traps are optional. Bury the traps such that the top of the sediment

TABLE 17.3

Sample Data Sheet for Sediment Colonization Experiment

Stream name _____

Stream location _____   Investigator _____

Date traps in _____   Date traps out _____

	Number of taxa	Number of individuals
Downstream trap		
Trap A		
Trap B		
Trap C		
Upstream trap		
Trap A		
Trap B		
Trap C		
Subsurface trap		
Trap A		
Trap B		
Trap C		
Aerial trap		
Trap A		
Trap B		
Trap C		
Control trap		
Trap A		
Trap B		
Trap C		

in each trap is flush with the streambed. The aerial traps are not buried, but rather are allowed to float at the water's surface. Placement of the traps should be such that water velocity at all traps is similar.

3. Leave the traps in place for a minimum of 7 days. A longer period of up to 30 days, however, is preferable to ensure complete colonization of the substratum. A minimum of 3 weeks usually is necessary if aerial traps are used. The netting must be periodically cleaned to allow free flow of water through the traps.

4. At the end of the colonization period, raise the traps slightly and slip a 250-μm mesh bag around them to prevent loss of animals before lifting the traps from the streambed. Preserve the sediment in each trap with ethanol, label the sample, and add Rose Bengal.

5. In the laboratory, remove, identify, and count all macroinvertebrates in each sample. Record data on Table 17.3.

6. Calculate the mean number of taxa and number of individuals that colonized each set of traps. Use an analysis of variance or Kruskal–Wallis test to test the null hypothesis that there was no significant difference in the number of taxa or individuals colonizing by the different mechanisms. Use a multiple comparison test (e.g., the Scheffe or the Tukey procedure) to partition significant differences among the means.

F. Exercise 6: Emergence of Adult Aquatic Insects

The objective of this exercise is to determine if differences exist in the numbers and species of adults that emerge from different habitats within a stream. A second, optional, objective is to determine if the numbers and species emerging differ between day and night.

1. Anchor emergence traps over the primary habitats in a stream. Typical habitats include riffles, pools, glides, or debris accumulations. Record the characteristics that differentiate the habitats (e.g., particle size or type, water velocity, water depth).

2. After 24 h, remove all insects from the sample bottle and netting and preserve them with 70% ethanol. Label the sample.

3. *Optional.* Rather than sampling different habitats, divide sampling into day and night periods, recording the number of hours that the traps were in place during both periods.

4. In the laboratory, identify and count the number of insects collected in each trap. Record data on Table 17.4.

5. Express the results as the mean number of taxa or individuals emerging per square meter per hour from each habitat or during the day and night. Use a *t* test or analysis of variance (or corresponding nonparametric

TABLE 17.4
Sample Data Sheet for Adult Insect Emergence

Stream name _____   Date _____

Stream location _____   Investigator _____

Type of trap _____   Trap sampling area _____

Habitat sampled	No. of taxa	No. of individuals
Habitat 1		
Trap A		
Trap B		
Trap C		
Habitat 2		
Trap A		
Trap B		
Trap C		
Habitat 3		
Trap A		
Trap B		
Trap C		

test) to test the null hypothesis that there was no significant difference in the number of taxa or individuals emerging from the different habitats or during the day vs the night.

IV. QUESTIONS

1. Are the species collected by the drift sampler representative of the entire benthic macroinvertebrate community in the stream?

2. What differences might occur in drift densities and the species drifting as stream size (order) increases? Would you expect seasonal differences in drift densities? Why?

3. What differences might occur in the drift densities of different-sized (e.g., large vs small) macroinvertebrates in streams with and without fish? Might the presence of fish have a different effect on day versus night drift densities compared to in a fishless stream?

4. Large numbers of macroinvertebrates can be transported downstream by drift each day. Why don't the upstream reaches of streams become depleted of macroinvertebrates?

5. Did the colonization curves reach an asymptote? If not, what does this suggest about the colonization process for your substratum?

6. What are some of the characteristics of the substrata used in your colonization study that may have been responsible for the differences in the numbers and species of macroinvertebrates that colonized each type of substratum?

7. To what extent was the downstream movement of macroinvertebrates compensated for by upstream movements? By movement up from the subsurface sediments or from aerial colonization?

8. What factors might be responsible for differences in the numbers of emerging insects among different substrata? What factors might cause day-to-day variation in the numbers of emerging insects?

9. Benthic sampling devices (e.g., Chapter 16) and emergence traps often provide very different estimates of the species composition of stream macroinvertebrate communities. What are some of the reasons for these differences?

V. MATERIALS AND SUPPLIES

Field Materials

Bottles or sealable bags
Buckets (20-liter)
Colonization traps
Colonization trays
Current velocity meter
Drift nets
Emergence traps
Ethanol (70%)
Forceps
Labeling paper and pencils
Meter sticks
Rose Bengal
Sieves (series of mesh sizes)

Laboratory Materials and Equipment

Aluminum weighing pans
Drying oven (optional)
Electronic balance (optional)
Ocular micrometer (optional)
Stereomicroscopes

REFERENCES

Allan, J. D. 1978. Trout predation and the size composition of stream drift. *Limnology and Oceanography* **23:**1231–1237.

Allan, J. D. 1984. The size composition of invertebrate drift in a Rocky Mountain stream. *Oikos* **43:**68–76.

Benzie, J. A. H. 1984. The colonization mechanisms of stream benthos in a tropical river (Menik Ganga: Sri Lanka). *Hydrobiologia* **111:**171–179.

Brittain, J. E., and T. J. Eikeland. 1988. Invertebrate drift—A review. *Hydrobiologia* **166:**77–93.

Ciborowski, J. J. H., P. J. Pointing, and L. D. Corkum. 1977. The effect of current velocity and sediment on the drift of the mayfly *Ephemerella subvaria* McDunnough. *Freshwater Biology* **7:**567–572.

Corkum, L. D., and P. J. Pointing. 1979. Nymphal development of *Baetis vagans* McDunnough (Ephemeroptera: Baetidae) and drift habits of large nymphs. *Canadian Journal of Zoology* **55:**1970–1977.

Cushing, C. E., and W. L. Gaines. 1989. Thoughts on recolonization of endorheic cold desert spring-streams. *Journal of the North Amercan Benthological Society* **8:**277–287.

Davies, I. J. 1984. Sampling aquatic insect emergence. Pages 161–227 *in* J. A. Downing and F. H. Rigler (Eds.) *A Manual on Methods for the Assessment of Secondary Productivity in Fresh Waters.* IBP Handbook 17. Blackwell Scientific, Oxford, UK.

Delucchi, C. M. 1989. Movement patterns of invertebrates in temporary and permanent streams. *Oecologia* **78:**199–207.

Ernst, M. R., and K. W. Stewart. 1985. Growth and drift of nine stoneflies (Plecoptera) in an Oklahoma Ozark foothills stream, and conformation to regression models. *Annals of the Entomological Society of America* **78:**635–646.

Gore, J. A. 1982. Benthic invertebrate colonization: Source distance effects on community composition. *Hydrobiologia* **94:**183–193.

Gray, L. J., and S. G. Fisher. 1981. Postflood recolonization pathways of macroinvertebrates in a lowland Sonoran desert stream. *American Midland Naturalist* **106:**249–257.

Kovalek, W. P. 1979. Day–night changes in stream benthos density in relation to current velocity. *Archiv für Hydrobiologie* **87:**1–18.

Krueger, C. C., and E. F. Cook. 1981. Life cycles, drift and standing stocks of some stoneflies (Insecta: Plecoptera) from streams in Minnesota, USA. *Hydrobiologia* **83:**85–92.

Malmqvist, B., and P. Sjostrom. 1987. Stream drift as a consequence of disturbance by invertebrate predators. *Oecologia* **74:**396–403.

Merritt, R. W., V. H. Resh, and K. W. Cummins. 1996. Design of aquatic insect studies: Collecting, sampling and rearing methods. Pages 12–28 *in* R. W. Merritt and K. W. Cummins (Eds.) *An Introduction to the Aquatic Insects of North America,* 3rd ed. Kendall/Hunt, Dubuque, IA.

Minshall, G. W. 1984. Aquatic insect-substratum relationships. Pages 358–400 *in*

V. H. Resh and D. M. Rosenberg (Eds.) *The Ecology of Aquatic Insects.* Praeger, New York, NY.

Müller, K. 1974. Stream drift as a chronobiological phenomenon in running water ecosystems. *Annual Review of Ecology and Systematics* **5:**309–323.

Müller, K. 1982. The colonization cycle of freshwater insects. *Oecologia* **52:**202–207.

Otto, C. 1976. Factors affecting the drift of *Potamophylax cingulatus* (Trichoptera) larvae. *Oikos* **27:**292–301.

Sedell, J. R., G. H. Reeves, F. R. Hauer, J. A. Stanford, and C. P. Hawkins. 1990. Role of refugia in recovery from disturbances: Modern fragmented and disconnected river systems. *Environmental Management* **14:**711–724.

Shaw, D. W., and G. W. Minshall. 1980. Colonization of an introduced substrate by stream macroinvertebrates. *Oikos* **34:**259–271.

Sheldon, A. L. 1984. Colonization dynamics of aquatic insects. Pages 401–429 *in* V. H. Resh and D. M. Rosenberg (Eds.) *The Ecology of Aquatic Insects.* Praeger, New York, NY.

Söderström, O. 1987. Upstream movements of invertebrates in running waters—A review. *Archiv für Hydrobiologie* **111:**197–208.

Thorp, J. H., E. M. McEwan, M. F. Flynn, and F. R. Hauer. 1985. Invertebrate colonization of submerged wood in a cypress-tupelo swamp and blackwater stream. *American Midland Naturalist* **113:**56–68.

Wallace, J. B., G. J. Lugthart, T. F. Cuffney, and G. A. Schnurr. 1989. The impact of repeated insecticidal treatments on drift and benthos of a headwater stream. *Hydrobiologia* **179:**135–147.

Walton, O. E., Jr., S. R. Reice, and R. W. Andrews. 1977. The effects of density, sediment particle size and velocity on drift of *Acroneuria abnormis* (Plecoptera). *Oikos* **28:**291–298.

Waters, T. F. 1972. The drift of stream insects. *Annual Review of Entomology* **17:**253–272.

Wiley, M., and S. L. Kohler. 1984. Behavioral adaptations of aquatic insects. Pages 101–133 *in* V. H. Resh and D. M. Rosenberg (Eds.) *The Ecology of Aquatic Insects.* Praeger, New York, NY.

Williams, D. D. 1980. Temporal patterns in recolonization of stream benthos. *Archiv für Hydrobiologie* **90:**56–74.

Williams, D. D., and H. B. N. Hynes, 1976. The recolonization mechanisms of stream benthos. *Oikos* **27:**265–272.

Wise, D. H., and M. C. Molles. 1979. Colonization of artificial substrata by stream insects: Influence of substratum size and diversity. *Hydrobiologia* **65:**69–74.

CHAPTER 18

Fish Community Composition

HIRAM W. LI* AND
JUDITH L. LI†

*Oregon Cooperative Fishery Research Unit
Oregon State University
†Department of Fisheries and Wildlife
Oregon State University

I. INTRODUCTION

The fish community is an assemblage of species inhabiting a prescribed area. It has the following properties: (1) richness (numbers of species), (2) diversity (relative composition of species abundances), (3) morphological and physiological attributes, and (4) trophic structure. The community is an important unit of study because community interactions influence the stability and flow of materials and energy through the ecosystem. Full understanding of any community can be broken down into two general sequential steps: (1) community description and (2) community processes and interactions. This chapter is concerned with community description, an important aspect of study because it leads to hypotheses on how the community may function.

Membership in the stream fish community is determined historically and ecologically (Gorman 1992). The size and composition of the available species pool of a zoogeographical province is governed by historical events of geologic change and evolution. Changes in pool membership arise from natural invasion caused by changes in climate, breakdown of geographical barriers, and human introduction. The number and kinds of species found locally can be ascribed to several ecological mechanisms: dispersal, physiological tolerances, biological interactions among species, and periodic environmental disturbances (Matthews and Heins 1987, Bayley and Li 1992). Colonists disperse during seasonal migrations from the available species pool. Some of these migrations are short, measured in tens of meters. Other migrations are extensive, extending over many kilometers (Welcomme 1985). The limitations to dispersal are often dictated by physical tolerances to habitat quality (e.g., temperature, pH, dissolved oxygen, current, availability of substrata or cover). Therefore, part of the explanation for the membership in an assemblage of fishes is physical attributes of the particular stream and its basin (Matthews 1987). Physiological tolerance of the physical setting does not ensure membership in the assemblage. Biotic interactions such as predation, competition, mutualism, parasitism, and disease are also important in shaping communities, especially in stable and predictable environments. Periodic, recurrent disturbances such as flash floods or droughts can cause local, short-term changes in community structure (Matthews 1987, Matthews et al. 1988). The effects of flooding in particular are especially harsh on young-of-the-year fishes (Harvey 1987). The severity of these disturbances can be ameliorated when the habitat is complex, because habitat complexity confers refugia (Schlosser 1987, Sedell et al. 1990, Pearsons et al. 1992). In summary, biological interactions may be very important community structuring agents in physically stable and complex streams, whereas the ability to disperse and colonize may be dominant factors in streams subject to harsh recurrent disturbances (Schlosser 1987).

In general, species richness increases as a function of habitat volume and habitat complexity (Angermeier and Schlosser 1989). Larger watersheds or catchment basins will have more species, all other things being equal. Two types of species distribution patterns have been observed in studies of fish communities (Rahel and Hubert 1991). First, when environmental gradients such as stream power or temperature change gradually, fish communities slowly change through the addition of species downstream. Second, when transitions are abrupt, the stream's fish fauna segregates into distinct zonal communities (e.g., trout zone, minnow zone) or faunal subunits.

Species composition may be used as an index of habitat conditions, if habitat requirements of several species are well understood (Karr et al. 1986). However, this is a very crude estimate because species interactions such as competition, predation, and disease influence species composition profoundly. Strong patterns of association and segregation may be clues to the nature of these interactions, but these clues must be treated as hypotheses for future experiments. To paraphrase Johannes and Larkin (1961), conclusions drawn from descriptive studies are subject to the same problems as inferring the plot of a mystery novel by reading only the last paragraph. It is worth noting that exotic species (i.e., those that are not native to the watershed) can disrupt native fish assemblages through competitive displacement, predation, as carriers of new diseases, and by altering food webs (Spencer *et al.* 1991, Li and Moyle 1993). A site that is dominated by exotic or nonindigenous species often reflects a disturbed environment (Fausch *et al.* 1990).

Community membership for a specified locale may be restricted by fish morphology. For instance, certain fishes may be constrained from using particular portions of streams because of size (Webb and Buffrénil 1990) or shape (Scarnecchia 1988). Analyses of fish morphologies should provide a better understanding of community function. Trophic (Gatz 1979) and microhabitat specializations (Douglas and Matthews 1992) within the fish community may be inferred from fish morphologies.

The kinds and abundances of fish will vary according to a wide variety of physical habitat differences such as habitat size, temperature, stream flow, and pollution. The purpose of this chapter is to examine how a particular factor may influence fish assemblages, by choosing sampling sites that differ markedly with respect to that characteristic. For example, if differences in temperature tolerance are to be examined, coolwater sites are compared to warmwater ones (presuming that all other things are equal or at least very similar) Contrasts to measure from the fish assemblage include the presence or absence of taxa, abundances, ages and sizes of each species, morphological attributes, habitat specializations, and trophic roles of fishes (e.g., predators, herbivores, or insectivores; see Chapter 22). Comparative study maximizes the breadth of information that can be derived from a descriptive examination of fish communities.

Using the exercises in this chapter, researchers will learn to conduct a comparative study of fish assemblages by (1) recognizing and measuring differences in physical habitats, (2) collecting and enumerating fishes, (3) measuring morphological attributes of the assemblages, (4) calculating measures of abundance and diversity for each assemblage, and (5) comparing species memberships and morphological diversity among assemblages.

II. GENERAL DESIGN

A. Site Selection

It is important to think of stream habitats at a number of scales (see Chapters 1–3). Thus, to place the fish community in context with its environment, study sites should be spatially referenced using topographic maps. Measuring habitat factors on large and small scales help meet the assumption that sites are comparable and valid for the contrast of interest. It is especially important at the watershed scale to note: (1) position of the sites within the watershed, (2) watershed conditions upstream and downstream from each study site (e.g., barriers to dispersal), and (3) factors associated with stream power (e.g., discharge, gradient, incoming tributaries).

B. Indices of Community Structure

Community structure can be analyzed in many ways. Structure can be examined from the perspective of classical Linnean taxonomy or from the perspective of functional groups. In either case, structure may be represented as the number of categories represented (richness) or the dispersion of categories (diversity, which combines richness and relative frequency). Functionally, the number of niches occupied may be far richer than the number of species. Through ontogeny, fishes change ecological function. This is due in large measure to shifts in size and allometry (nonlinear changes in one body part with respect to the entire anatomy). Changes in diet and habitat use occur through ontogeny, thus shifting niche requirements and resulting in different "life history niches" for a species. The task is to examine community structure from different points of view.

C. Habitat Inventory

The habitat inventory will consist of a number of related tasks, as follows. Inventory the habitat types available for fishes in the reaches to be sampled (see Chapter 2). Sketch these habitats, noting the distribution of large woody debris, undercut banks, substrata composition, and depth. Conduct a hydrological survey of these sites (see Chapter 3); obtain area, volume, and discharge. Collect data on temperature, light, dissolved oxygen, and turbidity (see Chapters 5 and 7). These factors are known to be as important to the distribution of fishes; however, other variables may be equally or more important than those listed (e.g., pesticide runoff, industrial discharge). Therefore, other measurements may be required, particularly in agriculturally or industrially developed watersheds. Many of these tasks may be minimal if you sample areas from which you already have physical data.

D. Data Analysis

The following analyses will be performed (details to be listed below): (1) graphically determine population abundances using the depletion method, (2) calculate and compare species richness of two communities, (3) calculate species diversity of two communities, (4) calculate richness and diversity of different trophic types as determined by mouth position and mouth size, (5) plot the fineness ratios (i.e., streamlining—see below) of fishes versus frequency of occurrence within the community, and (6) plot mouth gape and mouth width of a species versus standard length.

III. SPECIFIC EXERCISES

A. Exercise 1: Fish Collection

There are several methods by which to census fish communities. Visual estimation by snorkeling divers is a good technique where streams are highly transparent, especially when water is chest deep or higher. It is also the best technique to census small juvenile fishes. However, snorkeling may not be possible in all streams because of turbidity, or insufficient depth (see Helfman 1983 for snorkeling methods). In that case, there are two other capturing methods: seining and electrofishing. Electrofishing is an excellent tool when the habitat has many snags that can foul a net, but fish escape capture when water is thigh deep or higher. Also, electrofishing is biased for large fish. Seines are excellent for large streams with deep channels, slow current, and sandy, muddy, or pebbly substrata. Seines with small mesh nets (3/16″) capture small fish efficiently; however, they are not desirable gear when the current is swift (>50 cm/s) or when snags are present. Regardless of gear, accurate assessments of community structure require good estimates of population abundances of all species. Estimations of relative composition from any single sampling method can lead to biased results.

A standard method of estimating population size is the removal or depletion estimate. This technique requires that the upper and lower borders of the sampling reach of stream are blocked to prevent movement of fishes in or out of the study area. The length of the stream reach to be sampled depends upon stream size and the scale of interest. For example, in the Pacific Northwest (USA), sampling units are usually at the level of the habitat unit (e.g., riffles, runs, pools) and range between 3 and 40 m in length (e.g., Li *et al.* 1994). Larger streams of the Midwest may require sampling distances of 200 m or more before representative samples of the community can be obtained (Matthews 1990). The protocols listed below will work for either seining or electroshocking.

1. Block the upstream and downstream margins of the habitat to be sampled with block seines (3/16″ stretch mesh). Be sure to weight the bottom of the net with rocks to prevent fish escape.

2. Using a seine or electroshocking gear make three passes along the length of the blocked stream segment. This approach assumes that less fish will be caught at each successive pass given equal collecting effort.

3. After each pass with your collecting gear, put fish into a bucket or basket marked with a pass number.[1]

4. After each pass identify, count, and measure the length of all species. Preserve a representative subsample of the fishes (i.e., covering the range of size classes of each species) and retain for morphological measurements in the laboratory (see protocol for preservation below). To minimize your impact on the community, ethics dictate that you euthanize only those numbers of fishes necessary for your analysis. Release all other fishes outside of the block nets delineating the sampling area to prevent recounting errors in subsequent samples.

5. Narcotize those fishes to be preserved in a lethal solution of MS-222 (tricaine methanesulfonate, 200 mg/liter) and a sodium bicarbonate buffer (500 mg/liter). This is a humane technique for euthanasia. Once euthanized, the collection can be brought back to the laboratory in containers filled with 10% formalin[2] or in containers filled with ice. Be sure to label each jar with the pass number and the collection location. In the laboratory identify, count, and measure all fishes. If fishes are to be preserved for dietary analysis (see Chapter 22), be sure to make an incision along the abdominal cavity of large fish ($\geq$50 mm) so that the formalin will preserve the digestive tract.

6. For those fishes to be preserved for museum curation, allow the collection to fix in the formalin for 24 h. Decant the formalin and soak the collection in water for a second 24 h. Decant and repeat until the formalin odor is gone. Preserve the collection in 50% isopropyl alcohol. Use your institution's procedures for handling and disposing of toxic chemicals.

B. Exercise 2: Morphological Measurements

1. Determine the Fineness Ratio for each species by conducting the following measures. Measure the standard length (mm). This is the distance from the tip of the snout to the end of the vertebral column, which is approximately at the base of the caudal fin.

2. Measure maximum body depth (mm). This is frequently near the anterior insertion of the dorsal fin.

3. The Fineness Ratio is a measure of streamlining and is calculated as follows:

[1]Make sure to keep the water cool and aerated. Change the water frequently if a bucket is used. Alternatively, use a wicker laundry basket with a tight weave, which is kept in the stream. Be sure that a net or tarp is draped over the top to prevent fish from escaping.

[2]Both formalin and MS-222 are toxic and carcinogenic. The utmost care should be followed in their use. Gloves and protective eyewear should be used when handling these substances and exposed skin should be protected using a silicon lotion.

$$\text{Fineness Ratio} = \text{standard length/maximum body depth.} \quad (18.1)$$

A Fineness Ratio of 4.5 is optimal for hydrodynamic efficiency (see Scarnecchia 1988). Generally speaking, fishes inhabiting swifter water will be more streamlined. Fish with a Fineness Ratio above 4.5 appear "eel-like" and fishes below 4.5 are gibbose or stocky and less hydrodynamically efficient.

C. Exercise 3: Mouth Specialization

1. Note the position of the mouth for each species. Top feeders, such as killifish, have mouths in the superior position. Fishes that feed in the water column, such as golden shiners and rainbow trout, have mouths in the terminal position. Bottom feeders, like suckers, have mouths in an inferior position. Score superior mouths with an S, a terminal mouth with a T, and an inferior mouth with an I.

2. Measure the maximum width of the mouth (greatest distance between the left and right corners of the lower jaw) and the gape of the mouth. The gape is measured by gently opening the upper (the premaxillae and maxillae) and lower (mandible) jaws to the largest extent. The greatest distance between the upper and lower jaws when maximally distended is the gape of the mouth. Calipers will greatly facilitate measurements. Clearly, the bigger the mouth the larger the prey that can be ingested.

D. Optional Exercises

Some suggestions for additional work are (1) compare species richness and diversity as a function of habitat size (e.g., pool volume), (2) examine temporal shifts in assemblage or community structure of a habitat, (3) examine species richness and diversity along the longitudinal gradient in the stream, and (4) determine the variation in assemblage structure of fishes by sampling several habitats of similar size and quality.

E. Data Analysis

Graphical Estimate of Population Size   Estimates of population abundance for each species will be used in the assessment of species diversity. The basis of this technique is that each successive pass removes less fish than the previous one because the population is being depleted. There are statistically more sophisticated methods of making this estimate involving maximum likelihood indicators, but this involves much computation (Riley and Fausch 1992). The difference between the two techniques is that the maximum likelihood method makes allowances for different probabilities of capture among different passes, whereas the graphical method assumes that probability of capture is constant. The graphical method is primitive but simple and illustrates the concepts of the removal method.

1. Plot the catch on the y-axis and the sum of all catches on the x-axis.
2. Visually regress the catches (y-axis) on the sum of all catches (x-axis). The population estimate is where the line crosses the x-axis (x-intercept). For example see Table 18.1 and Fig. 18.1.

Morphological Analysis

1. Determine the relative frequency of different trophic functional groups as determined by mouth position.
2. Plot the frequency of occurrence within the fish assemblage of community (y-axis) vs fineness ratio (x-axis).
3. For each species, plot mouth area (y-axis) versus standard length (x-axis). The formula of an ellipse will be used to estimate mouth area, where mouth gape is assumed to be the major axis and mouth width the minor axis. The area of an ellipse is calculated as: (major axis × 0.5) × (minor axis × 0.5).

Taxa Richness and Diversity   Basic descriptors of biological assemblages are abundance, taxa richness, and measures that combine both, generally categorized as diversity indices. A few of these measures are used in this exercise; selected on their ability to discriminate between groups and ease of interpretation. Abundance is straightforward, either as a summation of all organisms or as numbers represented by particular groups. Taxa richness can be expressed as the number of taxa (as species, genera, or functional groups as defined by your study) or as proportions of total abundance.

1. Calculate species richness for each of two communities using Margalef's index (D_{Mg}), a simple measure of species richness that has good discriminating ability, but is very sensitive to the sample size,

$$D_{Mg} = (S - 1)/\ln n, \qquad (18.2)$$

where S is the total number of taxa in the sample and $\ln n$ is the natural logarithm of the total number of individuals.

2. Compare calculated values of Margalef's index to: (1) Hill's Diversity index (Hill 1973) of taxa richness (H') and (2) Alpha Diversity index (α), for each of two communities expressed as number of taxa. Theoretically H' is based on the actual number of taxa (we can only approximate it using

TABLE 18.1
Data Worksheet for Making Depletion Population Estimate

Page _____ of _____

Date: _____   Stream: _____   Basin: _____

County: _____   Township: _____   Site: _____

Habitat type (e.g., riffle, pool, other) _____

Species: _____

Axis	Pass 1	Pass 2	Pass 3
y (catch)			
x (sum of catches)			

Species: _____

Axis	Pass 1	Pass 2	Pass 3
y (catch)			
x (sum of catches)			

Species: _____

Axis	Pass 1	Pass 2	Pass 3
y (catch)			
x (sum of catches)			

Species: _____

Axis	Pass 1	Pass 2	Pass 3
y (catch)			
x (sum of catches)			

Species: _____

Axis	Pass 1	Pass 2	Pass 3
y (catch)			
x (sum of catches)			

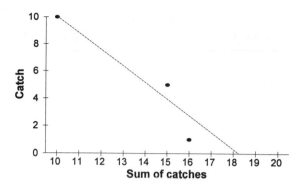

FIGURE 18.1 Graphical approach to the removal estimate of population size.

our estimates from samples of the entire habitat). Hill's Diversity index is among the easiest to interpret ecologically and is calculated as

$$H' = -\sum_{i=1}^{S} [(n_i/n)\ln(n_i/n)], \tag{18.3}$$

where n_i is the number of individuals in taxon i; and n is the total number of individuals in the sample.

Alpha Diversity index (α) emphasizes taxa richness and has a low sensitivity to sample size; this second property may make it very useful for comparing samples of different sizes. This diversity measure tends to be a good descriptor even when the actual sample may not be strictly a log series (Magurran 1988). The calculation is simple, and determined iteratively,

$$\alpha = [n(1-x)]/x, \tag{18.4}$$

where x is determined iteratively by

$$S/n = [(1-x)/x][-\ln(1-x)], \tag{18.5}$$

where x is usually more than 0.9 and always less than 1.0. If $n/S > 20$, x will be >0.99 *Note*. To solve for x, substitute S and n. Using a calculator, substitute for x until both sides of the equation are equal.

Alpha increases as diversity increases and is the log series coefficient in which a species abundance curve takes the form αx, $\alpha x^2/2$, $\alpha x^3/3$, ...

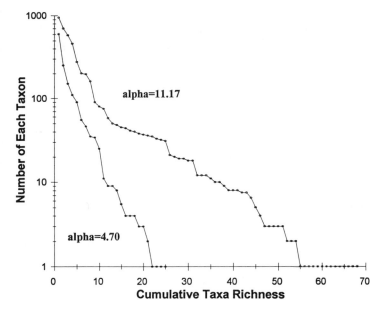

FIGURE 18.2 Log series curves of species abundance to taxa richness.

$\alpha x^n/n$, where αx is the number of species predicted to have one individual, $\alpha x^2/2$ with two, and so on. Figure 18.2 is an example of log series curves.

3. Calculate richness and diversity of different trophic functional groups as determined by mouth position and mouth size.

Comparing Communities   There are many instances where an investigator wishes to compare communities (e.g., Matthews 1986, Matthews *et al.* 1988). For instance, Matthews (1986) compared the effect of a catastrophic flood on an Ozark stream community by comparing their structure through time (i.e., before the flood and two periods after the flood). There are a wide variety of similarity indices that can be used to compare assemblages. Two indices used in many studies are suggested for this exercise.

1. Compare two communities using Jaccard's index. Jaccard's index (C_J) is simple, but does not account for abundances of species (Southwood 1978),

$$C_J = j/(a + b - j),\qquad\qquad (18.6)$$

where j is the number of taxa found in both sites; a, the number of taxa in Site A; and b, the number of taxa in Site B.

2. Compare two communities using the Morisita–Horn index. The Morisita–Horn index takes into account both taxa richness and abundance; however, it is highly sensitive to the abundance of the most abundant taxa (Wolda 1981),

$$C_{MH} = [2 \cdot \Sigma(an_i bn_i)]/[(da + db)\, aN \cdot bN],$$

where aN is the number of individuals in Site A; bN, the number of individuals in Site B; an_i, number of individuals of the ith species in Site A; bn_i, the number of individuals of the ith species in Site B; $da = \Sigma an_i^2/aN^2$; and $db = \Sigma bn_i^2/bN^2$.

3. Compare the performance of all indices with respect to the patterns they describe concerning the communities.

Other Analyses There are other useful statistical tools for examining patterns exhibited by fish communities. Spearman's Rank Correlation can be used to examine changes in the ranking of relative abundance of taxa for two sample comparisons. Kendall's W can be used to compare the relative ranking of several taxa (Siegal 1956, Grossman *et al.* 1982, Ross *et al.* 1985). A host of multivariate statistical tools such as cluster analyses and ordination techniques (e.g., PCA, DCA, DFA) can be applied to examine how fishes are distributed within their habitat (see Felley and Felley 1987, Ross *et al.* 1987, Whittier *et al.* 1988, for examples these analytical techniques).

IV. QUESTIONS

1. Why should the condition of the watershed upstream and downstream of the collection site be of concern to the interpretation of fish community composition?

2. What indices or community descriptors are the best discriminators between your study sites? Are there species that inhabit one site exclusively?

3. Why is it important to obtain estimates of population size before you calculate species diversity? Test this assumption by calculating species richness and species diversity based on catches on the first pass. Compare these measures with those based upon the population estimates.

4. Do morphological attributes of fishes reflect differences between sites or in physical gradients represented by sites?

5. Consult a regional guide concerning the fish fauna in your area. Classify the collection of fishes into trophic groups as suggested by the

guide. What are the relative abundances of various trophic groups? What are the ratios of one group to the next? What percentage of the fauna are piscivorous? Relate this to the concept of the trophic pyramid.

6. Overlay plots of mouth area versus body length for various species within a functional and trophic group (e.g., bottom feeding insectivores). What patterns do you see? If competition for food is related to mouth size and competition for space is related to fineness ratio and size (bigger fish are found in faster and deeper water), which size classes of various species might be subject to interspecific competition? Which size classes may be less subject to interspecific competition?

7. Some fishes are known to be highly piscivorous. If you capture piscivores in your sample, which size classes of each species are invulnerable to predation? Is this reflected in their relative abundances?

8. What happens when you exclude rare species from diversity indices?

V. MATERIALS AND SUPPLIES

Field Collecting Equipment

10% Formalin
50% Isopropyl alcohol
Anesthetic solution (MS-222 and sodium bicarbonate mixture)
Block seines (3/16″ mesh, height and length adequate for regional cir-
    cumstances)
Buckets for sorting and transporting fish
Collection labels
Either a common sense haul seine with a bag end (3/16″ mesh, height
    and length adequate for regional circumstances) or a backpack
    electroshocker, safety gear (electrician gloves, rubber insulated
    waders), polarized sun glasses, and dip nets
Ice chest, ice, large heavy duty plastic bags, and ties (optional fish
    storage method)
Large nalgene, wide-mouth jars or sealable buckets for storing samples
Live baskets (perforated holding tanks placed in the stream to hold
    fishes live)
Scalpels

Measuring Equipment

50-m Survey tape and wood stakes (for determining habitat area)
Calibrated gaging staff (for measuring water depth)
Calipers

Fish measuring boards
Stainless steel rule

Data Recording and Miscellaneous Supplies
Colored pencils for sketching maps
Data forms on water-resistant paper
Graph paper
Regional guide or key to fishes

REFERENCES

Angermeier, P. L., and I. J. Schlosser. 1989. Species-area relationships for stream fishes. *Ecology* **70:**1450–1462.

Bayley, P. B., and H. W. Li. 1992. Riverine fishes. Chapter 12, pages 251–281 *in* P. Calow and G. E. Petts (Eds.) *The Rivers Handbook, Hydrological and Ecological Principles,* Vol. 1. Blackwell Scientific, Oxford, UK.

Douglas, M. D., and W. J. Matthews. 1992. Does morphology predict ecology? Hypothesis testing within a freshwater stream fish assemblage. *Oikos* **65:**213–224.

Fausch, D. K., J. Lyons, J. R. Karr, and P. L. Angermeier. 1990. Fish communities as indicators of environmental degradation. *American Fisheries Society Symposium* **8:**123–144.

Felley, J. D., and S. M. Felley. 1987. Relationships between habitat selection by individuals of a species and patterns of habitat segregation among species: fishes of the Calcasieu drainage. Pages 61–68 *in* W. J. Matthews and D. C. Heins (Eds.) *Community and Evolutionary Ecology of North American Stream Fishes.* Univ. of Oklahoma Press, Norman, OK.

Gatz, A. J., Jr. 1979. Community organization in fishes as indicated by morphological features. *Ecology* **60:**711–718.

Gorman, O. T. 1992. Evolutionary ecology and historical ecology: Assembly, structure, and organization of stream fish communities. Pages 659–688 *in* R. L. Mayden (Ed.) *Systematics, Historical Ecology and North American Freshwater Fishes.* Stanford Univ. Press, Palo Alto, CA.

Grossman, G. D., P. B. Moyle, and J. O. Whitaker, Jr. 1982. Stochasticity in structural and functional characteristics of an Indiana stream fish assemblage: A test of community theory. *American Naturalist* **120:**423–454.

Harvey, B. C. 1987. Susceptibility of young-of-the-year fishes to downstream displacement by flooding. *Transactions of the American Fisheries Society* **116:**851–855.

Helfman, G. S. 1983. Underwater methods. Pages 349–370 *in* L. A. Nielsen and D. L. Johnson (Eds.) *Fisheries Techniques.* American Fisheries Society, Bethesda, MD.

Hill, M. O. 1973. Diversity and evenness: A unifying notation and its consequences. *Ecology* **54:**427–432.

Johannes, R. E., and P. A. Larkin. 1961. Competition for food between redside shiners *(Richardsonius balteatus)* and rainbow trout *(Salmo gairdneri)* in two British Columbia lakes. *Journal of the Fisheries Research Board of Canada* **18:**203–220.

Karr, J. R., K. D. Fausch, P. L. Angermeier, P. R. Yant, and I. J. Schlosser. 1986. *Assessing Biological Integrity in Running Waters. A Method and Its Rationale.* Special Publication 5, Illinois Natural History Survey, Champaign, IL.

Li, H. W., and P. B. Moyle. 1993. Management of introduced fishes. Pages 287–307 *in* C. C. Kohler and W. A. Hubert (Eds.) *Inland Fisheries Management in North America.* American Fisheries Society, Bethesda, MD.

Li, H. W., G. A. Lamberti, T. N. Pearsons, C. K. Tait, J. L. Li, and J. C. Buckhouse. 1994. Cumulative effects of riparian disturbances along high desert trout streams of the John Day Basin, Oregon. *Transactions of the American Fisheries Society* **123:**627–640.

Magurran, A. E. 1988. *Ecological Diversity and Its Measurement.* Princeton Univ. Press, Princeton, NJ.

Matthews, W. J. 1986. Fish faunal structure in an Ozark stream: Stability, persistence, and a catastrophic flood. *Copeia* **1986:**388–397.

Matthews, W. J. 1987. Physicochemical tolerance and selectivity of stream fishes as related to their geographic ranges and local distributions. Pages 111–120 *in* W. J. Matthews and D. C. Heins (Eds.) *Community and Evolutionary Ecology of North American Stream Fishes.* Univ. of Oklahoma Press, Norman, OK.

Matthews, W. J. 1990. Fish community structure and stability in warmwater midwestern streams. *United States Department of the Interior, United States Fish and Wildlife Service Biological Report* **90:**16–17.

Matthews, W. J., and D. C. Heins (Eds.). 1987. *Community and Evolutionary Ecology of North American Stream Fishes.* Univ. of Oklahoma Press, Norman, OK.

Matthews, W. J., R. C. Cashner, and F. P. Gelwick. 1988. Stability and persistence of fish faunas and assemblages in three midwestern streams. *Copeia* **1988:**947–957.

Pearsons, T. N., H. W. Li, and G. A. Lamberti. 1992. Influence of habitat complexity on resistance to flooding and resilience of stream fish assemblages. *Transactions of the American Fisheries Society* **121:**427–436.

Rahel, F. J., and W. A. Hubert. 1991. Fish assemblages and habitat gradients in a Rocky Mountain-Great Plains stream: Biotic zonation and additive patterns of community change. *Transactions of the American Fisheries Society* **120:**319–332.

Riley, S. C., and K. D. Fausch. 1992. Underestimation of trout population size by maximum-likelihood removal estimates in small streams. *North American Journal of Fisheries Management* **12:**768–776.

Ross, S. T., W. J. Matthews, and A. A. Echelle. 1985. Persistence of stream fish assemblages: effects of environmental change. *American Naturalist* **126:**24–40.

Ross, S. T., J. A. Baker, and K. E. Clark. 1987. Microhabitat partitioning of Southeastern stream fishes: Temporal and spatial predictability. Pages 42–51 *in* W. J.

Matthews and D. C. Heins (Eds.) *Community and Evolutionary Ecology of North American Stream Fishes.* Univ. of Oklahoma Press, Norman, OK.

Scarnecchia, D. L. 1988. The importance of streamlining in influencing fish community structure in channelized and unchannelized reaches of a prairie stream. *Regulated Rivers: Research and Management* **2:**155–166.

Schlosser, I. J. 1987. A conceptual framework for fish communities in small warmwater streams. Pages 17–24 *in* W. J. Matthews and D. C. Heins (Eds.) *Community and Evolutionary Ecology of North American Stream Fishes.* Univ. of Oklahoma Press, Norman, OK.

Sedell, J. R., G. H. Reeves, F. R. Hauer, J. A. Stanford, and C. P. Hawkins. 1990. Role of refugia in recovery from disturbances—modern fragmented and disconnected river systems. *Environmental Management* **14:**711–724.

Siegal, S. 1956. *Nonparametric Statistics for the Behavioral Sciences.* McGraw–Hill, New York, NY.

Southwood, T. R. E. 1978. *Ecological Methods with Particular Reference to the Study of Insect Populations,* 2nd ed. Chapman and Hall, London, UK.

Spencer, C. N., B. R. McClelland, and J. A. Stanford. 1991. Shrimp stocking, salmon collapse, and eagle displacement: Cascading interactions in the food web of a large aquatic ecosystem. *BioScience* **41:**14–21.

Webb, P. W., and D. V. Buffrénil. 1990. Locomotion in the biology of large aquatic vertebrates. *Transactions of the American Fisheries Society* **119:**629–641.

Welcomme, R. L. 1985. *River Fisheries.* FAO Fisheries Technical Paper 262, Food and Agricultural Organization, Rome, Italy.

Whittier, T. R., R. M. Hughes, and D. P. Larsen 1988. Correspondence between ecoregions and spatial patterns in stream ecosystems in Oregon. *Canadian Journal of Fisheries and Aquatic Sciences* **43:**1264–1278.

Wolda, H. 1981. Similarity indices, sample size and diversity. *Oecologia* **50:**296–302.

SECTION D

Community Interactions

CHAPTER 19

Plant–Herbivore Interactions

GARY A. LAMBERTI[*] AND
JACK W. FEMINELLA[†]

*Department of Biological Sciences
University of Notre Dame
†Department of Zoology and Wildlife Science
Auburn University*

I. INTRODUCTION

Benthic environments in streams comprise zones of high biological activity in which processes such as primary production, consumption, and decomposition occur. Plants and animals interact in this zone as they do in all ecosystems: plants become established, grow, and reproduce while herbivorous animals consume plant tissue to likewise grow and reproduce. Primary producers in streams consist of autotrophic bacteria, algae, bryophytes (mosses and liverworts), and vascular plants. In most small streams, however, benthic algae are the dominant primary producers (Bott 1983) and will grow on virtually any submerged surface, inorganic or organic, living or dead. Benthic algae commonly found in streams include diatoms, filamentous and nonfilamentous green algae, blue-green algae (cyanobacteria), and sometimes red algae and other algal groups (see Chapter 13). The entire attached microbial community is considered *periphyton* (also called "aufwuchs" or "biofilm"), of which algae are usually the main component.

Herbivory (or *grazing*) is the consumption of living plants or their parts by animals. Many aquatic animals consume periphyton, either for most of their energy intake (as with invertebrate *scrapers;* see Chapter 21) or as a variable portion of their diet (as with *omnivores*). The diversity of lotic herbivores spans a broad range of taxonomic groups, but insects, mollusks, and crustaceans are particularly important (Lamberti and Moore 1984). Among the more conspicuous benthic herbivores in streams are caddisflies (Trichoptera), mayflies (Ephemeroptera), and snails (Gastropoda), and much work has been done on their grazing ecology (reviewed by Lamberti 1993, Feminella and Hawkins 1995, Steinman 1996). More recently, the ecological importance of grazing by crayfish (Hart 1992, Creed 1994), fish (Power and Matthews 1983, Power *et al.* 1988, Wootton and Oemke 1992), and larval amphibians (Lamberti *et al.* 1992) also has been recognized. Regardless of the specific organisms, it is clear that many aquatic animals consume benthic plants (Gregory 1983) and that, for some, their growth and development is linked directly to algal production (e.g., Feminella and Resh 1990, 1991).

The organic matter synthesized by primary producers in streams (*autochthonous* production) is a major energy source for benthic food webs. In some streams with limited riparian shading and inputs of deciduous vegetation, algal production can dominate the annual energy budget (Minshall 1978). In most streams, but particularly third-order and higher, autochthonous production constitutes a significant proportion of the energy budget (Lamberti 1996). Middle-order (orders 3–6) streams frequently are autotrophic because light levels are high (influence of riparian shading is restricted to stream margins), water is shallow and clear (allowing light penetration to the streambed), and temperature and nutrient levels usually are suitable for benthic algal growth (Vannote *et al.* 1980, Minshall *et al.* 1985). In large rivers, internal production usually shifts from benthic algae to phytoplankton because of increased depth and turbidity, which limit light penetration to the riverbed. However, the shallow margins of large rivers can have substantial benthic primary production and abundant herbivores (Thorp and DeLong 1994). Even in small, heavily shaded streams with low algal standing crops, algae can support abundant herbivore populations by their rapid turnover and high nutritional value (i.e., possessing low $C:N$ ratios) relative to other carbon sources (e.g., McIntire 1973, Cummins and Klug 1979, Mayer and Likens 1987) and can strongly influence the structure of entire food webs (Lamberti 1996).

Given the energetic importance of benthic algae and its utilization by a diverse array of herbivores, it is reasonable to postulate that herbivores have a major impact on plant assemblages in many streams. Indeed, many structural and functional attributes of benthic algae are altered by grazers, but their effects are not consistent in direction or magnitude across streams, time, algal assemblages, or grazer type (Gregory 1983, Lamberti and Moore

1984, Feminella and Hawkins 1995, Steinman 1996). For example, biotic factors such as herbivore species, abundance, or size (Lamberti *et al.* 1987a, Steinman 1991) and algal successional state (Dudley *et al.* 1986, DeNicola *et al.* 1990, McCormick and Stevenson 1991) can each influence plant responses to herbivory. However, the strength and outcome of the algal–grazer interaction also is dependent on many abiotic factors such as light (Steinman 1992), nutrients (Rosemond 1993), substratum (Dudley and D'Antonio 1991), flow (DeNicola and McIntire 1991), season (Rosemond 1994), and disturbance (Feminella and Resh 1990). For example, a low standing crop of algae can result from heavy grazing pressure, low light or nutrient concentrations (poor growing conditions), recent disturbance, or some combination of these and other factors. Causal factors responsible for periphyton abundance patterns are impossible to identify using descriptive or observational approaches alone. Only controlled experiments, those done under field *(in situ)* conditions being best, can be used to evaluate the separate and combined effects of herbivory and other factors on periphyton assemblages.

   In this exercise, we describe two field experimental approaches to assess the impact of grazers on benthic algal assemblages in streams. Both experiments involve the manipulation of grazer abundances over time, using either grazer exclosures or enclosures. Optional exercises are presented to measure grazer colonization, depletion of algal biomass, and growth of grazers. The specific objectives of this chapter are to (1) provide an introduction to the plants and animals involved in lotic herbivory; (2) quantify reciprocal interactions between algae and herbivores in streams; (3) assess variability in herbivory within and among stream sites; and (4) illustrate the advantages and limitations of field experiments used to quantify lotic herbivory.

   These experiments are designed to be conducted in low- to middle-order streams (orders 2–5) where benthic herbivores often predominate; however, they can be modified for use along the margins of large rivers and even in littoral zones of lakes. Experiments with stream algae and grazers also may be conducted within laboratory artificial streams, where a high level of control is possible (see review by Lamberti 1993). However, field experiments typically provide more realistic conditions and responses (i.e., higher accuracy) than laboratory experiments. Unfortunately, although more accurate, experiments in natural streams often are prone to more variable response (i.e., lower precision) than those in the laboratory and so may require larger numbers of replicates to achieve the same level of precision. Logistical constraints and the possibility of unanticipated events (e.g., floods, drought, vandalism, etc.) also need to be considered when designing field experiments. Naturally, in both public and private waterways, permission from appropriate officials or landowners should be obtained before conducting field experiments.

II. GENERAL DESIGN

A. Site Selection

Small- to moderate-sized (order 2–5) streams are preferable for the exercises in this chapter because they typically contain benthic herbivores and productive algal assemblages, and often have the high water clarity necessary to allow visual estimates of grazer abundance (see below). If possible, several stream reaches, grouped by contrasting riparian canopy (shading) but all with similar channel form, should be selected. Here, each reach (and its associated habitat units such as pools, runs, or riffles; see Chapter 2) is considered a separate replicate. One group of reaches should have high irradiance (little or no shading) and the other should have low irradiance (heavy shading). In homogenous streams that have little variation in shading, researchers instead may select reaches that differ in current velocity (e.g., low- and high-flow classes) or other factors thought to influence herbivory (e.g., streamwater nutrient levels). Alternatively, simpler, but statistically flawed, experimental designs involve selection of two reaches that differ in some environmental feature or use of a single stream reach. In both cases, reaches are subdivided into replicate habitat units over which experiments are conducted. Use of two reaches and their nonindependent habitat units as spatial replicates to examine effects of selected environmental factors is considered pseudoreplication (sensu Hurlbert 1984) and thus is less desirable than using stream reaches as replicates. A single stream reach can be used if the study question concerns only grazing and treatments are interspersed (although environmental factors can be measured and used as statistical covariates). These latter two experimental designs may be the only practical approach in some studies of herbivory, although they have limited extrapolational power to other streams or stream reaches (Hurlbert 1984).

B. Field Experiments

Herbivore Exclusion Benthic herbivores that do not swim or that exhibit low drift rates (e.g., cased caddisflies, snails, etc.) can be excluded by elevating artificial substrata above the stream bottom, which are then mostly inaccessible to those herbivores (Lamberti and Resh 1983, Feminella *et al.* 1989). A "platform" supporting algal substrata (stream rocks or unglazed clay tiles) is erected in each replicate habitat unit and a control plot is placed directly on the streambed adjacent to each platform (Fig. 19.1). Sampling of grazers and periphyton on the treatment and control plots is conducted ≥30 days later. This design allows comparison of an "ambient" level of grazing (periphyton in control plot) with a "reduced" level of grazing depending on whether some or most herbivore species are excluded (periphyton in platform plot). Artifacts of caging are minimized with this design (see Exercise 2), but uncontrolled differences in depth, light, current velocity, or other factors still may exist between each control and platform pair, which may affect experimental results. This bias may be minimized by selecting plots that exhibit minimal environmental variation in all but the variables of interest. It is also important to note that such designs *rarely* exclude all grazers. Some swimming or drifting nontarget species (e.g., mayflies), as well as occasional target grazers, may accumulate on platforms. These will require manual removal periodically during the experiment. If this is not feasible, it may be necessary to estimate abundance of these grazers during the experiment and consider it as a covariate in statistical analyses.

Herbivore Enclosure Alternatively, grazer species and density can be manipulated directly within stream enclosures or "cages." These can be made from simple materials, stocked with known densities of grazers, and submersed or floated in the stream (e.g., Lamberti *et al.* 1987b, Feminella and Hawkins 1994). One set of cages is deployed in each stream section (e.g., shaded vs open reaches) and sampling of all treatments is conducted ≥30 days later. This design allows comparisons among known levels of grazing, which can range from none to high, while exerting more control over grazing pressure than provided by the platform design. It is also possible to measure grazing by particular species or size classes, which may exert very different impacts on periphyton. However, some consideration must be given to potential cage effects that result from altered flow, increased sedimentation, and colonization by unwanted grazers (see Walde and Davies 1984, Cooper *et al.* 1990); each of these may alter the effectiveness of grazer manipulations and confound interpretation of results. In addition, more maintenance generally will be required for cages than for platforms. In some instances, however, such as when conducting intraspecific grazing experiments or when quantifying effects of individual grazers on periphyton, cages may be the only suitable design.

Optional Exercises The above exercises are designed primarily to determine the effects of grazers on benthic algae. However, both exercises can be expanded to assess specific effects of algae on grazer populations. The exclosure exercise can be expanded to determine the rate of grazer

FIGURE 19.1 Grazer exclusion using platform design. Quarry tiles (7.5 × 7.5 cm) are shown on raised Plexiglas plates (ungrazed platforms) and on the streambed (grazed controls) at the beginning (A) and end (B) of the experiment. Tiles missing from platform and control plots in (B) were sampled before photograph was taken and objects on control tiles are grazing caddisflies. An improved design over that in the photograph is to use a Plexiglas plate underneath tiles for both platform *and* control treatments (from Feminella *et al.* 1989).

colonization to algal patches and the depletion rate of those algae by grazers. The enclosure exercise can be used to measure grazer growth rates by determining starting and ending weights of grazers, which may be used to test for grazer intraspecific or interspecific competition.

Grazer manipulations in streams are not limited to the two approaches presented above; certain grazers or groups of grazers are manipulated effectively using other techniques. For example, densities of certain sedentary or sessile grazers (e.g., fifth-instar hydroptilid caddisflies) can be altered by direct removal of larvae from rock surfaces (e.g., McAuliffe 1984, Hart 1985). In another approach, insecticides can be mixed with agar in diffusing substrata to deter some grazers (e.g., chironomid larvae) from colonizing substrata (Gibeau and Miller 1989, Peterson *et al.* 1993). However, care must be taken when using insecticides to avoid deleterious effects on nontarget organisms. A new technique that has been developed recently is the use of electricity to discourage lotic animals from entering small areas. Pringle and Blake (1994) excluded atyid shrimp from electrified "fence" enclosures (small in-stream hoops with wires hooked to battery-powered fence chargers) in a Puerto Rican stream. The continuous pulse of electricity within the electrified hoop excluded shrimp and allowed measurement of fine sediment accumulation in the absence of shrimp activity. This technique holds promise in excluding benthic grazers as well, while minimizing the artifactual effects of other techniques. However, unlike inert sediments, benthic algae are living organisms that likewise may respond to electric fields in unanticipated ways. Before electricity can be used as a herbivore deterrent, its effects on algae should be determined.

D. Laboratory Analyses

In both exercises below, periphyton is sampled and analyzed for biomass and chlorophyll *a* content at the end of the experiment (see methods in Chapter 14). In Exercise 1, macroinvertebrate community structure and density are determined for platform (ungrazed) and control (grazed) plots. For Exercise 2, growth rates of grazers in enclosures are determined by weighing grazers at the beginning and end of the experimental period.

III. SPECIFIC EXERCISES

A. Exercise 1: Grazer Exclusion Using Platforms

Platform Construction

1. Construct 10–20 grazer-exclusion platforms, the exact number depending on the experimental design (see below). Platforms are made of J-shaped metal or aluminum supporting rods and square 1/4″ Plexiglas plates,

the latter of which are used to hold periphyton substratum tiles. The resulting "platform" should look similar to the one shown in a stream in Fig. 19.1.

2. Cut Plexiglas into ≈400-cm² plates (approximate dimensions = 20 × 20 × 0.4 cm). Drill a 3/8″ hole midway through one of the four sides, about 1 cm from the edge; this allows the plate to be attached to the supporting rod. Using a drill press or metal lathe to secure the supporting rod, drill a 5/16″ hole (ca. 1″ long) into one end of the rod. Next, using a 3/8″ tap, thread the drilled hole to accept a 3/8″ hex bolt. Bend the threaded rods into the necessary J-shape with a metal jig in a bench vise, with the threaded end on the shorter end of the bend. Screw a hex bolt into the hole to ensure that it is seated properly. Cut a second set of plates for use in the control plots.

Initial Field Work

1. Three experimental design options exist. Study goals along with logistical and economic considerations will determine which design is most appropriate. *Design 1*. Locate two groups of replicate sites (reaches) to receive experimental plots, each group having some physical features in common (e.g., similar depth, substratum, etc.) but showing strong contrasts in others (e.g., shaded vs unshaded reaches, high vs low current velocity; see Section II, Site Selection). It is best, but not essential, to identify at least three habitat units within each site for plots, which collectively provide a reach-specific mean. This is a truly replicated design. *Design 2*. Locate two reaches that strongly differ in shading (or some other environmental factor of interest, such as upstream and downstream of a nutrient source). Deploy experimental plots within similar types of habitat units (i.e., pools, riffles, or runs) within each reach. Here, the habitat unit (rather than stream reach) is considered a spatial replicate. The environmental factor is not truly replicated in this design. *Design 3*. Locate a single stream reach that has relatively uniform environmental conditions (e.g., similar shading, flow, substratum, nutrient levels). Identify multiple habitat units in which to deploy experimental plots (one plot per habitat unit). This is a simple grazing study. Environmental factors can be measured for each plot and can be used as covariates in the statistical analyses (but not as main effects).

2. Incubate unglazed clay tiles on the streambed for at least 2 weeks (and preferably for 1 month) to allow algal colonization; the number of tiles incubated should be enough to supply all plots plus an additional 15% to allow for loss. A tile size of 7.5 × 7.5 cm is appropriate, four of which can be cut from one standard 6″ × 6″ tile using a tile cutter or masonry saw. Tiles should be incubated in a single microhabitat (similar depth, flow,

and shading) so that similar periphyton assemblages are present on all tiles across all treatments at the beginning of the experiment.

Installation of Platforms

1. Install supporting rods for platform plates in the selected sites within the stream. We recommend use of reinforcement bar (of similar diameter to that supporting rods) and a small sledge hammer to make a pilot hole in the substratum. Once embedded into the streambed about 30 cm, carefully remove the pilot bar and in its place insert the supporting rod; the rod is then tapped in place with the sledge. Next, secure four precolonized tiles to each platform and control plate with two heavy-duty rubber bands; these will prevent accidental dislodgement of tiles during the experiment. Make sure to subtract the area covered by rubber bands on each tile when estimating periphyton abundance later on, as tile areas underneath rubber bands will not maintain algae. Attach the platform plate to the supporting rod with a washer and hex bolt; use a wrench to tighten firmly but do not overtighten or plate may break. Place the control plot (plate with secured tiles accessible to grazers) on the streambed close to the platform.

2. In each habitat unit, install one replicate platform and one control plot, while matching current velocities, shading, and, if possible, depth. The latter may be achieved by using natural bed contours to minimize the platform distance above the substratum, while maintaining sufficient elevation to prevent colonization by crawling grazers. Alternatively, in more homogeneous-bottomed streams, bricks or cinder blocks may be used underneath control plates, which serve to match elevation (depth) with paired platforms, yet allow grazer access to tiles (Feminella et al. 1989). It is best to place each platform and control plot side-by-side (cf. upstream–downstream) so they will have minimal flow influences on each other.

3. Measure physical parameters for each platform and control plot including water depth, current velocity, irradiance, and nutrients (if relevant). Current velocity over the substrata can be measured with a current meter or by releasing dye with a pipet and measuring time-of-travel (see Chapter 4). Irradiance can be measured with a portable light meter held close to the water surface (or with an underwater probe; see Chapter 5), or estimated indirectly by measuring overhead canopy with a spherical densiometer (Lemmon 1957, Feminella et al. 1989). One set of four canopy measurements (i.e., facing upstream, downstream, and right and left banks) can be taken for both platform and control plot at each site; the four readings are then averaged for a single estimate for that site.

Sampling Platforms

1. Decide on how many tiles to sample from each plot (platform or control) at the end of the experiment ($\geq$30 days). Because each plot represents either (1) one reach "subsample" (i.e., one of three to five subsamples per reach) in the multiple-reach experimental design or (2) an actual replicate in the two-reach or single-reach designs, it is necessary to sample at least one tile per platform and control plot. However, for the two- and single-reach designs, because of natural variability it may be wise to sample two or more tiles per plot and then pool the sampled material before analysis. The same tiles can be used to sample macroinvertebrates and periphyton. However, extra tiles may be needed for additional analyses such as algal taxonomic composition or primary production (see below).

2. Sample macroinvertebrates from tiles at the end of the experiment. Dislodge macroinvertebrates from a tile into a downstream net (mesh size $\leq$250 μm) and then empty the net contents into a labeled heavy-duty plastic sample bag (e.g., Whirl-pak) or jar; preserve with 80% ethanol. Be sure to sample undersides of tiles for mobile invertebrates (e.g., mayflies, stoneflies, etc.). Manually remove any sessile invertebrates such as black flies, chironomid midges, and caddisflies. The latter two groups often attach their organic (algal-rich) cases to tiles, which may remain after larvae are removed. Thus, investigators should indicate whether they removed cases prior to sampling or left them in place as part of the periphyton sample.

3. Sample periphyton from tiles after invertebrates are removed. In the field, periphyton can be scraped or brushed from the tile into water and placed in a darkened container on ice. Alternatively, the entire tile can be placed in a plastic container, stored on ice in the dark, and scraped in the laboratory on the same day. Analyze periphyton for biomass (AFDM) and chorophyll *a* content (see Chapter 14). *Optional.* Determine algal taxonomic structure (see Chapter 13) and primary production (see Chapter 25).

B. Exercise 2: Grazer Density Manipulation Using Enclosures

Enclosure Construction

1. In-stream enclosures can be constructed from various materials, or prefabricated containers can be modified for use. All designs should be fitted with mesh on upstream and downstream sides to allow for water exchange. Enclosure (and mesh) size should be scaled to the size and density of grazers used in the study. Because animals vary greatly in their ability to behave "normally" within an enclosure, there is no steadfast rule governing the size of the cage relative to the size of the animal used. However, the best choice in a cage size is one that collectively (1) provides the best possible control of the variable of interest; (2) is economically and

logistically feasible to build and deploy; and (3) has the lowest potential for cage-related artifacts.

2. Construct cages ($n = 20$–30) suitable for the grazer of interest. For small grazers (e.g., small caddisflies or snails), plastic food containers with two or more sides replaced by window screening may be suitable (Fig. 19.2A). Cut window screening (1-mm mesh) to size and secure to containers with silicone aquarium sealant or chemically inert hot glue. Lightly roughen container edges with sandpaper prior to attaching screen to maintain the adhesive bond for a longer period. Allow at least 24 h for adhesive to cure before immersing the enclosure. For large grazers (e.g., limnephilid caddisflies, crayfish, tadpoles, small fish), larger enclosures made of hardware cloth, porous plastic containers, PVC pipe cut longitudinally (Fig. 19.3; see Feminella and Hawkins 1994), or other inexpensive materials may be used. These larger, more durable materials also allow experiments to be done in fast-flowing stream sections, such as in riffles or runs, where rheophilic grazers (e.g. hepatgeniid mayflies, glossosomatid caddisflies) predominate.

3. Enclosures may float or be submerged. To float small enclosures at the stream surface, cut "collars" from 1/2″ to 3/4″ thick styrofoam panels (Fig. 19.2A). Leave enough of a styrofoam border around the complete enclosure (i.e., containing substratum tile, grazers, and any additional material, such as sand for caddisfly casebuilding materials) so that it remains buoyant and stable when placed in the stream. Covers for individual enclosures are optional, but are not recommended in herbivory studies if they reduce irradiance. Sets of enclosures may be held in place within rectangular wooden racks, which are predrilled and strung tennis racquet-style with monofilament line (30–60 lb. test) to secure individual enclosures and collars (Fig. 19.2B). Racks can be floated by attaching styrofoam blocks to each of the corners and held in place by tethering to rebar stakes or trees on the bank. Some play in the tethers is desirable in case water level changes. A wooden frame fitted with a chicken-wire screen and placed over the rack may protect enclosures from falling debris or disturbance from small animals. Alternatively, individual enclosures can be tethered to stakes or bricks in the channel with monofilament, and allowed to "free float" in the stream.

4. Submersed enclosures are placed directly on the streambed. In swift current, it may be necessary to attach them with hose clamps, twist ties, or other materials to reinforcement bar pounded into the streambed. For larger enclosures, stream cobbles lining enclosure bottoms may be used as ballast. Unless enclosure walls are considerably higher than that of the stream surface, enclosures must be completely covered so that grazers are effectively isolated and experimental treatments can be maintained.

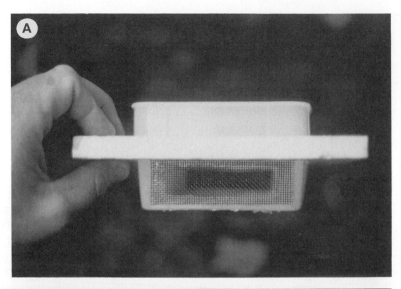

FIGURE 19.2 Grazer enclosures used to manipulate densities of small benthic grazers. (A) Each enclosure is ≈500 cm^3 with sides screened with 1-mm^2 mesh and fitted with a styrofoam collar for flotation. (B) A floating rack holding 27 enclosures is placed in the stream and secured (see Lamberti *et al.* 1987).

FIGURE 19.3 Photograph of 12 large, submersed enclosures (dimensions: 103 cm L ×
32.5 cm D) used to study behavior of grazing tadpoles inhabiting high-gradient streams at
Mount St. Helens, Washington. Enclosures contain algal-covered stream cobbles that are used
both as grazeable substrata for tadpoles and as ballasts for enclosures in fast riffles. Coarse
mesh covers used to isolate and protect animals have been removed. Arrow indicates direction
of flow (from Feminella and Hawkins 1994).

Initial Field Work

1. Locate a study site that will receive enclosures. It less critical to
replicate the experiment over several sites (cf. platform design), because
a full range of experimental treatments and replicates may be interspersed
within a stream reach. Hence, there are fewer concerns about spatial pseudoreplication, which may be problematic in the platform experimental
design (see Exercise 1). In the enclosure study, the main effect is herbivore
density and the response is various periphyton variables.

2. Incubate unglazed clay tiles on the streambed, as described previously (Exercise 1 under Initial Field Work).

Installation of Enclosures

1. Survey the stream reach to identify the numerically dominant large,
benthic herbivore(s) in the stream (e.g., snails, caddisflies, etc.). Measure
the ambient density of these herbivores on the streambed by making visual
counts from replicate quadrats ($N = 15–30$) placed randomly in all habitat
types (e.g., riffles and pools) throughout the study reach, as well as in

specific microhabitats where grazers are most common (e.g., cobbles, boulders). These two separate estimates will yield a full range of grazer densities from low (across all habitat units) to high (within preferred habitat); doing so will yield a large operational range of grazer abundances in which to bracket experimental treatments and will provide greater extrapolational power to experimental results. Quadrat size may be scaled to the size, density, and spatial distribution of the grazer studied. For example, small quadrats (<100 cm^2) may be precise enough to characterize small, abundant grazers such as mayflies, sessile midges, or small caddisflies, whereas much larger quadrats ($\geqslant 100$ cm^2) may be required for larger, more patchily distributed grazers such as snails, tadpoles, or large caddisflies.[1]

2. Choose treatments and number of replicates of each treatment. At a minimum, use ambient grazer density and no grazers, with three replicates of each treatment. Optionally, include treatments of one-half ambient density and double ambient density, or add replicates to increase statistical power. In a more complex design, shading can be added to some replicates to achieve a two-factor experimental manipulation (i.e., grazing and irradiance). This can be achieved by placing a shade screen over the top of some enclosures.

3. Select a relatively uniform site in the study reach in which to install enclosures. Place precolonized tiles in enclosures and suspend in the rack or anchor to the stream bottom (see above). Collect individuals of the selected herbivore from the stream. Randomize the collected grazers and place the appropriate number of herbivores into each replicate enclosure.[2] Establish a block in each identified microhabitat (velocity range, depth, etc.) and place one replicate of each treatment in each block. Ideally, the number of blocks should correspond to the number of treatments so that treatments are equally represented across blocks.

Sampling Enclosures

1. At the end of the experiment, sample all enclosures containing tiles and grazers; the same tiles can be used to quantify both macroinvertebrates and periphyton abundance. However, extra replicate enclosures may be

---

[1]Visual counts are generally only useful in estimating densities of large grazers that are easily seen on the streambed surface. This approach may not be feasible in streams with high turbidity (low water clarity) or for grazers that cannot be quantified visually, such as those species occurring in subsurface habitats or exhibiting low activity (i.e., visibility) during daylight hours. In these cases, direct substratum sampling (e.g., Surber or Hess samplers) or nighttime estimates (Feminella and Hawkins 1994) may be more appropriate (see Chapter 16, Resh et al. 1984).

[2]Minimize position effects (e.g., velocity or depth gradients) by using a randomized block design (Zar 1984).

needed for additional analyses such as algal taxonomic composition or primary production (see below).

2. Manually remove grazers and all other macroinvertebrates from enclosures at the end of the experiment ($\geq$30 days) and preserve all in 80% ethanol. Count the remaining target grazers; identify and count other macroinvertebrates that have colonized the enclosures. Quantify the difference between starting and ending grazer abundance to determine mortality. Optionally, measure grazer growth (see below).

3. Sample periphyton chlorophyll *a* and biomass from tiles at the end of the experiment, as described briefly above (see Exercise 1, Sampling Platforms) and in more detail in Chapter 14.

4. *Optional*. Estimate growth rates of grazers in enclosures to determine if growth is density dependent. If target grazers are approximately the same size (e.g., synchronously developing insect cohort), then subsample animals at the beginning of the experiment to determine average starting individual mass. In the laboratory, measure as tissue dry mass or ash-free dry mass (see Chapters 14, 16). At the end of the experiment, remove and preserve grazers from enclosures, and estimate individual grazer mass (determined individually or batch-weighed) for each replicate enclosure. Determine individual mass of grazers from the stream bottom, collected on the same day as those from enclosures; this will allow an assessment of grazer growth patterns attributable to enclosures themselves (i.e., cage effects). If cage effects are minimal, growth of grazers in enclosures containing "ambient" densities will be equivalent to those on the stream bottom. If grazers are slow-growing or occur as multiple cohorts, then growth of tagged individuals will need to be determined. We have used individually numbered "bee" tags, glued carefully to the shells of snails or cases of caddisflies, to monitor individual growth (see Lamberti *et al.* 1989).

C. Exercise 3: Manipulation of Periphyton Abundance

1. Algal patches can be created on tiles and then exposed to grazers to determine response of those grazers to modified resource levels. Platforms or nongrazed enclosures can result in substantial algal growth on tiles. These algal-rich tiles can then be returned to the streambed to measure grazer colonization, spatial distribution, and algal depletion rates (see Lamberti and Resh 1983). Different assemblages and densities of algae can be created by exposing tiles to various densities or species of grazers within enclosures or modified by partial scraping (see Kohler 1984). These tiles then can be placed on the streambed and used to study grazer patch use (see below), food preference, and growth (see Feminella and Resh 1991).

2. Grazer colonization of algal "patches" can be observed at various time intervals and compared with that of grazed tiles. This can be documented by visual counts of grazers on site or from photographs (Lamberti and Resh 1983, Resh *et al.* 1984), or with direct substratum sampling for less conspicuous grazers (see Exercise 2, Installation of Enclosures). Direction of movement and travel rates of grazers across patches also can be determined (see Hart 1981, Kohler 1984).

3. Algal depletion rates can be determined by sampling a subset of tiles at the beginning of the experiment and then at various time intervals. Periphyton biomass or chlorophyll *a* content can be measured as described above.

4. Enclosures can be used to examine the outcome of intraspecific or interspecific interactions among grazers. The effects of different grazer densities of one population on individual growth of those grazers will provide information on intraspecific interactions and possible competition (see Lamberti *et al.* 1987). Combinations of two or more grazer species in enclosures will provide information on interspecific interactions (see Feminella and Resh 1991, Kohler 1992).

D. Data Analysis

Platform Experiments   Parametric statistics can be used to analyze data, provided that assumptions of normality, homogeneity of variance, and sample unit independence are met (Zar 1984). For the platform study, if a single stream reach is used with only two grazing levels (i.e., platform vs control plots), then a paired *t* test (parametric test) or Wilcoxon Signed Ranks test (nonparametric test) should be used. These tests examine the *difference* between the paired control and platform plots for a specific response variable. This approach will also remove the variation associated with unmeasured differences among paired plots. Differences between platform and control plots can be compared for each variable (both treatment and response) measured (e.g., periphyton biomass, chlorophyll *a*, grazer density, etc.). If two stream reaches were used, then platform–control differences can be compared *between* reaches using an independent, two-sample *t* test (or nonparametric Wilcoxon test). Differences in grazing treatments between the two reaches can be determined, but one *cannot* attribute differences to any other environmental variables measured (Hurlbert 1984). If multiple stream reaches were used (that varied in a specific environmental factor), then a two-factor ANOVA that includes a fixed factor (manipulated grazing level) and a random factor (unmanipulated environmental variable) can be used. This "mixed-model" ANOVA (Zar 1984) will allow an assessment of effects on periphyton that result from grazing, light (or other factor), or an interaction between these factors.

Enclosure Experiments   If only grazing (≥ 2 levels) was manipulated, then a one-way ANOVA or its nonparametric equivalent (Kruskal–Wallis test) is appropriate (Sokal and Rohlf 1995). If, in addition to grazing, a second factor (e.g., irradiance) was manipulated, then a two-factor ANOVA should be used. If ANOVA reveals significant differences among treatments, a posteriori multiple contrasts (e.g., Tukey HSD test) can be used to identify where specific differences reside.

Regression Approaches   As an alternative to the above analyses, regression can be used to examine the relationship between grazer density, light, or current (independent variables) and periphyton abundance or grazer growth (dependent variable). This analysis may be appropriate if densities or abiotic factors vary continuously across the full range of treatment replicates or change substantially over time for individual replicates. Construct graphs of treatment-specific grazer density, irradiance, or current (x-axis) against corresponding periphyton abundance or grazer growth (y-axis) and choose the most appropriate regression model (e.g., linear, exponential, power, etc.) that accounts for the highest amount of variation in the data (Sokal and Rohlf 1995). If grazer density was the only "fixed" factor in experiments, then it may be most appropriate to treat measured environmental variables as covariates in the analyses (Zar 1984). Multiple regression may also be used to examine collectively the predictive power of all independent variables on periphyton abundance.

IV. QUESTIONS

1. Was the grazer exclusion effective at eliminating some or all grazers? Did you notice any colonization by invertebrates other than the ones you were trying to exclude? Were there any confounding effects of the exclusion?

2. Were there significant effects of grazing on periphyton abundance? If tested, were there significant effects of canopy or current velocity on periphyton within or among grazed treatments (i.e., grazer × canopy/current interactions)? What can you conclude about the relative importance of light, current, and grazing on periphyton abundance in your stream?

3. If the platform grazer exclusion resulted in significant periphyton accrual, what does this say about the overall importance of grazing in your stream? Can you conclude that all of the streambed would accumulate that much periphyton in the absence of grazing? Why or why not? Would you expect this same result during all seasons or in all other nearby streams?

4. If multiple grazer treatments were used in enclosures, what was the relationship between grazer density and periphyton abundance? If grazer growth was measured, how did this vary with density, and what can you conclude about intraspecific or interspecific interactions among grazers?

5. Did the biomass/chlorophyll *a* ratio differ among grazing treatments? Among stream reaches? Why might this occur?

6. Do you believe that periphyton is a limiting resource for grazers in your stream? Why? Is it possible for the biomass of benthic herbivores to exceed that of the primary producers? How might this occur? If periphyton was indeed limiting and grazers were competing for periphyton, how might you test for the specific competitive mechanism (i.e., exploitation vs interference)?

V. MATERIALS AND SUPPLIES

Platform Materials

1/4″ Plexiglas squares (20 × 20 cm)
1/2″ Aluminum bar
3/8″ Drill bit, drill, and drill press (or metal lathe)
3/8″ Hex bolts (1/2″ long), washers, and nuts
3/8″ Tap for hex bolt
Form or jig used to bend aluminum bar into J-shape
Heavy-gauge rubber bands
Reinforcement bar (used for pilot hole)
Sledge hammer
Unglazed clay tiles (7.5 × 7.5 cm)

Enclosure Materials

Aquarium (silicone) sealant
Duct tape
Fiberglass mosquito netting (≈1-mm mesh size)
Monofilament nylon (30- to 60-lb. test)
Plastic food containers (pint size or larger) or similar enclosure
Queen bee tags (optional)
Reinforcement bar
Rope for rack tethers
Styrofoam blocks for rack flotation (ca. 1 foot^3); four per rack
Styrofoam sheets to make collars to float enclosures within rack; one
    per enclosure
Utility knife

Wood frame and hardware (for floating rack), large enough to fit all enclosures

Field and Laboratory Materials (see Chapter 14 for materials to conduct periphyton analyses)

80% Ethanol
Forceps
Net or sieve ($\leq$250-μm mesh)
Paper labels
Squirt bottles
Whirl-pak sample bags

Field Equipment

Current velocity meter (optional)
Light meter (optional)

Laboratory Equipment

Desiccator
Dissecting microscope
Drying and ashing ovens
Electronic balance ($\pm$0.1 mg)
Filtration apparatus (vacuum pump, filter funnel, Erlenmeyer filter flask, tubing)
High-speed centrifuge
Spectrophotometer or fluorometer

REFERENCES

Bott, T. L. 1983. Primary productivity in streams. Pages 29–53 *in* J. R. Barnes and G. W. Minshall (Eds.) *Stream Ecology.* Plenum, New York, NY.

Cooper, S. D., S. J. Walde, and B. L. Peckarsky. 1990. Prey exchange rates and the impact of predators on prey populations in streams. *Ecology* **71:**1503–1514.

Creed, R. P., Jr. 1994. Direct and indirect effects of crayfish grazing in a stream community. *Ecology* **75:**2091–2103.

Cummins, K. W., and M. J. Klug. 1979. Feeding ecology of stream invertebrates. *Annual Review of Ecology and Systematics* **10:**147–172.

DeNicola, D. M., C. D. McIntire, G. A. Lamberti, S. V. Gregory, and L. R. Ashkenas. 1990. Temporal patterns of grazer–periphyton interactions in laboratory streams. *Freshwater Biology* **23:**475–489.

DeNicola, D. M., and C. D. McIntire. 1991. Effects of hydraulic refuge and irradiance on grazer–periphyton interactions in laboratory streams. *Journal of the North American Benthological Society* **10:**251–262.

Dudley, T. L., S. D. Cooper, and N. Hemphill. 1986. Effects of macroalgae on a stream invertebrate community. *Journal of the North American Benthological Society* **5**:93–106.

Dudley, T. L., and C. M. D'Antonio. 1991. The effects of substrate texture, grazing, and disturbance on macroalgal establishment in streams. *Ecology* **72**:297–309.

Feminella, J. W., and C. P. Hawkins. 1995. Interactions between stream herbivores and periphyton: A quantitative analysis of past experiments. *Journal of the North American Benthological Society* **14**:465–509.

Feminella, J. W., and C. P. Hawkins. 1994. Tailed frog tadpoles differentially alter their feeding behavior in response to non-visual cues from four predators. *Journal of the North American Benthological Society* **13**:310–320.

Feminella, J. W., M. E. Power, and V. H. Resh. 1989. Periphyton responses to grazing and riparian canopy in three Northern California coastal streams. *Freshwater Biology* **22**:445–457.

Feminella, J. W., and V. H. Resh. 1990. Hydrologic influences, disturbance, and intraspecific competition in a stream caddisfly population. *Ecology* **71**:2083–2094.

Feminella, J. W., and V. H. Resh. 1991. Herbivorous caddisflies, macroalgae, and epilithic microalgae: Dynamic interactions in a stream grazing system. *Oecologia* **87**:247–256.

Gibeau, G. G., Jr., and M. C. Miller. 1989. A micro-bioassay for epilithon using nutrient-diffusing artificial substrata. *Journal of Freshwater Ecology* **5**:171–176.

Gregory, S. V. 1983. Plant–herbivore interactions in stream systems. Pages 157–189 *in* J. R. Barnes and G. W. Minshall (Eds.) *Stream Ecology*. Plenum, New York, NY.

Hart, D. D. 1981. Foraging and resource patchiness: Field experiments with a grazing stream insect. *Oikos* **37**:46–52.

Hart, D. D. 1992. Community organization in streams: The importance of species interactions, physical factors, and chance. *Oecologia* **91**:220–228.

Hart, D. D. 1985. Causes and consequences of territoriality in a grazing stream insect. *Ecology* **66**:404–414.

Hurlbert, S. H. 1984. Pseudoreplication and the design of ecological experiments. *Ecological Monographs* **54**:187–211.

Kohler, S. L. 1984. Search mechanism of a stream grazer in patchy environments: The role of food abundance. *Oecologia* **62**:209–218.

Kohler, S. L. 1992. Competition and the structure of a benthic stream community. *Ecological Monographs* **62**:165–188.

Lamberti, G. A. 1993. Grazing experiments in artificial streams. *Journal of the North American Benthological Society* **12**:337–342.

Lamberti, G. A. 1996. The role of periphyton in benthic food webs. Pages 533–572 *in* R. J. Stevenson, M. L. Bothwell, and R. L. Lowe (Eds.) *Algal Ecology: Freshwater Benthic Ecosystems*. Academic Press, San Diego, CA.

Lamberti, G. A., L. R. Ashkenas, S. V. Gregory, and A. D. Steinman. 1987a. Effects of three herbivores on periphyton communities in laboratory streams. *Journal of the North American Benthological Society* **6**:92–104.

Lamberti, G. A., J. W. Feminella, and V. H. Resh. 1987b. Herbivory and intraspecific competition in a stream caddisfly population. *Oecologia* **73:**75–81.

Lamberti, G. A., and J. W. Moore. 1984. Aquatic insects as primary consumers. Pages 164–195 *in* V. H. Resh and D. M. Rosenberg (Eds.) *The Ecology of Aquatic Insects.* Praeger, New York, NY.

Lamberti, G. A., and V. H. Resh. 1983. Stream periphyton and insect herbivores: An experimental study of grazing by a caddisfly population. *Ecology* **64:**1124–1135.

Lamberti, G. A., S. V. Gregory, L. R. Ashkenas, A. D. Steinman, and C. D. McIntire. 1989. Productive capacity of periphyton as a determinant of plant–herbivore interactions in streams. *Ecology* **70:**1840–1856.

Lamberti, G. A., S. V. Gregory, C. P. Hawkins, R. C. Wildman, L. R. Ashkenas, and D. M. DeNicola. 1992. Plant–herbivore interactions in streams near Mount St. Helens. *Freshwater Biology* **27:**237–247.

Lemmon, P. E. 1957. A new instrument for measuring forestry overstory canopy. *Journal of Forestry* **55:**667–668.

Mayer, M. S., and G. E. Likens. 1987. The importance of algae in a shaded headwater stream as food for an abundant caddisfly (Trichoptera). *Journal of the North American Benthological Society* **6:**262–269.

McAuliffe, J. R. 1984. Competition for space, disturbance, and the structure of a benthic stream community. *Ecology* **65:**894–908.

McCormick, P. V., and R. J. Stevenson. 1991. Mechanisms of benthic algae succession in lotic environments. *Ecology* **72:**1835 1848.

McIntire, C. D. 1973. Periphyton dynamics in laboratory streams: a simulation model and its implications. *Ecological Monographs* **43:**399–420.

Minshall, G. W. 1978. Autotrophy in stream ecosystems. *BioScience* **28:**767–771.

Minshall, G. W., K. W. Cummins, R. C. Petersen, C. E. Cushing, D. A. Bruns, J. R. Sedell, and R. L. Vannote. 1985. Developments in stream ecosystem theory. *Canadian Journal of Fisheries and Aquatic Sciences* **42:**1045–1055.

Peterson, B. J., L. Deegan, J. Helfrich, J. E. Hobbie, M. Hullar, B. Moller, T. E. Ford, A. Hershey, A. Hiltner, G. Kipphut, M. A. Lock, D. M. Fiebig, V. McKinley, M. C. Miller, J. R. Vestal, R. Ventullo, and G. Volk. 1993. Biological responses of a tundra river to fertilization. *Ecology* **74:**653–672.

Power, M. E., and W. J. Matthews. 1983. Algae-grazing minnows (*Campostoma anomalum*), piscivorous bass (*Micropterus* spp.), and the distribution of attached algae in a small prairie-margin stream. *Oecologia* **60:**328–332.

Power, M. E., A. J. Stewart, and W. J. Matthews. 1988. Grazer control of algae in an Ozark Mountain stream: Effects of short-term exclusion. *Ecology* **69:**1894–1898.

Pringle, C. M., and G. A. Blake. 1994. Quantitative effects of atyid shrimp (Decapoda: Atyidae) on the depositional environment in a tropical stream: use of electricity for experimental exclusion. *Canadian Journal of Fisheries and Aquatic Sciences* **51:**1443–1450.

Resh, V. H., G. A. Lamberti, E. P. McElravy, J. R. Wood, and J. W. Feminella. 1984. *Quantitative Methods for Evaluating the Effects of Geothermal Energy Development on Stream Benthic Communities at The Geysers, California.* California Water Resources Center Contribution 190, University of California, Davis, CA.

Rosemond, A. D. 1993. Interactions among irradiance, nutrients, and herbivores constrain a stream algal community. *Oecologia* **94:**585–594.

Rosemond, A. D. 1994. Multiple factors limit seasonal variation in periphyton in a forest stream. *Journal of the North American Benthological Society* **13:**333–344.

Sokal, R. R., and F. J. Rohlf. 1995. *Biometry,* 3rd ed. Freeman, New York, NY.

Steinman, A. D. 1991. Effects of herbivore size and hunger level on periphyton communities. *Journal of Phycology* **27:**54–59.

Steinman, A. D. 1992. Does an increase in irradiance influence periphyton in a heavily-grazed woodland stream? *Oecologia* **91:**163–170.

Steinman, A. D. 1996. Effects of grazers on freshwater benthic algae. Page 341–373 *in* R. J. Stevenson, M. L. Bothwell, and R. L. Lowe (Eds.) *Algal Ecology: Freshwater Benthic Ecosystems.* Academic Press, San Diego, CA.

Thorp, J. H., and M. D. DeLong. 1994. The riverine productivity model: An heuristic view of carbon sources and organic processing in large river ecosystems. *Oikos* **70:**305–308.

Vannote, R. L., G. W. Minshall, K. W. Cummins, J. R. Sedell, and C. E. Cushing. 1980. The river continuum concept. *Canadian Journal of Fisheries and Aquatic Sciences* **37:**130–137.

Walde, S. J., and R. W. Davies. 1984. Invertebrate predation and lotic prey communities: Evaluation of in situ enclosure/exclosure experiments. *Ecology* **65:**1206–1213.

Wootton, J. T., and M. P. Oemke. 1992. Latitudinal differences in fish community trophic structure, and the role of fish herbivory in a Costa Rican stream. *Environmental Biology of Fishes* **35:**311–319.

Zar, J. H. 1984. *Biostatistical Analysis,* 2nd ed. Prentice–Hall, Englewood Cliffs, NJ.

CHAPTER 20

Predator–Prey Interactions

BARBARA L. PECKARSKY

Department of Entomology
Cornell University

I. INTRODUCTION

Riffle and pool communities in streams may be viewed as part of an open, nonequilibrium system (Caswell 1978), having multiple patches connected by migration (Cooper *et al.* 1990, see also Chapter 17). Historically, ecologists did not consider predation an important determinant of the structure of stream communities (Allan 1983a, b, 1995, Peckarsky 1984), because most theory describing predator effects on prey communities had been developed for closed, equilibrium systems (e.g., Slobodkin 1961). Caswell (1978), however, developed a model predicting that predator and prey populations can coexist over the long term in open nonequilibrium systems. Furthermore, his model suggested that predation could be a major determinant of prey community structure in such systems, underscoring the value of studying predator–prey interactions in streams. Below is a brief review of the types of effects that predators can have on prey communities and a description of some proximate mechanisms that can explain such effects, as background for studies on predator–prey interactions in streams.

In streams the predominant predators (organisms that consume animal prey) are fish, which feed on the benthos or on drift in the water column (Hyatt 1979, see also Chapters 18 and 22). Prey items include representatives of many orders of benthic insects such as dragonflies and damselflies (Odonata), stoneflies (Plecoptera), hellgrammites (Megaloptera), caddisflies (Trichoptera) and true flies (Diptera) (Peckarsky 1982, Allan 1995, see Chapters 16 and 21), and other macroinvertebrates such as crayfish. Because most investigations of predation in streams have been conducted on either fish (e.g., trout) that feed on invertebrate drift (Metz 1974, Allan 1981, Healey 1984) or benthic-feeding stoneflies (e.g., Malmqvist and Sjostrom 1980, Allan 1982, Molles and Pietruszka 1983, 1987, Peckarsky 1985, Walde and Davies 1987), little is known about the natural history of other predators in streams. The effects of predators on prey populations depend on their predation rate compared to the prey *exchange rate*, the rate at which prey move in and out of areas where predators are feeding (Cooper *et al.* 1990, Lancaster *et al.* 1991, Forrester 1994, Sih and Wooster 1994). In stream reaches with high background rates of prey migration (e.g., fast flowing riffles with high invertebrate drift rates, see Chapter 17), effects of predation may not be measurable if they are swamped by prey immigration. In stream reaches with low rates of background prey migration, or where predators induce high rates of prey emigration, predators may have more substantial impacts on prey communities. Likewise, the impact of predators may vary on different prey species depending on relative prey vulnerability to predation, migration rates, and tendency to emigrate from patches where predators are foraging (Peckarsky 1985, Lancaster *et al.* 1991, Forrester 1994).

Predator–prey interactions also can have effects on prey communities that extend well beyond direct predator-induced mortality (Kerfoot and Sih 1987). Two general classes of predator impacts other than direct lethal effects on prey populations (terminology here as in Strauss 1991) are: (1) indirect community level effects, such as "top-down" or cascading trophic effects (Carpenter *et al.* 1987, Power 1990) and (2) nonlethal or sublethal effects on prey populations (Peckarsky *et al.* 1993). The first class refers to situations in which top predators can influence trophic levels below that of their prey, and competitors of their prey, if direct prey reductions cause indirect increases in the abundance of the prey's resources. The second class involves direct interactions between predators and prey that do not result in prey death, but have negative consequences on prey population growth. This may occur if predator-avoidance behavior is costly to prey in terms of lost feeding time, shifting to unfavorable food patches, or shifting to less favorable feeding times. Thus, the impacts of predation on prey populations can be studied from two general perspectives: (1) effects of predator-induced mortality on prey populations and communities and (2) consequences of anti-predatory behavior on prey fitness.

The mechanisms or specific causes of predator effects on prey communities or on prey fitness often depend on whether predators are selective (i.e., consume certain prey types disproportionate to their densities in the environment). Community ecologists are interested in measuring whether selective predation can alter the relative proportions of prey in communities (Paine 1966, Connell 1975). Behavioral or evolutionary ecologists often focus on the mechanisms underlying observed patterns of selective predation (Allan et al. 1987a,b), the significance of differential prey defenses (Cooper 1985, Greene 1985, Peckarsky and Penton 1989), or on the life history consequences of predator-induced changes in prey behaviors (Crowl and Covich 1990, Peckarsky *et al.* 1993, 1994). Selective predation may result from concentration of predator search in the preferred habitat of the prey, selection of prey types most frequently encountered, active rejection of some encountered prey individuals, or differential prey vulnerability. Selection that occurs because of differential prey rejection is considered "active" behavioral selection (Greene 1985), and is a function of predator choice (Sih 1987). Selectivity that occurs because of differential prey vulnerability is more a function of stereotypical predator behavior ("fixed" behavioral selection; Greene 1985), prey defensive attributes (passive selection; Allan and Flecker 1988), or microhabitat differences among prey that affect vulnerability (Fuller and Rand 1990). These potential mechanisms of selective predation can be differentiated by measuring predator—prey encounter rates, attacks per encounter, and captures per attack, which are the major components of the predator—prey interaction.

The purpose of this chapter is to introduce researchers to the study of predator—prey interactions from both community and behavioral perspectives. More specifically, methods are presented to (1) calculate electivity indices comparing field measurements of predator gut contents to estimates of prey availability to generate hypotheses on selective predation at the community level (Exercise 1), (2) test field-generated hypotheses by conducting predation experiments to confirm whether predators feed selectively on certain prey species (Exercise 2), and (3) conduct behavioral observations to determine which components of predator—prey interactions explain observed patterns of selective predation (Exercise 3).

II. GENERAL DESIGN

A. Site and Species Selection

The feasibility and specifics of these exercises will depend on the vicinity of the researchers to low (first—fourth) order rocky-bottom streams with riffle habitats containing abundant populations of large predatory stoneflies (Plecoptera: families Perlidae, Perlodidae, or large Chloroperlidae) and potential mayfly prey species (Ephemeroptera: families Baetidae, Leptophlebiidae, Heptageniidae, Ephemerellidae). While it is possible to substitute other predatory taxa (e.g., benthic fish; see Kotila 1987, dragonflies, or hellgrammites), these exercises were written specifically for stonefly–mayfly interactions and, thus, have the highest probability of succeeding if those taxa are used. Exercise 1 involves field collection of predators and prey to compare proportions of prey in predator diets and the environment. The most abundant predator species should be used for the electivity analysis, such that each researcher obtains several predators per sample.

Predation experiments in Exercise 2 can be carried out in enclosures placed in very shallow (<10 cm), moderately flowing (15–20 cm/s) riffles in the field, if such habitats are available and will not be disturbed overnight. Likewise, behavioral observations (Exercise 3) can be done in enclosures *in situ,* but with less concern for disturbance, since they will not be left unattended. Alternatively, Exercises 2 and 3 can be carried out in the laboratory if the researchers have access to dechlorinated water that can be distributed to replicate enclosures. For these exercises researchers should select one predator species and, ideally, three abundant alternative prey species; one of which is overrepresented in predator diets, one underrepresented in predator diets, and one eaten in proportion to its availability in the predator's habitat (as determined in Exercise 1).

B. Field-Derived Electivity Indices—Hypothesis Generation

The most common method for measuring patterns of differential predation is to compare the proportion of prey remains found in a predator's gut to the relative prey abundances measured in the predator's habitat (Chesson 1978). Although gut content data may provide an accurate record of undigested prey parts, there are many potential limitations to this method. Differential gut clearance time of different prey species (Hildrew and Townsend 1982) may lead to overestimation of prey with heavily sclerotized parts compared to soft-bodied prey. Partial consumption of prey may leave heavily sclerotized parts uneaten, as is the case with some predatory aquatic insects (Martin and Mackay 1982, Peckarsky and Penton 1985). Furthermore, ingestion of prey fragments or prey masceration may also constrain our ability to identify and quantify predator diets from gut contents. Also, regurgitation may occur during preservation, or preservatives may alter gut contents. Thus, gut contents show only part of what has been eaten and could result in misinterpretation of the relative consumption rates of different prey species.

Calculation of prey preferences also depends on the ability to estimate prey availability, and assumes that (1) samples of prey populations accurately reflect their relative densities, (2) prey are encountered by predators at a rate commensurate with measured prey density, and (3) the predator's perception of available prey is the same as that of the investigator. There is a large literature concerning potential problems with the accuracy of benthic samples (Resh 1979 and see Chapter 16). Furthermore, little is known about actual predator–prey encounter rates (Peckarsky *et al.* 1994) or predator perception of available prey in streams (O'Brien and Showalter 1993), since it is difficult to observe stream predators in their natural habitat. Consequently, hypotheses of differential predation based on data obtained by the gut content approach should be tested using other methods (see below).

Two general approaches can be used to compare the composition of a predator's diet to the composition of prey collected from the predator's habitat. Both approaches standardize the relative importance of each prey item by indicating its proportion of the total of all food items. Percentage or fractional composition can be calculated for each prey species in predator gut contents (r_i), and habitat (p_i), which indicate the relative abundances of each food type in the diet and in the environment, respectively. The first method (correlation) involves the ranking of prey types in the predator guts and in the habitat and comparing those ranks using Spearman's rank correlation analysis (Siegel 1956). A significant positive correlation (Fig. 20.1A) indicates no selectivity (ranks of prey items in the diet and in the environment are similar). No correlation (Fig. 20.1B) or a significant negative correlation (Fig. 20.1C) indicate selective predation (i.e., feeding that is weakly or strongly disproportionate to availability of prey items in the environment).

A second method (electivity indices) can be used to compare the relative proportions of each prey species in the predator diet and the environment. A wide variety of such indices have been developed (Ivlev 1961, Jacobs 1974, Chesson 1978), including situation-specific modifications

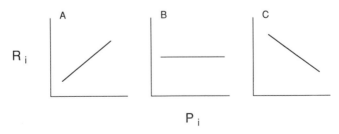

FIGURE 20.1 Plots of the possible relationships between the proportions of prey species i in the habitat (p_i) and in predator diets (r_i). (A) Positive correlation. (B) No correlation. (C) Negative correlation.

(Johnson 1980, Lechowicz 1982). All are variations of the forage ratio (in simplest form), which is the proportion of a specific prey item in the predator's gut (r_i) divided by the proportion of that prey item in the habitat (p_i). In this case, positive values indicate apparent preference for that prey species, negative values suggest avoidance or prey unavailability, and values around zero show that a prey item is being consumed in the same proportion as it is found in the environment. This method generally provides no significance tests (but see Lechowicz 1982). The higher the positive or negative electivity index, the stronger the selection or avoidance, respectively. Finally, recall that electivity indices or correlation coefficients can be used only to hypothesize positive or negative selection for certain prey species and that further tests are necessary to determine the reason that specific patterns were observed.

C. Predation Experiments—Tests of Hypotheses

One simple way to test hypotheses on selective predation generated from field data is to conduct a predation experiment in the field or the laboratory. Known numbers of alternative prey of differing abundance in predator guts can be presented to predators in replicate enclosures that do not allow immigration or emigration. After a prescribed time period, prey disappearance rates can be determined and compared to numbers of prey disappearing from control enclosures containing the same prey numbers but no predators. Prey mortality rates (Dodson 1975) can be calculated and compared among alternative prey species; significance tests (analysis of variance) can be applied to determine whether predators selectively consumed one or more prey species (Peckarsky and Penton 1989). The advantage of this method over observations of predator guts and field samples is that experiments can provide data to examine cause and effect questions. However, enclosures can cause predator or prey responses that are artifacts of the experiment rather than being indicative of natural responses (Hulberg and Oliver 1980, Peckarsky and Penton 1990). For example, mesh cages can slow stream flow and cause deposition of fine sediments, altering the behavior of predators or prey (Peckarsky 1985, Walde 1986). Researchers must be aware of these potential artifacts and interpret experimental data accordingly. Correspondence between field observations and experimental data provide a powerful tool for answering questions about selective predation. If data from the two methods disagree, the investigator is then challenged to identify the artifacts biasing one or both methods.

D. Behavioral Experiments—Tests of Mechanisms

Prey that are positively selected, avoided, or eaten in proportion to availability can be observed in enclosures to determine whether selection is active or passive (i.e., which components of the predator–prey interaction are responsible for the observed pattern). The biggest challenge in this approach is to design an enclosure that is as close to the natural environment of predators and prey as possible, yet allows accurate viewing of interactions (e.g., Peckarsky *et al.* 1994). When observing benthic organisms that are often nocturnal and hidden beneath substrata by day, compromises may have to be made and data should always be interpreted with caution. Removal of stream organisms from natural conditions and the presence of an observer can affect their performance (Peckarsky 1983, Wiley and Kohler 1984). With this in mind, observers can conduct timed, replicated trials with known densities of alternative prey species and one predator. The numbers of predator–prey encounters (contact), attacks, and captures per trial can be compared among prey species using significance tests (analysis of variance). Behavioral experiments enable researchers to determine whether prey taxa are selected or avoided on the basis of differences in encounter rates, attacks per encounter, or captures per attack. Thus, data indicate whether prey selection is due to active choice by the predator or a passive consequence of prey attributes or behavior (Peckarsky and Penton 1989).

III. SPECIFIC EXERCISES

A. Exercise 1: Electivity Indices

Field Protocols Researchers should collect field samples using an aquatic D-frame net, Surber Sampler, or some other sampler designed for collecting invertebrates in stream riffles. Sampling methods should be standardized such that the same microhabitat and sampling effort is used by each investigator.

1. Using methods described in Chapter 16, collect macroinvertebrate samples from a prescribed surface area of substrata (which should include large cobbles) of a shallow (<30 cm) riffle with moderate flow (20–30 cm/s). The size of the area disturbed depends on the productivity of the stream (i.e., take smaller samples in more productive streams) with a goal of collecting at least 100 benthic invertebrates.

2. Place contents of the collecting net in a shallow pan. Remove all large predatory stoneflies and preserve with 70% ethanol. If no predatory stoneflies are obtained, the sample should be discarded and another collected until at least one predator (preferably several) is collected.

3. Remove as much detritus and inorganic sediment as possible from the remainder of the sample to improve the efficiency of sorting invertebrates from debris in preserved samples.

Laboratory Sorting, Counting, and Identification Protocols

1. Remove macroinvertebrates from detritus and inorganic matter in benthic samples. Set aside all predatory stoneflies. The species of predatory stonefly most abundant in the samples should be used in the following analysis.[1]

2. Tally the numbers of individuals collected of each potential prey taxon on data sheet Table 20.1. Most potential prey taxa need only be identified to order (see Chapter 16), with the exception that among the Diptera, blackflies (Simuliidae) and midges (Chironomidae) should be differentiated, and mayflies should be identified to family (especially Baetidae, Leptophlebiidae, Heptageniidae, and Ephemerellidae).

3. Prepare a reference collection of the invertebrates found at the stream to facilitate this process and so that errors in identification are minimized.

4. Calculate the total numbers of each taxon in all samples combined and the proportion of the total individuals in all samples combined (p_i) (see Table 20.1).

Protocol for Stomach Content Analyses

1. Use two pair of forceps to pull the head from the prothorax of individuals from the selected, predatory stonefly taxa. The foregut, which should remain intact and attached to the head, can then be dissected and examined for recognizable prey parts. If the foregut does not remain attached to the head, dissect the thorax (through the ventrum) and anterior abdomen to extract the foregut. Since large predatory stoneflies swallow their prey whole, prey should be identifiable, provided a short time has elapsed since the predator's last meal.[2]

2. Use the reference collection of potential prey taxa or taxonomic references to identify prey in the predator's foregut. Stoneflies primarily eat chironomid midges, black flies, mayflies (primarily the four families listed above), and caseless caddisflies (Stewart and Stark 1988). If prey are fragmented, compare fragments (claws, mandibles, head capsules, etc.) to whole specimens to make identifications.

3. Tally numbers of each prey taxon found in each predator gut, and calculate totals for each taxon, and the proportion of the total prey individuals in all of the predator guts for the whole group combined (r_i) (see Table 20.1).

[1]Only samples with predators are processed.

[2]For best results, samples should be taken in the morning, because most predatory stoneflies are nocturnal feeders (Peckarsky 1982) and food items in the gut will be fresher.

TABLE 20.1
Data from Predator Guts and Benthic Samples for Calculating Ivlev's Electivity Index and Spearman Rank Correlation Coefficient

Sample	No. of prey in predator guts						Rank:		No. of prey in benthic samples						Rank:	
Taxon (common name)	1	2	3	...n	Total	r_i	Diet	1	2	3	...n	Total	p_i	Habitat	Ivlev's E_i
Baetidae															
Ephemerellidae															
Heptageniidae															
Leptophlebiidae															
Other Ephemeroptera (mayflies)															
Plecoptera (stoneflies)															
Trichoptera (caddisflies)															
Chironomidae (midges)															
Simuliidae (black flies)															
Other Diptera (true flies)															
Total					Grand total							Grand total			

Data Analysis

1. Using the data from the whole group, calculate a Spearman Rank Correlation Coefficient (Siegel 1956) to determine whether there is a significant correlation ($\alpha = 0.05$) between the ranks of the potential prey taxa in the diets and in the habitat of the stoneflies (see Table 20.1).

2. Also using the group data (Table 20.1) compare the fractional composition of each item (i) found in the guts of the stoneflies (r_i) to its fractional composition in the available food supply (p_i) using Ivlev's Electivity Index (1961):

$$E_i = (r_i - p_i)/(r_i + p_i). \tag{20.1}$$

Values of E_i can range from -1 to $+1$, indicating avoidance to preference, with values near zero indicating that the prey item is eaten in the same proportion that it was collected in the environment.

3. Record the electivities of each prey taxon on data sheet (Table 20.1), and use the data from the whole group to prepare a bar graph illustrating the electivities of each taxon, placing the prey taxa on the horizontal axis in order of decreasing electivity.

B. Exercise 2: Predation Experiments

Protocols for Field or Laboratory Trials

1. Collect predators and prey in the field and hold predators in aerated, cooled (10–15°C), or flowing water without prey for at least 24 h to standardize hunger levels. For best results, minimize handling of the insects that will be used in trials. Predators should be handled with soft forceps, and prey species can be transferred between containers using large-mouthed plastic pipetts.

2. Set up at least six enclosures (single prey trials), two each per three prey species containing 15–20 prey and either one predatory stonefly (predator treatment) or no stonefly (control).[3]

> a. Choose the three prey species from the available taxa identified in Exercise 1. Preferably, they should include one overrepresented in stonefly diets (positive electivity), one underrepresented in stonefly diets (negative electivity), and one eaten in proportion to its availability (electivity around zero).

---

[3]If this exercise is used for a class, there should be as many replicates of treatments and controls as there are investigators.

   b. If possible, use mayfly species, because they are easier to handle
      and manipulate than dipterans or caddisflies, which tend to slip
      through meshes (midges) or spin silken threads in which stoneflies
      get tangled (black flies and caddisflies).
   c. If it is not feasible to use three prey species, this exercise can be
      accomplished with two prey species.

3. Field enclosures should be rectangular with upstream and downstream ends covered with mesh (i.e., size of openings small enough to retain
prey but large enough to minimize clogging).

   a. Fill enclosures with natural stream substrata (from which all inver-
      tebrates but not algae have been carefully removed) to provide
      refuges for predators and prey and anchor enclosures to the
      streambed.
   b. A suggested design for enclosures is a 25-cm-long × 10-cm-
      wide × 10-cm-high Plexiglass box (e.g., Fig. 20.2) containing 10
      cobbles ranging from 5 to 15 cm in diameter with the same
      substrata distribution in each enclosure. Cheaper materials may
      be used to construct enclosures, such as Rubbermaid shoe boxes,
      with openings cut in the sides and screened with Nitex® attached
      to walls with hot-melt glue.

4. Laboratory enclosures can be circular (10–15 cm diameter), which
reduces edge effects, and made of Plexiglass (e.g., Fig. 20.3) or modified
cylindrical food containers, and powered by water (Walde and Davies
1984, Peckarsky and Cowan 1991) or air pressure (Wiley and Kohler 1980,
Mackay 1981). Air pressure necessitates recirculation of water and some
type of refrigeration; water-powered chambers can use cold running water

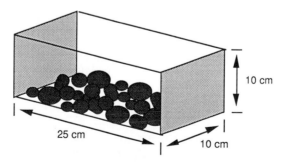

FIGURE 20.2 Example of rectangular chambers that could be used for Exercise 2 (predation choice trials) in the field (from Peckarsky and Penton 1989).

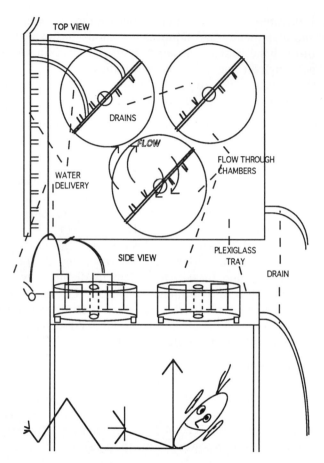

FIGURE 20.3 Example of circular chambers that could be used for Exercise 2 (predation choice trials) in the laboratory (from Peckarsky *et al.* 1994).

and central mesh-covered standpipes to regulate water levels. These designs can be modified depending on facilities available, but cool temperatures (10–15°C) and good oxygenation are essential for stonefly foraging. Natural substrata can be collected from the stream and used as refuges for predators and prey.

5. Allow predators to feed overnight or for 24 hr. It is advisable to conduct a pilot trial to determine the time during which predators eat detectable numbers of prey but do not deplete prey in any chambers (about 10–50% prey consumption is optimal).

6. After the end of the trial remove the contents of the chambers, and tally numbers of prey remaining in each chamber (Table 20.2).

Data Analysis

1. Record on Table 20.2 the numbers of prey dead or missing from control chambers containing each of the three prey species, and calculate a correction factor for losses of prey due to factors other than predation for each prey species.

2. Calculate instantaneous prey mortality rates (m) for each predator and prey species tested using the equation

$$N_f = N_o e^{-mt}, \tag{20.2}$$

or rewritten,

$$m = [\ln N_o - \ln N_f]/t, \tag{20.3}$$

where N_f represents final density of live prey in chambers (corrected for average number lost from all controls of that species, see Table 20.2); N_o, initial prey density (15–20 individuals); and t, duration (days) of the trial (Dodson 1975). The units of this parameter (m) are prey mortality per prey per predator per day, which takes into account exploitation of prey over the time of the trial.

3. Using the data from the entire group recorded on Table 20.2, prepare a bar graph of the mortality rate m, showing mean ± SE for each of the three prey species. Use a one-way analysis of variance and multiple comparisons tests (e.g., Sokol and Rolff 1995) to test whether there were significant differences between predation rates among the three different prey species. Compare these results to those predicted by hypotheses generated from the field data (Exercise 1).

C. Exercise 3: Behavioral Experiments

Field or Laboratory Trials

1. Using the same combinations of predators and prey as in Exercise 2, conduct behavioral trials to determine which components of the predator–prey interaction are responsible for observed patterns of selective predation.

2. Containers used in Exercise 2 can be used for these trials, except that substrata will have to be modified for viewing of behavior. Ideally, circular Plexiglass chambers with natural substrata can be viewed by two

TABLE 20.2
Data for Calculating Correction Factors and Predator-Induced Prey Mortality (m)

Investigator (Replicate No.)	Mortality in controls			Mortality in treatments			Mortality due to predation		
	Prey sp. 1	Prey sp. 2	Prey sp. 3	Prey sp. 1	Prey sp. 2	Prey sp. 3	Prey sp. 1	Prey sp. 2	Prey sp. 3
1									
2									
3									
4									
5									
6									
7									
8									
9									
10									
..n									
	*Correction factors								
Mean	*	*	*						
SE									
n									

observers from the top and bottom if they are placed in elevated Plexiglass trays (Fig. 20.4). If Plexiglass chambers are not available, use of gravel or sand substrata into which prey and predators cannot burrow is an acceptable alternative.

3. Each investigator should conduct three 10-min behavioral trials observing one 24-h starved predator with 15–20 (same density as in Exercise 2) individuals of each prey species one at a time (i.e., single prey species trials). Mixed prey species combinations can be used here and in Exercise 2, but statistical analyses become complicated, necessitating the use of MANOVA (Peckarsky and Penton 1989).

4. Observe each predator and prey individual only once, using the same careful handling techniques that were outlined in Exercise 2.

5. Record all predator–prey encounters, attacks, and captures that take place during the 10-min trials.

6. For each trial, record the number of encounters, attacks per encounter, and captures per attack on Table 20.3. If there are no encounters, a new predator should be observed because there will be no useful data obtained from an inactive predator that does not encounter any prey. In contrast if there are encounters but no attacks, the trial is useful, but captures per attack are undefined.

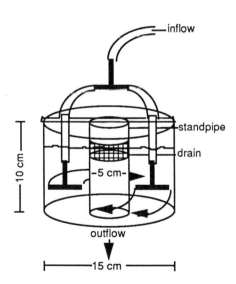

FIGURE 20.4 Example of Plexiglass observation arenas that could be used for Exercise 3 (behavioral observations of predator–prey interactions) in the lab or field (from Peckarsky *et al.* 1994).

TABLE 20.3
Data on Number of Encounters, Attacks Per Encounter, and Captures Per Attack

Investigator (Replicate No.)	No. of encounters			Attacks per encounter			Captures per attack		
	Prey sp. 1	Prey sp. 2	Prey sp. 3	Prey sp. 1	Prey sp. 2	Prey sp. 3	Prey sp. 1	Prey sp. 2	Prey sp. 3
1									
2									
3									
4									
5									
6									
7									
8									
9									
10									
..n									
Mean									
SE									
n									

Data Analysis

1. Using the data from all Exercise 3 observations (Table 20.3), compare each of the three parameters (i.e., encounters, attacks per encounter, and captures per attack) among the three different prey species using a one-way ANOVA and multiple comparisons tests, as in Exercise 2.

2. Prepare three bar graphs, one each for encounters, attacks per encounter, and captures per attack, to illustrate and compare and the mean ± SE values (Table 20.3) for each of the three species.

IV. QUESTIONS

1. What are the strengths and limitations of field-generated electivity indices? Predation experiments? Behavioral observations?

2. What hypotheses were suggested by the electivity indices? By correlations between gut contents and benthic data?

3. What can you conclude from the predation experiments regarding selective predation by stoneflies?

4. What did the behavioral observations reveal about the importance of encounter rates, attacks per encounter, and captures per attack as mechanisms explaining patterns of selective predation by stoneflies?

5. Do the behavioral observations suggest that prey selection by stoneflies is active or passive? Explain.

6. Are the data from all the different methods consistent? Describe any discrepancies.

7. If the data are not consistent what conclusions would you draw? Do you trust some methods more than others? Why?

8. Why should investigators include controls and replication when designing experiments?

9. How would you redesign this exercise to make it more successful (if it fails!!)?

V. MATERIALS AND SUPPLIES

Field Equipment

70% Ethanol
Collecting net
Forceps
Enclosures
Holding chambers (for predators and prey, Exercise 2)

Observation chamber (Exercise 3)
One small collecting jar (for predators)
One wide-mouthed jar (for prey)
Petri dishes for sorting samples
Plastic eyedroppers
Shallow sorting pan
Stopwatch
Water or air source (for circulating flow in chambers if trials are done
    in the laboratory)

Laboratory Equipment

Dissecting microscope
Invertebrate identification guide (see Appendix 16.1)

REFERENCES

Allan, J. D. 1981. Determinants of diet of brook trout *(Salvelinus fontinalis)* in a
    mountain stream. *Canadian Journal of Fisheries and Aquatic Sciences* **38:**
    184–192.
Allan, J. D. 1982. Feeding habits and prey consumption of three setipalpian stoneflies
    (Plecoptera) in a mountain stream. *Ecology* **63:**26–34.
Allan, J. D. 1983a. Food consumption by trout and stoneflies in a Rocky Mountain
    stream, with comparisons to prey standing crop. Pages 371–390 *in* T. D. Fon-
    taine III and S. M. Bartell (Eds.) *Dynamics of Lotic Ecosystems.* Ann Arbor
    Science Publishers, Ann Arbor, MI.
Allan, J. D. 1983b. Predator–prey relationships in streams. Pages 191–229 *in* J. R.
    Barnes and G. W. Minshall (Eds.) *Stream Ecology.* Plenum, New York, NY.
Allan, J. D. 1995. Predation and its consequences. Pages 163–186 *in* J. D. Allan
    (Ed.) *Stream Ecology.* Chapman and Hall, London, UK.
Allan, J. D., A. S. Flecker, and N. L. McClintock. 1987a. Prey size selection by
    carnivorous stoneflies. *Limnology and Oceanography* **32:**864–872.
Allan, J. D., A. S. Flecker, and N. L. McClintock. 1987b. Prey preference of stone-
    flies: Sedentary vs. mobile prey. *Oikos* **49:**323–331.
Allan, J. D., and A. S. Flecker. 1988. Preference in stoneflies: a comparative analysis
    of prey vulnerability. *Oecologia* **76:**496–503.
Carpenter, S. R., J. F. Kitchell, J. R. Hodgson, P. A. Cochran, J. J. Elser, M. M.
    Elser, D. M. Lodge, D. Dretchmer, X. He, and C. N. von Ende. 1987. Regula-
    tion of lake primary productivity by food web structure. *Ecology* **68:**1863–
    1876.
Caswell, H. 1978. Predator-mediated coexistence: A nonequilibrium model. *Ameri-
    can Naturalist* **112:**127–154.
Chesson, J. 1978. Measuring preference in selective predation. *Ecology* **59:**211–215.
Connell, J. H. 1975. Some mechanisms producing structure in natural communities:

A model and evidence from field experiments. Pages 460–490 *in* M. L. Cody and J. M. Diamond (Eds.) *Ecology and Evolution of Communities.* Belknap Press of Harvard University, Cambridge, MA.

Cooper, S. D. 1985. Prey preference and interactions of predators from stream pools. *Verhandlungen der Internationalen Vereinigung für Theoretische und Angewandte Limnologie* **22**:1853–1857.

Cooper, S. D., S. J. Walde, and B. L. Peckarsky. 1990. Prey exchange rates and the impact of predators on prey populations in streams. *Ecology* **71**:1503–1514.

Crowl, T. A., and A. P. Covich. 1990. Predator-induced life-history shifts in a freshwater snail. *Science* **247**:949–51.

Dodson, S. I. 1975. Predation rates of zooplankton in Arctic ponds. *Limnology and Oceanography* **20**:426–433.

Forrester, G. E. 1994. Influences of predatory fish on the drift dispersal and local density of stream insects. *Ecology* **75**:1208–1218.

Fuller, R. L., and P. S. Rand. 1990. Influence of substrate type on vulnerability of prey to predaceous aquatic insects. *Journal of the North American Benthological Society* **9**:1–8.

Greene, C. H. 1985. Planktivore functional groups and patterns of prey selection in pelagic communities. *Journal of Plankton Research* **7**:35–40.

Healey, M. 1984. Fish predation on aquatic insects. Pages 255–288 *in* V. H. Resh and D. M. Rosenberg (Eds.) *The Ecology of Aquatic Insects.* Praeger, New York, NY.

Hildrew, A. G., and C. R. Townsend. 1982. Predators and prey in a patchy environment: a freshwater study. *Journal of Animal Ecology* **51**:797–815.

Hulberg, L. W., and J. S. Oliver. 1980. Caging manipulations in marine soft-bottom communities: Importance of animal interactions or sedimentary habitat modifications. *Canadian Journal of Fisheries and Aquatic Sciences* **37**:1130–1139.

Hyatt, K. D. 1979. Feeding strategy. Pages 71–119 *in* W. S. Hoar, D. J. Randall, and J. R. Brett (Eds.) *Fish Physiology,* Vol. 3. Academic Press, New York, NY.

Ivlev, V. S. 1961. *Experimental Ecology of the Feeding of Fishes.* Yale Univ. Press, New Haven, CT.

Jacobs, J. 1974. Quantitative measurement of food selection. A modification of the forage ratio and Ivlev's electivity index. *Oecologia* **14**:413–417.

Johnson, D. H. 1980. The comparison of usage and availability measurements for evaluating resource preference. *Ecology* **61**:65–71.

Kerfoot, C. W., and A. Sih (Eds.). 1987. *Predation: Direct and Indirect Impacts on Aquatic Communities.* Univ. Press of New England, Hanover, NH.

Kotila, P. M. 1987. Using stream fish to demonstrate predator–prey relationships and food selection. *The American Biology Teacher* **49**:104–106.

Lancaster, J., A. G. Hildrew, and C. R. Townsend. 1991. Invertebrate predation on patchy and mobile prey in streams. *Journal of Animal Ecology* **60**:625–641.

Lechowicz, M. J. 1982. The sampling characteristics of electivity indices. *Oecologia* **52**:22–30.

Mackay, R. J. 1981. A miniature laboratory stream powered by air bubbles. *Hydrobiologia* **83**:383–386.

Malmqvist, B., and P. Sjostrom. 1980. Prey size and feeding patterns in *Dinocras cephalotes* (Plecoptera). *Oikos* **35**:311–316.

Martin, I. D., and R. J. Mackay. 1982. Interpreting the diet of *Rhyacophila* larvae (Trichoptera) from gut analysis: An evaluation of techniques. *Canadian Journal of Zoology* **60**:783–789.

Metz, J. P. 1974. The invertebrate drift on the surface of a prealpine stream and its selective exploitation by rainbow trout *(Salmo gairdneri). Oecologia* **14**:247–267.

Molles, M. C., Jr., and R. D. Pietruszka. 1983. Mechanisms of prey selection by predaceous stoneflies: Roles of prey morphology, behavior and predator hunger. *Oecologia* **57**:25–31.

Molles, M. C., Jr., and R. D. Pietruszka. 1987. Prey selection by a stonefly: The influence of hunger and prey size. *Oecologia* **72**:473–478.

O'Brien, W. J., and J. J. Showalter. 1993. Effects of current velocity and suspended debris on the drift feeding of Arctic grayling. *Transactions of the American Fisheries Society* **122**:609–615.

Paine, R. T. 1966. Food web complexity and species diversity. *American Naturalist* **100**:65–75.

Peckarsky, B. L. 1982. Aquatic insect predator–prey relations. *BioScience* **61**:932–943.

Peckarsky, B. L. 1983. Use of behavioral experiments to test ecological theory in streams. Pages 196–254 *in* J. R. Barnes and G. W. Minshall (Eds.) *Stream Ecology.* Plenum, New York, NY.

Peckarsky, B. L. 1984. Predator–prey interactions among aquatic insects. Pages 196–254 *in* V. H. Resh and D. M. Rosenberg (Eds.) *Ecology of Aquatic Insects.* Praeger, New York, NY.

Peckarsky, B. L. 1985. Do predaceous stoneflies and siltation affect the structure of stream insect communities colonizing enclosures? *Canadian Journal of Zoology* **63**:1519–1530.

Peckarsky, B. L., and C. A. Cowan. 1991. Consequences of larval intraspecific interference to stonefly growth and fecundity. *Oecologia* **88**:277–288.

Peckarsky, B. L., C. A. Cowan, M. A. Penton, and C. R. Anderson. 1993. Sublethal consequences of stream-dwelling predatory stoneflies on mayfly growth and fecundity. *Ecology* **74**:1836–1846.

Peckarsky, B. L., C. A. Cowan, and C. R. Anderson. 1994. Consequences and plasticity of the specialized predatory behavior of stream-dwelling stonefly larvae. *Ecology* **75**:166–181.

Peckarsky, B. L., and M. A. Penton. 1985. Is predaceous stonefly behavior affected by competition? *Ecology* **66**:1718–1728.

Peckarsky, B. L., and M. A. Penton. 1989. Mechanisms of prey selection by stream-dwelling stoneflies. *Ecology* **70**:1203–1218.

Peckarsky, B. L., and M. A. Penton. 1990. Effects of enclosures on stream microhabitat and invertebrate community structure. *Journal of the North American Benthological Society* **9**:249–261.

Power, M. E. 1990. Effects of fish in river food webs. *Science* **250**:87–90.

Resh, V. H. 1979. Sampling variability and life history features: Basic considerations on the design of aquatic insect studies. *Journal of Fisheries Research Board of Canada* **36**:290–311.

Siegel, S. 1956. *Nonparametric Statistics for the Behavioral Sciences.* McGraw–Hill, New York, NY.

Sih, A. 1987. Predators and prey lifestyles: An evolutionary and ecological overview. Pages 203–224 *in* C. W. Kerfoot and A. Sih (Eds.) *Predation: Direct and Indirect Impacts on Aquatic Communities.* Univ. Press of New England, Hanover, NH.

Sih, A., and D. E. Wooster. 1994. Prey behavior, prey dispersal, and predator impacts on stream prey. *Ecology* **75:**1199–1207.

Slobodkin, L. B. 1961. *The Growth and Regulation of Animal Populations.* Holt, Rinehart and Winston, New York, NY.

Sokal, R. R., and F. J. Rohlf. 1995. *Biometry,* 3rd ed. Freeman, New York, NY.

Strauss, S. Y. 1991. Indirect effects in community ecology: Their definition, study and importance. *Trends in Ecology and Evolution* **6:**206–209.

Stewart, K. W., and B. P. Stark. 1988. *Nymphs of North American Stonefly Genera (Plecoptera),* vol. 12. Thomas Say Foundation, The Entomological Society of America.

Walde, S. J. 1986. Effect of an abiotic disturbance on a lotic predator–prey interaction. *Oecologia* **69:**243–247.

Walde, S. J., and R. W. Davies. 1984. The effect of intraspecific interference on *Kogotus nonus* (Plecoptera) foraging behaviour. *Canadian Journal of Zoology* **62:**2221–2226.

Walde, S. J., and R. W. Davies. 1987. Spatial and temporal variation in the diet of a predatory stonefly (Plecoptera: Perlodidae). *Freshwater Biology* **17:**109–116.

Wilcy, M. J., and S. L. Kohler. 1980. Positioning change of mayfly nymphs due to behavioral regulation of oxygen consumption. *Canadian Journal of Zoology* **58:**618–622.

Wiley, M. J., and S. L. Kohler. 1984. Aquatic insect behavior. Pages 101–133 *in* V. H. Resh and D. M. Rosenberg (Eds.) *Ecology of Aquatic Insects.* Praeger, New York, NY.

CHAPTER 21

Trophic Relations of Macroinvertebrates

RICHARD W. MERRITT[*] AND KENNETH W. CUMMINS[†]

*Department of Entomology
Michigan State University
†Ecosystem Research and Implementation Department
South Florida Water Management District

I. INTRODUCTION

A major observation resulting from studies of aquatic invertebrate feeding (e.g., Berrie 1976, Cummins and Klug 1979, Anderson and Cargill 1987, Palmer *et al.* 1993a, Wotton 1994) is that, based on food ingested, essentially all aquatic insects are omnivorous. For instance, insects that chew leaf litter in streams, termed *shredders,* ingest not only the leaf tissue and associated microbiota, (e.g., fungi, bacteria, protozoans, microarthropods), but also diatoms and other algae that may be attached to the leaf surface, as well as very small macroinvertebrates (e.g., first-instar midge larvae). For this reason, the trophic level analysis pioneered by Lindeman (1942), and used extensively in investigations of trophic relationships in marine and terrestrial communities, does not lend itself well to simple trophic categorization of stream macroinvertebrates.

An alternate classification technique involves the functional analysis of invertebrate feeding, based on morphobehavioral mechanisms of food acquisition. This technique, described 18 years ago (Cummins 1973), has been modified in some detail since then (e.g., Cummins 1974, Cummins and Klug 1979, Wallace and Merritt 1980, Merritt *et al.* 1984, Cummins and Wilzbach 1985, Merritt and Cummins 1996). The macroinvertebrate *functional feeding group* method is based on the association between a limited set of feeding adaptations found in freshwater invertebrates and their basic nutritional resource categories. The same morphobehavioral mechanisms can result in the ingestion of a wide range of food items, the intake of which constitutes herbivory (i.e., living plants; Gregory 1983, Lamberti and Moore 1984, Webster and Benfield 1986), detritivory (i.e., dead organic matter; Anderson and Sedell 1979, Wallace and Merritt 1980, Webster and Benfield 1986, Cummins *et al.* 1989, Palmer *et al.* 1993b), or carnivory (i.e., live animal prey; Allan 1983, Peckarsky 1984).

Although food type intake would be expected to change from season to season, habitat to habitat, and with growth stage, limitations in food acquisition mechanisms have been shaped over evolutionary time and these are relatively more fixed. As an example, a mayfly scraper such as *Stenonema* (Heptageniidae), adapted for shearing off food attached to surfaces, especially diatoms, can ingest a variety of organic substrata within the restriction of its food-harvesting morphology and without any change in behavior. As the animal gets larger, the microhabitat being scraped for food may change or the items scraped from surfaces may differ (Cummins 1980). Comparable (analogous) morphological structures that enable insects to scrape off attached periphyton can be found in the very similar mandibles of taxonomically diverse groups such as the trichopteran families Glossosomatidae (saddle-case makers) and Helicopsychidae (snail-case makers), and the coleopteran family Psephenidae (water penny beetles). This similarity of structure is a striking example of convergent evolution.

The basic food resource categories in stream ecosystems are: (1) *CPOM,* coarse particulate organic matter (particles greater than 1 mm in size); represented by litter accumulations consisting of leaves, needles, bark, twigs, and other plant parts, large woody debris (i.e., large branches and logs), and macrophytes including macroalgae and rooted and floating vascular plants (see Chapter 11); (2) *FPOM,* fine particulate organic matter (particles less than 1 mm and greater than 0.5 μm in size) generally composed of unattached living or detrital material including that created through physical and biological reduction of CPOM and associated microbiota (see Chapter 10); (3) *periphyton,* predominantly attached algae (especially diatoms) and associated material growing on rock, wood, and plant surfaces (see Chapters 13 and 14); and (4) *prey,* all invertebrates captured by predators, predominantly small species and early instars of large species (see also Chapters 16 and 20).

These four nutritional resource categories related to food acquisition mechanisms were chosen on the basis of: (1) size range of the material (coarse and fine) and (2) general location, such as attached to surfaces (periphyton), suspended in the water column (seston), deposited in the sediments, found in litter accumulations, or in the form of live invertebrates. This categorization also reflects: (1) biochemical differences in the nutritional resources, such as presence of living chlorophyll in periphyton or microorganisms on CPOM and (2) the major source of the food, such that whether it was either *autochtonous* (produced within the aquatic system; see Chapter 25) or *allochthonous* (produced from the stream-side or riparian terrestrial area; see Chapter 27).

A general classification system for aquatic insect trophic relations is presented in Table 21.1. This functional group classification distinguishes invertebrate taxa within aquatic ecosystems according to the different morphological–behavioral adaptations used to harvest nutritional resources. These feeding mechanisms determine the food resources that are processed: *shredders* feed on CPOM, *collectors* on FPOM, *scrapers* on periphyton, and *predators* on prey. The functional groups described in this classification are analogous to *guilds,* which are groups of organisms using a particular resource class (Root 1973, Georgian and Wallace 1983, Hawkins and Mac-Mahon 1989); thus, function is defined as use of similar resource classes.

Use of the functional feeding group approach is advantageous in that it allows an assessment, numerically or by standing crop biomass, of the degree to which the invertebrate biota of a given aquatic system is dependent upon a particular food (nutritional) resource. It also makes more apparent the linkages that exist between food resources and insect morphological–behavioral adaptations (Cummins 1974). As the relative dominance of various food resource categories changes, there often is a corresponding shift in the ratios of the different functional feeding groups. In this manner, invertebrate functional group analysis is sensitive to both the normal pattern of geomorphic and the concomitant biological changes that occur along river systems from headwaters to lower reaches (e.g., Vannote *et al.* 1980), as well as to alterations in these patterns resulting from human impact (Cummins 1992, 1993). The objective of the exercise in this chapter is to demonstrate how functional feeding group ratios can be used as a tool for assessing the ecological state of running water communities.

II. GENERAL DESIGN

The general procedure described here focuses on identifying key functions in stream invertebrate communities that can be determined at succeeding levels of taxonomic resolution. The technique is particularly useful for macroinvertebrate groups for which the state of taxonomic knowledge is presently incomplete. For example, determinations of community struc-

TABLE 21.1

General Classification System for Aquatic Insect Trophic Relations[a]

Functional group (general category based on feeding mechanism)	Subdivision of functional group		Examples of taxa	General particle size range of food (in micrometers)
	Dominant food	Feeding mechanism		
Shredders	Living vascular hydrophyte plant tissue	Herbivores—chewers and miners of live macrophytes	Trichoptera: Phryganeidae, Leptoceridae	$>10^3$
	Decomposing vascular plant tissue and wood—coarse particulate organic matter (CPOM)	Detritivores—chewers, wood borers, and gougers	Diptera: Tipulidae, Chironomidae	
Collectors	Decomposing fine particular organic matter (FPOM)	Detritivores—filterers or suspension feeders	Trichoptera: Hydropsychidae Diptera: Simuliidae	$<10^3$
		Detritivores—gatherers or deposit (sediment) feeders (includes surface film feeders)	Ephemeroptera: Ephemeridae Diptera: Chironomidae	
Scrapers	Periphyton—attached algae and associated material	Herbivores—grazing scrapers of mineral and organic surfaces	Trichoptera: Glossosomatidae Coleoptera: Psephenidae Ephemeroptera: Heptageniidae	$<10^3$
Predators (Engulfers)	Living animal tissue	Carnivores—attack prey, pierce tissues and cells, and suck fluids	Hemiptera: Belostomatidae	$>10^3$
	Living animal tissue	Carnivores—ingest whole animals (or parts)	Odonata, Plecoptera: Perlidae	$>10^3$

[a]Modified from Merritt and Cummins (1996).

ture for many groups of macroinvertebrates are based on measures of *taxa* richness or diversity (i.e., a combination of family and generic identifications), *not* species richness or diversity. However, using functional feeding group analysis enables macroinvertebrate communities to be evaluated at a range of levels of taxonomic resolution. The objective of this approach is to maximize the ecological information obtained for the taxonomic effort expended. For example, determination of functional feeding groups in the Odonata (dragonflies and damselflies) is achieved by separation to order alone. In contrast, such determinations for some subfamilies of Chironomidae (midges) or genera of ephemerellid mayflies may require species identifications.

A. Site Selection

Sites should be selected to ensure that three basic habitats are covered: (1) coarse sediments of riffles (golf ball- to soccer ball-sized cobbles), (2) accumulations of organic litter and small woody debris (handful amounts), and (3) fine sediments in depositional zones (one scoop with a small, fine-mesh aquarium net or equivalent). Because the results are presented as dimensionless ratios, they are relatively independent of sample size. The overall evaluation of the stream or river site should be weighted for the amount of the given habitat type, since the three habitats naturally favor different functional groups (e.g., cobble, scrapers and filtering collectors; leaf litter, shredders; fine sediments, gathering collectors).

B. Collection and Processing of Samples

All sampling can be accomplished using a D-frame net or equivalent (250–500 μm). The net is held downstream and below a rock or leaf pack as it is lifted from the stream or the net is used to scoop surface sediments ($\approx$2-cm depth) from depositional habitats. Samples may be returned to the laboratory for processing (see Chapter 16) or it may be convenient to conduct some initial processing of the sample in the field by washing it into a white enamel tray to sort the specimens into separate containers[1] by functional categories.

Specimens should be identified, separated into functional category, counted, and tallied. (Detailed taxonomic identification may be made using good taxonomic reference texts such as Thorp and Covich (1991) and Merritt and Cummins (1996).) Specimens can be placed in vials, labeled, and preserved in 70% ethanol. Record data on Table 21.2.

The seasonal timing of functional feeding group sampling is critical, as

---

[1]Wells of a muffin tin, plastic ice-cube tray, or styrofoam egg carton work well for this purpose.

TABLE 21.2
Macroinvertebrate Functional Feeding Groups Data Sheet

Team: _____

Location: _____ Date: _____

    Substratum: Cobble ▢   Litter ▢   Fine Sediment ▢

Tally

	Collectors	
Shredders	Gathering collectors	Filtering collectors

Scrapers	Predators

Totals

  Shredders _____

  Collectors _____

    Gathering collectors _____

    Filtering collectors _____

  Scrapers _____

  Predators _____

    Grand total _____

Ratios

  Scrapers/shredders + total

    collectors _____

Shredders/total collectors _____

  Filtering collectors/

    gathering collectors _____

Scrapers + filtering collectors/

  shredders + gathering collectors

  _____

    Predators/total of all

     other groups _____

Common representatives

  Shredders:

_____

Collectors

  Gathering collectors:

Filtering collectors:

_____

Scrapers:

_____

Predators:

is also true when determining any taxonomically based numerical index. It is important to sample when the greatest number of taxa are in feeding stages and are as large as possible. This means avoiding periods of maximum recruitment (e.g., egg laying and hatching). In general, collecting in the spring should be done before water temperatures exceed 10°C and in the autumn before water temperatures fall below 10°C. Because there are definite winter and summer communities of stream macroinvertebrates, at least two samplings per year are required to adequately characterize the functional feeding groups (Cummins *et al.* 1989).

Biomass data, which are preferable to numerical data, but more time-consuming to obtain, can be estimated by measuring *biovolume.* Use a small graduated cylinder (5–10 ml) to measure volumetric displacement of composited functional group categories. The specimens in each functional group are added to the graduated cylinder containing a known volume of water. The number of milliliters of water displaced by the functional group biomass is recorded. Functional group ratios are calculated using the volumes of each group. Biomass also can be estimated in the laboratory from measurements of the lengths of specimens using length–mass relationships (e.g., Smock 1980).

C. Functional Group Designations

The key that appears below (Appendix 21.1) emphasizes higher level taxonomic separations that permit reliable categorization of functional feeding groups. The key is organized in two levels of resolution. The first level can be done in the field with a minimum of taxonomic skill, usually resulting in an accurate separation of 80–85% of the material. The second level should be done in the laboratory and may increase the resolution 5–10% by categorizing those macroinvertebrates that either do not readily fit the first level grouping or are likely to be misclassified (e.g., the organic case trichopteran *Brachycentrus,* which is a filtering collector, would initially be classified as a shredder). The amount of taxonomic effort and skill required and the need for the use of a microscope increases with the second level of resolution.

D. Determination and Analysis of Functional Feeding Group Ratios

Examples of functional feeding group ratios used as indicators of stream ecosystem attributes are summarized in Table 21.3. The ratios shown can serve as indicators of the relative importance of stream ecosystem autotrophy or heterotrophy, the size categories and relative amounts of coarse CPOM and FPOM in transport and in storage, and the stability of the channel (Table 21.4). Studies have shown that functional feeding group ratios may be used as indicators of stream ecosystem attributes (Vannote

TABLE 21.3
Examples of Functional Feeding Group Ratios as Indicators of Stream Ecosystem Attributes

Ecosystem parameter	Symbols	Functional feeding group ratios	General criteria ratio levels[a]
Autotrophy to Heterotrophy Index or Gross Primary Production to Community Respiration Index	Auto/Hetero or P/R	Scrapers to Shredders + Total Collectors	Autotrophic >0.75
Coarse Particulate Organic Matter (CPOM) to Fine Particulate Organic Matter (FPOM) Index	CPOM/ FPOM	Shredders to Total Collectors	Normal Shredder association linked to functioning riparian system >0.25
FPOM in Transport (suspended) to FPOM in Storage in Sediments (deposited in benthos)	TFPOM/ BFPOM	Filtering Collectors to Gathering Collectors	FPOM Transport (in suspension) greater than normal particulate load in suspension >0.50
Substrate (channel) Stability	Stable channel	Scrapers + Filtering Collectors to Shredders + Gathering Collectors	Stable Substrates (e.g., bedrock, boulders, cobbles, large woody debris) plentiful >0.50
Top-down Predator Control	Top-down control	Predators to Total of all other groups	Normal Predator to Prey balance <0.15

[a]General ratio ranges given are for numerical or biomass data taken when most species are in mid- to late larval instars or aquatic adults (see discussion under field sampling).

et al. 1980, Minshall *et al.* 1983). Because the functional feeding group method is responsive to changes in food resource base (e.g., algae, litter, fine organics, prey), it is sensitive to both general and site-specific influences on riparian–watershed–land use. For example, the localized input of a toxic effluent in the form of FPOM from a paper mill might be a site-specific influence, while increased sediment or reduced litter inputs resulting from altered land-use patterns would be a general influence.

The example shown in Table 21.4, for a second-order woodland stream in the Allegheny National Forest, Pennsylvania, includes ecosystem evaluations based on functional group ratios. The stream in question is heavily shaded by a riparian zone dominated by hemlock. The shaded nature of

TABLE 21.4

Example of Functional Feeding Group Ratios as Indicators of Stream Ecosystem Attributes[a]

Ecosystem parameter	Functional feeding group ratios	Calculated ratios		General criteria ratio levels	Evaluation
		Numbers	Biomass		
Auto/hetero or P/R	Scrapers to Shredders + Total Collectors	0.34	0.24	Autotrophic >0.75	Heterotrophic site, dependent on allochthonous organic matter inputs
CPOM/FPOM	Shredders to Total Collectors	0.32	0.64	Normal shredder association linked to functioning riparian system >0.25	A summer shredder stream; species dependent mainly on slow processing rate of litter[b]
TFPOM/BFPOM	Filtering Collectors to Gathering Collectors	0.87	1.76	FPOM transport (in suspension) enriched; unusual particulate loading >0.50	High FPOM loading (presence of Philopotamidae indicates very fine FPOM)
Stable Channel	Scrapers + Filtering Collectors to Shredders + Gathering Collectors	1.08	1.03	Stable substrates (e.g., bedrock, boulders, cobbles, large woody debris) plentiful >0.50	Channel stability high with numerous attachment sites for macroinvertebrates
Top-down Control	Predators to Total of all other groups	0.22	0.14	Typical predator to prey balance <0.15	Typical predator to prey ratio

[a]Data from a second-order stream in the Allegheny National Forest, Pennsylvania, July 1993.
[b]For example, see Cummins et al. 1989.

the stream is reflected in the low number and biomass of scrapers. The stream is distinctly heterotrophic (see also Chapters 25 and 28) and dependent on allochthonous organic matter from the riparian zone, as indicated by the dominance of shredders and collectors that use detritus as a food resource. Significant numbers of shredders (e.g., Plecoptera: Peltoperlidae; Trichoptera: Lepidostomatidae) indicate that the system is a "summer-shredder stream" (Cummins *et al.* 1989). The shredders are likely dependent upon litter that requires a long conditioning time (see Chapter 27), such as hemlock needles. *Conditioning* is the time period required for plant litter to be sufficiently colonized by stream microbes to render it a food resource usable by the invertebrates. The significant amount of FPOM available, indicated by the abundant filtering-collectors, is consistent with the supply of organic fragments derived from processed autumn–winter litter and organic soils of the riparian zone. The conclusion that the supply of FPOM is not related to land-use disturbance is supported by the high channel stability rating. This, in turn, is indicated by the ratio of those groups requiring nonshifting surfaces (scrapers and filtering-collectors) to those that occupy the interstices of sediments and litter accumulations (shredders and gathering collectors). The normal abundance of predators observed at the site indicates that there is a balance between prey species with long- and short-term life cycles (i.e., a high top–down ratio would indicate a dominance of prey species having short life cycles and populations that turn over rapidly). Thus, in this example (Table 21.4), the measurements of functional group ratios are consistent with the observations of ecosystem properties of the sampling site.

III. SPECIFIC EXERCISES

A. Exercise 1: Determining Macroinvertebrate Functional Feeding Groups

Field Exercise

1. Establish sampling teams of two or three individuals.
2. Sample from each of the three general habitat types (rock, litter, fine sediments) using a 500-μm mesh net or take composite samples, for example with a Surber, Hess, D-frame net, or kick-net as described in Chapter 16, that are representative of a wider range of microhabitat types. Place the collected material in a bucket and subsample from this composite. The goal is to have at least three estimates of the functional group ratios

associated with each habitat type. If the samples are to be processed in the field go to step 3. If the samples are to be processed in the laboratory, then go to Laboratory Exercise (below).[2]

3. Wash each sample into an enamel tray.

4. Remove specimens from the tray with forceps. Use the key (Appendix 21.1) to sort by functional group. Accumulate specimens belonging to the same functional group as described above.

5. Using the data sheet (Table 21.1), tally numbers in each functional feeding group, and enter taxa of common representatives for each of the three habitat types (rock, litter, fines).

6. Calculate functional group ratios from totals for each of the three habitat types.

7. For each habitat type, calculate average functional group ratios using all data.

8. Summarize stream ecosystem condition based on criteria given in Table 21.2 and examples given in Table 21.3 for each of the three habitat types. See also Chapter 2 if more extensive physical habitat assessment is desired.

Optional Addition to Field Exercise

1. After steps 1–5 above, determine the biovolume by volumetric displacement of each of the composite functional groups in a 5- to 10-ml graduated cylinder.

2. Complete steps 6–8 above using volume as well as numerical ratios.

Laboratory Exercise

1. Samples for laboratory analysis can be taken as under Field Exercise above.

2. After collecting field samples, each team should preserve material from each habitat collection in sample jars or bags containing a final concentration of 70% ethanol or 4% formalin.[3] Each jar should be labeled with habitat type, team designation, site, and date. Write the information in pencil on a white paper label and place into jar or bag.

3. Repeat steps 3 and 4 of the Field Exercise and, under a dissecting microscope, identify specimens using both levels of resolution in the key

---

[2]Detailed taxonomic and functional feeding group analysis must be done in the laboratory using a dissecting binocular microscope for specimen identification.

[3]*Caution.* Formaldehyde is a known carcinogen; use with great care and always wear eye and skin protection.

provided. Additional resolution can be achieved by using the keys and ecological tables provided in Merritt and Cummins (1996).

4. Repeat steps 5–8 under Field Exercise above.

Optional Addition to Laboratory Exercise

1. Conduct steps 1–3 under Laboratory Exercise above.

2. After specimens are sorted, identified, and categorized by functional group, measure and record total length of each specimen to the nearest millimeter under a dissecting microscope with a clear plastic millimeter ruler placed on the stage or with a calibrated ocular micrometer.

3. Convert lengths of each specimen to an estimate of biomass using the regression equations in Chapter 17, Table 17.3 (from Smock 1980).

4. Repeat steps 5–8 under Field Exercise above.

IV. QUESTIONS

1. How could the timing (season) of your sampling have influenced your estimates of functional group ratios?

2. When would biomass yield a better estimate of functional organization of a stream macroinvertebrate community than numbers of individuals?

3. What types of human perturbations (e.g., increased sediment inputs resulting from agricultural land use) would tend to change the different functional group ratios and how would these changes influence the ratios?

4. What are some of the advantages and disadvantages of the use of functional group ratios over other methods (e.g., diversity indices) for assessing the ecological state of running water systems?

5. What does the ratio of scrapers : shredders or shredders : total collectors tell you about the food resources of the stream?

6. What type of collectors, filterers or gatherers, were most abundant? Why?

V. MATERIALS AND SUPPLIES

Field Materials

95% Ethanol (70% ethanol final concentration)
Calculator
Enamel sorting pans
Forceps
Graduated cylinder (10 ml graduated in 0.1 ml)

Hand lens
Hand tally counter
Kick-net net, D-frame aquatic net, Surber, or Hess sampler
Multiple-cup container (for separating specimens in categories prior
   to counting)
Sieve (250–500 μm)
Vials, sample jars, or zip-lock bags

Laboratory Materials and Equipment

Clear plastic ruler (in mm) or ocular micrometer
Dissecting microscope and light
Petri dishes to examine insects
Taxonomic guides (e.g., Appendix 21.1; Appendix 16.1; Merritt and
   Cummins 1996)

APPENDIX 21.1: SIMPLIFIED KEYS TO THE FUNCTIONAL FEEDING GROUPS OF LOTIC MACROINVERTEBRATES

KEY TO FUNCTIONAL FEEDING GROUPS

⊢⊣ ⊢—————————⊣ indicates size or range of sizes

1. ANIMALS IN HARD SHELL (Phylum Mollusca)

    a. LIMPETS (Class Gastropoda)

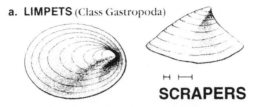

SCRAPERS

    b. SNAILS (Class Gastropoda)

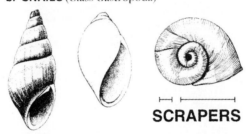

SCRAPERS

Snails are generalized (facultative) feeders and can also function as Shredders.

    c. CLAMS OR MUSSELS (Class Pelecypoda)

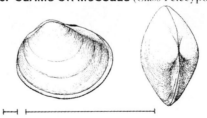

FILTERING COLLECTORS

APPENDIX 21.1—*continued*

2. SOW BUG OR SHRIMP-LIKE ANIMALS
(Class Crustacea)

SHREDDERS

Generalized, can also function as Gathering Collectors.

3. LARVAE IN PORTABLE CASE OR "HOUSE"
Go to KEY 2

4. LARVAE IN FIXED RETREAT
WITH CAPTURE NET

Note: Care must be taken when collecting to observe nets.

Go to KEY 3

5. WITHOUT CASE OR FIXED RETREAT

a. WORM-LIKE LARVAE
WITHOUT JOINTED LEGS

Go to KEY 4

b. NYMPHS OR ADULTS
WITH JOINTED LEGS

Go to KEY 5

6. DOES NOT FIT KEY 5 EXACTLY. GO TO KEY 6

APPENDIX 21.1—*continued*

KEY 2

FIRST LEVEL OF RESOLUTION

LARVAE IN PORTABLE CASE
Caddisflies (Order Trichoptera)

CASES ORGANIC
Leaf, stick, needle, bark

CASES MINERAL
Sand, fine gravel

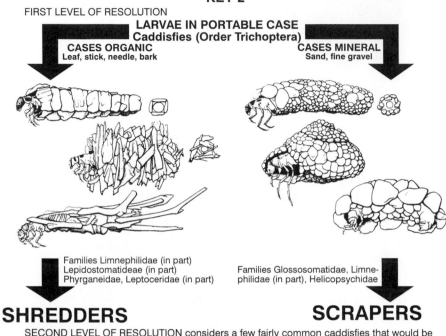

Families Limnephilidae (in part)
Lepidostomatideae (in part)
Phyrganeidae, Leptoceridae (in part)

Families Glossosomatidae, Limnephilidae (in part), Helicopsychidae

SHREDDERS   SCRAPERS

SECOND LEVEL OF RESOLUTION considers a few fairly common caddisflies that would be
misclassified above on the basis of case compositon alone.

CASES ORGANIC   **CASES MINERAL**

Cases square in cross section
and tapered, with no bark or
flat leaf pieces included. Front
attached to substrate. Larvae
extend legs and filter the current.

Cases long, slender, and
tapered, made of plant material

Cases long, slender, and
tapered (mostly fine sand)
or cases ovoid and very flat
in cross section

Foreleg with
filtering hairs

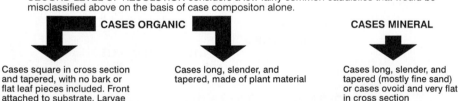

Family Leptoceridae (in part)

Family Brachycentridae

Family Leptoceridae (in part)

FILTERING   GATHERING   GATHERING
COLLECTORS   COLLECTORS   COLLECTORS

APPENDIX 21.1—*continued*

KEY 3

FIRST LEVEL OF RESOLUTION

LARVAE WITH FIXED RETREAT AND CAPTURE NET

Note: Care must be taken when collecting to observe nets.

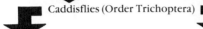

Caddisflies (Order Trichoptera)

True Flies (Order Diptera)

COARSE NET IN "SCAFFOLDING"

FLATTENED SOCK-LIKE OR TRUMPET-SHAPED NET OF FINE MESH

TUBE WITH SILK STRANDS STRUNG BETWEEN TERMINAL PRONGS

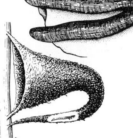

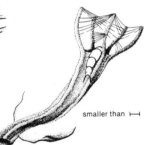

smaller than ⊢—⊣

True Midges (Family Chironomidae)

Families Hydropsychidae, Philopotamidae, Polycentropodidae

FILTERING COLLECTORS

SECOND LEVEL OF RESOLUTION separates from free living larvae those net spinning caddisflies that may have been inadvertently collected without being associated with their nets.

NET SPINNING CADDISFLIES
Frequently separated from their nets

FREE LIVING CADDISFLIES
Non net spinning

HEAD AS WIDE AS THORAX

HEAD LONG, SMALL, AND NARROWER THAN THORAX

Especially Philopotamidae (bright yellow) and Hydropsychidae (bright green or brown)

Rhyacophilidae (often bright green)

FILTERING COLLECTORS

PREDATORS

APPENDIX 21.1—*continued*

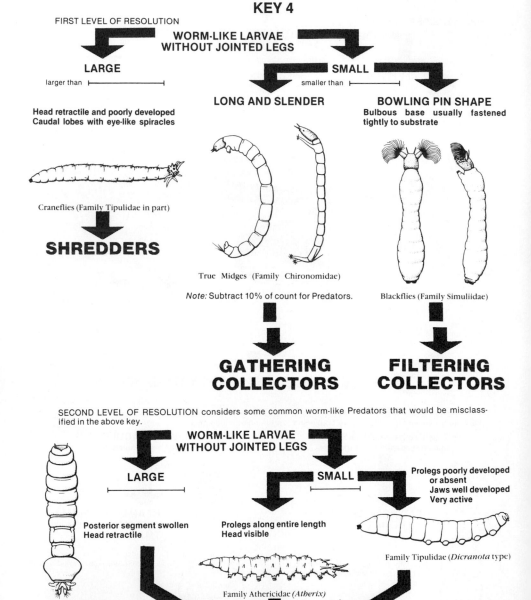

KEY 4

FIRST LEVEL OF RESOLUTION

**WORM-LIKE LARVAE
WITHOUT JOINTED LEGS**

LARGE

larger than |⟷|

SMALL

smaller than |⟷|

LONG AND SLENDER

BOWLING PIN SHAPE

Head retractile and poorly developed
Caudal lobes with eye-like spiracles

Bulbous base usually fastened
tightly to substrate

Craneflies (Family Tipulidae in part)

SHREDDERS

True Midges (Family Chironomidae)

Note: Subtract 10% of count for Predators.

Blackflies (Family Simuliidae)

GATHERING
COLLECTORS

FILTERING
COLLECTORS

SECOND LEVEL OF RESOLUTION considers some common worm-like Predators that would be misclassified in the above key.

**WORM-LIKE LARVAE
WITHOUT JOINTED LEGS**

LARGE

|⟷|

SMALL

|⟷|

Prolegs poorly developed
or absent
Jaws well developed
Very active

Posterior segment swollen
Head retractile

Prolegs along entire length
Head visible

Family Tipulidae (*Dicranota* type)

Family Tipulidae (*Eriocera* type)

Family Athericidae (*Atherix*)

PREDATORS

APPENDIX 21.1—*continued*

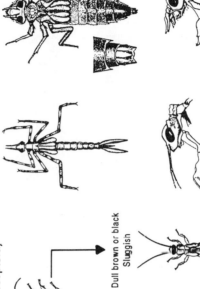

KEY 5

FIRST LEVEL OF RESOLUTION

NYMPHS WITH JOINTED LEGS

3 (or 2) TAILS (FILAMENTS) AT BACK
NO EXTENDIBLE LOWER LIP (LABIUM)

3 FLAT PADDLES OR POINTS AT BACK
EXTENDIBLE LOWER LIP

3 (OR 2) TAILS WITH
LATERAL ABDOMINAL GILLS
Mayflies (order Ephemeroptera)

2 TAILS WITHOUT
LATERAL ABDOMINAL GILLS
Stoneflies (order Plecoptera)

3 FLAT PADDLES AT BACK

POINTS AT BACK

Body shape cylindrical
Round in cross section

Body shape ovoid
Flat in cross section

Bright color pattern
Very active

Dull brown or black
Sluggish

Dragonflies
(suborder Anisoptera)

Damselflies
(suborder Zygoptera)

Families Heptageniidae,
Ephemerellidae (in part)

Families Baeridae, Leptophlebiidae,
Ephemerellidae (in part),
Ephemeridae

Setipalpian Stoneflies

Filipalpian Stoneflies

SCRAPERS

**GATHERING
COLLECTORS**

PREDATORS

SHREDDERS

PREDATORS

APPENDIX 21.1—*continued*

KEY 6

SECOND LEVEL OF RESOLUTION considers some fairly common insects that do not fit in the above key or would be misclassified on the basis of body shape alone.

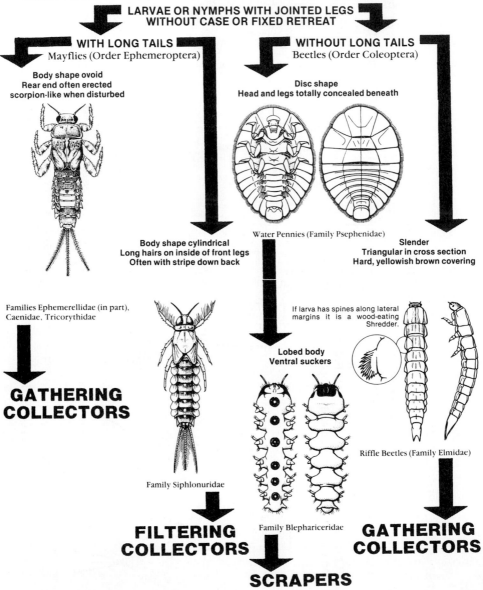

**LARVAE OR NYMPHS WITH JOINTED LEGS
WITHOUT CASE OR FIXED RETREAT**

WITH LONG TAILS
Mayflies (Order Ephemeroptera)

WITHOUT LONG TAILS
Beetles (Order Coleoptera)

**Body shape ovoid
Rear end often erected
scorpion-like when disturbed**

**Disc shape
Head and legs totally concealed beneath**

Water Pennies (Family Psephenidae)

**Body shape cylindrical
Long hairs on inside of front legs
Often with stripe down back**

**Slender
Triangular in cross section
Hard, yellowish brown covering**

Families Ephemerellidae (in part),
Caenidae, Tricorythidae

If larva has spines along lateral
margins it is a wood-eating
Shredder.

**Lobed body
Ventral suckers**

**GATHERING
COLLECTORS**

Riffle Beetles (Family Elmidae)

Family Siphlonuridae

Family Blephariceridae

**FILTERING
COLLECTORS**

**GATHERING
COLLECTORS**

SCRAPERS

REFERENCES

Allan, J. D. 1983. Predator–prey relationships in streams. Pages 191–229 *in* J. R. Barnes and G. W. Minshall (Eds.) *Stream Ecology: Application and Testing of General Ecological Theory.* Plenum, New York, NY.

Anderson, N. H., and A. S. Cargill. 1987. Nutritional ecology of aquatic detritivorous insects. Pages 903–925 *in* F. Slansky, Jr., and J. G. Rodriguez (Eds.) *Nutritional Ecology of Insects, Mites and Spiders.* Wiley, New York, NY.

Anderson, N. H., and J. R. Sedell. 1979. Detritus processing by macroinvertebrates in stream ecosystems. *Annual Review of Entomology* **24:**351–377.

Berrie, A. D. 1976. Detritus, microorganisms and animals in fresh water. Pages 323–338 *in* J. M. Andersen and A. MacFadyen (Eds.) *The Role of Terrestrial and Aquatic Organisms in Decomposition Processes.* Blackwell Scientific, Oxford, UK.

Cummins, K. W. 1973. Trophic relations of aquatic insects. *Annual Review of Entomology* **18:**183–206.

Cummins, K. W. 1974. Structure and function of stream ecosystems. *BioScience* **24:**631–641.

Cummins, K. W. 1980. The multiple linkages of forests to streams. Pages 191–198 *in* R. H. Waring (Ed.) *Forests: Fresh Perspectives from Ecosystem Analysis.* Proceedings 40th Annual Biology Colloquium, Oregon State Univ. Press, Corvallis, OR.

Cummins, K. W. 1992. Invertebrates. Pages 234–250 *in* P. Calow, and G. E. Petts (Eds.) *The Rivers Handbook: Hydrological and Ecological Principles.* Blackwell Scientific, London, UK.

Cummins, K. W. 1993. Bioassessment and analysis of functional organization of running water ecosystems. Pages 155–169 *in* S. L. Loeb and A. Spacie (Eds.) *Biological Monitoring of Aquatic Systems.* Lewis, Boca Raton, FL.

Cummins, K. W., and M. J. Klug. 1979. Feeding ecology of stream invertebrates. *Annual Review of Ecology and Systematics* **10:**147–172.

Cummins, K. W., and M. A. Wilzbach. 1985. *Field Procedures for the Analysis of Functional Feeding Groups in Stream Ecosystems.* Appalachian Environmental Laboratory, Contribution 1611, University of Maryland, Frostburg, MD.

Cummins, K. W., M. A. Wilzbach, D. M. Gates, J. B. Perry, and W. B. Taliaferro. 1989. Shredders and riparian vegetation. *BioScience* **39:**24–30.

Georgian, T., and J. B. Wallace. 1983. Seasonal production dynamics in a guild of periphyton-grazing insects in a southern Appalachian stream. *Ecology* **64:**1236–1248.

Gregory, S. V. 1983. Plant–herbivore interactions in stream systems. Pages 157–190 *in* J. R. Barnes and G. W. Minshall (Eds.) *Stream Ecology: Application and Testing of General Ecological Theory.* Plenum, New York, NY.

Hawkins, C. P., and J. A. MacMahon. 1989. Guilds: The multiple meanings of a concept. *Annual Review of Entomology* **34:**423–451.

Lamberti, G. A., and J. W. Moore. 1984. Aquatic insects as primary consumers. Pages 164–195 *in* V. H. Resh and D. M. Rosenberg (Eds.) *The Ecology of Aquatic Insects.* Praeger Scientific, New York, NY.

Lindeman, R. L. 1942. The trophic–dynamic aspect of ecology. *Ecology* **23:**399–418.

Merritt, R. W., and K. W. Cummins (Eds.). 1996. *An Introduction to the Aquatic Insects of North America,* 3rd ed. Kendall/Hunt, Dubuque, IA.

Merritt, R. W., K. W. Cummins, and T. M. Burton. 1984. The role of aquatic insects in the processing and cycling of nutrients. Pages 134–163 *in* V. H. Resh and D. M. Rosenberg (Eds.) *The Ecology of Aquatic Insects.* Praeger Scientific, New York, NY.

Minshall, G. W., R. C. Petersen, K. W. Cummins, T. L. Bott, J. R. Sedell, C. E. Cushing, and R. L. Vannote. 1983. Interbiome comparison of stream ecosystem dynamics. *Ecological Monographs* **53:**1–25.

Palmer, C., J. O'Keeffe, A. Palmer, T. Dunne, and S. Radloff. 1993a. Macroinvertebrate functional feeding groups in the middle and lower reaches of the Buffalo River, eastern Cape, South Africa. I. Dietary variability. *Freshwater Biology* **29:**441–453.

Palmer, C., J. O'Keeffe, and A. Palmer. 1993b. Macroinvertebrate functional feeding groups in the middle and lower reaches of the Buffalo River, eastern Cape, South Africa. II. Functional morphology and behaviour. *Freshwater Biology* **29:**455–462.

Peckarsky, B. L. 1984. Predator–prey interactions among aquatic insects. Pages 196–254 *in* V. H. Resh and D. M. Rosenberg (Eds.) *The Ecology of Aquatic Insects.* Praeger Scientific, New York, NY.

Root, R. B. 1973. Organization of a plant-arthropod association in simple and diverse habitats: the fauna of collards (*Brassica oleracea*). *Ecological Monographs* **43:**95–124.

Smock, L. E. 1980. Relationships between body size and biomass of aquatic insects. *Freshwater Biology* **10:**375–383.

Vannote, R. L., G. W. Minshall, K. W. Cummins, J. R. Sedell, and C. E. Cushing. 1980. The river continuum concept. *Canadian Journal of Fisheries and Aquatic Sciences* **37:**130–137.

Wallace, J. B., and R. W. Merritt. 1980. Filter-feeding ecology of aquatic insects. *Annual Review of Entomology* **25:**103–132.

Webster, J. R., and E. F. Benfield. 1986. Vascular plant breakdown in freshwater ecosystems. *Annual Review of Ecology and Systematics* **17:**567–594.

Wotton, R. S. 1994. Particulate and dissolved organic matter as food. Pages 235–288 *in* R. S. Wotton (Ed.) *The Biology of Particles in Aquatic Systems.* Lewis Publishers, Boca Raton, FL.

CHAPTER 22

Trophic Relations of Stream Fishes

FRANCES P. GELWICK[1] AND
WILLIAM J. MATTHEWS

University of Oklahoma Biological Station
University of Oklahoma

I. INTRODUCTION

Viewed across all species and life-stages, fishes feed from every category of prey available, from both aquatic and terrestrial, from within or near streams, including microscopic organisms, (e.g., algae, bacteria, plankton), macrophytes, macroinvertebrates (e.g., insects, crustaceans, mollusks), fish, and other vertebrates. Descriptions of food use by individual species are compiled in review volumes (Carlander 1969, 1977), state references (Robison and Buchanan 1984), and workshop proceedings (Stouder *et al.* 1994). Conceptually, trophic relations of fishes begin with foods and feeding behavior of individuals or species, but include feeding guilds (i.e., species or life-stages using similar foods or feeding behaviors) and community food webs.

Researchers may address management concerns, evolution of trophic adaptations, or theoretical predictions about species interactions (e.g., Polis *et al.* 1989) within the context of experimental approaches. Such studies may require data on weight, volume or size of prey items, size of predator, morphology of trophic structures, behavior, or physical habitat variables

---

[1]Current address: Department of Wildlife and Fisheries Science, Texas A&M University, College Station, TX 77843.

(e.g., Adams and Bric 1990, Noakes and Baylis 1990, Wainwright and Lauded 1992). Studies for describing the structure or function of stream ecosystems may require various levels of enumeration or taxonomic distinction of both predators and their prey (Karr *et al.* 1986, Flecker 1992, Gelwick and Matthews 1992, Roell and Orth 1994). Such choices can have important consequences for conclusions drawn from the study (Winemiller 1990).

The role of fish as predators in stream ecosystems has long been recognized. Fishes feed on many types of organisms and, in turn, are food for many aquatic and terrestrial predators, including humans. Thus, fishes provide important links within aquatic, and between aquatic and terrestrial, food webs. However, determination of the effects of fishes on stream ecosystem function has only just begun (Grimm 1988, Flecker 1992, Gelwick and Matthews 1992). Such effects are largely due, directly and indirectly, to trophic activities of fishes. Thus, the objectives in this chapter are: (1) to provide the researcher with tools to answer questions about trophic relations of fishes and (2) to provide information helpful in understanding the functional role of fishes in stream ecosystems. In this chapter we present basic and optional methods for collecting and analyzing data for diets, trophic morphology, habitat use, and feeding behavior of fishes. We include examples of mensurative studies and optional manipulative experiments, and discuss the suitability of various methods for particular questions.

II. GENERAL DESIGN

The basic exercises in this chapter require one-half day at a stream to obtain fish, and one-half day in the laboratory to identify stomach contents. Optional exercises require more laboratory and field time with multiple trips to the stream to set up experiments, collect samples, and analyze data.[2]

Exercises in this chapter could be combined with those for Chapters 2 to 4, which describe stream flow and geomorphology. By so doing, quantitative habitat variables can be related to diet. By including protocols in Chapters 16 to 18 and 23, researchers could relate overall body shape, and trophic morphology of fish to habitat use and diet (see also Chapter 20). Food availability can be estimated from collections of stream biota and

---

[2]Every state requires a permit for collecting fishes for scientific purposes, and local authorities should be notified of collecting activities. The research coordinator should ensure that the research group can access the field site and that permission is granted to work in the stream. This is especially important if sensitive species of concern for conservation are in the area. It is always good practice to review information on the expected distribution of fishes in the study area.

used to calculate an electivity index for fish diet. The effect of predators on habitat use and diet of prey species could be demonstrated by the optional exercise for short-term field manipulations. The optional exercise for longer-term pen experiments that exclude fish could be combined with studies in most of the other chapters to demonstrate the potential effects of fish on stream communities and ecosystem processes.

Groups of at least three persons are ideal for field sampling. However, if two or more groups are working together, they should avoid working in the same habitat unless activities are well coordinated. Otherwise, fish may be excessively disturbed, seek refuge, leave the area, or be difficult to collect. Behavioral observations, laboratory procedures, and data compilation and analysis can be carried out entirely by individuals or allocated within or among groups.

A. Site Selection

The stream reach should be wadeable and include both pools and riffles as well as other categories of in-stream structural complexity (e.g., backwaters, incised banks, boulders, woody debris) and possess a variety of fishes (e.g., sunfish, suckers, minnows, darters). Select a site with clear water if behavior of fish is to be observed. A comfortable area on the streambank protected from wind is needed for the observations in these exercises, unless underwater methods are used (see Helfman 1983). For the optional field exercises, pools should be large enough to accommodate the experimental pens, and should contain piscivorous fishes for manipulation. Experimental pens or fences require fence posts to be driven securely into the substrata and lower edges of fencing to be buried. Pools about 20 to 30 cm deep with velocity of 10 to 15 cm/s provide good experimental conditions.

B. General Procedures

Behavioral Observations   Researchers should be thoroughly familiar with the fishes of the study site before attempting this kind of data collection. However, even if the species are well known in the hand, their correct identification in the field by visual observation from above the water is often difficult and imprecise. If stream conditions are appropriate, snorkeling is effective. Underwater methods greatly improve the quality of data in regard to identification and feeding behavior of fishes. A video camera mounted on a tripod or enclosed in a waterproof case underwater may be used to record activity. Video recordings may save field time, but require time later when data are retrieved from the tapes.

Researchers should select observation points sufficiently elevated above the stream that a view of the entire pool is available, yet near enough to view individual fishes (either unaided or with binoculars). Some cover for the observer is helpful; however, most fishes return to normal behavior after 15 to 20 min if observers remain still.

Collecting Fishes   Fish have various diel schedules of feeding and habitat use. By sampling discrete habitat units (e.g., riffle edge or center; backwater; pool water column, edge, or bottom) and keeping collections separate in prelabeled containers, diet can be analyzed with respect to habitat use.

The time of collection influences the abundance and kinds of food items found in the stomach or gut of fishes. Fish may feed extensively throughout the day, for only a few hours, or only at night, dusk, or dawn. Some feed infrequently, especially if they eat large prey items that are digested slowly. Furthermore, when water temperatures are very cold (winter) or very warm (mid-summer), feeding is often greatly reduced and may even cease entirely in some systems. Therefore, not all fish collected will have food items in their stomachs. Differential digestion of prey items and postcapture digestion may result in reduction or loss of dietary information. In addition, some collection techniques can result in regurgitation if fish are stressed severely (e.g., rotenone, electroshocking). For this and other technical and safety considerations, we recommend seining for collecting fish. A large (30–40 cm diameter) zooplankton net may be used in late spring and summer to collect larval fishes (<10 mm total length, TL). A small aquatic dip net or fine mesh (ca. 4 mm) seine is adequate to collect young-of-the-year (10–20 mm TL) in summer and early fall.

Collecting fishes at the peak of their feeding activity, and placing them immediately into ice water, reduces stress to the fish (ASIH *et al.* 1987) and effectively slows digestion. This is an efficient and safe method for initial field preservation.[3] Fish should be frozen as soon as possible if they are not to be dissected immediately. Specimens used for morphological or museum studies should be preserved in 10% formalin.[4]

Depending on the research goals, fish of different ages, sizes, and sexes should be sampled from various habitats to adequately describe the feeding ecology of a species. Differences may relate to mouth size (i.e., relating to

[3]Before preserving fish, narcotize them with 2-phenoxyethanol or tricaine methanesulfate (MS-222) (ASIH *et al.* 1987; see also Chapter 18).

[4]Formalin is a known carcinogen and should be handled only with appropriate precautions. Use gloves and eye shields when handling formalin. If formalin touches skin or eyes, flood the exposed area with clean water. If necessary, seek further medical attention after returning from the field. Before formalin-preserved fish are handled, rinse them in water until no formalin odor remains (see also Chapter 18). Avoid direct contact with formalin solution or breathing fumes. To maintain preserved fish after examination, transfer them to 50% isopropanol or 70% ethanol.

vulnerable prey size), gut morphology, physiology and behavior (i.e., relating to switching foods, such as zooplankton to fish), nutritional needs (e.g., for egg development), or restricted habitat use in a nesting territory.

C. Experimental Manipulations

The optional procedures include experimental manipulations of predatory fish and construction of mesh pens. Polyethylene plastic mesh (5 mm, commonly used in aquaculture) allows drifting invertebrates to pass through, yet prevents the passage of most fishes (Gelwick and Matthews 1992). Larger-mesh fences installed across riffles prevent the movement of larger predatory fish, but not smaller prey fishes during experimental manipulations. Fences can be secured with wire or plastic bands to T-style metal fence posts. A fence post driver is very useful for hammering posts into the substratum. The top edge of the fencing should extend well above the water. The lower edge should be buried approximately 15 cm. In our experience, pens about 4 m on a side allow fish room to swim freely (Fig. 22.1), and are readily constructed from a single roll (ca. 18 m) of fencing.

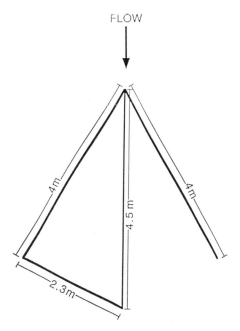

FIGURE 22.1 Pen design (top view) for Exercise 3. Fish are free to enter the open section of the pen from downstream.

Fences should be checked periodically and brushed clean of accumulated detritus.

Clay tiles (15 × 15 cm, commercially available) are convenient sampling units for colonization by algae and invertebrates. Unglazed tiles mimic natural cobble substrata and allow accurate estimation of surface area for calculating standing crop of biota (Lamberti and Resh 1985). Fences and tiles are vulnerable to being swept downstream during stage rises. Thus, we recommend short-term experiments, carried out during periods of stable flow.

D. Analysis

Morphological Structures   Trophic behavior and foods often are associated with morphological adaptations for differential feeding. Some examples are: (1) scraping mouth parts for grazing periphyton (*Campostoma anomalum, Acrocheilus alutaceus*), (2) long or finely spaced gill rakers of planktivores (*Lepomis macrochirus, Dorosoma cepedianum*), (3) stout pharyngeal jaws (*Lepomis gibbosus*) or pharyngeal teeth (*Moxostoma carinatum*) to crush snail or clam shells, and (4) long coiled guts associated with detritivory and herbivory (*Notropis nubilis, Hybognathus placitus*). Care should be taken to avoid damage of these structures if the study includes their examination and measurement. This work requires a dissecting microscope, especially for small fish.

Removing, Identifying, and Quantifying Gut Contents   Stomach contents can be evaluated by volume, weight, or occurrence and compared to that of other individuals or species (diet overlap) or to availability of food in the environment (diet breadth, electivity). The objectives of the study guide the choice of techniques for collecting and analyzing data (see methods of Bowen 1983, Hyslop 1980). For large piscivores (e.g., black basses), a smooth-sided plastic or PVC tube (appropriately sized for the fish) can be inserted into the stomach, and filled with water. By covering the end of the tube with the palm of one's hand, a gentle vacuum can be created as the tube is removed, and the stomach contents are transferred to a plastic bag. The fish can be returned quickly and in good condition to the stream. Anal and stomach flushing can be used for smaller fishes (Culp *et al.* 1988). However, this requires a special apparatus, fewer animals can be sampled during limited field time, and reliability must be determined for each species. Alternatively, dissection allows the gut to be examined for fullness and removal of all prey, and morphology of trophic structures can be evaluated.

Even recently ingested food items may be ground by pharyngeal jaws and teeth, making recognition difficult. Consequently, digestion-resistant parts of prey (exoskeletons, head capsules and shields, tarsal claws) are most useful. We highly recommend the use of a reference collection of invertebrates from the stream site to aid in prey identification (see Chapter 16). Macrophytes often have characteristic leaf shapes or edges, and algal cells are often found intact (see Chapter 13). The level of identification should be determined by the researcher's skill, time available, and information needed.

Calculating the Frequency of Occurrence of Foods This is the proportion of fish that contains a given food type. It is the fastest approach by which to quantitatively analyze fish diets, because only the presence or absence of a food item needs to be recorded. This analysis indicates the extent to which fish in the sample functioned as a single feeding unit (i.e., high frequencies of occurrence for food types used by many individuals; low frequencies for most foods when individual fish specialize). Incidental ingestion of some items (e.g., sediments or algae by benthivorous fish) also may result in their occurrence at high frequency. Alternatively, the occurrence of even a single prey item in a stomach is recorded as positive, even if it is a small percentage of the total contents by weight or volume. Consequently, researchers should emphasize the uniformity of the diet, among fishes analyzed, rather than the importance of individual food types when interpreting results.

Calculating the Percentage Composition of Food by Number This is the number of items of a given food type expressed as a percentage of the total number of food items of all types counted in the stomach of a fish. If food items are disarticulated or digested, a characteristic part (best if found once per prey) is counted as one food item. For amorphous foods taken in bites (e.g., detritus, vascular plants), the percentage by volume can be used. Small, yet abundant prey often need to be subsampled from a known volume of stomach contents (50 to 100 of each prey-type should be counted). Although mean values summarize the prey in the sample, the food value among prey types cannot be compared using this approach (e.g., 1 amphipod vs 10,000 bacteria). Numerical counts combined with estimates of prey weight and volume, or feeding rate, can be used to calculate bioenergetic relationships, relative prey biomass removed, or rate of prey removal.

Calculating the Percentage Composition of Foods by Weight Wet (blotting to remove excess moisture) or dry (desiccation at 105°C) mass can be assessed if items are large enough to be handled individually and have been digested only slightly (Bowen 1983). Mass of disarticulated prey must be extrapolated. For prey too small for direct weighing, biovolume may be calculated (Cummins and Wuycheck 1971, see also

Chapter 10). Because detritus can make up a large portion of some diets, stomach contents can be combusted (at 550°C in a muffle furnace) and the ash-free dry mass (AFDM) calculated as dry weight minus ash weight (see Chapter 14). Most prey items can be considered 100% organic matter for these calculations. The weight of mollusks should be adjusted for shell weight. Percentage composition by mass is roughly proportional to food value.

Calculating Ecological Indices   These indices are based on ecological theory of the niche. Electivity indices can be used to compare the diet to the resources available in the natural habitat (Strauss 1979, 1982). Electivity values indicate the extent to which the observed diet differs from a diet selected at random (Ivlev 1961). However, the accurate determination of the relative abundances of all potential prey is difficult (Ready *et al.* 1985). Indices that quantify overlap, breadth, or percentage similarity in diet among individuals or taxa were developed to study competition and re-source partitioning. The index of Levins (1969) was one of the first to be used to calculate breadth and ranges from 1.0 to n, where n is the number of resource categories. Overlap calculated by Pianka's (1973) index ranges from zero (no overlap) to 1.0 (complete overlap).

The mathematical and statistical properties, and biological relevance of the various indices have been debated (Hurlbert 1978, Jumars 1980, Wallace 1981, Smith 1985). Overall, it is best to chose a simple index that is relevant to the question addressed and to be aware of its limitations when interpreting the analysis (Loreau 1990). Overlap indices are not a true statistic, but they place relationships between species pairs into a relative context. Values greater than 0.75 have been considered to indicate high overlap while values below 0.30 are considered to have a very low overlap (Matthews *et al.* 1982). Overlap or similarity in diet or trophic morphology can be used to suggest mechanisms of ecological interactions (e.g., competition, predation) that potentially affect community structure (Matthews *et al.* 1992).

III. SPECIFIC EXERCISES

A. Exercise 1: Basic Collection and Trophic Analysis

Field Procedures

1. Research teams of three persons, two to seine and one to carry the collection jar and assist those seining, should sample particular homogeneous habitat patches. If there is more than one research team, then they

should stay separated such that sediments disturbed by seining have minimal effect on other teams.

2. Sample specific habitats with appropriate combinations of seining techniques (see protocols in Chapter 18).

3. Place all fish into plastic jars (prelabeled on or in the jar) one-third full of ice water, unless formalin-preserved specimens are needed. If using the ice water method with same-day dissection, place collection jars into insulated coolers with ice. Researchers may choose to evacuate stomach contents of large piscivores in the field. The fish can then be released alive to another section of the stream so that it is not collected again.

4. Record the characteristics of the habitat seined (e.g., substratum size, presence of macrophytes, woody debris) on a data sheet.

Laboratory Procedures   Using a dissecting microscope, examine stomach contents of fish from each species collected.

1. Using appropriately sized scissors or scalpel, make a longitudinal cut on the ventral side of the fish from just behind the isthmus of the gills posteriorly to the anal fin. Then make two transverse cuts at each end of the first cut to open the coelom and expose the viscera.

2. Using sharp-pointed scissors, sever the esophagus, the last few millimiters of the intestine, and the mesentery at its dorsal point of attachment. This allows the visceral mass to be lifted out of the coelom for more detailed examination and manipulation.

3. Separate the digestive tract from the other visceral organs. Note the coiled gut (e.g., of herbivores) or gastric caecae (finger-like projections from the stomach of piscivores) of some fishes. Sever the stomach (or anterior compartment for species without a true muscular stomach) from the rest of the gut. This section will contain the most recently ingested prey.

4. Open the stomach or gut segment by making a shallow slit (be careful to not cut prey) lengthwise with fine scissors. For piscivores, whole prey items can be lifted directly from the stomach. For smaller prey, it is often useful to hold the slit segment with forceps over a petri dish and wash out the contents with a small amount of water from a squirt bottle or pipet. The food also may be extruded by sliding a blunt probe along the length of the segment; this may extrude much of the gut mucosa as well, which should not be mistaken for part of the diet (Bowen 1983).

5. For each individual fish, sort, identify, and count prey items and visually estimate the percentage of the total volume of the stomach contents made up of each prey category (this is especially appropriate if detritus or algae is included in the diet). Record the data for each fish.

B. Exercise 2: Predator Manipulation Experiments (Optional)

1. About 2 to 4 days before the group study, install fences across shallow riffles above and below selected pools.

2. Remove predatory fishes by seining (or with proper safety precautions, electroshocking) from two pools and place these in two other pools. Adjust predator abundances between pools to create "removed" and "enhanced" predator treatments. This is a demonstration only, with low statistical power.

3. Proceed as in Exercise 1.

C. Exercise 3: Fish Exclosure Experiments (Optional)

1. One month before the planned study date, place 80 unglazed clay tiles (15 × 15 cm) into the stream to be colonized by algae and invertebrates. Extra tiles are included to compensate for loss or breakage.

2. One week before the study, build two triangular double (open/closed) pens (Fig. 22.1), taking care that the lower edge of the fence is buried in the substrata.

3. Remove any fish that have entered the closed pens during construction. Place 15 stream-conditioned tiles into each side (open/closed) of the pens.

4. Return periodically (every 1–2 days) to maintain experimental conditions.[5]

5. Proceed as in Exercise 1.

6. Visually assess and record the macroscopic types of algae and epilithon (filamentous green, thin layer diatoms, detritus, or silt) on tiles and natural substrata in open and closed sides of each pen.

7. Record the thickness (mm) of the epilithon and the length of the longest filament (mm) on each tile. These values can be related to habitat use by fishes and other stream fauna.

8. Sample tiles and substrata for algae and invertebrates according to protocols in Chapters 14 and 16.

D. Optional Laboratory Procedures

1. Describe or sketch the location, orientation, and morphology of the mouth.

2. Measure the approximate gape size by gently extending the lower jaw and measuring the largest interior diameter of the mouth with a ruler or calipers.

---

[5]We found that week 1 is sufficient for response variables to begin to change. This will vary with stream conditions and fish densities used.

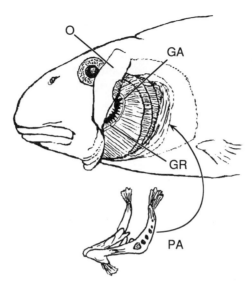

FIGURE 22.2 Relationships of the gill arches under the operculum and the pharyngeal arches (under tissue) in upper figure. Lower figure shows orientation of the pharyngeal arches and muscle attachments. GA, gill arch; GR, gill rakers; O, opercular flap; PA, pharyngeal arches and teeth.

3. Cut and lift the operculum away from the underlying gills (Fig. 22.2).

4. Examine the structure of the gill rakers located on the buccal side of the first gill-arch; make a sketch for each species examined.

5. The last pharyngeal arch lies behind the last gill-bearing arch (push gill arches anteriorly) and is embedded in tissue (Fig. 22.2). The arch is removed by first running a sharp probe or scalpel around the posterior edge of the arch, and severing the muscles at the dorsal and ventral attachment points.[6] Note how the arch was oriented in the fish. Take care to not break the teeth.

6. Carefully remove any flesh from around the teeth and describe or sketch their morphology. Often teeth are broken off during dissection; look for evidence of this. Number of teeth and their morphology vary among species (Robison and Buchanan 1984) and are related to trophic habits.

7. Go to Exercise 1, above, and conduct Laboratory Procedures.

[6]The pharyngeal (throat) teeth are primarily on the ventral–anterior side of the arch; therefore, hold the dorsal end of the arch with forceps (use fine No. 5 watchmaker forceps, especially for minnows) and gently pull the arch away.

8. Measure the width (second longest axis) of the largest prey item (this measure is not appropriate for fishes that are primarily detritivores or algivores). If prey are disarticulated use an estimate from head capsule or shield width. Record the measurements on a data sheet. This can be used to analyze the relationship of prey size to gape width.

E. Data Analysis

1. To determine the effect of habitat on diet, calculate the frequency distribution of the dominant prey item for all species in each habitat category. Use a G test (Sokal and Rohlf 1995) to compare the observed frequency distribution of dominant prey for each species, in each habitat, versus the expected distribution of prey. The expected distribution is based on the frequency distribution of prey items across all fish pooled across all habitats.

2. If measure of gape size and prey size were made, plot the gape size versus largest prey size for each species (or taxon of interest). Calculate the linear relationship (transform data if necessary to achieve normality) using regression analysis (e.g., Matthews et al. 1982), or use analysis of covariance (Sokal and Rohlf 1995) to compare these relationships among taxa.

3. If gill-raker morphology was measured (optional laboratory procedure), calculate the frequency distribution of the dominant prey item (i.e., the prey item in highest proportion in one fish) for all fish in each category of gill-raker morphology. Use a G test to compare the distribution of dominant prey taxa for each gill-raker morphology versus the expected distribution of prey. The expected distribution is based on the frequency distribution of prey items combined across all fish examined. Similarly, this could be done to test the effect of pharyngeal-tooth morphology.

4. The effect of predators can be determined by comparing the frequency distribution of dominant prey of predator-enhanced versus predator-removed treatments (Exercise 2). The expected distribution is based on the frequency distribution of prey items across all fish in all pools combined.

5. Determine if diets are more similar among or within families by calculating a percentage similarity index. First, list all the prey categories that occurred in either species being compared (zero–zero matches do not affect the index). Next, calculate the mean percentage (i.e., the average percentage, by number or volume, for each prey category across all individuals) of each species being compared. Then choose the smaller mean percentage for each prey category, and sum them across all categories. Repeat this procedure for each species pair within and between families. Use a t test to compare these averages for within-family versus between-family similarities.

6. Determine the effect of predators on diet, by comparing the average similarity among species in predators-present versus predators-removed treatments (Exercise 2).

7. If data for invertebrate abundance at the study site are available and adequately describe the availability of prey items in the habitat, an index of electivity can be calculated for invertivores. Use the odds ratio (Jacobs 1974)

$$O = \frac{p_1 q_2}{p_2 q_1}, \tag{22.1}$$

where p_1 represents percentage of diet comprised by a given prey taxon; p_2, percentage of diet comprised by all other prey taxa; q_1, percentage of food complex in the environment comprised by the given taxon; and q_2, percentage of food complex in the environment comprised by all other taxa.

The natural log of the odds ratio (L) is symmetrically distributed about a mean of 0 and ranges to ($+$) infinity in case of positive selection and ($-$) infinity in case of negative selection. The standard error of L can be calculated as

$$SE\,(L) = \sqrt{\frac{1}{n_1 \, p_1 \, q_1} + \frac{1}{n_2 \, p_2 \, q_2}}, \tag{22.2}$$

where n_1 is the total number of prey in the diet sample; n_2, total number of food organisms in the environmental sample; and p_1, p_2, q_1, and q_2 are as previously defined.

Because L has a log-normal distribution, the null hypothesis that an observed L is not significantly different from 0 (i.e., prey are consumed nonselectively and in proportion to their occurrence in the environment) can be tested. The difference is expressed in terms of standard normal deviates,

$$Z = \frac{L_{observed} - L_{expected}}{SE\,(L)}, \tag{22.3}$$

where $L_{expected}$ in this case is 0. The value can be compared in a table of areas of the normal curve (z distribution) to determine the probability of obtaining such a difference. Where the ratios of p_1 to p_2 are the same, the larger the absolute difference between these values the greater the absolute value of L.

6. If the exclosure exercise (3) was done, test the effect of excluding fish for selected ecosystem variables (e.g., invertebrate abundance, algal standing crop). Compare the values for each response variable measured in the two pen sides (fish/no fish) by *t* tests or ANOVA, with necessary data transformation (Sokal and Rohlf 1995).

IV. QUESTIONS

1. Were diets more similar among species within than between families? Were diets more similar among species within than between morphological groups? Would you predict potentially stronger competitive trophic interactions among larger (generally adults) or smaller (generally juveniles) fishes? How could this affect taxonomic structure and relative importance of trophic levels in a fish community? In a benthic macroinvertebrate community?

2. If invertivore diet electivity was calculated, did species appear to choose prey nonrandomly? If analyses were done by habitat, did fish feed more nonrandomly in some habitats than others? How might the presence of a predator affect the electivity of each species? What was the effect of predators? Was the effect different among fish of different body size (see Werner and Gilliam 1984)?

3. What was the effect of habitat on distribution (dominance) of prey? Was this different for some species or body sizes? Do predators force species into common habitats (see Savino and Stein 1989)? What was the effect of predators on prey dominance and diet similarity? Were some species or body sizes more affected than others?

4. Was the relationship between gape size and prey size stronger within a species or a family? Was the strength of the relationship between gape size and prey size related to mean prey size? Under what conditions would you predict that a species would be gape limited?

5. Diagram a simplified food web for your stream community. Draw an arrow from each prey category to each fish species and make the width of the arrow roughly proportional to the percentage of each prey in the predator's diet. Draw a box around each fish taxon reflecting its proportion in your collections. In the food web, did some prey categories appear to be used more than others? Is this concordant with the abundance of the prey in the environment or are some fish specializing on similar prey, independent of fish abundance or habitat use (see Mendelson 1975)? Do your analyses of diet support or refute this? How would a food web in this stream differ, regarding relative values for each parameter, if rather than

species, size or age classes of fish were used? If you did Exercise 2, how do piscivorous fish influence food web structure (see Power *et al.* 1985, Power 1990)?

6. How do fish influence each ecosystem component? Which fish appear to have the greatest effect due to abundance or trophic habits (see Gelwick and Matthews 1992, Vaughn *et al.* 1993)?

V. MATERIALS AND SUPPLIES

Field Equipment

Fish collecting permit
Gallon plastic jars (one per habitat) with large screw-on lids
Heavy, rag or waterproof paper (for labeling collections)
Insulated coolers with ice (for preserving fish)
Pencils or water-proof ink pens (for labeling collections)
Seines: 1.3 m deep × 1.8 m wide and 1.3 m deep × 4.6 m wide (5 mm mesh size, heavy leaded bottom lines)

Optional Field Equipment

1.3 m deep × 1.8 m wide (4 mm mesh), for postlarval and juvenile fish
Backpack electroshocker, lineman's rubber gloves, long-handled dip nets (for electroshocking fish)
Canvas shoulder bags (to accommodate the collecting jars and free hands while collecting)
Concentrated formalin (diluted to 10% in the field for preserving fish)
Polarized sunglasses
Zooplankton net for collecting larval fish

Optional Field Equipment for Observations

Binoculars for observations from the stream bank
Snorkel and mask or SCUBA gear
Video camera

Optional Field Equipment for Predator Manipulation Experiments

Fence post driver
Flexible wire or plastic straps (to secure fence to posts)
Polypropylene fencing (dimensions adequate to block large piscivorous fish)
T-style fence posts (to stabilize plastic fencing across riffles)

Optional Field Equipment for Fish Exclosure Experiments

Meter stick

Polypropylene fencing (6-mm mesh, comes as approximately 15-m to
    a roll)

Unglazed clay tiles (80 15 cm × 15 cm)

Laboratory Equipment for Gape Analysis

Blunt and pointed dissecting probes

Dissecting microscope and light source

Fine-tipped scissors

Large flat trays (for sorting and dissecting fish)

Number 5 "watchmaker" straight tipped-forceps

Ocular micrometer (for estimating prey size)

Petri dishes (for stomachs and prey items)

Pipets (for sorting and counting prey)

Plastic millimeter ruler (150 mm)

Sieve (for rinsing formalin or detritus out of fish collections)

Vernier-type calipers

Wash bottles (for rinsing prey from guts)

REFERENCES

Adams, S. M., and J. E. Bric. 1990. Bioenergetics. Pages 389–415 *in* C. Schreck
    and P. Moyle (Eds.) *Methods for Fish Biology.* American Fisheries Society,
    Bethesda, MD.

American Society of Ichthyologists and Herpetologists (ASIH), American Fisheries
    Society, and American Institute of Fisheries Research Biologists. 1987. Guide-
    lines for the use of fishes in field research. *Fisheries* **13:**16–23.

Bowen, S. 1983. Quantitative description of the diet. Pages 325–336 *in* L. A. Nielsen
    and D. L. Johnson (Eds.) *Fisheries Techniques.* American Fisheries Society,
    Bethesda, MD.

Carlander, K. D. 1969. *Handbook of Freshwater Fishery Biology,* Vol. 1. Iowa State
    Univ. Press, Ames, IA.

Carlander, K. D. 1977. *Handbook of Freshwater Fishery Biology,* Vol. 2. Iowa State
    Univ. Press, Ames, IA.

Culp, J. M., I. Boyd, and N. E. Glozier. 1988. An improved method for obtaining
    gut contents from small, live fishes by anal and stomach flushing. *Copeia*
    1988:1079–1082.

Cummins, K. W., and J. C. Wuycheck. 1971. Caloric equivalents for investigations
    in ecological energetics. *International Association for Theoretical and Applied
    Limnology, Special Communication* **18:**1–158.

Flecker, A. S. 1992. Fish trophic guilds and the structure of a tropical stream: Weak direct vs. strong indirect effects. *Ecology* **73:**927–940.

Gelwick, F. P., and W. J. Matthews. 1992. Effects of an algivorous minnow on temperate stream ecosystem properties. *Ecology* **73:**1630–1645.

Grimm, N. B. 1988. Feeding dynamics, nitrogen budgets, and ecosystem role of a desert stream omnivore, *Agosia chrysogaster* (Pisces: Cyprinidae). *Environmental Biology of Fishes* **21:**143–152.

Helfman, G. S. 1983. Underwater methods. Pages 349–370 *in* C. Schreck and P. Moyle, editors. Methods for fish biology. American Fisheries Society, Bethesda, MD.

Hurlbert, S. H. 1978. The measurement of niche overlap and some relatives. *Ecology* **59:**67–77.

Hyslop, E. J. 1980. Stomach contents analysis—A review of methods and their application. *Journal of Fish Biology* **17:**411–429.

Ivlev, V. S. 1961. *Experimental Ecology of the Feeding of Fishes.* Yale Univ. Press, New Haven, CT.

Jacobs, J. 1974. Quantitative measurement of food selection: A modification of the forage ratio and Ivlev's electivity index. *Oecologia* **14:**413–417.

Jumars, P. A. 1980. Rank correlation and concordance tests in community analyses: An inappropriate null hypothesis. *Ecology* **61:**1553–1554.

Karr, J. R., K. D. Fausch, P. L. Angermeier and P. R. Yant. 1986. *Assessing Biological Integrity in Running Waters: A Method and Its Rationale.* Illinois Natural History Survey, Special Publication 5, Champaign, IL.

Lamberti, G. A., and V. H. Resh. 1985. Comparability of introduced tiles and natural substrates for sampling lotic bacteria, algae and macroinvertebrates. *Freshwater Biology* **15:**21–30.

Levins, R. 1969. *Evolution in Changing Environments: Some Theoretical Explorations.* Princeton Univ. Press, Princeton, NJ.

Loreau, M. 1990. The Colwell–Futuyma method for measuring niche breadth and overlap: A critique. *Oikos* **58:**251–253.

Matthews, W. J., J. R. Bek, and E. Surat. 1982. Comparative ecology of the darters *Etheostoma podostemone, E. flabellare,* and *Percina roanoka* in the upper Roanoke River drainage, Virginia. *Copeia* **1982:**805–814.

Matthews, W. J., F. P. Gelwick, and J. J. Hoover. 1992. Food of and habitat use by juveniles of species of *Micropterus* and *Morone* in a Southwestern reservoir. *Transactions of the American Fisheries Society* **121:**54–66.

Mendelson, J. 1975. Feeding relationships among species of *Notropis* (Pisces: Cyprinidae) in a Wisconsin stream. *Ecological Monographs* **45:**199–230.

Noakes, D. L. G., and J. R. Baylis. 1990. Behavior. Pages 555–583 *in* C. Schreck and P. Moyle (Eds.) *Methods for Fish Biology.* American Fisheries Society, Bethesda, MD.

Pianka, E. R. 1973. The structure of lizard communities. *Annual Review of Ecology and Systematics* **4:**53–74.

Polis, G., C. A. Myers, and R. D. Holt. 1989. The ecology and evolution of intraguild predation: Potential competitors that eat each other. *Annual Review of Ecology and Systematics* **20:**297–330.

Power, M. E. 1990. Effects of fish in river food webs. *Science* **250:**811–814.

Power, M. E., W. J. Matthews, and A. J. Stewart. 1985. Grazing minnows, piscivorous bass, and stream algae: Dynamics of a strong interaction. *Ecology* **66:**1448–1450.

Ready, R. C., E. L. Mills, and J. L. Confer. 1985. A new estimator of, and factors influencing, the sampling variance of the linear index of food selection. *Transactions of the American Fisheries Society* **114:**258–266.

Robison, H. W., and T. M. Buchanan. 1984. *Fishes of Arkansas.* Univ. of Arkansas Press, Fayetteville, AR.

Roell, M. J., and D. J. Orth. 1994. The roles of predation, competition, and exploitation in the trophic dynamics of a warmwater stream: A model synthesis, analysis and application. *Hydrobiologia* **291:**157–178.

Savino, J. F., and R. A. Stein. 1989. Behavior of fish predators and their prey: Habitat choice between open water and dense vegetation. *Environmental Biology of Fishes* **24:**287–293.

Smith, E. P. 1985. Estimating the reliability of diet overlap measures. *Environmental Biology of Fishes* **13:**125–138.

Sokal, R. R., and F. J. Rohlf. 1995. *Biometry,* 3rd ed. Freeman, New York, NY.

Stouder, D. J., K. L. Fresh, R. J. Feller (Eds.). 1994. *Theory and Application in Fish Feeding Ecology.* Belle W. Baruch Library in Marine Science No. 18. Univ. of South Carolina Press, Columbia, SC.

Strauss, R. E. 1979. Reliability estimates for Ivlev's electivity index, the forage ratio, and a proposed linear index of food selection. *Transactions of the American Fisheries Society* **108:**344–352.

Strauss, R. E. 1982. Influence of replicated subsamples and subsample heterogeneity on the linear index of food selection. *Transactions of the American Fisheries Society* **111:**517–522.

Vaughn, C. C., F. P. Gelwick, and W. J. Matthews. 1993. Effects of algivorous minnows on production of grazing stream invertebrates. *Oikos* **66:**119–128.

Wainwright, P. C., and G. V. Lauded. 1992. The evolution of feeding biology in sunfishes (Centrarchidae). Pages 472–491 *in* R. Mayden (Ed.) *Systematics, Historical Ecology and North American Freshwater Fishes.* Stanford Univ. Press, Palo Alto, CA.

Wallace, R. K. 1981. An assessment of diet-overlap indices. *Transactions of the American Fisheries Society* **110:**72–79.

Werner, E. E., and J. F. Gilliam. 1984. The ontogenetic niche and species interactions in size-structured populations. *Annual Review of Ecology and Systematics* **15:**393–425.

Winemiller, K. O. 1990. Spatial and temporal variation in tropical fish trophic networks. *Ecological Monographs* **60:**331–367.

CHAPTER 23

Habitat Use and Competition among Stream Fishes

CHRISTOPHER A. FRISSELL[*] AND DAVID G. LONZARICH[†]

*Flathead Lake Biological Station
The University of Montana
†Department of Biology
University of Wisconsin at Eau Claire

I. INTRODUCTION

Species with similar morphology, life histories, and ecological requirements coexist in many river and stream systems. For example, one may
wonder how so many species of trout and salmon can persist in a single
Pacific coastal stream (Glova 1986), or how three species of darters can
live together on the bed of a single stream on the Mississippi Plateau
(Fisher and Pearson 1987). Over biogeographic history, fishes that originally
evolved in isolation within divergent river or lake basins have subsequently
gained access to the same drainages, and today they can share habitats.
The result is that species with similar phylogeny, morphologies, behavior,
and ecological requirements are commonly found together in streams and
there is potential for competition among these taxa. Some fishes are highly
sensitive to the presence of potentially competing species, especially when
there has been limited opportunity for coevolution. For example, native
salmonids often exhibit substantial shifts in habitat use and sometimes

complete exclusion where nonnative species have been introduced (Fausch 1988). Fishery managers and conservation biologists are vitally concerned about the exclusion and possible local extinction of native or naturalized fish populations when potential competitors are introduced or invade.

One of the important ways that such species coexist is through habitat partitioning. By exploiting different habitat or microhabitat patches at different times, potentially competing species can find the opportunity to avoid competitive exclusion and thereby coexist (Connell 1980). The differential use of habitats by closely related or similar species is also an important component of riverine biodiversity in that it promotes spatial complexity in community structure at various scales and may contribute to resilience of fish assemblages when habitats are altered. Habitat partitioning is certainly of consequence in mediating many other biotic interactions, such as predation and disease transfer, in complex ways that are as of yet poorly understood.

Stream ecosystems are spatially heterogeneous, such as in the habitat diversity offered by pools, riffles, and morphological features, or the convergence of flow velocity, depth, substratum, and temperature conditions that define different microhabitat patches within a single pool. Stream fishes can exhibit habitat segregation at either or both of these scales, depending on species, life stage, and season (e.g., Gorman 1987). Temporal heterogeneity, such as in the variations in flow or temperature over time, can also afford time-variant niches among which species are differentiated, thus reducing or directly avoiding sustained biotic interactions. Temporal variability in the environment can curtail competitive interactions between species and promote their continued coexistence regardless of overlap in their ecological niches (Connell 1980). For example, despite their similarity in many aspects of juvenile and adult morphology and behavior, the preadaption of brook trout *(Salvelinus fontinalis)* for fall spawning and rainbow trout *(Oncorhynchus mykiss)* for spring spawning may prevent either species from monopolizing a small stream where floods damaging to eggs and fry can strike alternately in spring or winter (Seegrist and Gard 1972).

Morphological similarity, including body size, shape, and structure of the fins and mouthparts, can be a useful indicator of the potential of fish taxa for overlapping habitat requirements (Douglas 1987, see also Chapter 18). Habitat structure and available food resources tend to be highly correlated in stream environments, so the evolution of stereotypic morphologic patterns in fishes is not surprising. Some taxonomic groups, such as the sunfishes (Centrarchidae), however, demonstrate much overlap in these *ecomorphological* features, and sunfish species commonly display dramatic shifts in use of habitats depending on the presence of congeneric species (Werner and Hall 1976).

Among fishes that are presumably more highly coevolved, such as assemblages of native minnows in the midwestern and eastern United States, habitat selection is relatively rigid and responses to the presence or absence of a possible competitor may be less apparent (Douglas 1987, Gorman 1987). The fine-scale partitioning evident in such minnow assemblages might, in theory, reflect the "Ghost of Competition Past" (Connell 1980), whereby strong and consistent natural selection among species that have coexisted for many generations has led to stereotyped behavioral and ecological avoidance of potential competitors. Introduced species may have much different success and effects in a tightly coevolved assemblage than in a more loosely coevolved group of fishes (Herbold and Moyle 1986, Ross 1991).

Researchers can gain important information about the degree to which ongoing interspecific competition is a process shaping fish assemblages by experimentally removing or adding species and then observing changes in distribution and behavior of the remaining species. However, it is often more practical to take advantage of "natural experiments," where quirks of nature or history have resulted in different species combinations in similar streams. We can examine the ecology of a target species in the absence and presence of potential competitor species, seeking evidence of shifts in habitat use, behavior, or microdistribution that might be attributed to competitive exclusion or interactive segregation (e.g., Finger 1982, Fisher and Pearson 1987). If we find evidence of niche shifts or dynamic segregation in an observational field study, further experimental work involving direct manipulation of presence or abundance of each species might be helpful in confirming and better describing the mechanisms of interaction (e.g., Finger 1982, Reeves et al. 1987, Greenberg 1988). Experimental manipulation of either field or laboratory systems can be fruitful, but interpretation of the results can be ambiguous or confounding unless they are closely tied to experiences with natural communities.

II. GENERAL DESIGN

This chapter presents a basic field survey using a design intended to allow measurement of shifts in habitat use among possible competitor species by sampling in streams where different combinations of target species occur. Additional options provide for field sampling at a finer level of resolution and describe some possible experimental follow-up studies. In each of the exercises, researchers will be actively involved in all aspects of field sampling, from surveys of physical habitat to collection and analysis of fish data. These exercises require work on two different streams with similar physical conditions, but different species complements. Researchers could be divided into one team for each stream or the same team can conduct surveys of the two streams on different days. Teams will probably function best with three to five members, splitting specialized tasks among the team members. Reaches can be subdivided among multiple teams for simultaneous sampling of habitats and fishes if sufficient gear are available, reducing overall field time necessary at each stream. In the lab, each team should summarize its raw physical and biological data, and then all data should be shared among teams for analysis and interpretation.

A. Site Selection

The study streams should be small- to medium-sized and easily wadable. Streams with naturally high fish densities are most likely to yield interesting information using the techniques described in this chapter. Low densities may require more careful and much more extensive sampling. The streams should be similar in elevation, gradient, width, flow, thermal regime, chemistry, and riparian vegetation, but should have different complements of potentially interacting species. Results from Chapters 1, 2 and 18 can facilitate the classification process to identify streams and reaches of high physical similarity (see also Frissell *et al.* 1986). Approximate information about fish assemblages may be obtained from previous surveys, museum collections, and other existing sources, but should be confirmed with sampling at the sites where the exercises will be conducted.

B. Identification of Target Taxa

A select subset of the fish assemblage should be the target of sampling and analysis. This subset should comprise two or three species that are morphologically, behaviorally, and/or taxonomically similar, so that there is reason to expect the potential for competition for habitat (or spatial segregation that suggests such competition). The study streams should differ in the presence of the target species, so that one key species is abundant in one stream but absent in the other. The simplest case of two ecologically similar taxa in stream A, but only one species in stream B, is likely to be the most tractable. The cooccurrence of three or more similar species allows multiple and interacting relationships that can be far more difficult to sort out with simple field data. Target species may include native or introduced fishes. Likely fishes best suited to these methods include young-of-the-year or juveniles of mid-water species such as salmonids or cyprinids or small benthic species such as darters and sculpins. Sampling and analysis should be targeted on species of similar age and size classes. If the field exercises are conducted in conjunction with other exercises from this book, researchers are encouraged to sample the entire fish assemblage and then focus on an appropriate subset of species to evaluate for habitat segregation and possible competition. However, if exercises about competition are the primary objective, sampling can be made more efficient with better likelihood of success by targeting the sampling methods on the selected species.

C. Field Methods for Habitat Surveys

Researchers should classify the study reach into pools, riffles, and other habitat types, following the methods described in Chapter 2. Habitats should be mapped freehand, and using meter tapes or meter sticks, the length, width, and maximum depth of each habitat unit measured and recorded on the map. In addition, throughout habitat and biological sampling, users should measure and record water and air temperature hourly. A long enough reach should be surveyed such that each team can identify 10 or more similar habitats in each of three general categories: pools, riffles, and, where appropriate, off-channel habitats. This portion of the field work will take from 2 to 8 h, depending on stream size, reach length, accessibility, and researcher skills. Next, use a random numbers table or other unbiased method to select three to five habitats (pools, riffles, side channels, etc.) in each of these three general categories for biological sampling.[1]

III. SPECIFIC EXERCISES

A. Exercise 1: General Habitat Survey and Habitat Associations

Field Methods for Biological Sampling of Pools and Riffles The objective of sampling is to quantify absolute or relative abundance of the target species subset within each habitat unit selected. Depending on conditions, target species, and available equipment, pools or riffles can be sampled with seine or electrofishing gear using *removal to depletion* methods (see also Chapter 18). Electrofishing may be more effective in shallow waters and where cover is abundant; seining is effective in deeper waters lacking wood debris, roots, or other obstructions. Often neither method is satisfactory for highly benthic species (e.g., sculpins), except in very small streams. Turbidity, temperature, and water chemistry can also affect the efficiency of these methods. Visual observation from the banks or by divers with mask and snorkel may be effective in some situations (see below, Exercise 2), and involve less physical disturbance of fishes than more intrusive physi-

---

[1]One way to simplify and shorten this exercise is to restrict sampling to one type of habitat, such as pools.

cal sampling techniques. Nonintrusive visual methods are highly preferred where threatened or endangered species are present. Preparatory work may be necessary to confirm the effectiveness of the proposed sampling technique, and researchers must be trained in fish identification and safe capture, handling, and release procedures.

1. Prior to sampling, set blocking nets at the upstream and downstream ends of the habitat to be sampled.

2. Record local water temperature, turbidity, and weather conditions.

3. Using seines or electrofishing gear, sample each habitat. Repeat sampling passes, allowing at least 5 min between passes, until two full passes are completed with no individuals of the target species captured.[2]

4. Hold all captured fish in buckets or a secure net pen in a nearby portion of the stream. Fish sampling and postcollection data recording will take about 30–60 min for each habitat sampled, or 3–6 team-hours to sample five pools and five riffles in a stream reach. Difficult sampling conditions or larger streams will increase the time necessary.

5. Within each habitat sampled, record size and species of each fish captured. It may not be necessary to anesthetize fish since minimal handling will be required. Proper use of fish anesthetics can be tricky, and is probably best avoided by untrained users. The result of this biological sampling is a record of number and sizes of all individuals of each fish species captured in each sampled habitat unit.

Population Estimation   The Zippin removal method (Zippin 1958, Everhart *et al.* 1975) of population estimation is a simple method to analyze data from multiple-pass sampling methods. In this method, the rate of removal of fish (without replacement) from the population per increment of sampling effort is simply extrapolated to determine the total catch if sufficient effort were applied to completely deplete the population. This method assumes that the population is closed, that all fish in the population have equal chance of capture, and that the chance of capture remains equal from sample to sample.

1. Plot on standard graph paper the cumulative catch from each sampling pass on the *y*-axis against the cumulative catch from all previous passes on the *x*-axis. (The cumulative previous catch will be zero for the first pass, of course.)

---

[2]If no individuals of the target species are captured in the first two passes, make a third pass before moving on to the next habitat.

2. Fit a straight line to the three points. Where this line intercepts the *x*-axis is the estimate of population size (for total cumulative catch) had the population been sampled to depletion (*y*-axis = 0).[3]

3. Apply this procedure to each species separately in each sampled habitat unit to obtain an estimate of total abundance and aerial density (population estimate divided by habitat area in square meters).

4. Calculate mean density (or abundance) and variances for each major habitat category (e.g., pools, riffles, side channels, etc.). If multiple size or age classes are evident, separate population estimates should be made for each.

Visual estimates from snorkeling or bank observations, or sampling of benthic fishes using kick nets or other techniques, may be less amenable or unsuitable for the removal method of population estimation. For these methods, researchers can assume that for a given sampling effort proportional to the area and complexity of habitat available, counts are a sufficient estimate of relative abundance and distribution of fishes among habitat types. In this case, specific care should be given to ensure that sampling effort is even and appropriate among habitats. In some cases it may be best to combine counts from different techniques. For example, diver counts may provide a good measure of fish in the deeper parts of the water column, whereas observers walking along the shallow channel margins may provide better counts of smaller fishes in those habitats.

B. Exercise 2: Microhabitat Associations and Niche Analysis

Objectives and Design   Even if two species appear to occupy the same habitats in a stream, spatial segregation may occur at the finer scale of microhabitat patches. This is especially true among smaller and less mobile fishes (e.g., sculpins (Finger 1982) and darters (Greenberg 1988)), but even free-swimming, mid-water fishes demonstrate fine-scale partitioning of microhabitats based on water velocity, depth, cover types, and food availability (e.g., Fausch and White 1981, Glova 1986, Gorman 1987). The purpose of the following exercise is to describe patterns of microhabitat use by target species of fishes in areas of sympatry and allopatry, using a design similar to that in Exercise 1, but with sampling at a finer scale of resolution.

Microhabitat association analysis is usually best conducted in conjunction with, or as a complement to, sampling at the coarser scale of pool and riffle habitat units. Microhabitat sampling can be conducted by a separate

---

[3]If the third pass (and particularly if both the second and third passes) produce no fish of the target species, complete removal is assumed and the population estimate is calculated as equivalent to the total catch.

team in the same stream reaches at the same time as Exercise 1, or as a follow-up study at a later date. However, note that visual sampling methods (e.g., snorkeling) will require careful planning to avoid upstream activities fouling water clarity or frightening fishes for downstream observations. This will probably require separation of snorkeling observations and other sampling activities in time and/or space.

Microhabitat Sampling Methods   Various methods are available for describing microhabitat use and resource segregation in stream fishes (e.g., Finger 1982, Gorman 1987, Ross *et al.* 1992, Kessler and Thorp 1993). We suggest two approaches, recognizing that the appropriate method will depend on physical conditions of the surveyed streams, habits of the target species, and type of gear to be used.

1. Characterize and quantify the availability of microhabitat types in the study reach.

2. Using wire flags or other markers, establish grids of appropriate size so that each pool and riffle in both sympatric and allopatric zones are completely subdivided into 1 × 1-m quadrats (larger quadrats may be appropriate in streams more than 8 m wide).

3. Draw a map of each habitat, and on a mylar overlay sheet draw the quadrats and enter an index number to identify each quadrat. It may help in keeping track of data to label each wire flag with the code number of the quadrat that lies to its right and downstream.

4. At the center of each quadrat, measure and record the average depth, relative current velocity, substrata class, and cover type (Table 23.1).

TABLE 23.1

Suggested Categories and Codes for Substrata, Cover, and Velocity Measurements in Microhabitat Studies.

Substrate class	Cover class	Velocity class
1, Boulder (>100 cm)	1, Smooth and featureless	1, Still
2, Cobble (25–100 cm)	2, Cobble crevice cover	2, Slow
3, Large gravel (5–25 cm)	3, Boulder crevice cover	3, Moderate
4, Small gravel (0.1–5 cm)	4, Aquatic vegetation	4, Fast
5, Sand (0.1–1 mm)	5, Leaf litter and fine debris	5, Very fast
6, Silt (<0.1 mm)	6, Woody debris	

Note. Substrata dimensions are diameter of second-longest axis of particles; substrate class reflects areally dominant particle size category within the quadrat. Cover categories reflect presence of suitable cover of the noted type in quadrat or dominant cover type in quadrats where two or more types occur. Velocity classes are relative and should be scaled to full range of velocities present in stream reaches studied.

If a flowmeter is available, researchers can conduct precise current measurements (surface, bottom, mean water-column) instead of the relative visual categories provided.

5. After allowing at least 1 h for fish to recover from any physical microhabitat surveys (1 day or more might be better when possible), snorkel from the downstream to the upstream end of each habitat unit.

6. Survey each quadrat within a single habitat unit (pool or riffle) before moving on to the next habitat.

7. For every fish observed, the diver should report to a streambank recorder the following information: quadrat number, species, length class, depth at which the fish appears to be holding, total depth of the water column at the fish's location, distance to the nearest streambank (or water's edge), distance to the next nearest fish, and distance to the nearest obvious cover. Depending on conditions and the behavior of the fish, these observations may be best made at the point where the diver first observed the fish, or at the point to which the fish returns and holds after it is first disturbed, assuming the fish shows no signs of acute stress, shock, or fright.

8. Divers must be very careful to minimize disturbance of the habitat in order to observe fish engaged in natural behavior and positions, rather than sulking in hiding cover. Trial runs in nearby reaches will be necessary to train new researchers in the nuances of "snorkel sneaking" with minimal disturbance to fish. Disturbance of the stream by even careful divers can obscure visibility, sometimes necessitating that divers move upstream. When it works well, however, this method allows unfettered direct observations of nuances of microdistribution and behavior that no other field method can provide.

9. *Benthic species option.* For benthic species, fixed-net samples (e.g., kick nets or frame nets) may be used when snorkeling is not possible or effective (e.g., Finger 1982). Using the grid of quadrats described above, place a kick or frame net (about 1 m wide, mesh size as appropriate for target species and age classes) immediately downstream of the quadrat to be sampled. Sample the quadrat by thoroughly kicking and disturbing the gravel for 20 s. A backpack electroshocker may be used as an alternative to this approach, with appropriate adjustments in the gear to ensure a restricted zone of effectiveness.

10. All fish collected in the net should be identified, measured, and held until the entire habitat is sampled.

11. Either sample all quadrats in this manner, or if it appears that the sampling method likely interferes with adjacent quadrats, sample every other quadrat, or a randomly selected subset of them. Unlike snorkeling, this method does not allow accurate point estimates, such as distance to nearest neighboring fish or distance to nearest cover element. However,

these values can be roughly estimated by using the center of the quadrat as a reference point for all fish found within the quadrat. Because a relatively swift current is required to drive fish into the net, this method is most suitable for shallow, fast-water habitats. It can be used where high turbidity precludes visual methods.

C. Experimental Options

Habitat or microhabitat surveys alone cannot provide conclusive evidence for or against competition. Only manipulative experiments can allow scientists to fully separate competitive effects from other factors that can determine the distribution and abundance of stream fishes. For this reason, manipulative experiments are common in studies of competition in fish assemblages (e.g., Fausch and White 1981, Finger 1982, Reeves et al. 1987, Greenberg 1988). Experiments provide the opportunity to measure the influence of various interspecific interactions on distribution, growth, survival, and behavior. However, experimental environments and treatments also can introduce artifacts that cause their results to be unrealistic or misleading. This is grounds for forethought and critical evaluation in experimental design and development, and for care in nesting experiments in the context of good field studies.

Properly designed experiments can be elaborate and difficult unless special facilities and plenty of time are available to researchers. In this section we briefly describe the utility of laboratory and field experiments in competition studies of stream fishes. For a more thorough presentation of this topic, see Gelwick and Matthews (1993). The appropriateness of any of these methods will depend on the target species, the character of their natural environment and interactive relationships, and the specific questions of interest. Ideally, experiments allow researchers to control for all or most factors that could confound the search for competitive effects (e.g., food availability, temperature, current, presence of predators). In competition experiments, densities of potential competitors are manipulated and the response of the other species is measured in terms of distribution, growth, or behavior. Experiments can range from manipulation of specific conditions in natural streams, with relatively little control over extraneous variables, but substantial realism, to creation of artificial habitats in the laboratory with a high degree of control over the environment and organisms, but with the loss or severe simplification of some elements and processes that may be important in natural systems. Experimental setups can provide the rare opportunity for close visual observation of the behavior of cryptic fishes or species that normally inhabit highly turbid waters, which might allow some experiments that would be virtually impossible in the field.

Natural Stream Experiments Experimental manipulation of natural streams provides naturally complex conditions, such as light, food, temperature, cover, and spatial scale, but establishing true replicates and controls can be difficult due to the wide range and multidimensionality of natural variation in these factors. Two general experimental approaches are species removals from open stream habitats (e.g., Greenberg 1988) and establishment of enclosures within which densities are altered. If a species can be effectively removed from an area of sympatry, surveys can be made to measure the microhabitat use response of the second species. Complete removal of a species can be very difficult, however, especially for a species that frequently moves across habitats or is hard to capture because of their ability to escape. With such a design, it may be several days or longer before a response occurs or is detectable. Enclosure experiments have the benefit of a known species pool, control of emigration and immigration, and more control over the habitat conditions included. They can also be designed to produce a rapid response from the test subjects. For example, D. Stouder (personal communication) constructed 1 × 1-m enclosures across a depth gradient and measured habitat selection by fish on the basis of 1-h observations.

Laboratory Tank Experiments Laboratory tank experiments provide the best opportunity to perform competition experiments. Although they are easy to set up and control, they suffer from obvious oversimplification and represent poor caricatures of running-water habitats. Some stream fishes are difficult to maintain in small tanks, and others may exhibit abnormal behavior. However, small species adapted to slow-water habitats can sometimes be used successfully. For example, Gorman (1988) used 75-liter (20-gallon) aquaria to examine the influence of interspecific competition on the vertical distribution of small fishes in the water column. He introduced fishes into aquaria 1 day prior to each experimental trial and made several repeat observations at intervals of 10 to 30 s to record the position of each fish. Treatments were composed of several densities and species combinations. Fish size, lighting, substratum composition, depth, and temperature were controlled in an effort to mimic natural conditions as closely as the setup would allow.

Greenberg (1988) examined competition for shelter between two benthic fishes in indoor "Living Streams," which are commercially produced circulating tanks measuring 145 L × 60 W × 50 D cm. Artificial shelters constructed of PVC tubes were placed in streams where the water was 16 cm deep. Greenberg augmented the current using hoses, with flows ranging from 0 to 18.5 cm/s. He used this setup to examine use of shelters by each species in allopatry at low (4 fish per tank) and high (8 fish) densities, and in sympatry (4 individuals of each species). Smaller, conventional aquaria with flow-through arrangements might be suitable for similar work in the laboratory using small fishes, such as sculpins or darters.

Some commercially available tropical fish species would also be suitable for laboratory tank experiments, and might be easier to maintain under indoor conditions than many native fishes. However, it is likely to be difficult for beginning researchers to relate the results of experiments using unfamiliar species to natural stream ecosystems without access to and experience with the natural habitats. Compiling background research on the species used, their native habitats, and what is already known about their habits and possible interactions could overcome some of these limitations. Likewise, a laboratory experiment using exotic aquarium fishes may be especially worthwhile if there is no comparable opportunity to work with native systems and species.

Artificial Stream Experiments   Large artificial streams, located indoors or outdoors, offer excellent opportunities for control over environmental conditions, the species pool, and emigration or immigration. They can be replicated and yet provide a reasonably complex and semirealistic proxy to the natural stream environment (Gelwick and Matthews 1993). If artificial streams are available, laboratory exercises using them could include short-term studies of habitat use and behavior (e.g., Taylor 1992), or long-term studies of growth (e.g., Reeves *et al.* 1987). Because a wide variety of manipulations are possible with experimental stream channels (e.g., light, temperature, food), the appropriate design for a lab exercise will depend on linking the laboratory work to the results of prior field exercises, and will also depend on the particular nature and capabilities of the artificial stream system available.

D. Analytic Methods

Habitat Associations   The Strauss (1979) linear selection index, originally developed to measure dietary selectivity, is a simple and useful tool for comparing the distribution of fishes among habitat types. It measures the degree to which fish select a particular category or subset of a resource relative to the range of other categories that is theoretically available to them in the study area. For each species and habitat, the selection index (L) is calculated as

$$L = r_i - p_i, \tag{23.1}$$

where r_i represents the relative proportion of the total population of that species found in habitat type (i); and p_i, the relative proportion of habitat type (i) in the stream survey reach.

The index ranges from a minimum of near -1.00, indicating a habitat that is extremely abundant but which is completely avoided by fish, to a maximum approaching $+1.00$, indicating a habitat that is extremely rare but is used exclusively by the species of concern. A value of 0 indicates that the habitat is used roughly in proportion to its abundance in the stream. Examine this index to determine whether the target species exhibit similar or dissimilar habitat associations, and whether these associations vary in the presence or absence of potentially interacting species. Selection indices can be calculated at coarse (pool and riffle) or fine (microhabitat) scales, depending on the availability of data, but requires that habitats be classified categorically (e.g., continuous variables such as depth must be converted to discrete depth classes).

Microhabitat Associations and Comparisons   Univariate analysis of the associations of species with microhabitat gradients is a good starting point for analysis of microhabitat data from field or laboratory studies. Construct histograms of the relative frequency of occurrence of each species in each stream within categories of each microhabitat variable. Continuous data, such as depth, should be subdivided into discrete classes for categorical analysis. Chi-square contingency tests (see Zar 1984) are a simple way of testing whether observed associations differ significantly from the distributions expected, assuming that fish were randomly distributed among habitat categories. Nonparametric tests (e.g., the Mann–Whitney U test and Kruskal–Wallis test) can be useful and are easily performed to examine whether observed microhabitat associations differ between streams with different species complements. Parametric tests (analysis of variance) are sometimes applicable, but require more careful attention to assumptions about data structure, and most data will require transformation before such analysis is appropriate (Zar 1984).

Multivariate analysis can be useful for exploring microhabitat data, but these methods can be complex, especially when they require combining data of categorical and continuous nature and varying or unknown normality and variance. We refer readers interested in multivariate analysis to specialized texts such as Green (1979), Gauch (1982), or Ludwig and Reynolds (1988) for further information.

Niche Analysis   Analysis of niche structure is a useful way of examining patterns in microhabitat use and drawing inferences about the magnitude and effects of possible competition (Schoener 1970). Niche breadth (β)

indicates the scope of resource or habitat use by a single species along one environmental gradient or niche dimension and is calculated as

$$\beta = \frac{1}{\sum p_i^2},$$ (23.2)

where p_i is the proportion of records for the species in each category (i) of a particular resource or habitat variable.

Niche breadth is low when the species occurs in just a single category of the resource or habitat variable of interest and is high when the species is spread equally over many or all categories of the variable. Presumably, the effects of interspecific competition would include compression of niche breadth of one or both species through competitive exclusion from some microhabitats. Keep in mind, however, that a species can simultaneously have low niche breadth for one microhabitat dimension (e.g., depth) and high niche breadth for another dimension (e.g., cover).

Niche overlap measures the overlap of microhabitat use between species. A high value of niche overlap indicates substantial potential for cooccurrence and interaction among the species in the assemblage. A low value of overlap, on the other hand, may indicate that past or ongoing competition or other factors have resulted in a high degree of partitioning of habitat among coexisting species. The principle question is whether niche overlap between species changes in the presence of another species and whether overlap differs among different habitat factors. The most widely used measure of niche overlap (O_{jk}) is from Schoener (1970),

$$O_{jk} = 100 \times \left[1 - \left(\frac{1}{2} \times \sum |p_{ij} - p_{ik}| \right) \right],$$ (23.3)

where the overlap is measured between species j and species k; p_{ij}, the proportion of species (j) found in category (i) of a particular habitat dimension; and p_{ik}, the same measured for species (k).

Niche overlap ranges from 0% if the species are completely separated among microhabitats to 100% if they are distributed equally among the same microhabitat types. These niche dimensions can be calculated for any set of species and species pairs, depending on the specific question or hypothesis of interest.

IV. QUESTIONS

1. Did any of the target species show associations with particular habitat types? Were these associations positively correlated between species, or were they complementary?

2. Did the pattern or strength of habitat type associations appear to shift in the presence of a potential competitor? What kind of behavioral response might this shift (if any) indicate to a potential competitor?

3. Did habitat associations and possible shifts in response to a competitor species vary by size class of the fishes of concern? Do your results and interpretations change if you conduct separate analyses by size classes?

4. Can you infer from these data whether any changes in habitat association reflect direct competitive interactions for food or space, behavioral avoidance, predation, or some other interaction between the species? What criteria might you use to make such a judgment? What kind of additional data might you need? What ancillary information on life history, behavior, reproduction, food preferences, and other aspects of the biology of these fishes is helpful in determining the possible importance of competition among them?

5. If you calculated variance among habitat patches, and/or confidence limits on your population size estimates, is potential error in these estimates great enough to confound your results and interpretations? How might you reduce this source of error to improve your ability to discern patterns in habitat use?

6. Does the use of any one microhabitat variable change in the presence of the potential competitor? Is this reflected in any other microhabitat variable? Can you sort out which factors might be "causitive," and which might be simply correlative?

7. Does the breadth or overlap of microhabitat use change in the presence of the potential competitor?

8. Does microhabitat selectivity change in the presence of the potential competitor (i.e., niche shifts)?

9. Are there size-based differences, in addition to species-specific differences, in microhabitat use and habitat segregation? Do your results for the effects of the potential competitor on niche breadth, overlap, and selectivity change if you account for size class variation?

10. If you detect evidence of niche shifts, do the data suggest whether the shift is driven by competitive interactions for habitat, or perhaps by some other interaction such as predation? What evidence can you point to? Did you note anything about behavior of these fishes during field observations that might be helpful to answer this question?

11. Do the results of microhabitat-level analyses appear to be consistent with those from coarser-scale (pool and riffle) analyses? If not, how can you explain any discrepancies? Is it possible for competition and other interspecific interactions to have scale-specific effects? What does this mean for interpretation and design of ecological studies?

12. Did any laboratory studies conform to expectations generated by field studies? If not, why not? What spatial scale of habitat do the laboratory systems compare to in the field, and how might limitations of scale affect the results?

V. MATERIALS AND SUPPLIES

Sampling at Pool and Riffle Level

Bag seine of a length, height, and mesh size appropriate to study streams; or

Blocking nets (two)

Buckets (two perforated to allow free flow of stream water for fish recovery before release)

Data forms for recording sampling effort and catch

Electrofishing gear (optional)

Fish measuring board

Insulated minnow buckets with battery-operated aerators for holding containers

Small aquarium net (for transferring fish)

Thermometer

Microhabitat Level Sampling

Current velocity meter

Meter sticks for recording depth and velocity

Permanent ink markers or wax pencils for labeling markers with grid location

Wet- or dry suits, masks, and snorkels (two sets)

Wire flags or other visible markers (250 to 500 will be required)

Benthic Species Option

Kick net or frame net, or electrofishing gear, in lieu of diving gear; otherwise same as microhabitat list above

REFERENCES

Connell, J. H. 1980. Diversity and the coevolution of competitors, or the ghost of competition past. *Oikos* **35**:131–138.

Douglas, M. E. 1987. An ecomorphological analysis of niche packing and niche dispersion in stream fish clades. Pages 144–149 *in* W. J. Matthews and D. C.

Heins (Eds.) *Community and Evolutionary Ecology of North American Stream Fishes.* Univ. of Oklahoma Press, Norman, OK.

Everhart, W. H., A. W. Eipper, and W. D. Youngs. 1975. *Principles of Fishery Science.* Cornell Univ. Press, Ithaca, NY.

Fausch, K. D. 1988. Tests of competition between native and introduced salmonids in streams: What have we learned? *Canadian Journal of Fisheries and Aquatic Sciences* **45:**2238–2246.

Fausch, K. D., and R. J. White. 1981. Competition between brook trout *(Salvelinus fontinalis)* and brown trout *(Salmo trutta)* for position in a Michigan stream. *Canadian Journal of Fisheries and Aquatic Sciences* **38:**1220–1227.

Finger, T. R. 1982. Interactive segregation among three species of sculpins *(Cottus).* *Copeia* **1982:**680–694.

Fisher, W. L., and W. D. Pearson. 1987. Patterns of resource utilization among four species of darters in three central Kentucky streams. Pages 69–76 *in* W. J. Matthews and D. C. Heins (Eds.) *Community and Evolutionary Ecology of North American Stream Fishes.* Univ. of Oklahoma Press, Norman, OK.

Frissell, C. A., W. J. Liss, M. D. Hurley, and C. E. Warren. 1986. A hierarchical framework for stream habitat classification: Viewing streams in a watershed context. *Environmental Management* **10:**199–214.

Gauch, H. G. 1982. *Multivariate Analysis in Community Ecology.* Cambridge Univ. Press, New York, NY.

Gelwick, F. P., and W. J. Matthews. 1993. Artificial streams for studies of fish ecology. *Journal of the North American Benthological Society* **12:**313–384.

Glova, G. J. 1986. Interaction for food and space between experimental populations of juvenile coho salmon *(Oncorhynchus kisutch)* and coastal cutthroat trout *(Salmo clarki)* in a laboratory stream. *Hydrobiologia* **132:**155–168.

Gorman, O. T. 1987. Habitat segregation in an assemblage of minnows in an Ozark stream. Pages 33–41 *in* W. J. Matthews and D. C. Heins (Eds.) *Community and Evolutionary Ecology of North American Stream Fishes.* Univ. of Oklahoma Press, Norman, OK.

Gorman, O. T. 1988. An experimental study of habitat use in an assemblage of Ozark minnows. *Ecology* **69:**1239–1250.

Green, R. H. 1979. *Sampling Design and Statistical Methods for Environmental Biologists.* Wiley, New York, NY.

Greenberg, L. A. 1988. Interactive segregation between the stream fishes *Etheostoma simoterum* and *E. rufilineatum. Oikos* **51:**192–202.

Herbold, B., and P. B. Moyle. 1986. Introduced species and vacant niches. *American Naturalist* **128:**751–760.

Kessler, R. K., and J. H. Thorp. 1993. Microhabitat segregation of the threatened spotted darter *(Etheostoma maculatum)* and the closely related orangefin darter *(E. bellum). Canadian Journal of Fisheries and Aquatic Sciences* **50:**1084–1091.

Ludwig, J. A., and J. F. Reynolds. 1988. *Statistical Ecology: A Primer on Methods and Computing.* Wiley, New York, NY.

Reeves, G. H., F. H. Everest, and J. D. Hall. 1987. Interactions between the redside shiner *(Richardsonius balteatus)* and the steelhead trout *(Salmo gairdneri)* in western Oregon: The influence of water temperature. *Canadian Journal of Fisheries and Aquatic Sciences* **44:**1603–1613.

Ross, S. T. 1991. Mechanisms structuring stream fish assemblages: are there lessons from introduced species? *Environmental Biology of Fishes* **30:**359–368.

Ross, S. T., J. G. Knoht, and S. D. Wilkins. 1992. Distribution and microhabitat dynamics of the threatened bayou darter, *Etheostoma rubrum. Copeia* **1992:**658–671.

Schoener, T. W. 1970. Non-synchronous spatial overlap of lizards in patchy habitats. *Ecology* **51:**408–418.

Seegrist, D. W., and R. Gard. 1972. Effects of floods on trout in Sagehen Creek, California. *Transactions of the American Fisheries Society* **101:**478–482.

Strauss, R. E. 1979. Reliability estimates for Ivlev's electivity index, the forage ratio, and a proposed linear index of food selection. *Transactions of the American Fisheries Society* **108:**344–352.

Taylor, E. B. 1992. Behavioral interaction and habitat use in juvenile chinook *Oncorhynchus tshawytscha,* and coho, *Oncorhynchus kisutch. Animal Behavior* **42:**729–744.

Werner, E. E., and D. J. Hall. 1976. Niche shifts in sunfishes: experimental evidence and significance. *Science* **191:**404–406.

Zar, J. H. 1984. *Biostatistical Analysis,* 2nd ed. Prentice–Hall, Englewood Cliffs, NJ.

Zippin, C. 1958. The removal method of population estimation. *Journal of Wildlife Management* **22:**82–90.

CHAPTER 24

Stream Food Webs

ANNE E. HERSHEY* AND
BRUCE J. PETERSON†

*Department of Biology
University of Minnesota-Duluth
†The Ecosystems Center
Marine Biological Laboratory

I. INTRODUCTION

An appreciation for stream food webs is essential for integrating studies of organic matter processing and community interactions. Food webs differ in structure and function among stream types, although they will all have some common elements. Here we describe gut contents analyses and ^{15}N tracer studies that can be used to describe food web structure and nitrogen flow through food web components in a variety of stream types.

When we study stream food webs, the initial objectives are to identify the principle sources of organic matter, assign consumers to trophic levels within the web, and identify the major food sources for each of these consumers. Most streams have approximately three or four trophic levels, but occasionally fewer or more may be present. Primary producers, including algae, bryophytes, and vascular macrophytes, occupy the lowest trophic level. However, detritus also belongs to this lowest trophic level. Defining the higher trophic levels gets increasingly difficult. There are groups of macroinvertebrates and some vertebrates that we readily characterize as grazers and detritivores, apparently a primary consumer trophic level. However, both the producers and especially the detritus are intimately associated with heterotrophic microbes, which more clearly belong to the primary consumer trophic level. Thus, macroconsumers that feed on aquatic plants or plant detritus also ingest stream microbes and thereby function as both primary and secondary consumers. We could say they are somewhere between trophic levels 2 and 3. This places primary predators between levels 3 and 4. However, many of them also have mixed diets of detritus, diatoms, and animal prey. Finally, predators that feed on other predators nearly always have mixed diets to include algivores and detritivores, as well as other predators. To what trophic level should these predators be assigned? Detritivores complicate this web even further since there are several functional feeding groups of detritivores (shredders, collector–filterers, collector–gatherers; see Chapter 21). The detritus fed upon by these groups will vary tremendously in both the mass proportion of the microbial component as well as the importance of this microbial component in the detritivore's diet. This is true not only among functional feeding groups, but also among species within a group.

Species comprising stream food webs are constrained by many factors, which then determine the structure and function of the food web of a particular stream. Such factors include biogeography (geographic distributions of species), geomorphology, substratum characteristics, and gradient (Gregory et al. 1991), riparian characteristics (Cummins et al. 1989), temperature (Ward and Stanford 1982), and interspecific interactions (Power 1990). The food web in any particular stream will reflect all of these factors, and among streams, a wide variation in food webs can be found.

Two factors that greatly influence the organic matter sources for the food web, and thus the structure and function of the web, are canopy cover and gradient. Although often interrelated in a particular geographic region, the relationships change dramatically across regions, and thus it is important to consider these factors separately. For example, in some forested regions first-order streams may be high gradient with closed canopy, but in other forested regions they may be low gradient with closed canopy. Alternatively, prairie, arctic, or desert first-order streams usually have an open canopy. Low-gradient streams often have unconsolidated substrata (sand, peat, etc.), whereas high-gradient systems are typically characterized by consolidated substrata (bedrock, boulders, cobble). Substratum type has a strong effect on the fauna and flora present and, therefore, on the food web structure. Canopy, riparian zone, and watershed characteristics determine the light, nutrient, and organic matter inputs, which are major determinants of the energy base for the food web.

Three general approaches might be used to measure food web structure and dynamics: (1) gut analyses, (2) carbon or energy budgets (see Chapter 28), and (3) stable isotope tracer studies. Note that these approaches are generally complementary and to some extent overlapping, but provide slightly different types of information. Thus, in undertaking food web studies in a stream system the approach used will depend on the questions of interest and the resources available. Here, we focus on approaches (1) and (3).

Approaches to studying food webs may be either qualitative or quantitative. A qualitative approach identifies components and linkages or connections within the web, without ascribing strengths to the interactions. Various properties of a food web assembled in this manner may be calculated to reveal patterns within the web (Pimm 1982). This approach has only recently been applied to a lentic food web (Martinez 1991), and has not been adopted by stream ecologists, likely because of the high species diversity, high trophic overlap, and difficulties with taxonomic resolution that are characteristic of lotic food webs. However, knowledge of the structure of the food web can lend insight into the function of the food web by providing a template for developing testable hypotheses about specific aspects of community or ecosystem function.

A first step in any food web study is to identify trophic levels and linkages. Sampling of organic matter sources and consumer components across habitats is essential regardless of the approach used. Qualitative and quantitative sampling approaches for primary producer (Chapters 13 and 14), detrital (Chapters 10, 11, and 28), microbial (Chapter 12), and macroconsumer (Chapters 16, 17, and 18) pools are described elsewhere in this volume and will not be reiterated here. Quantitative study of food webs typically requires years of intensive work for even one stream and involves many different types of approaches, including production studies (Chapters 25 and 26), and development of organic matter or energy budgets (Chapter 28). In addition to an intensive sampling effort over at least one annual cycle, quantification of food web transfers often also involves feeding experiments or measurements of metabolic or digestion rates. In this chapter, it is our objective to outline approaches used to identify linkages, focusing on gut content analyses, natural abundances of stable isotopes in food web components, and experimental use of ^{15}N as a food web tracer. We do not attempt to illustrate quantification of food web linkages.

Gut analyses are often conducted to determine major food sources for consumers. Although much information can be gained in this manner, there are also some cautionary notes to be considered. First, for most stream consumers, gut contents will underestimate both the biomass consumed as well as the variety of components, since some diet items may go unrecognized, or only soft parts may be ingested. Thus, this approach provides a minimum estimate of the diet of a consumer. Second, consumer diets may change dramatically with seasonal availability of food, ontogeny, or even on a diel basis, requiring a long-term and comprehensive study to fully characterize consumer diet items. Third, many consumers ingest material that is difficult to identify. Even a cursory gut analysis effort will result in a diet category labeled "other," or "amorphorus detritus," or both. Amorphorus detritus likely includes material of DOM origin that has become incorporated into the FPOM pool, as well as thoroughly processed FPOM that is impossible to characterize as to source. Fourth, depending on the techniques used, gut analyses can overlook important information. For example, bacteria are not evident in guts unless properly preserved and stained, yet may be extremely important numerically and nutritionally (see Chapter 12). Many fish and some predatory insects swallow prey whole. In these cases, examination of guts provides a good indication of the items consumed, but unless digestion rate is determined and size of prey measured, these data do not quantify the linkages between a predator and its prey community.

Even with these constraints, gut analyses can be used to construct food web diagrams. Pimm (1981) refers to such diagrams as "caricatures of nature," but notes that they still contain much information that is of value. As well as providing information on linkages within a system, food web diagrams can be used to construct hypotheses about population, community, or ecosystem processes and dynamics that can then be tested by manipulating components of the web. For example, if we hypothesized that a stream food web was constrained by habitat heterogeneity, then a test of this hypothesis could be to experimentally increase habitat heterogeneity (e.g., add woody debris) and then resample food web components some time later and reconstruct the food web. Does the food web look similar after habitat manipulation than it did before? Comparison of food webs from different streams is also very important in construction of hypotheses regarding differences between streams, and/or controls of food web structure and function within one stream of interest. To return to the heterogeneity example, say we had two streams that differed in food web complexity, one of which had considerable woody debris and another that was clear of woody debris. We could hypothesize that the food web disparity was due to this difference in debris between the streams and then test the hypothesis by either adding or removing debris from one (or both) streams, and observing changes in food web complexity over time. A common exercise used in field courses is to compare the biota of a perturbed stream (e.g., by urban runoff, agricultural runoff, forest clearcutting, etc.) to that of a pristine stream. Food web diagrams can then be constructed from these data, and hypotheses can be generated regarding the mechanisms contributing to observed differences. Frequently, however, these hypotheses are difficult to test explicitly because the streams may differ in ways other than the perturbation in question.

Stable isotope analyses are an additional independent way of following the transfers of organic carbon and nitrogen from plant and detrital sources to primary and secondary consumers (Peterson and Fry 1987, Fry and Sherr 1984, Minagawa and Wada 1984; see Section II.C.2 below, for definition and discussion of isotope terminology). In many ecosystems the organic matter sources have different $^{13}C:^{12}C$ and $^{15}N:^{14}N$ ratios, and the diets of animals can be inferred from the isotopic ratios in the animal tissue. The reason for this is simply that you are what you eat. If algae with $\delta^{13}C = -30$‰ and $\delta^{15}N = 0$ are the sole food of an animal, the isotopic composition of the animal is predicted to be $\delta^{13}C = -30$ to -29 and $\delta^{15}N = +2.5$ to $+3.5$. The ranges given illustrate that there are some small and variable isotopic shifts (or fractionation) due to animal metabolism of carbon and nitrogen compounds. Animal tissues are usually just slightly enriched (1.0‰ or less) in ^{13}C relative to their food, but significantly enriched (2.0 to 4.0‰) in ^{15}N relative to their food. The trophic enrichment in ^{15}N is sufficiently predictable that it is often used as one indicator of trophic level (Minagawa and Wada 1984, Fry 1991). For example, an animal with a $\delta^{15}N$ value of 6 would be considered to occupy a higher trophic level than one with a $\delta^{15}N$ value of 3.5. Since there is little trophic transfer shift in carbon isotopes, but a large and predictable shift in nitrogen isotopes, the combination of C and N isotopes is frequently used as an aid to determine both pathways of organic matter transfer and trophic structure in ecosystems.

Ecosystems sometimes contain natural isotopic distributions that allow an easy differentiation of organic matter sources for different consumers. For example, in a grassland stream, insects might be utilizing either detritus from streamside grasses, or epilithic diatoms as food. If the grass has a $\delta^{13}C$ value of -14 and the diatoms have a value of -30, it will be easy to distinguish the food source. Probably some species of insects will be close to -14 (the detritivores) while others will be close to -30 (grazers), and still others will be in between, reflecting a mixed diet. Top predators are frequently intermediate in ^{13}C values because they will likely feed on both grazers and detritivores. One might ask why isotopes are useful if you can tell on morphological bases (the functional feeding group concept; see Merritt and Cummins 1996, and Chapter 21) that one species is a grazer and another is a detritivore. The answer is that feeding types based on morphology do not always agree with animal diets. Also consider the concepts that diatom detritus is potentially mixed with grass detritus in FPOM on the river bottom and in transport and also that dissolved organic matter leaching from the grassland into the stream will become incorporated into the epilithic biofilm where the grazer is feeding. Thus, what initially looks like a clear-cut situation may in fact be more complex. The isotopic composition of the dissolved organic matter leaching from the grasses or from diatoms will likely reflect the isotope ratio of the different sources quite well. On the other hand, microscopic analyses of the grazer gut contents would be unlikely to detect dissolved or colloidal organic matter of higher plant origin. For reasons such as this a combination of approaches is most powerful.

Some ecosystems have several organic matter sources that have almost identical isotopic compositions. In these situations, the determination of sources assimilated by consumers from natural isotopic abundance is probably impossible. In such a case, one option is to deliberately introduce an isotopic signal that can be followed throughout the food web. One example would be to add corn leaf detritus ($\delta^{13}C = -13$) to a stream reach that normally receives oak leaves ($\delta^{13}C = -27$). While this might be an interesting experiment, it suffers the objection that corn is exotic to the system and may not cycle like oak detritus. Another approach would be to add an inorganic nutrient that is highly enriched or depleted in heavy isotope. For example, if you add isotopically enriched dissolved ammonium to a stream, the algae and bryophytes that assimilate this nitrogen will be isotopically enriched and very soon the consumers utilizing these components will also become enriched. Conversely, detritus entering from the riparian zone will not be enriched and detritivores that specialize in using this material may be only slightly enriched. This type of experiment can yield information about nitrogen biogeochemistry and spiralling as well as trophic structure. When applied to a small unpolluted, low-nitrogen stream, the reasonable cost of the isotope and the ease of sampling make this a viable alternative for determining N flow pathways, which are clearly related to trophic structure. One disadvantage is that this experiment requires more equipment and time than the natural isotope distribution approach.

II. GENERAL DESIGN

Based on field samples from one or more streams, food web diagrams should be constructed by one or some combination of the methods outlined below. This will include taxonomic evaluation of samples and gut contents. If organic matter sources with distinct N or C signals are present in an available study site, this could be exploited for food web studies. Depending on resource availability, a stable isotope tracer experiment could be conducted to measure movement of nitrogen (or carbon) through the web.

A. Site Selection

For the study of stream food webs it makes the most sense to choose sites for which the maximum amount of data are available. Thus, if there are streams for which you have extensive nutrient or invertebrate data, food web studies on these streams would likely be more meaningful than in streams not previously studied. Choice of sites, however, obviously must depend on the question being asked. If two or more streams are used, they should be selected to illustrate the range of food web structures available. Suggestions for contrasting pairs of streams include low-order versus high-order, low-gradient versus high-gradient, forested versus prairie stream, and urban versus pristine stream. Unless cost is not a factor, inclusion of a low-order stream is essential if a whole-stream stable isotope tracer addition experiment is to be performed (Exercise 3 below), because low discharge is crucial to minimizing isotope costs while introducing a strong enough signal to be useful as a tracer.

B. General Procedures

Conventional Food Web Diagrams   You will construct a food web diagram based on your sampling of stream ecosystem components and gut analyses. Include about four trophic levels in your diagram. Diagrams should include organic matter sources, consumers (positioned according to trophic level), and linkages between source and consumer components.

Fractionation of Isotopes and Equations   The elements carbon and nitrogen both have heavy and light isotopes that can be used to follow the flow of these elements in ecosystems. The ratios of $^{15}N : {}^{14}N$ and $^{13}C : {}^{12}C$ in components of ecosystems vary in predictable ways as illustrated above. The ratios in environmental samples can be measured with great accuracy (a tenth of a part per thousand deviation from a standard) through the use of mass spectrometers that determine the isotope ratios in the gas from a combusted sample and compare it to a standard. The standards are carbon from carbonate rock (the Peedee Belemnite formation) and nitrogen in air, for C and N, respectively. Results are usually expressed as del or δ (which refers to deviation) values in parts per thousand (‰, also termed "per mil") difference between sample and standard ratios according to the formula

$$\delta^{13}C \text{ or } \delta^{15}N = [(R_{sample} - R_{standard})/R_{standard}] \times 1000, \qquad (24.1)$$

where $R = {}^{13}C/{}^{12}C$ or $^{15}N/^{14}N$. Thus, samples enriched in ^{13}C or ^{15}N are isotopically "heavy" and have higher δ values, whereas samples depleted in ^{13}C or ^{15}N are isotopically "light" (relatively rich in the lighter isotopes ^{12}C and ^{14}N) and have lower δ values.

If all environmental samples had identical stable isotope ratios, there would be little information (no signal) in isotope data. However, as carbon and nitrogen cycle in ecosystems the elements undergo fractionation during certain processes. This means that during those reactions or processes the light and heavy isotopes move at slightly different rates, with the result that the donor and recipient pools or components end up with different isotope ratios. The common example for carbon is the approximately 20‰ fractionation in CO_2 uptake by trees. The $\delta^{13}C$ value for trees of about -27 to -28‰ is less than the -8‰ value for atmospheric CO_2 because ^{13}C diffuses and reacts more slowly than ^{12}C during stomatal passage and photosynthesis. This fractionation accounts for the consistent large difference between the carbon isotope ratios in the atmospheric CO_2 pool and the terrestrial biota. In lakes and streams the $\delta^{13}C$ value of dissolved inorganic carbon (DIC) varies considerably because stream and lake waters are not usually in equilibrium with the atmosphere. Thus, in our example below we assign stream algae a $\delta^{13}C$ value of -35‰, which is feasible if the stream is supersaturated with CO_2 derived from decomposition of terrestrial detritus ($\delta^{13}C = -28$). DIC in such a stream might average -15‰, rather than the atmospheric value of -8‰, reflecting that it is derived from both atmospheric and respiratory CO_2 sources. Several factors can affect algal fractionation of C, but if, as an example, algal fractionation of C is 20‰ the ^{13}C signal should be -15–$20 = -35$‰. In contrast, nitrogen fixation by microbes and plants often exhibits little fractionation and it is not uncommon for plants to have $\delta^{15}N$ values close to the 0‰ atmospheric value. However, microbial processes such as nitrification and denitrification, and animal metabolism, fractionate nitrogen isotopes sufficiently such that all ecosystems contain components with significant (>1‰) variation in N isotope ratios.

An example of a hypothetical stream food web analyzed using stable isotopes is shown in Fig. 24.1. Initial study of this stream has shown that it receives large amounts of leaf detritus, but also has sufficient light and nutrient input to support a benthic diatom community. Samples of detritus, epilithic algae, insects, and fish have been collected and analyzed for C and N isotope ratios. As expected, the tree leaf detritus had a $\delta^{13}C$ value of -28 and a $\delta^{15}N$ value of 0. Absolutely clean samples of diatom cells are almost impossible to collect in the field as the cells grow in an epilithic or epibenthic matrix of microbial slime and detritus, but for illustrative purposes we assign them a $\delta^{13}C$ value of -35 and $\delta^{15}N$ value of $+2$‰. Assuming a literature-based trophic transfer shifts of $+0.5$‰ for C and $+3$‰ for N, the predicted values for insects with contrasting feeding modes and predators is shown in Fig. 24.1. Note the wide separation in $\delta^{13}C$ values for consumers specializing on diatoms versus detritus, clearly indicating their organic matter sources. Also note that predators have higher $\delta^{15}N$ values than their prey, and filter feeders that use both algal and detrital components have $\delta^{13}C$ values that are intermediate between algae and detritus, and $\delta^{15}N$ values that reflect a greater trophic shift than would be expected based on

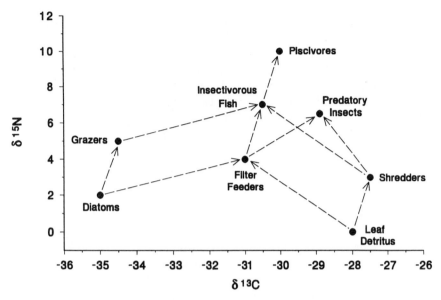

FIGURE 24.1 Sample plot of hypothetical values for $\delta^{13}C$ versus $\delta^{15}N$ for various components of a typical stream food web. Arrows indicate hypothesized trophic transfers back on a fractionation of approximately 3‰ for N and 1‰ for C for each trophic level.

detritus alone (Fig. 24.1). The system has approximately four trophic levels. While this system is oversimplified for illustrative purposes, it is not very different from what we see in many stream ecosystems (Fry 1991). In actual practice, there are likely to be more than two organic matter sources and many consumers are likely to have more generalized diets (e.g., filter feeders in Fig. 24.1), leading to less clear isotopic separation and more ambiguity in interpretation of organic matter transfer through the food web. This point emphasizes the importance of having additional information from gut contents, feeding studies, and morphological studies.

III. SPECIFIC EXERCISES

A. Exercise 1: Construction of Food Web Diagrams Based on Sampling Data and Gut Analyses

This exercise is a cursory study of organic matter sources and several specimens of a few dominant consumers to be completed by researchers working in groups for 6–8 h. A detailed, quantitative analysis would take one person several months.

1. Collect samples of food resources and consumers. In representative stream habitats, sample primary producer groups (e.g., algae, vascular macrophytes, bryophytes) in each habitat (see Chapters 13 and 14), detrital CPOM and FPOM components (see Chapters 10 and 11), invertebrate macroconsumers (Chapters 16 and 21), and fish (Chapters 18 and 22).

2. Sort invertebrate samples into functional feeding groups (Merritt and Cummins 1996, and Chapter 21).

3. Determine gut contents of representatives from each consumer group, or choose those groups that are most common if samples are very diverse.

   a. To perform gut analyses, first dissect guts from animals. The ease of this dissection is variable among invertebrates, but, with patience, can be accomplished for even small invertebrates. For many invertebrates, especially insects, the entire gut can often be extracted by holding the abdomen firmly (but not squeezing) with fine-tipped forceps, then pulling the head away from the body using another pair of fine-tipped forceps. Generally the gut will come out attached to the head. When this does not work, the gut can sometimes be pulled out of the abdomen once the head is removed, especially if an incision is made longitudinally along the ventral side of the abdomen. Use a fine, sharp dissecting pin for the incision. Care should be taken to remove the gut intact whenever possible. For taxa where the foregut is distinct from the midgut, it might be preferable to use only the foregut.

   b. Once the intact gut has been dissected, place it on a clean slide in a drop of mounting media. Tease gut contents from the gut or foregut with a fine dissecting pin, and disperse in the mounting media. Place a coverslip over the preparation, and press gently to spread the gut material, then place the slide on a slide warmer to dry.

   c. Examine the gut contents and identify components as well as possible. This is not unlike solving a puzzle, since gut contents are often fragmented. Knowledge of the stream flora and fauna, and prior microscopic examination of FPOM, is very helpful. Diatoms and filamentous algae are often in good condition, and may be identified to genus or even species using appropriate sources (see Chapter 13). Detrital material is usually very difficult to identify, but may be characterized as vascular plant detritus, animal detritus, amorphous detritus (e.g., FPOM of DOM origin), or other categories that you can recognize. Invertebrate diet items may or may not be intact depending on the consumer, but often fragments including such things as head capsules or claws, may be identifiable if the stream fauna is well known. This often involves matching fragments against comparable fragments of taxa that have been previously keyed from the habitat. Enter your findings on a data sheet such as the one shown in Table 24.1.

4. Sort and identify fish. Remove stomachs and conduct stomach analyses on these specimens (see Chapter 22), similar to invertebrate gut analyses. For predatory fish, dietary items are often intact or nearly so, and individuals can be keyed using the appropriate references (see Pennak 1989). Herbivorous and detritivorous fish are more challenging. For preliminary work, disperse stomach contents in a small amount of water in a petri dish. Pick out large items and identify using a dissecting scope, or mount on slides as necessary. Smaller amorphous material can be subsampled with a Pasteur pipet and mounted on slides. You may want to prepare several slides per animal. Tally your results on Table 24.1.

5. Based on collections of organic matter and gut analyses, categorize each animal taxon according to the specific food items consumed, and according to the appropriate trophic category. Prepare a summary table as in Table 24.2.

6. Construct a food web diagram from your data. Organize the data vertically with organic matter sources at the bottom, predators at the top, and primary consumers in between. For organic matter sources or consumer groups at the same trophic level, organize these laterally in the web at the same respective vertical level.

B. Exercise 2: Analysis of Food Webs Using Stable Isotopes

1. Prepare samples of components collected in Exercise 1 for isotope analyses, and send to a commercial laboratory for analysis of ^{13}C and ^{15}N (see Section V below). Look for a "natural experiment," either by choosing two streams in Exercise 1 that are sufficiently different to provide a different signal in the resource base, or where there is an upstream–downstream comparison involving an input of C or N (e.g., different isotopic ratio downstream than upstream).

2. Collect samples for stable isotope analyses consisting of several milligrams dry mass of leaves, mosses, algae, or animal tissues. Animals should be held in clean filtered streamwater for at least several hours to help clear their guts. Crustaceans, snails, or bivalves should be removed from carbonate shells as carbonate will have a different δ^{13}C value than animal body tissue. Care should be taken to separate individuals to species level if possible since even closely related species will often have quite different diets. Frequently, it is both necessary and desirable to pool several

TABLE 24.1
Example Data Sheet for Gut Contents Analyses

Stream or stream reach:

Habitat:

Collection date:

Investigator:

Notes:

		Gut contents for each specimen				
Specimen	Animal prey	Algae	Plant detritus	Amorphous detritus	Other	Comments
1. *Rhyacophila*	3 *Prosimulium* hypostomas	3 *Cladophora* filaments	Rare	Common		Some gut material lost during dissection.
	1 *Cricotopus* head capsule					
	2 *Tanytarsus* head capsule					
	1 *Simulium* head					
	2 *Baetis* ? claws					
2.						
etc.						

Note. For each specimen, identify the specimen and itemize each of the items found in its gut (or stomach for fish). Food items should be identified to the extent possible, and each type should be enumerated. An example is given. For items that cannot be enumerated, categorize them as abundant, common, or rare (establish criteria for these categories and use them consistently, e.g., rare, <5%; common, 5–20%; abundant, >20%).

TABLE 24.2
Summary of Food Web Data from Exercise 1

Trophic category	Taxa represented in stream	Diet items from gut analyses
Producer sources		
Detrital sources		
Primary consumers	Grazer taxa	
	Shredder taxa	
	Collector-gatherers	
	Collector-filterers	
Predators	Primary predators	
	Secondary predators	

to many individuals of a species for a single analysis. While it is interesting to
know the individual variability, the sample quantity and cost considerations
usually require pooling. Furthermore, pooling many individuals may give
a better estimate of the mean value for the population from a site. For
larger organisms, it is possible to analyze specific tissues, but in this exercise
we will use either whole body analysis or muscle tissue (for fish or crayfish).

3. Place clean samples in glass scintillation vials and dry in an oven at
60°C. Dried samples can be held indefinitely in a dessicator. Alternatively,
samples can be frozen for later dissection and drying, or can be preserved
in ethanol. Ethanol dissolves lipids from animal tissues, but is generally
satisfactory since it evaporates completely on drying, leaving little if any
detectable alteration of C or N isotope ratios.

4. Dried samples can be sent to any of several commercial or university
laboratories specializing in mass spectrometry. Sample costs are becoming
less as machines improve. Current costs range from $10 to $30 per sample
for ^{13}C, and $15 to $45 per sample for ^{15}N. The range in costs reflects the
differing instruments as well as different financial bases of the various
laboratories. By shopping around, a cost of $30 or less per sample for both
C and N ratios is achievable. If you plan to do serious work, send a known
blind sample as a standard with each shipment to increase your confidence
in the resulting data. Because isotopic analyses are expensive, it is prudent
to analyze a few preliminary samples of either organic matter sources or
various types of consumers to discover whether or not your system contains
naturally contrasting $\delta^{13}C$ and $\delta^{15}N$ values. If the system turns out to have
large signals and is well poised isotopically for your question, then analyze
additional samples.

5. Your data will arrive on a data sheet from the isotope laboratory
and will look something like those in Table 24.3. First plot all the data for
^{13}C and ^{15}N separately to look for the range of values and the pattern of

TABLE 24.3
Sample Stable Isotope Data from Commercial Laboratory

Code	Sample ID	$\delta^{13}C$	$\delta^{15}N$	Comments
1	Lab Creek oak detritus	−28.0	0.0	
2	Lab Creek "diatoms"	−35.0	2.0	Small and dirty
3	Lab Creek black flies	−31.8	4.0	
etc.				

grouping of ^{13}C and ^{15}N values (see Fry 1991 for examples). A rule of thumb is that values that are different by 1‰ or less should not be considered different. While the laboratory analyses may be more precise than that, the variability in the ecosystem and in sampling is larger. Next construct an isotope cross-plot of C versus N and draw in your hypothesized food web pathways. Do they make sense or does the isotopic data suggest a different pattern? Do the gut contents and isotope approaches agree? Do the isotopic distributions support your morphological observations and functional group classifications?

6. Usually the answers to these questions do not come easily because natural food webs are complex, with temporal and spatial variability, as well as species changes that confound any simple interpretation. Equally important is that the different approaches to food web analysis give different kinds of information. For example, the identity of the species eaten can be determined via skillful gut content analysis, whereas the isotopic composition of the diet is determined via mass spectrometry. Gut contents indicate who eats whom, but isotopes help identify the original source of the organic matter and how many trophic links are involved in the transfer from source to ultimate consumer. Except in unusually simple systems, isotopes tell little about the species composition of diets. It is best to use either a combination of these complementary approaches or to use the approach that comes closest to answering your specific question. You will be frustrated if you expect isotopic analysis to define all the links in a complex web because there are too many unknowns for two tracers (^{13}C and ^{15}N) to resolve.

C. Exercise 3: Experimental Manipulation of ^{15}N in a Small Stream

The use of stable isotopes to analyze ecosystems is not restricted to naturally occurring signals. It is often best to address specific questions by introducing an isotopic signal that allows easier measurement of C and N flows than is naturally available. One interesting natural experiment occurs when spawning salmon return to reproduce, die, and then decompose in freshwater lakes and streams. The movement of this detrital and mineralized nitrogen throughout the river food web can be followed because it is isotopically enriched in ^{15}N compared to N inputs from land (Kline *et al.* 1985, Mathison *et al.* 1988, Schuldt and Hershey 1995). Likewise, one might experimentally introduce detritus with contrasting ^{13}C or ^{15}N content to discover which insects rely more heavily on allochthonous detritus than autochthonous production.

Another approach is to label the inorganic nitrogen in a stream by dripping ^{15}N-enriched ammonium or nitrate into the stream. This approach involves labeling the components of the ecosystems most heavily dependent on autochthonous production because stream algae and bryophytes rapidly assimilate ammonium and nitrate. Enriched compounds are available commercially (see Section V for vendors) and the purchase cost of adding a readily detectable isotopic signal to a small stream can be as low as $50 to $500, depending on mean discharge and nutrient concentration. The experiment involves the following steps:

1. Calculate the feasibility of tracer addition by determining the amount of ^{15}N required to elevate the δ value of in-stream N by 100‰ or more for several days or weeks (e.g., Table 24.4). Check calculations carefully.

TABLE 24.4

Sample Calculation for a Stream with Discharge of 100 liters/s (0.1 m³/s) and NH₄⁺-N Concentration of 5 mg/liter (0.36 mM) or 5 mg/m³

(a) Daily flux of nitrogen is 0.1 m³/s × 60 s/min × 60 min/h × 24 h/day × 5 mg/m³ = 43,200 mg/day or 43.2 g/day.

(b) The daily flux of ^{15}N due to the natural levels of approximately 1 g of ^{15}N for each 273 g of N is found by dividing the NH₄⁺ flux by 273. Thus the flux of ^{15}N is 43.2 g/day divided by 273, or about 0.158 g/day of ^{15}N.

(c) To obtain a target 100‰ enrichment of ^{15}N (not ^{14}N or total N!), we want the ^{15}N flux to be increased from 0.158 g/day to 0.158[(100/1000)+1] = 1.1 × 0.158 = 0.174 g/day. The difference of 0.174 − 0.158 or 0.016 g/day is the amount of ^{15}N isotope needed for the experiment. If the experiment is to continue for three weeks, 21 days × 0.016 g/day = 0.336 g of ^{15}N will be needed.

(d) Commercial compounds enriched in isotopes are usually sold by the gram of compound, not element. For example, for purchasing ammonium chloride, the quoted price is for grams ammonium chloride, not grams ^{15}N. Also note that commercial firms offer different enrichment levels: 5% enrichment means that 5% of the N atoms are ^{15}N as compared to 1/273 or 0.3663% occurring naturally. The lower percentage enrichments are satisfactory for these experiments and often less costly.

If we choose to buy 5% ^{15}N ammonium chloride, we will need to calculate how much ^{15}N it contains per gram in order to purchase the correct amount. A mole of NH₄Cl weighs 53.45 g, approximately 14 g of which is N. Therefore, 1 g of NH₄Cl contains (1 g)(14 g/mol ^{14}N)/(53.45 g/mol NH₄Cl) = 0.262 g N. If a product is enriched 5% with ^{15}N, then 5% of its N atoms will be ^{15}N rather than ^{14}N, or (0.05) (0.262) = 0.0131 g ^{15}N per gram NH₄Cl. To purchase the needed 0.336 g of ^{15}N, we would need to purchase 0.336 g ^{15}N per g NH₄Cl = 25.7 g NH₄Cl that is 5% enriched. Because 5% enriched NH₄Cl costs about $2 to $3 per gram, the cost of the isotope tracer would be about $50–$75. At higher discharge or higher NH₄⁺ concentration in the water, more would be needed. For example, if stream discharge were 0.3 m³/s, rather than 0.1, and NH₄⁺ concentration were 10 mg/liter rather than 5, you would need six times more isotope to stay at the 100‰ target enrichment level, and the cost would be $300–$450.

2. Determine the optimum study reach and conduct baseline sampling. Plan to have an upstream control reach and a downstream reach of from 100 m in a small (0.01 m³/s) stream to 1 km in a slightly larger stream (0.1 m³/s). The baseline sampling is the same as in the natural abundance exercise above but should be conducted along upstream–downstream transects to uncover any large natural isotopic gradients that might be due to spatial variation in nitrogen inputs or in-stream processes.

3. Test the pumping apparatus for reliability, as continuous and accurate delivery is important. During a season of active algal growth and low flood frequency, deliver the tracer to the stream by means of a continuous drip at a turbulent site for a period of several days to several weeks. Introduction can be via Marriote bottle (see Chapter 8) or peristaltic pump. The reason for a continuous addition is that it takes a period of several days for producer components to become well labeled and a pulse addition would result in too small a signal at the upper trophic levels. A continuous addition for several weeks allows the investigator to track the strength of the signal via sequential sampling to determine the rate of N uptake and the asymptotic δ value of many ecosystem components.

4. Once the ^{15}N addition has started, sample algae, insects, and small fish at a series of stations downstream of the dripper. Also sample one or more stations in the upstream control reach. Locate the first experimental station as close to the dripper as you expect the isotope to be well mixed. This could be examined beforehand using a dye such as rhodamine or fluoresccin to observe mixing in a stream riffle. Locate the last station well downstream of the point where you think most of the tracer will have been taken up. Sample three to five stations in between but more closely spaced near the dripper. Uptake distances for ammonium or nitrate in typical unpolluted streams range from a few meters in the smallest first-order stream to several kilometers in third- or fourth-order streams. Uptake distances at any site are variable depending on discharge, depth, temperature, nutrient concentration, and biotic activity.

5. As soon as you have taken your first set of samples (between 2 and 5 days after starting the ^{15}N dripper), select a type of sample you expect to be quickly labeled (algae or grazing insects) and submit a "rush" set to an isotope laboratory by express mail. Request a special rapid turnaround. Plot these data immediately. These components will likely not be at their maximum δ^{15}N value, but will show the distribution pattern. Adjust your sampling locations if necessary to sample the full travel distance of the introduced signal.

6. Sample important ecosystem components weekly until 3 weeks after the dripper is turned off. The last 2 weeks of data will show how rapidly stream organisms and detrital compartments lose incorporated N. After

sampling is complete, inventory all samples and select your most interesting and complete series for isotope analysis. Sampling costs for enriched ^{15}N samples requiring relatively low precision can be as low as $5 to $15 per sample.

7. Plot the ^{15}N data for each type of sample versus distance downstream and versus time.

IV. QUESTIONS

1. Considering your food web diagrams, what types of additional information would you need to quantify linkages between components? How would you go about collecting these data?

2. Does the food web you constructed using stable isotope data differ from that based on gut analyses? Develop testable hypotheses to resolve any discrepancies.

3. How variable are gut analyses for individual detritivores of the same species? Invertebrate predators? Fish? Based on your answers, how many specimens should you collect to get a good idea of the average diet of each of the dominant species sampled?

4. Do dominant consumers show considerable diet overlap or minimal diet overlap? Answer this question based on gut analyses data, then answer the same question based on ^{15}N data.

5. If you conducted Exercise 3, how far did the dissolved nutrient travel downstream? What species of consumers contain the highest amount of ^{15}N? Do all species have similar shaped distributions in space or in time? Why or why not?

V. MATERIALS AND SUPPLIES

Food Web Diagrams

Algae, invertebrate, and fish taxonomic keys
Dissecting and compound microscopes
Fine dissecting pins
Fine-tipped forceps
Microscope slides and coverslips
Mounting media
River samples of detrital and producer components and macroconsumers
Slide warmer

All ^{15}N Studies

Drying oven
Glass vials (scintillation vials work well)
River samples of organic matter components and macroconsumers

Additional Supplies for ^{15}N Enrichment Study

Discharge data
NO_3 and NH_4 data from small stream
Peristaltic pump or Marriott bottle, tubing

APPENDIX 24.1 LABORATORIES FOR STABLE ISOTOPE ANALYSES AND ISOTOPE VENDORS

Commercial laboratories
  Geochron Laboratory
  711 Concord Ave.
  Cambridge, MA 02138
  (617)876-3691

  Global Biogeochemistry Corporation
  6919 Eton Ave.
  Canoga Park, California 91303
  (818)992-4103

University laboratory (on a research specific and space available basis only)
  Water Research Center, University of Alaska—Fairbanks
  Institute of Northern Engineering
  P.O. Box 755900
  460 Duckering Building
  Fairbanks, AK 99775-1760
  (907)474-7777

Isotope vendors
  Cambridge Isotope Laboratories
  20 Commerce Way
  Woburn, MA 01801
  (617)938-0067

  Isotec, Inc
  3858 Benner Road
  Miamisburg, OH 45345
  (513)859-1808

REFERENCES

Cummins, K. W., M. A. Wilzbach, D. M. Gates, J. B. Perry, and W. B. Taliaferro. 1989. Shredders and riparian vegetation. *BioScience* **39:**24–30.

Fry, B. 1991. Stable isotope diagrams of freshwater food webs. *Ecology* **72:**2293–2297.

Fry, B., and E. Sherr. 1984. $\delta^{13}C$ measurements as indicators of carbon flow in marine and freshwater ecosystems. *Contributions in Marine Science* **27:**13–47.

Gregory, S. V., F. J. Swanson, W. A. McKee, and K. W. Cummins. 1991. An ecosystem perspective of riparian zones. *BioScience* **41:**540–551.

Kline, T. C., J. J. Goering, O. A. Mathisen, and P. H. Poe. 1990. Recycling of elements transported upstream by runs of pacific salmon. 1. ^{15}N and ^{13}C evidence in Shashin Creek, southeastern Alaska. *Canadian Journal of Fisheries and Aquatic Sciences* **47:**136–144.

Martinez, N. D. 1991. Artifacts or attributes? Effects of resolution on the Little Rock Lake food web. *Ecological Monographs* **61:**367–392.

Mathison, O. S., P. L. Parker, J. J. Goering, T. C. Kline, P. H. Poe, and R. S. Scalan. 1988. Recycling of marine elements transported into freshwater by anadromous salmon. *Verhandlungen der Internationalen Vereinigung für Theoretische und Angewandte Limnologie* **23:**2249–2258.

Merritt, R. W., and K. W. Cummins. 1996. *An Introduction to the Aquatic Insects of North America,* 3rd ed. Kendall/Hunt, Dubuque, IA.

Minagawa, M., and E. Wada. 1984. Stepwise enrichment of ^{15}N along food chains: Further evidence and the relation between $\delta^{15}N$ and animal age. *Geochimica et Cosmochimica Acta* **48:**1135–1140.

Pennak, R. W. 1989. *Freshwater Invertebrates of the United States,* 3rd ed. Wiley, New York, NY.

Peterson, B. J., and B. Fry. 1987. Stable isotopes in ecosystem studies. *Annual Review of Ecology and Systematics* **18:**293–320.

Pimm, S. L. 1982. *Food Webs.* Chapman Hall, New York, NY.

Power, M. E. 1990. Effects of fish in river food webs. *Science* **250:**811–814.

Schudt, J. A., and A. E. Hershey. 1995. Impact of salmon carcass decomposition on Lake Superior tributary streams. *Journal of the North American Benthological Society* **14:**259–268.

Ward, J. V., and J. A. Stanford. 1982. Thermal responses in the evolutionary ecology of aquatic insects. *Annual Review of Entomology* **27:**97–117.

SECTION E

Ecosystem Processes

CHAPTER 25

Primary Productivity and Community Respiration

THOMAS L. BOTT

Stroud Water Research Center
The Academy of Natural Sciences

I. INTRODUCTION

Primary producers provide the base of the food web for the biosphere. Even though fluxes of energy into a particular aquatic system include reduced chemical forms (e.g., detritus, DOM), primary productivity usually is a significant energy input. Aquatic primary producers include algae, Cyanobacteria (blue-green algae), bryophytes (mosses and liverworts), and vascular macrophytes. Other photosynthetic and chemosynthetic bacteria are not considered here because they are found only in specialized habitats and do not carry out oxygen-evolving photosynthesis.

Primary productivity is defined as the rate of formation of organic matter from inorganic carbon by photosynthesizing organisms and thus represents the conversion of solar energy to reduced chemical energy. Some of this fixed energy is lost through plant (autotrophic) respiration (R_a); the portion stored in biomass is termed net primary productivity (NPP), and the total (respired plus stored) is gross primary productivity (GPP). Thus,

$$GPP = NPP + R_a. \qquad (25.1)$$

Primary productivity measurement techniques have their basis in the equation for photosynthesis:

$$6\ CO_2 + 12\ H_2O \rightarrow 6\ O_2 + C_6H_{12}O_6 + 6\ H_2O. \qquad (25.2)$$

In practice, one measures changes in dissolved O_2 or CO_2 concentrations, with O_2 being most often used, or the rate of uptake of added [^{14}C]bicarbonate, which is used as a radioactive tracer of inorganic carbon. Gas change procedures allow the determination of both GPP and community respiration. The ^{14}C uptake method provides an estimate of something between GPP and NPP (Vollenweider 1974), regulated in part by environmental conditions and shifting from GPP to NPP with increasing length of incubation (Dring and Jewson 1982). The ^{14}C uptake procedure is the more sensitive technique and is useful where production rates are expected to be low. Productivity is a rate and thus measurements carry the units of mass area^{-1} time^{-1} or mass volume^{-1} time^{-1}.

The *in situ* dissolved O_2 change technique was introduced by Odum (1956) and has been used by many others to measure community metabolism in streams (e.g., Flemer 1970, Hall 1972, Hornberger *et al.* 1977). Wright and Mills (1967) used the method in principle but measured CO_2 change. All *in situ* methods are based on the premise that the change in dissolved gas concentration (Q) is related to photosynthesis (P), respiration (R), and gas exchange with the atmosphere (E) as long as accrual from surface and groundwater inputs is negligible; thus,

$$Q_{(\text{dissolved }O_2)} = P - R \pm E. \qquad (25.3)$$

Measurements made on the open system require accurate determination of gas exchange with the atmosphere. Instream measurements may be done at one location (single station method), in which case it is assumed that changes in O_2 concentration are identical throughout the reach, or at two stations, one at either end of a study reach (upstream–downstream method) which allows estimation of metabolism in a parcel of water flowing downstream.

Two parameters are directly measurable by gas change procedures: net oxygen change in the light (the balance of photosynthesis and respiration) and respiration in the dark. Other metabolic parameters are derived from these data. Because respiration rates include the metabolism of heterotrophs (R_h) such as microbes and insects as well as autotrophs (R_a) they are termed community respiration (CR). Measures made over 24 h are analyzed as diel curves (often referred to as diurnal curves) as exemplified in Fig. 25.1. Since CR can be determined only in darkness, daily respiration (CR_{24}) is estimated by extrapolating average night-time respiration through the daylight hours. Estimates of GPP will be subject to error from (1) differences between dark and light respiration including the inability to

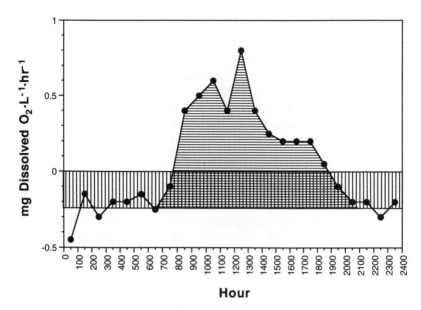

FIGURE 25.1 Rate of change curve derived from changes in dissolved O_2 concentration in a respirometer chamber (13 liter water volume) containing a benthic sample (290 cm^2) from Deadman Hole, Salmon River, ID, July 1977. Horizontal lines, GPP, 6.75 mg/liter × 13 liters × 1/290 cm^2 = 0.303 mg cm^{-2} day^{-1}. Vertical lines, CR$_{24}$, 5.52 mg/liter × 13 liters × 1/290 cm^2 = 0.247 mg cm^{-2} day $^{-1}$.

account for photorespiration by plants (light-dependent O_2 uptake and CO_2 evolution and synthesis of glycolate, some of which is excreted), (2) the inability to measure autotrophic respiration separately, and (3) any asynchrony between photosynthesis and associated metabolic costs (Bott *et al.* 1985). Net daily metabolism (NDM) can be estimated from the difference between GPP and CR$_{24}$; thus,

$$NDM = GPP - CR_{24}. \qquad (25.4)$$

NDM is free from errors associated with the estimation of daytime respiration. If GPP exceeds CR$_{24}$ for a given 24-h period, there is a net addition of energy to the system, NDM is a positive number and the GPP/CR$_{24}$ ratio is >1. If the reverse occurs, there is a net loss of energy from the system, NDM is a negative number and GPP/CR$_{24}$ is <1. NDM has been referred to by others as net ecosystem productivity or net community productivity.

Measurements performed on the open system determine total system metabolism. In small to mid-size streams community biomass is greater in the benthos than in the water column and benthic metabolism dominates system activity. In large rivers and reservoirs planktonic primary productivity and respiration can be greater than benthic sources. In some systems macrophyte productivity may be significant and can be measured using harvest techniques (Westlake 1974).

McIntire *et al.* (1964) introduced the use of respirometer chambers equipped for water recirculation for measuring benthic community metabolism without the need to determine reaeration. In the past 25 years chambers have been used in numerous studies of lotic primary productivity (e.g., Sumner and Fisher 1979, Bott *et al.* 1985). While free from the need to correct for gas exchange, concerns related to the use of chambers center on nutrient limitation, alteration of flow regime from that *in situ*, change in the ratio between sample surface area and water volume from that *in situ*, and other aspects of chamber design that may affect metabolic rates (Uehlinger and Brock 1991). Nonetheless, working with chambers allows researchers to compartmentalize the environment and relate metabolic parameters to standing crops of organisms on particular substrata with considerable confidence. However, in order to estimate metabolism of the whole system, it is important to measure the activity on all substrata with adequate replication.

Despite widespread use in measuring lentic and marine phytoplankton productivity, the ^{14}C uptake procedure has been used less frequently with benthic stream communities (Naiman 1976, Bott and Ritter 1981, Hornick *et al.* 1981, Hill and Webster 1982). If used, the procedure must include correction for the adsorption of isotope to surfaces and for the proportion of photosynthate excreted by the phytoplankton as labeled organic matter. The technique must be applied using closed systems.

The specific objectives of this chapter are to: (1) provide researchers with instructions for measuring primary productivity and community respiration and for analyzing data, (2) acquaint researchers with necessary considerations in applying particular methods, and (3) provide criteria used for evaluating the relative importance of these processes in stream ecosystem energetics. Additional information concerning methods can be found in Vollenweider (1974), Hall and Moll (1975), Wetzel and Likens (1991), and APHA *et al.* (1992).

II. GENERAL DESIGN

A. Site Selection

Stream selection will be strongly influenced by logistic considerations, especially if the chamber technique is used. The reach should be located in an area safe for personnel making night-time measurements and for equipment. For illustration of techniques, third- or fourth-order streams are best for this work because light reaches the benthos and periphyton growths can be luxuriant. The study reach should be wadeable, have no entering tributaries or significant groundwater inputs, a fairly uniform flow and a water surface unbroken by protruding rocks or logs because they generate turbulence and alter reaeration characteristics. If possible, use a reach for which geomorphologic parameters have already been determined (see Chapters 2 and 3). Most pronounced diel changes will occur during warm weather in an open reach where large standing crops of algae are found. However, very active communities can rapidly supersaturate water with O_2 and if chambers are used, gas bubbles collect on the lids.

B. Methods Selection

Depending on time and resources, researchers can choose from among several experimental protocols that are presented for measuring gas changes in the open system or in respirometers or for measuring ^{14}C uptake. If fieldwork is not possible, a problem using real data (Exercise 4) has been included to illustrate the principles of data analysis for O_2 change in chambers. Although a few researchers, and even an individual, can carry out a particular protocol, a larger team will allow division of labor for diel measurements and perhaps comparison of: (1) different methods on one stream reach or (2) activity in two or more reaches using the same procedure. Teams will be needed to measure streamflow parameters and reach geomorphologic characteristics and to obtain water samples for dissolved gas determinations over 24 h. If chambers are used, teams can compare the activity associated with different substrata in replicate chambers.

Responses to the following questions will aid in selecting the most appropriate method for the system under study. (1) Is the system nutrient rich with large algal standing crops or nutrient poor with sparse growth? Gas change methods will work best with moderate to large algal standing crops, whereas the ^{14}C uptake technique[1] is useful where algal densities are low or if algal responses to light, temperature, or nutrient manipulations are under study. (2) Are respirometer chambers with pumps for water recirculation available? For a chamber design which can be modified as desired, see Bott *et al.* (1978). (3) If an instream gas change method is to be used, is there a sufficiently long reach with uniform flow to do the upstream–downstream technique or is it best to use a single station method?

---

[1] Any use of radioisotopes must be in strict compliance with licensing regulations, including those affecting use at field sites. Researchers must be thoroughly familiar with the safe handling of radioisotopes.

For *in situ* techniques, pick a single work station or a site at the top and bottom of the reach for sample collection. The upstream–downstream procedure can be best illustrated when metabolic rates upstream of the study reach are very different from those in the study reach (e.g., when a stream enters a meadow after traversing a woodlot). Here the open-water technique is treated as a single station procedure, although researchers may want to collect samples at two stations and use the upstream–downstream analysis if differences are large enough and average the data for a single station analysis if they are not. The required length of stream for a predetermined change in dissolved gas can be estimated using the approach described in Bott *et al.* (1978). Detailed directions for the upstream–downstream analysis of the data are found in Owens (1974).

If chambers are used, it is desirable for autotrophic and heterotrophic components to be fairly uniformly distributed so that measures on replicate samples will be representative of the stream reach. If biomass is unevenly distributed, the best approach is to map the distribution of substrata in the reach, measure the metabolism of each substratum type, and extrapolate to the reach using the areal proportions of each component.

The protocols measure changes in dissolved O_2 over 24 h, although shorter times can be used if gas change measurements are conducted in chambers because darkened-chamber controls can be used to provide separate simultaneous measures of respiration. Likewise, the ^{14}C uptake procedure is generally performed over shorter times using darkened-chamber controls to measure heterotrophic uptake and sorption of ^{14}C to surfaces. Dissolved O_2 can be measured using the Winkler titration (see Chapter 5) or with commercially available probes and meters, which also can be linked to strip chart recorders or electronic data loggers for continuous monitoring. Probes must be calibrated according to manufacturers directions and should match changes in dissolved O_2 measured with Winkler titrations. Corollary measurements of photosynthetically active radiation (see Chapter 5), water chemistry (Chapter 9, APHA *et al.* 1992), algal biomass or chlorophyll *a* (Chapter 14), and ash free dry mass of detritus (Chapter 10) allow metabolism rates to be related to solar energy input, nutrient status, and periphyton biomass, and thus aid in interpreting results.

Techniques that have been used to determine the gas exchange coefficient include measuring the evasion of a radioisotopically labeled gas or volatile hydrocarbon (e.g., propane) (Kilpatrick *et al.* 1989, Marzolf *et al.* 1994) and monitoring O_2 exchange under a domc floated on the water surface (Copeland and Duffer 1974). These empirical methods are beyond the scope of work presented in this chapter. Odum (1956) calculated a diffusion constant from changes in dissolved gas concentrations during predawn and postsunset periods. However, since respiration rates change overnight this method may produce erroneous estimates. In this chapter, estimates are obtained from stream morphometry and hydrology because these are independent of metabolic activity. Two options are presented based on two models of reaeration: (a) the surface exchange model (Owens 1974) and (b) the energy dissipation model (Tsivoglou and Neal 1976, APHA *et al.* 1992).

C. General Basis of Data Analysis

A data sheet is provided for each exercise, each containing sample calculations. Data should be entered on those sheets and analyzed as directed. O_2 saturation status can be determined by comparing dissolved O_2 concentration at a given temperature and barometric pressure (if working at significant elevation) to a table of saturation values (e.g., found in APHA *et al.* 1992) or an O_2 saturation nomogram (e.g., Wetzel 1983). Rates of change in dissolved O_2 concentration are determined for sequential sampling intervals, adjusted for exchange of O_2 with the atmosphere for open system measurements, and integrated over the 24-h period. Respiration rates for the 24-h period are estimated by extrapolating the average nighttime respiration value through the daylight hours. Recall the limitations associated with this approach and note that more refined approaches are available (Hall and Moll 1975).

Data obtained in O_2 units can be converted to carbon (or joules) as follows. For photosynthesis we will assume a photosynthetic quotient (PQ, mol O_2 released during photosynthesis/mol CO_2 incorporated) of 1.2. Then,

$$gC = gO_2 \times \frac{1}{PQ} \times \frac{12}{32}, \qquad (25.5)$$

where 12 = the atomic weight of C and 32 = the molecular weight of O_2. Conversion to calories (cal) is accomplished using the conversion factor of 11.4 cal/mg C (Platt and Irwin 1973) and to joules using 4.2 joules/cal. For respiration, a respiratory quotient (RQ, mol CO_2 released/mol O_2 consumed) of 0.85 is employed. Then,

$$gC = gO_2 \times RQ \times \frac{12}{32}. \qquad (25.6)$$

Data analysis for the ^{14}C uptake procedure is based on the premise that organisms use ^{12}C dissolved inorganic carbon in the same proportion as the added [^{14}C]bicarbonate tracer. Results are corrected for excreted dissolved organic matter, dark uptake, and sorption of ^{14}C to surfaces.

III. SPECIFIC EXERCISES

A. Exercise 1: GPP and CR Assessed from Dissolved Oxygen Changes *in Situ*

Determination of Reach Characteristics   Using procedures described in Chapters 2, 3, and 4, determine the reach geomorphologic and hydrologic characteristics required for computation of the gas exchange coefficient. For the energy dispersion model determine streamwater velocity, discharge, and slope of the reach. Slope can be estimated from a topographic map (7.5-min series quadrangle). The surface renewal model requires determination of streamwater velocity, discharge, and mean depth. Determine the vertical and lateral variability in dissolved O_2 concentrations before proceeding with the diel curve measurements.

Diel Curves—Field Protocols

1. At either a single station or at upstream and downstream stations, collect water samples in triplicate for dissolved O_2 determinations at least once every 2 h over a 24-h period.[2]

2. Measure the water temperature at each sampling time.

3. Determine dissolved O_2 concentrations using Winkler titrations (Chapter 5) or with probes and meters in either the field or the laboratory.

4. Determine the gas exchange coefficient using *either* Option A *or* Option B below and analyze data accordingly.

Option A: Data analysis for a single station study with determination of gas exchange coefficient by the energy dissipation model (Tsivoglou and Neal, 1976; use Table 25.1).

a. Determine gas exchange coefficient, $K_{2(20°C)}$ at 20°C, using the equation

$$K_{2(20°C)} = K' \times \frac{\Delta H}{\Delta X} \times V, \qquad (25.7)$$

where $\Delta H/\Delta X$ is the slope expressed as m/1000 m; V, velocity in m/s; and K' varies with stream flow as in Table 25.2 (APHA *et al.* 1992). K_2 has the unit d^{-1}.

---

[2]Hourly determinations or continuous monitoring is desirable. If the upstream–downstream method is used, sampling can be done simultaneously at each station, adjusting for the time required for water to traverse the distance between stations when analyzing the data, but it is easier to set the downstream sampling time at the upstream time plus time of travel.

TABLE 25.1
Calculation of Data for a Single Station Curve with Reaeration Calculated from the Energy Dissipation Model

	A	B	C	D	E	F	G	H	I	J	K	L
	Time (h)	Dissolved O_2 concentration (C) (mg/liter)	Temperature (°C)	Plot time (h)	O_2 Rate of change (mg liter^{-1} h^{-1})	Saturation concentration (mg/liter)	Saturation concentration at altitude (Cs) (mg/liter)	Saturation deficit/surplus (C-Cs)	$K_{2(°C)}$	Gas exchange (mg liter^{-1} h^{-1})	Ave. gas exchange (mg liter^{-1} h^{-1})	Exchange corrected rate of change (mg liter^{-1} h^{-1})
	0000	7.00	17.9			9.487	7.802	-0.80	23.756			
	0100	6.95	17.6	0030	-0.05	9.545	7.850	-0.90	23.559	-0.79	-0.84	-0.89
	0200	7.00	17.4	0130	0.05	9.585	7.882	-0.88	23.476	-0.88	-0.87	-0.82
	0300	6.95	17.0	0230	-0.05	9.665	7.948	-1.00	23.255	-0.97	-0.92	-0.97
1												
2												
3												
4												
5												
6												
7												
8												
9												
10												
11												
12												
13												
14												
15												
16												
17												
18												
19												
20												

Your site data	Elevation =		m
	Mean depth =		m
	Discharge =		m³/s
	Velocity =		m/s
	Slope =		m/km
	$K_{2(20)}$ =		/day

For the example data calculation shown: elevation, 1600 m; mean depth, 0.76 m; discharge, 24.3 m/s; velocity, 0.51 m/s; slope, 3.2 m/km; and $K_{2(20°C)}$, 24.97.

TABLE 25.2
Estimated Variation in K' With
Stream Flow

Discharge (m³/s)	K' (s m^{-1} day^{-1})
0.028–0.28	28.3×10^3
0.28–0.56	21.3×10^3
>0.56	15.3×10^3

Note. From APHA et al. 1992.

b. Record field data in columns (Col.) A, B, and C.
c. Calculate the mid-point between each sampling time. Enter as plot time in Col. D.
d. Calculate the rate of change in dissolved O_2 concentration for each time interval and enter (with sign) in Col. E to coincide with plot time.
e. Determine O_2 saturation concentration for each time and enter in Col. F.
f. If working at elevation >1000 m adjust saturation concentration for barometric pressure and enter in Col. G.
g. Determine the O_2 saturation deficit or surplus ($C–C_s$) at each sampling time and enter in Col. H.
h. Adjust $K_{2(20°C)}$ for the streamwater temperature at each time using the following equation (Elmore and West 1961) and enter in Col. I:

$$K_{2(t°C)} = K_{2(20°C)} \times 1.024^{(t-20)}. \qquad (25.8)$$

i. Calculate gas exchange at each time by multiplying the saturation deficit or surplus in Col. H by $K_{2(t°C)}$ (Col. I) and enter in Col. J:

$$\text{gas exchange (mg liter}^{-1}\,\text{h}^{-1}) = \frac{(C - C_s) \times K_{2(t°C)}}{24}. \qquad (25.9)$$

Division by 24 converts the daily rate to an hourly rate. When a deficit exists O_2 diffuses into the water and masks the true metabolic rate, and gas exchange has a ($-$) sign so that when algebraically added to the observed rate of change, the "true" respiration rate is obtained. When water is supersaturated, O_2 diffuses out of the system, minimizing the actual metabolic rate, and gas exchange has a (+) sign so that when algebraically added to the observed rate of change the "true" photosynthesis rate is obtained.

j. Average the hourly gas exchange values and enter in Col. K to coincide with plot time (Col. D).

k. Add Col. K to Col. E to obtain rate of oxygen change corrected for gas exchange and enter in Col. L.

l. Plot the corrected rate of oxygen change (Col. L) against plot time (Col. D).

m. On the plot, set the respiration line by extrapolating the mean overnight respiration rate through the daylight hours.

n. Determine area under the curve to obtain GPP. CR_{24} is determined by determining the area beneath the "0 rate of change" line or by multiplying the average hourly night-time respiration rate by 24. Areas can be determined from plots on graph paper by summing the number of squares and multiplying by the value of a square ("units on y-axis" $\times$ "units on x-axis"). Calculate NDM.

o. Multiply metabolic parameters by mean depth to obtain areal metabolism estimates.

Option B: Data analysis for a single station study with determination of gas exchange coefficient by the surface renewal model (Owens *et al.*, 1964; use Table 25.3).

a. Determine the reaeration coefficient from velocity (*V*, in cm/s) and mean depth (*H*, in cm) according to the following equation (Owens 1974):

$$f_{(20°C)} = 50.8 \ V^{0.67} \ H^{-0.85}. \tag{25.10}$$

This equation is suitable for streams with velocities from 3 to 150 cm/s and depths from 12 to 335 cm. Wilcock (1982) has examined the applicability of similar equations to systems of specific character. The quantity *f* is in units of cm/h and is easily converted to m/h by dividing by 100.

b. Record field data in columns (Col.) A, B, and C.

c. Calculate the mid-point between each sampling time. Enter as plot time in Col. D.

d. Calculate the rate of change in dissolved O_2 concentration for each time interval and enter (with sign) in Col. E to coincide with plot time.

TABLE 25.3
Calculation of Data for a Single Station Curve with Reaeration Calculated from the Surface Renewal Model

	A	B	C	D	E	F	G	H	I	J	K	L	M
	Time (h)	Dissolved O₂ concentration (C) (mg/liter)	Temperature (°C)	Plot time (h)	O₂ Rate of change (mg liter⁻¹ h⁻¹)	Saturation concentration (mg/liter)	Saturation concentration at altitude (Cs) (mg/liter)	Saturation deficit/surplus (C-Cs)	$f_{(2O°C)}$ (m/h)	Gas exchange (g m⁻² h⁻¹)	Average gas exchange (g m⁻² h⁻¹)	O₂ rate of change × depth (g m⁻² h⁻¹)	Exchange corrected rate of change (g m⁻² h⁻¹)
	0000	7.00	17.9			9.487	7.802	−0.80	0.170	−0.136			
	0100	6.95	17.6	0030	−0.05	9.545	7.850	−0.90	0.168	−0.152	−0.14	−0.038	−0.18
	0200	7.00	17.4	0130	0.05	9.585	7.882	−0.88	0.167	−0.148	−0.15	0.038	−0.11
	0300	6.95	17.0	0230	−0.05	9.665	7.948	−1.00	0.166	−0.166	−0.16	−0.038	−0.19
1													
2													
3													
4													
5													
6													
7													
8													
9													
10													
11													
12													
13													
14													
15													
16													
17													
18													
19													
20													

Your site data		
Elevation =		m
Mean depth =		m
Discharge =		m³/s
Velocity =		m/s
$f_{(2O°C)}$ =		m/H

Note For the example data calculation shown: elevation, 1600 m; mean depth, 0.76 m; discharge, 24.3 m³/s; velocity, 0.51 m/s; slope, 3.2 m/km; and $f_{(2O°C)}$ = 17.83 cm/h (or 0.178 m/h).

e. Determine O_2 saturation concentration for each time and enter in Col. F.

f. If working at significant altitude adjust saturation concentration for barometric pressure and enter in Col. G.

g. Determine the O_2 saturation deficit or surplus $(C-C_s)$ at each sampling time and enter in Col. H.

h. Adjust $f_{(20°C)}$ for the streamwater temperature at each time using the following equation (Elmore and West 1961) and enter in Col. I:

$$f_{(t°C)} = f_{(20°C)} \times 1.024^{(t-20)}. \tag{25.11}$$

i. Calculate O_2 exchange by multiplying the saturation surplus or deficit (Col. H) by f (Col. I), and enter in Col. J. Retain the sign, and refer to the comments in Option A, step i (above).

j. Average the gas exchange for each hour and enter in Col. K to coincide with plot time (Col. D).

k. Multiply the oxygen rate of change (Col. E) by depth and enter in Col. L.

l. Sum the hourly rate of change (Col. L) and gas exchange (Col. K) and enter in Col. M.

m. Plot Col. M against Col. D (plot time) and determine GPP, CR_{24}, and NDM as directed in Option A, step n (above). Note that data are in areal units.

B. Exercise 2: Metabolism Measurements on Benthic Communities Transferred to Chambers

Field Protocols

1. Transfer streambed sediments with minimal disturbance to trays that will fit in the chambers. If possible, do this several weeks prior to the experiment and place the trays in the streambed contiguous with the sediment surface for continued colonization.

2. For experiments, transfer trays of sediment or samples of other streambed substrata to the respirometer chambers filled with streamwater and placed *in situ*. Place lids on the chambers, start pumps to recirculate water, and release all gas bubbles.

3. Take initial (T_0) water samples from the respirometers immediately after completing step 2. Gently siphon water from the chamber into BOD bottles for O_2 determination and replace with fresh streamwater. Note water temperature at each sampling time. It is convenient to start measures just after sunset because overnight respiration will lower dissolved O_2 concentrations and retard the potential onset of supersaturation the next day.

4. If using a dissolved O_2 probe and meter, record the meter output continuously or take readings at frequent intervals. If a probe and meter are not used, sample at 2- to 3-h intervals. Turn off the pump, siphon water samples into BOD bottles for dissolved O_2 determinations, empty the chamber water with care to avoid loss of sample, and replace with fresh streamwater. Determine the dissolved O_2 in the new water immediately, replacing only the water removed to do so, continue incubation for another period, and repeat the sampling procedure. Sample for 24 h to generate a diel curve. If a shorter experiment is performed, incubate some chambers in the light to measure photosynthesis and cover replicates with black plastic to obtain a measure of respiration. At the end of the incubation period, drain the chamber carefully to obtain an estimate of water volume and sample the chamber contents for algal biomass (chlorophyll *a,* see Chapter 14) and detritus standing stock estimates (ash-free dry mass, see Chapter 10).

Data Analysis Protocols for Diel Curve in Respirometer (Use Table 25.4)

1. Enter field data in Col. A and Col. B.
2. Enter mid-point between each sampling time as plot time in Col. C.
3. Determine rate of change in O_2 concentration between successive intervals and enter in Col. D to coincide with plot time.
4. Plot the rate of change against plot time.
5. On plot, set the mean overnight respiration line and extrapolate through the daylight hours.
6. Determine the area under curve to obtain GPP. Determine the area beneath the "0 rate of change" line to obtain CR_{24} or multiply the average hourly respiration during the overnight period by 24. If shorter term incubations with covered and uncovered chambers were performed, net oxygen change is measured in the light chamber, CR in the dark chamber, and the sum of these provides an estimate of GPP, just as for the light bottle–dark bottle procedure. Multiply data (in mg/liter) by the water volume in the chamber (liters) and divide by the substratum surface area to obtain data in units of mg O_2 cm^{-2} day^{-1}.

C. Exercise 3: Primary Productivity from ^{14}C Incorporation

Field Protocols

1. Transfer, with minimal disruption, replicate samples of periphyton to screw-capped vials (5- to 15-ml capacity). A cork borer can be used to core samples from thick mats. More flocculent growths can be scraped from a known area, pooled, and transferred to experimental vials. Use several replicates for each incubation condition. Collect five to six additional repli-

TABLE 25.4
Calculation of Data for a Diel Curve Performed in a Respirometer Chamber

	A	B	C	D
	Time (h)	DO (mg/liter)	Plot time (h)	O_2 rate of change (mg liter^{-1} h^{-1})
	0000	6.90		
	0100	6.45	0030	−0.45
	0200	6.30	0130	−0.15
	0300	6.00	0230	−0.30
1				
2				
3				
4				
5				
6				
7				
8				
9				
10				
11				
12				
13				
14				
15				
16				
17				
18				
19				
20				
21				
22				
23				
24				
25				
26				
27				
28				
29				
	Chamber volume =	liters		
	Tray surface area =	cm^2		

cates for chlorophyll *a* analyses (see Chapter 14). Experiments can also be performed by transferring streambed substrata to respirometer chambers in which case periphyton samples are taken at the end of the experiment, pooled as appropriate, mixed thoroughly, and subsampled. Much more isotope is needed for an experiment of this scale.

2. Add site water to nearly fill the vial (e.g., add 9.6 ml water to a 10-ml vial, so that with additions of isotope and formalin the total volume will be 9.9 ml).

3. Cover several replicates with aluminum foil sleeves to completely block the light and use as "dark controls."

4. Add [^{14}C]bicarbonate to each vial to provide a final concentration in the range of 0.01–0.1 μCi/ml and seal immediately. Make additions in a convenient volume, such as 0.1 ml, to vials at regularly spaced intervals (e.g., every 30 s). You will stop incubations (step 8, below) at the same intervals to keep incubation times identical. Set up a table to note time of isotope addition and time of formaldehyde addition to each sample. Note sample number, isotope manufacturer, lot No., and specific activity in Table 25.5.

5. Incubate all samples at ambient temperature and light conditions for an estimate of *in situ* productivity or under a range of conditions (e.g., temperatures) if a manipulation experiment is desired.

6. Collect streamwater samples for pH and total alkalinity determination. Measure water temperature. Determine pH and total alkalinity (APHA *et al.* 1992).

7. Incubate samples for 1 h (0.5- to 4-h incubations are typically used).

8. Stop incorporation by the addition of formaldehyde to provide a final concentration of $\approx$0.4% in each vial. Note time of addition in Table 25.5.

Sample Processing in Laboratory

1. Measure excreted ^{14}C by filtering the entire sample or some of the water from each light incubation through a 0.45-μm pore size membrane filter at a vacuum of no more than 0.5 atmosphere. Collect the filtrate in a vial or test tube placed in the filter flask under the filter head. Dark incubations are not treated this way since photosynthesis was eliminated in them.

2. Acidify each filtrate to pH 2.5 with 3% H_3PO_4 and bubble with air or N_2 for 10 min in a hood to drive off inorganic ^{14}C.

3. Transfer subsamples of filtrate from each sample to liquid scintillation vials and add a suitable cocktail. Reserve the filtrate for additional subsampling if needed.

4. If the whole sample was filtered in step 1, remove the filtrate vial and rinse the filter and algae twice with 5 ml of water. If levels of incorporated

radioactivity are expected to be very high, resuspend the recovered periphyton and filter several aliquots of the suspension, noting both the total sample volume and aliquot volume. Use postfiltration rinses as described above.

5. Air dry filters.

6. Expose dried filters to fumes of HCl for 10 min to remove adsorbed [^{14}C]bicarbonate, which can be a problem with highly active communities or in sites with hard water.

7. Ideally, combust dried filters in a sample oxidizer and count the combusted sample in a liquid scintillation counter. If a sample oxidizer is not available, digest samples with tissue solubilizer and then add cocktail.

8. Count each sample twice in a liquid scintillation counter, displacing the duplicate counts in time.

Data Analysis (Use Table 25.5)

1. Decide whether data can be analyzed on a per sample basis or whether variability in weight or chlorophyll a is great, thereby requiring normalization. Table 25.5 is set up on a "per sample" basis, assuming little variation in algal biomass between vials. The analysis of the extra samples taken in Field Protocols, step 1, will provide a reasonable estimate of the biomass in all vials. If this is not the case, aliquots of each sample can be analyzed for chlorophyll a or dry weight, and each datum normalized accordingly. If this step is needed, remember to handle the samples as radioactive. Enter sample identification in Cols. A–C.

2. Enter data from the liquid scintillation counter for biomass in Cols. D and E. Average the duplicate counts and enter in Col. F. Calculate the DPM in the whole sample if DPM data were for aliquots and enter in Col. G.

3. Average the biomass data for dark incubations and enter in Col. H.

4. Correct light incubation biomass data for isotope adsorption by subtracting the mean of the dark controls (Col. H) and enter the corrected value in Col. I.

5. Enter data from the liquid scintillation counter for filtrates in Cols. J and K. Average these values and enter in Col. L. Extrapolate to the whole sample volume and enter total DPM excreted for each light sample in Col. M.

6. Sum the dark corrected biomass DPM (Col. I) and excreted DPM (Col. M) to get the total DPM metabolized for each light sample and enter in Col. N.

7. Calculate the percentage photosynthate excreted (Excreted DPM/total DPM metabolized) × 100 or (Col. M/Col. N) × 100 and enter in Col. O.

TABLE 25.5
Calculations for Determining Primary Productivity from ^{14}C Incorporation

	A	B	C	D	E	F	G	H	I
	Sample	Light or dark incubation	Rep	Biomass DPM-1	Biomass DPM-2	Biomass mean DPM/sample	Total DPM in biomass	Mean dark DPM in biomass	Light DPM in biomass (dark corrected)
	Ambient	L	1	8147	8153	8150	24697		19790
		L	2	8345	8360	8353	25311		20404
		L	3	10765	10729	10747	32567		27660
		D	1	1756	1740	1748	5297	4907	
		D	2	1501	1480	1491	4517		
1									
2									
3									
4									
5									
6									
7									
8									
9									
10									
11									
12									
13									
14									
15									
16									
17									
18									
19									
20									

Site data:		pH =			Isotope data:	Lot No. =		
	Total alkalinity =		mg/liter			Manufacturer =		
	Temperature =		°C			Specific activity =		
	Area sampled =		cm²					

Note. In the example computations, data are analyzed on a per sample basis. One-third of the biomass in each vial was filtered and counted, and 2.5 ml of the total 15 ml of filtrate. Samples were taken from an area of 0.95 cm² and incubated for 45 min in 1 µCi of [^{14}C]bicarbonate. Total alkalinity was 55 mg/liter and the correction factor was 0.27 (pH 7.3; T = 19°C), yielding 14.85 mg of available ^{12}C/liter.

TABLE 25.5 (*continued*)

J	K	L	M	N	O	P	Q
Excretion DPM-1	Excretion DPM-2	Excretion mean DPM/sample	Total DPM excreted	Total DPM metabolized (dark and excretion corrected)	Percentage excreted	Total DPM/h	mg [^{12}C] metabolized cm^{-2} h^{-1}
611	609	610	3660	23450	15.6	31267	0.0035
488	485	487	2919	23323	12.5	31097	0.0035
530	525	528	3165	30825	10.3	41100	0.0046

TABLE 25.6
Data from a Respirometer Suitable for Constructing an O_2 Rate of Change Curve

	A	B	C	D
	Time (h)	D.O. (mg/liter)	Plot time (h)	O_2 rate of change (mg liter^{-1} h^{-1})
1	0000	6.60		
2	0100	6.45		
3	0200	6.30		
4	0300	6.10		
5	0400	5.90		
6	0500	5.75		
7	0600	5.60		
8	0700	5.50		
9	0800	5.55		
10	0900	5.80		
11	1000	6.25		
12	1100	6.75		
13	1200	7.55		
14	1300	8.30		
15	1400	8.80		
16	1500	8.95		
17	1600	9.00		
18	1700	9.05		
19	1800	9.10		
20	1900	9.10		
21	2000	8.70		
22	2100	8.40		
23	2200	8.20		
24	2300	8.00		
25	2400	7.80		

Note. The respirometer volume was 13 liters and a tray (337 cm^2) of streambed substrata was introduced. Day length 0600 to 1900 h.

8. Normalize for incubation time (h) to derive DPM/h and enter in Col. P.

9. Determine available ^{12}C inorganic carbon from a table of pH, total alkalinity, and temperature (Saunders *et al.* 1962) or from a nomogram found in Vollenweider (1974) or a limnology text (e.g., Wetzel 1983).

10. Calculate primary productivity,

^{12}C metabolized $cm^{-2} h^1$ = (Total DPM metabolized h^{-1}) × (1/DPM added)
  × (mg ^{12}C available $liter^{-1}$) × (liters/incubation vessel)
  × (1/cm^2 substrate area) × 1.06,   (25.12)

where 1.06 equals a factor to correct for isotopic discrimination. Enter in Col. Q.

D. Exercise 4: Sample Data for Dissolved Oxygen Change in Respirometer Chambers *(Use Table 25.6)*

1. From data in Table 25.6, determine plot time and rate of change in dissolved O_2 concentration between successive intervals and enter to coincide with plot time.
2. Plot the rate of change against plot time.
3. On the plot, set the respiration line by extrapolating overnight respiration through daylight hours.
4. Determine metabolic parameters as described in step 6 of the data analysis for Exercise 2.

IV. QUESTIONS

1. What methodological considerations had greatest affect on your results? If you compared methods on the same reach, did they give similar results? If you compared reaches, were the results similar? Why or why not?
2. How will an error in the determination of reaeration affect experimental results? How did oxygen change from reaeration compare with metabolic rates in your system?
3. Why is it that gas change procedures do not provide an estimate of NPP? What sources of error are included in GPP estimates from gas change procedures?
4. What is NDM and the GPP/CR_{24} ratio for your system? What does this indicate about the energetics of your system? If GPP does not exceed CR_{24} and the condition persisted for a period of time, what would be required to maintain the system in a steady-state condition?
5. What factors are most likely to affect system productivity on a daily basis, and on a seasonal basis? If your experiments involved manipulations of light, temperature, or nutrients, what effects were observed?
6. What kinds of organisms were the predominant primary producers in the community you studied?

V. MATERIALS AND SUPPLIES

Dissolved O₂ Monitoring

BOD bottles (for Winkler determinations)
Dissolved O_2 probe and meter or chemicals for Winkler dissolved O_2
  determinations (see Chapter 5)
Strip chart recorder or electronic data logger if dissolved O_2 probe
  and meter are used
Thermometer or recording thermograph
Waste containers for samples titrated in the field

Instream Method

Materials for determinations of reach length, width, and depth (Chapter
  2) to allow calculation of reach surface area, discharge, and water
  retention time
Velocity measurement equipment (see Chapter 4)

Respirometer Method

BOD bottles (if probe and meter are not used) and tubing for siphon.
Clear acrylic or Plexiglas chambers equipped with pumps for water
  recirculation large enough to hold streambed samples
Graduated cylinders for water volume measurements.
Source of electricity (line current or generator) or storage battery
  (depending on pump requirements).

¹⁴C Uptake Method

[¹⁴C]bicarbonate to provide a final concentration of no more than 0.1
  μCi/ml in each incubation vial
Aluminum foil to cover dark incubations
Forceps for handling membrane filters
Formaldehyde
HCl
Liquid scintillation counter
Membrane filters (0.2- to 0.45-μm pore size)
pH meter and probe
Pipets and propipet
Plastic backed absorbent paper to protect work surfaces
Plastic gloves and disposable lab coat for protection when handling
  isotope
Reagents for total alkalinity determination
Sample oxidizer

Screw-capped vials with secure seals for incubation
Side arm filtering flasks and membrane filter holders
Straight topped test tubes
Syringe and needle for adding isotope
Test tube racks
Vacuum pump
Waste containers

REFERENCES

APHA, AWWA, and WEF. 1992. *Standard Methods for the Examination of Water and Wastewater.* 18th ed. American Public Health Association, Washington, DC.

Bott, T. L., J. T. Brock, C. E. Cushing, S. V. Gregory, D. King, and R. C. Petersen. 1978. A comparison of methods for measuring primary productivity and community respiration in streams. *Hydrobiologia* **60:**3–12.

Bott, T. L., J. T. Brock, C. S. Dunn, R. J. Naiman, R. W. Ovink and R. C. Petersen. 1985. Benthic community metabolism in four temperate stream systems: An inter-biome comparison and evaluation of the river continuum concept. *Hydrobiologia* **123:**3–45.

Bott, T. L., and F. P. Ritter. 1981. Benthic algal productivity in a piedmont stream measured by ^{14}C and dissolved oxygen change procedures. *Journal of Freshwater Ecology* **1:**267–278.

Copeland, B. J. and W. R. Duffer. 1964. Use of clear plastic dome to measure gaseous diffusion rates in natural waters. *Limnology and Oceanography* **9:**494–499.

Dring, M. J., and D. H. Jewson. 1982. What does ^{14}C uptake by phytoplankton really measure? A theoretical modeling approach. *Proceedings of the Royal Society of London, Series B, Biological Sciences* **214:**351–368.

Elmore, H. L. and W. F. West. 1961. Effect of water temperature on stream reaeration. *Journal of the Sanitary Engineering Division ASCE* **87:**59–71.

Flemer, D. A. 1970. Primary productivity of the north branch of the Raritan River, New Jersey. *Hydrobiologia* **35:**273–296.

Hall, C. A. S. 1972. Migration and metabolism in a temperate stream ecosystem. *Ecology* **53:**586–604.

Hall, C. A. S., and R. Moll. 1975. Methods of assessing aquatic primary productivity, Pages 19–53 *in* H. Lieth and R. H. Whittaker (Eds.) *Primary Productivity of the Biosphere.* Springer-Verlag, New York, NY.

Hill, B. H., and J. R. Webster. 1982. Periphyton production in an Appalachian river. *Hydrobiologia* **97:**275–280.

Hornberger, G. M., M. G. Kelly, and B. J. Cosby. 1977. Evaluating eutrophication potential from river community productivity. *Water Research* **11:**65–69.

Hornick, L. E., J. R. Webster, and E. F. Benfield. 1981. Periphyton production in an Appalachian mountain trout stream. *American Midland Naturalist* **106:**22–36.

Kilpatrick, R. F. A., R. E. Rathbun, N. Yotsukura, G. W. Parker, and L. L. DeLong. 1989. *Determination of Stream Reaeration Coefficients by Use of Tracers.* Techniques of water resources investigations, TWI: 03-A18. United States Geological Survey, Reston, VA.

Marzolf, E. R., P. J. Mulholland, and A. D. Steinman. 1994. Improvements to the diurnal upstream-downstream dissolved oxygen change technique for determining whole-stream metabolism in small streams. *Canadian Journal of Fisheries and Aquatic Sciences* **51:**1591–1599.

McIntire, C. D., R. L. Garrison, H. K. Phinney, and C. E. Warren. 1964. Primary production in laboratory streams. *Limnology and Oceanography* **9:**92–102.

Naiman, R. J. 1976. Primary production, standing stock and export of organic matter in a Mohave Desert thermal stream. *Limnology and Oceanography* **21:**60–73.

Odum, H. T. 1956. Primary production in flowing waters. *Limnology and Oceanography* **1:**102–117.

Owens, M. 1974. Measurements on non-isolated natural communities in running waters Pages 111–119 *in* R. A. Vollenweider (Ed.) *A Manual on Methods for Measuring Primary Production in Aquatic Environments,* IBP Handbook 12. Blackwell Scientific, Oxford, UK.

Owens, M., R. W. Edwards, and J. W. Gibbs. 1964. Some reaeration studies in streams. *International Journal of Air and Water Pollution* **8:**469–486.

Platt, T., and B. Irwin. 1973. Caloric content of phytoplankton. *Limnology and Oceanography* **18:**306–310.

Saunders, G. W., F. B. Trama, and R. W. Bachman. 1962. *Evaluation of a Modified C-14 Technique for Shipboard Estimation of Photosynthesis in Large Lakes,* Publication 8. Great Lakes Division, University of Michigan, Ann Arbor, MI.

Sumner, W. T., and S. G. Fisher. 1979. Periphyton production in Fort River, Massachusetts. *Freshwater Biology* **9:**205–212.

Tsivoglou, H. C., and L. A. Neal. 1976. Tracer measurement of reaeration. III. Predicting the reaeration capacity of inland streams. *Journal of the Water Pollution Control Federation* **489:**2669–2689.

Uehlinger, U., and J. T. Brock. 1991. The assessment of river periphyton metabolism: A method and some problems. Pages 175–181 *in* B. A. Whitton, E. Rott, and G. Friedrich (Eds.) *Use of Algae for Monitoring Rivers.* Institut for Botanik, Universität Innsbruck, Innsbruck, Austria.

Vollenweider, R. A. 1974. *A Manual on Methods for Measuring Primary Production in Aquatic Environments,* I.B.P. Handbook 12, 2nd ed. Blackwell Scientific, Oxford, UK.

Westlake, D. F. 1974. Macrophytes. Pages 126–130 *in* R. A. Vollenweider (Ed.) *A Manual on Methods for Measuring Primary Production in Aquatic Environments,* I.B.P. Handbook 12, 2nd ed. Blackwell Scientific, Oxford, UK.

Wetzel, R. G. 1983. *Limnology,* 2nd ed. Saunders College Publishing, Philadelphia, PA.

Wetzel, R. G., and G. E. Likens. 1991. *Limnological Analyses,* 2nd ed. Springer-Verlag, New York, NY.

Wilcock, R. J. 1982. Simple predictive equations for calculating stream reaeration coefficients. *New Zealand Journal of Science* **25:**53–56.

Wright, J. C., and I. K. Mills. 1967. Productivity studies on the Madison River, Yellowstone National Park. *Limnology and Oceanography* **12:**568–577.

CHAPTER 26

Secondary Production of Macroinvertebrates

ARTHUR C. BENKE

Department of Biological Sciences
University of Alabama

I. INTRODUCTION

Bioenergetics deals with the capture and fate (or flow) of energy in biological systems, whether dealing at the organism, population, or ecosystem level (e.g., Benke *et al.* 1988). Since bioenergetics involves energy *flows,* it is a dynamic property, and is an aspect of ecological function rather than structure (e.g., population density, community composition, etc.). This chapter is concerned with the bioenergetics of stream *macroinvertebrates,* the major group of primary and secondary consumers in streams, and it focuses primarily on growth and production. Other exercises dealing with bioenergetics are Chapter 25 (primary production and community respiration) and Chapter 12 (production of microbial heterotrophs). Although this book does not include an exercise for measuring fish production (but see Chapter 18), the methods for fish are very similar to those considered here for invertebrates (e.g., Waters 1977). Finally, an important aspect of bioenergetics is the transfer of energy and materials through food webs (i.e., herbivory, detritivory, predation). Various aspects of such feeding relationships are dealt with in Chapters 19 to 24.

Secondary production is the formation of heterotrophic biomass *through time* (e.g., Benke 1993), and its magnitude and dynamics are a direct function of other aspects of bioenergetics. Specifically, the relationship between production and other bioenergetic parameters can be represented by the familiar equations

$$I = A + F \tag{26.1}$$

and

$$A = P + R + U, \tag{26.2}$$

where I represents ingestion; A, assimilation; F, food that is defecated (egestion); P, production; R, respiration; and U, excretion (e.g., Calow 1992). In practice, assimilation is apportioned to production (P) and respiration (R), and excretion (U) is often ignored. Each of these terms are fluxes (or flows) of energy or carbon, with units of energy area^{-1} time^{-1}. For the individual organism, P represents growth, whereas for a population it represents the collective growth of all individuals. Obviously, how much an organism grows in an interval of time depends on how much it eats, but growth also depends on how efficiently that food is converted to new tissue. Two characteristics of an organism's bioenergetics determine this efficiency: assimilation efficiency (A/I) and net production efficiency (P/A). Among stream macroinvertebrates, assimilation is likely to be the most variable term, ranging from less than 5% for detritivores to almost 90% for carnivores (Benke and Wallace 1980). Net production efficiency for macroinvertebrates shows less variation and is often close to 50%. Thus, a detritivore might convert only 2% ($\approx 0.05 \times 0.5$) of its food to production, whereas a predator might convert as much as 45% ($\approx 0.9 \times 0.5$).

Historically, different kinds of units have been used to represent secondary production. Strictly speaking, bioenergetic analyses should use energetic units (e.g., kcal m^{-2} year^{-1} or kJ m^{-2} year^{-1}). However, most studies of aquatic animals have used mass units. For example, in studies of fish production, wet (live or preserved) mass is commonly used. In studies of macroinvertebrates, dry mass (or ash-free dry mass) is the norm. Carbon units, as in primary production studies (Chapter 25), are rarely used. Nonetheless, standard conversions are available (e.g., 1 g dry mass $\approx$ 6 g wet mass $\approx$ 0.9 g ash-free dry mass $\approx$ 0.5 g C $\approx$ 5 kcal $\approx$ 21 kJ) (Waters 1977).

The methods used in measuring animal secondary production contrast sharply with those used to measure bioenergetic aspects of heterotrophic and autotrophic microbes (Chapters 12 and 25). Instead of using a community level approach, animal production methods usually operate at the population level. The basic procedures for measuring the production of aquatic animals have been in existence for many decades, with the exception of the size–frequency method (see series of reviews by Edmondson and

Winberg 1971, Winberg 1971, Waters 1977, Downing and Rigler 1984, Benke 1984, 1993). However, the great majority of studies on macroinvertebrate production in streams have occurred only within the past 20 years (Benke 1993).

A. Biomass Turnover and the *P/B* Concept

In order to appreciate the concept of secondary production, it is important to understand the relationship between production and biomass. Biomass (*B*) is a measurement of how much living tissue mass for a population is present at one point in time (or averaged over several periods of time), and its units are mass (or energy) per unit area (e.g., g/m^2) (Benke 1993). Production, on the other hand, is a flow (e.g., $g \, m^{-2} \, year^{-1}$). Production divided by biomass (*P/B*) is therefore a rate, with units of inverse time (e.g., $year^{-1}$). Since any unit of time can be selected for a rate, we can calculate annual *P/B*, weekly *P/B*, daily *P/B*, etc. *P/B* is essentially a weighted mean value of biomass growth rates of all individuals in the population.

Annual *P/B* values of benthic invertebrates were once thought to vary from only about 1 to 10 (Waters 1977), and this is probably still true for several groups. However, much higher values (approaching or exceeding 100) have now been shown for at least some of the dipterans and mayflies (Benke 1984, 1993). High *P/B* values are also possible for microcrustaceans and other types of microinvertebrates (meiofauna, see Chapter 15). *P/B* values (as rates) are a direct function of the *cohort P/B* (not a rate), which is often relatively constant (about 5), and the development time of a population. The shorter the development time, the higher the annual *P/B*.

B. Usefulness of Secondary Production in Ecological Studies

The measurement of secondary production of stream invertebrates often does not require much more information than is collected in a basic life history analysis (i.e., quantitative samples taken at regular intervals). Thus, many production studies that provide this basic information (densities, growth rates, survivorship, and production) for one population (e.g., Waters and Crawford 1973, Elliott 1981) or several closely related or functionally similar species have been done (e.g., Freeman and Wallace 1984, Lauzon and Harper 1988). Production analyses are extremely useful in this regard in understanding how production and *P/B* relate to life history features, and in comparisons of production parameters among coexisting species or between populations of the same species from different systems (e.g., Short *et al.* 1987). If production can be estimated for all major invertebrate populations in a stream, the values can be combined into functional feeding groups (Chapter 21) to help describe total energy flow through the macroconsumer community (e.g., Benke *et al.* 1984, Huryn and Wallace 1987).

Secondary production analyses can also be used for testing various ecological hypotheses (Downing 1984, Benke 1984, 1993). Models of benthic invertebrates which consider how production or P/B is affected by organism size and system temperature have been developed (e.g., Plante and Downing 1989, Benke 1993). Production is useful in estimating the intensity of utilization of a population's food source (Ross and Wallace 1981, 1983, Wallace and O'Hop 1985, Benke and Jacobi 1994). The distribution of production of coexisting species through time can be used for analyzing resource partitioning (e.g., Georgian and Wallace 1983, Benke and Jacobi 1994). Finally, production is a comprehensive representation of a population's "success" because it is a composite of several other components of success: density, biomass, individual growth rate, reproduction, survivorship, and development time (Benke 1993). As such, it can be used as a response variable in experimental studies that deal with specific biotic interactions (e.g., Dudley *et al.* 1990, Vaughn *et al.* 1993) or whole ecosystem manipulations (e.g., Wallace and Gurtz 1986, Lugthart and Wallace 1992, Peterson *et al.* 1993).

II. GENERAL DESIGN

Most attempts to measure macroinvertebrate production involve sampling at least monthly for an entire year. For taxa with short life spans, more frequent samples are sometimes taken. For long-lived taxa (>1 year), less frequent samples are acceptable. The exercises presented here are modified from the usual procedure (i.e., sampling for an entire year) to enable the researchers to complete the exercise in a shorter time period. These modifications either require some simplifying assumptions, calculate production for short time intervals, or use previously collected samples. However, the exercises illustrate some of the basic methods in ways that easily can be expanded to a full year's study. Exercises 1 and 2 can be coordinated with Chapter 16 in which general procedures for benthic sampling are described in detail, particularly quantitative sampling devices, replication, processing, and identification.

A. Site Selection

Site selection is not limited by stream size, but rather by accessibility to a site at which it is possible to obtain quantitative samples. If a site contains more than one type of invertebrate habitat (e.g., cobble, bedrock, gravel, or submerged wood), each habitat can be sampled by different teams of researchers (two to three individuals per team). If sampling is restricted to a time period shorter than a year, it is extremely helpful if the leader is already familiar with the major macroinvertebrates in the stream and has some idea of the life history of species for which this exercise will work. Knowing the life history in advance will enable the leader to select species or taxonomic groupings for which a limited sampling regime will provide meaningful data. For example, many species of caddisflies and mayflies have relatively synchronous life histories in which a cohort can be followed for all or part of its life span. In some cases, this synchronous growth is limited to a span of only 3–6 months.

B. Review of Basic Methods in Secondary Production

The methods for estimating production can be divided into two basic categories: cohort and noncohort (Waters 1977, Benke 1984, 1993). Cohort techniques are often employed when it is possible to follow a cohort (i.e., individuals that hatch from eggs within a reasonably short time span and grow at about the same rate) through time. When this is not possible, a noncohort technique must be used.

Cohort Techniques From following a cohort through time, one can recognize a general decrease in density (N) due to mortality and increase in individual mass (W) due to growth (Fig. 26.1). Given such a situation, interval production (i.e., between two sampling dates) is most easily calculated directly from field data by the *increment-summation method* as the product of the mean density between two sampling dates ($\overline{N}$) and the increase in individual mass ΔW (i.e., $\overline{N} \times \Delta W$). Assuming there is only one generation per year, annual production is most easily calculated as the sum of all interval estimates:

$$P = \Sigma \overline{N} \Delta W. \qquad (26.3)$$

However, if one wants to examine production patterns throughout the year, mean daily production for the interval should be calculated by dividing $\overline{N}\Delta W$ by the days in the interval needed to achieve a true flow (i.e., g m^{-2} day^{-1}). A study of the stream caddisfly *Brachycentrus spinae* provides an especially clear example for illustrating the calculation of production using a cohort method, such as the increment-summation method (Table 26.1, modified from Ross and Wallace 1981). Ross and Wallace (1981) used the *instantaneous growth method* (see below), which provides production estimates very similar to those in Table 26.1. Note that the cohort *P/B* ratio is close to 5 (i.e., 5.8).

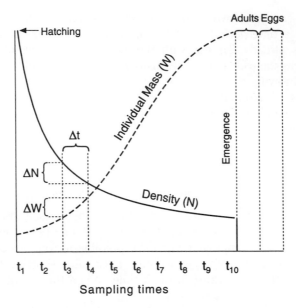

FIGURE 26.1 Hypothetical cohort of a stream insect showing curves of individual growth in mass (*W*) and population mortality (*N*) (modified after Benke 1984).

In addition to the increment-summation method, there are three closely related ways of calculating production using cohort data that should give very similar results (e.g., Waters 1977, Gillespie and Benke 1979, Benke 1984). The *removal-summation method* is most similar, but calculates production *lost* during the sampling interval as the product of the decrease in density (ΔN, Fig. 26.1) and the mean individual mass ($\overline{W}$) over the interval (i.e., $\overline{W} \times \Delta N$ rather than $\overline{N} \times \Delta W$). Adding the increase in biomass (ΔB) between sampling dates to the production loss equals interval production as calculated above (i.e., $\overline{W}\Delta N + \Delta B = \overline{N}\Delta W$ for any interval). The *instantaneous growth method* can also be used to calculate production during a sampling interval using the increase in individual mass to estimate an *instantaneous growth rate* (see below). The *Allen curve method* is a graphical approach (Fig. 26.2) in which the area under a curve of density vs mean individual mass approximates total production of a cohort (Allen 1951, Waters 1977, Gillespie and Benke 1979, Benke 1984). The Allen curve also illustrates the relationship between biomass (*B*), and the changes in numbers (ΔN) and individual mass (ΔW) over sampling intervals that are used in the tabular methods (e.g., Table 26.1). For example, $Y + Z$ (Fig. 26.2) $\cong \overline{N}\Delta W$ in the increment-summation method (Table 26.1).

TABLE 26.1
Calculation of Annual and Daily Production of *Brachycentrus spinae* (data from Ross and Wallace 1981) using the Increment-Summation Method

Date	Density (No./m²) N	Individual mass (mg) W	Biomass (mg/m²) $N \times W$	Individual growth (mg) $\Delta W = W_2 - W_1$	Mean $\overline{N}$ (No./m²) $(N_1 + N_2)/2$	Interval P (mg/m²) $\overline{N}\Delta W$	Daily P (mg m⁻² day⁻¹) $\overline{N}\Delta W/\Delta t$
18 May	282.9	0.021	5.9				
				0.036	254.9	9.17	0.66
1 Jun	226.8	0.057	12.9				
				0.031	204.4	6.33	0.53
13 Jun	181.9	0.088	16.0				
				0.085	160.2	13.62	0.85
29 Jun	138.5	0.173	24.0				
				0.179	123.9	22.17	1.58
13 Jul	109.2	0.352	38.4				
				0.588	98.4	57.83	4.45
26 Jul	87.5	0.940	82.3				
				0.266	73.9	19.64	0.89
17 Aug	60.2	1.206	72.6				
				0.590	54.3	32.01	2.46
30 Aug	48.3	1.796	86.7				
				0.025	42.6	1.06	0.07
15 Sep	36.8	1.821	67.0				
				1.378	32.0	44.03	2.45
3 Oct	27.1	3.199	86.7				
				0.358	20.1	7.18	0.17
14 Nov	13.0	3.557	46.2				
				1.074	10.9	11.71	0.51
7 Dec	8.8	4.631	40.8				
				2.222	6.2	13.83	0.27
27 Jan	3.8	6.853	26.0				
				1.624	3.1	5.08	0.24
17 Feb	2.6	8.477	22.0				
				3.071	2.2	6.60	0.28
13 Mar	1.7	11.548	19.6				
				3.252	0.9	2.76	0.06
27 Apr	0.0	14.800	0.0				

Cohort B = 43.4   Annual P = 253.03

Annual B = 39.8   Cohort P/B = 5.8

Annual P/B = 6.4

Note. P, production; B, mean biomass; $\overline{N}$, mean density between two consecutive dates. Mean biomass was estimated from monthly means since the sampling regime involved both monthly and bimonthly samples. Thus, mean cohort biomass was for 11 months and mean annual biomass for 12 months.

Noncohort Techniques—Size–Frequency Method When a population cannot be followed as a cohort from field data, it is necessary to use a noncohort technique to estimate secondary production. There are two basic approaches, both of which require independent approximations of either development time or biomass growth rates. The *size–frequency method* (Hynes and Colemen 1968, Hamilton 1969, Benke 1979) has been used more than any other production method for stream invertebrates (Benke 1993). It is based upon the assumption that a mean size–frequency distribution determined from samples collected throughout the year approximates

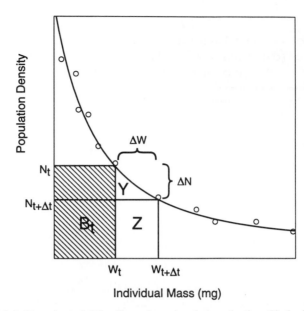

FIGURE 26.2 Hypothetical Allen Curve for estimating production. Circles indicate means of density and individual mass from samples. Curve is smoothed to provide an approximate fit to the points. Production is equal to the area under the curve. Note that W_t, N_t, ΔW, ΔN, and B_t correspond to the same terms in Table 26.1.

a mortality curve for an *average cohort*. A study of the stream mayfly *Tasmanocoenis tonnoiri* provides a good illustration of the size–frequency method (Table 26.2, modified from the data of Marchant 1986). The decrease in density (ΔN) from one size (i.e., length) category to the next is multiplied by the mean mass between size categories ($\overline{W}$), using the same rationale as for the removal-summation method. Before summation of the products (i.e., $\overline{W}\Delta N$) for each size class, each value should be multiplied by the total number of size classes (Table 26.2, final column). This must be done because it is assumed that there is a total development time of 1 year and that there are the same number of cohorts during the year as size classes (see Hamilton 1969 or Benke 1984 for rationale). Cohort P/B is equal to the sum of the biomass column (i.e., the true mean annual biomass) divided by the sum of the final column (i.e., production assuming a 1-year life span). In this particular case, the cohort P/B (9.5) is considerably higher than usually expected (5), due to the fact that a very small fraction of the population survived to the larger size classes.

   If development time is much different than 1 year, it is necessary to apply a correction factor to the basic size–frequency calculation, the sum

TABLE 26.2
Calculation of Annual Production of *Tasmanocoenis tonnoiri* (data from Marchant 1986) using the Size–Frequency Method

Length (mm)	Density (No./m²) N	Individual mass (mg) W	No. lost (No./m²) ΔN	Biomass (mg/m²) $N \times W$	Mass at loss (mg) $\overline{W} = (W_1 + W_2)/2$	Biomass lost (mg/m²) $\overline{W}\Delta N$	Times no. size classes $\overline{W}\Delta N \times 6$
0.5	706.0	0.001		0.71			
			−142.0		0.011	−1.49	(−8.95)[a]
1.5	848.0	0.02		16.96			
			730.0		0.050	36.50	219.00
2.5	118.0	0.08		9.44			
			72.0		0.130	9.36	56.16
3.5	46.0	0.18		8.28			
			42.0		0.265	11.13	66.78
4.5	4.0	0.35		1.40			
			3.7		0.435	1.61	9.66
5.5	0.3	0.52		0.16			
			0.3		0.520	0.16	0.94

Biomass   = 36.94   Production   = 352.5 (uncorrected)

Cohort P/B =   9.5
Annual P/B =   22.9   Annual P = 846.1 (Prod. × 12/5)

Note. The density column (the average cohort) is the mean value from samples taken throughout the year. P, production; B, mean biomass; $\overline{W}$, mean individual mass between two size classes.

[a]Negative value at top of table (far column) disregarded since it is probably an artifact caused by inefficient sampling of smallest size class or rapid growth through size interval.

of the final column of Table 26.2. This involves multiplication by 365/CPI where CPI (i.e., cohort production interval) is the mean development time from hatching to final size (Benke 1979). In the example of Table 26.2, Marchant estimated a mean CPI of 5 months based upon his interpretation of life histories from size–frequency histograms. Annual production is thus calculated as 352.5 × 12/5 = 846.1 mg m^{-2} year^{-1}, with an annual P/B of 22.9. These estimates are somewhat different than those found by Marchant since he used a geometric rather than a linear calculation of mean individual mass between size categories (mass at loss, sixth column of Table 26.2). The use of geometric vs linear means is usually a matter of investigator preference, where the former will generally provide only slightly lower values than the latter. Shorter CPIs (e.g., 30 days) require even greater corrections (i.e., 365/30 ≅ 12). If it is not possible to approximate CPI from field data, as done by Marchant, it is necessary to obtain this information from populations reared in the laboratory or in the field. A final point is that CPI is inversely related to biomass turnover rates (i.e., daily or annual P/B). For example, if CPI = 30 days (a relatively short time), then annual P/B = cohort P/B × 365/CPI ≈ 5 × 365/30 ≈ 60 (a relatively high value).

Noncohort Techniques—The Instantaneous Growth Method   The second noncohort technique is the *instantaneous growth rate method*. It involves the calculation of a daily instantaneous growth rate,

$$g = \frac{\ln(W_{t+\Delta t}/W_t)}{\Delta t}, \qquad (26.4)$$

where W_t is mean mass of an individual at time t; $W_{t+\Delta t}$, mean mass of an individual at time $t + \Delta t$; and Δt, length of the time interval. As a noncohort method, the growth rate is estimated from animals grown in the laboratory or under controlled conditions in the field (e.g., Huryn and Wallace 1986, Hauer and Benke 1987, 1991). The instantaneous growth rate method can also be used as a cohort method (see above) if the mean growth of individuals can be followed from sequential field samples. Daily production (P_d) is simply calculated as

$$P_d = g \times \overline{B}, \qquad (26.5)$$

where $\overline{B}$ is mean population biomass for two consecutive dates in units of g/m^2 (e.g., Benke and Parsons 1990, Benke and Jacobi 1994). The most accurate estimates will be obtained when it is possible to determine size- and temperature-specific growth rates. Production of the ith size class is $P_i = g_i \times \overline{B}_i$, where g_i and $\overline{B}_i$ are the growth rate and mean biomass of the ith size class, respectively, and total daily production is the sum of the production of all size classes:

$$P_d = g_1 B_1 + g_2 B_2 + \cdots + g_f B_f. \qquad (26.6)$$

C. Quantitative Sampling (See Also Chapter 16)

Production analysis requires quantitative sampling. Many types of samplers can be used, depending on substratum type. For example, a Surber or Hess sampler might be used in a cobble area or on a flat bedrock habitat. A corer might work best in shallow gravel (e.g., coffee can) or sand (e.g., Ogeechee River corer, Gillespie *et al.* 1985, also manufactured by Wildco, Inc.). As with any quantitative sampling, replication is necessary in order to obtain confidence limits of density and biomass, and every attempt should be made to obtain random samples. The estimation of confidence limits (or standard errors) can easily be done for density and biomass. Determination of confidence limits for production estimates has been attempted (see methods by Krueger and Martin 1980, Newman and Martin 1983), but it is somewhat controversial. The basic problem with attempting to apply statistical measures to production is that variance estimates of field growth rates or development times are very difficult to obtain.

One means of obtaining a more accurate picture of the distribution of production in a stream is by taking samples in different habitats and determining the relative production in each habitat. If the habitats themselves can be quantified, it is then possible to determine total system production by weighting the individual habitat-specific production calculations by the size of the habitat (e.g., Huryn and Wallace 1987, Smock *et al.* 1992).

D. Determining Relationship between Length and Mass

In order to calculate production by any method, it is essential that biomass values are determined. The product of length-specific mass (mg/individual) and density (No. individuals/m^2) yields an estimate of size-specific biomass (mg/m^2). The sum of biomass values for all size groups is equal to population biomass.

The relationship between individual length and mass for a given taxon is usually obtained from a length/mass equation developed by the investigator or from previously published studies. Nonpreserved (fresh) animals collected for this purpose provide the best results since preservation (especially in ethanol) results in shrinkage of soft body parts and losses of dry mass by leaching. Animals preserved in a formalin solution will provide estimates comparable to nonpreserved specimens. The procedure involves measuring the lengths of individual animals from a wide range of size categories under a dissecting microscope. The eyepiece must be fitted with a micrometer so that lengths can be measured to at least 0.1 mm. Subsequently, the measured individuals usually are dried in a drying oven for a minimum of 24 h at 60°C, cooled in a desiccator, and weighed on an analytical balance that is accurate to at least 0.01 mg. It is best to have at least 20 measurements. A linear regression is then developed of the form

$$\ln W = \ln a + b \ln L, \tag{26.7}$$

where W is individual mass; L, length; a, a constant; and b, the slope of the regression. This equation is the linear equivalent of a power curve, $W = aL^b$. Since we expect a cubic relationship between L and W, b should be reasonably close to 3.

In the absence of time or equipment to determine a length/mass regression, one can use literature values to obtain length-specific mass. Obviously, it is best to use a regression for the actual species or genus of interest. Smock (1980) has developed regressions for many insect species from streams in North Carolina and Meyer (1989) has done the same thing from a stream in southwestern Germany. Both authors also have provided equations for larger taxonomic groupings. Smock's order-level equations are presented in Table 26.3 for use in these exercises in the event that more specific equations are unavailable from the literature, or have not otherwise been determined by the researchers.

III. SPECIFIC EXERCISES

A. Field Collections and Laboratory Processing

The first two exercises below call for field collections of invertebrate populations and laboratory processing of these collections. The general procedures are as follows:

1. Select one or more sampling dates, depending on the exercise selected below. One or more teams (two to four researchers each) should collect at least three replicate samples on each date. If there is more than one team, they may be assigned to different habitats. For example, one team might sample cobble habitat, another bedrock, another gravel, etc.

2. Field sampling will require the use of a benthic sampler (e.g., Surber or Hess sampler) appropriate for the selected habitat. Other necessary

TABLE 26.3

Values for the Constants ln a and b Obtained from the Log-Transformed Regression Equation ln W = ln a + b ln L, where W is Dry Mass (mg) and L is Body Length (mm), According to Insect Order (from Smock 1980)

| | Regression constants | | |
Order	ln a	b	r
Coleoptera	−1.878	2.18	0.94
Diptera	−5.221	2.43	0.96
Ephemeroptera	−5.021	2.88	0.94
Hemiptera	−3.461	2.40	0.93
Megaloptera	−5.843	2.75	0.95
Odonata	−4.269	2.78	0.94
Plecoptera	−6.075	3.39	0.95
Trichoptera	−6.266	3.12	0.83

Note. r is the correlation coefficient.

supplies might include: (1) a scrub brush used for gently brushing animals off large substrata such as bedrock, (2) hard forceps for removing animals that are tightly adhering to substrata or hiding in crevices, and (3) plastic bags and buckets within which a sample can be temporarily stored. If samples can be transported back to the laboratory for immediate processing (i.e., separation from organic debris), no preservative is necessary in the field. If samples must be saved for more than a few hours (depending on storage temperature), then a preservative must be added to each sample.[1]

3. The technique used in sorting animals from the debris depends on the animals selected for analysis and the type of debris. If large taxa are selected, separate much of the fine sediment by rinsing through a 500-μm sieve. However, if small taxa are used, a finer sieve (100–300 μm) might be required (see discussion of sampling in Chapter 16). Additional sorting within a flat white pan or using a dissecting microscope may be necessary. If organisms are sorted when alive, their color and movement in the pan is extremely helpful in finding them. If they are sorted after preservation, the use of the stain should help to distinguish them from the organic debris. If samples were preserved, processing in the laboratory first requires that the preservative be rinsed from the sample. During sorting, it is convenient to place all animals in labeled, 4-dram vials containing 70% ethanol for permanent storage.

4. After animals are separated from the debris, identify them to the appropriate taxonomic level (see Chapter 16), measure, and then place into length categories (e.g., total length, excluding cerci) or some equally reliable width category (e.g., head width). Size categories (e.g., 1-mm length intervals) are usually determined by measurement under a dissecting microscope outfitted with an ocular micrometer.[2] A more rapid means of grouping (e.g., a template) may be used if the taxon is large enough.

5. For the taxon of interest, record the number of animals in each size class for each sample and calculate the mean density of each size class (Table 26.4). Determine individual mass for each size class (e.g., from equations in Table 26.3) and record in Table 26.4. Multiply the density in each size class by the size-specific individual mass to obtain size-specific biomass. Total densities and biomass values for each date are determined by summing all size class values (sums of columns shown in Table 26.4).

---

[1]Use either 70% ethanol (final concentration) or formalin with either Rose Bengal or Phloxine B stain added. If using formalin, approximately 1 part formalin to 9 parts water in the sample will provide the desired 4% formaldehyde concentration. *Caution.* Formaldehyde is a known carcinogen, use with great care and always wear eye and skin protection.

[2]*Note.* Measurement of total length may cause problems for taxa with soft abdomens since alcohol results in some shrinkage.

TABLE 26.4
Worksheet for Calculation of Mean Size-Specific Density, Total Mean
Density, and Total Mean Biomass for Each Date

Stream/habitat _____ Taxon _____ Date_____

Size class (i)	No. per sample				Density (N_i)	Indiv. Mass (W_i)	Biomass (B_i)
	Sample 1	Sample 2	Sample 3	Mean	No./m²	(mg)	(mg/m²)
1							
2							
3							
4							
5							
6							
7							
8							
9							
10							
Totals					$\Sigma N_i =$ _____	$\Sigma B_i =$ _____	

Note. Values for individual mass are obtained from a length/mass regression (e.g., Table 26.3) for each size (i.e., length) class.

Estimate the mean individual mass for a specific date by dividing population biomass by density. Size-specific densities (N_i) from Table 26.4 are used for the size–frequency method. Total mean density and biomass are used in the increment-summation method.

B. Exercise 1: Estimating Secondary Production—Single Date Production

The simplest exercise that one can do to estimate secondary production is to sample the population from the field on a single date and assume that the results are similar to what would be found from monthly sampling and obtaining an average value for the year.[3]

1. Apply the size–frequency method to these data. This exercise will work best if a population is selected that *does not grow as a cohort* and is

---

[3]It should be emphasized that this is only an illustrative exercise and it should not be employed in obtaining a true production estimate.

represented by a wide range of size classes. If it is difficult to obtain a wide range of size classes for a single population, a broad size distribution can be achieved by combining several taxa. In fact, the size–frequency method was originally conceived to apply to several species simultaneously (Hynes and Coleman 1968), but the assumptions required to do this are rarely obtainable. For this exercise, it might work best to combine all taxa within a major family (e.g., Chironomidae) or order (Ephemeroptera) to obtain densities across a wide range of sizes.

2. After animals are placed into size groups and mean densities for each size class are determined from worksheets (Table 26.4), enter the data into a size-frequency table (Table 26.5). Ideally, the densities will decline from the smallest to the largest animals. For each length category, calculate individual mass values from the appropriate length/mass equation (e.g., Table 26.3). Then follow the procedure described above for the *size– frequency method* to obtain a production estimate (Table 26.2). Any nega-

TABLE 26.5
Worksheet for Calculating Mean Biomass and Annual Production using the Size-Frequency Method

Size class (mm)	Density (No./m^2)	Indiv. mass (W, mg)	Biomass (g/m^2)	ΔN	Mass at loss ($\overline{W}$, mg)	$\overline{W}\Delta N$	$\overline{W}\Delta N \times$ No. sizes
	a	b	$c = a \times b$	$d = a_2 - a_1$	$e = (b_1 + b_2)/2$	$f = d \times e$	
1							
2							
3							
4							
5							
6							
7							
8							
9							
10							

Biomass = _____   P = _____

Cohort P/B = _____

Annual P/B = _____   Annual P = _____

Note. For purposes of this exercise, the size–distribution on a single date may be used. It is assumed that this distribution represents the mean size distribution of that taxon for the year and that the total development time equals 1 year.

tive value should be included in the summation, unless it occurs at the top of the table in the smallest size classes (Table 26.2).

3. If it is possible to sample the population(s) on additional dates throughout its period of occurrence at the site (this could be anywhere from 3 to 12 months), then a reliable production estimate can be calculated from the mean size-specific densities from all sampling dates. The calculations are identical to those for a single date.

C. Exercise 2. Estimating Secondary Production—Production Between Dates

To calculate production realistically, it is necessary to sample a population on more than one date or have independent estimates of growth rates. If a population grows synchronously and can be followed as a cohort (Fig. 26.1), it is possible to calculate production between sampling dates (even if only two dates) using a cohort method. Normally, the population would be sampled from the start of the cohort (at hatching) until the end of the cohort (at emergence for aquatic insects), which could be as short as a month to more than a year.

1. In order to effectively apply this method over a short period of time (e.g., 3 months), the presence of one or more species that are known to have cohort (synchronous) development must be determined in advance. Likewise, sampling must be conducted during a time when temperatures are high enough for growth to occur. Ideally, a species that has only one or two generations per year will work best. Many north-temperate populations have synchronous development and only one or two generations per year. In the south-temperate or subtropical zone, some groups are more likely to be multivoltine (several generations per year) and will be more difficult to follow (e.g., many dipterans and some mayflies). For example, the ephemerellid mayflies are a family that often has synchronous development and only a single generation per year. In colder climates, they are often considered summer species, but in warmer areas, they have a cold-season development period (e.g., Benke and Jacobi 1994). Several caddisfly species, including net-spinners (e.g., hydropsychids) and grazers (e.g., glossosomatids), often have relatively synchronous development. Although many stoneflies have generation times longer than 1 year, the small and widely distributed *Perlesta placida* (Perlidae) has a cohort that develops through the winter in the southeastern states. Since most species grow faster at high temperatures, it is advantageous if at least one of the sampling dates is when temperatures are relatively warm and measurable growth occurs.

2. This exercise should work best if the taxon can be identified to the species level. Researchers should know the most abundant species found

in the stream and select those that are most likely to show cohort growth during the sampling period. Mayflies, caddisflies, and stoneflies should work best, and chironomids should probably be avoided.

3. Place animals into size groups and determine total density and biomass from worksheets for each date (Table 26.4). Then enter data into a table for calculating cohort production using the increment-summation method (Table 26.6). Calculate individual mass (W) by dividing biomass by density. Then follow the procedure described above for *increment-summation method* to obtain a production estimate (Table 26.1). Note that for two consecutive dates, mean numbers between the dates ($\overline{N}$) and the increase in individual mass (ΔW) are calculated. They are then multiplied to obtain an interval production estimate ($\overline{N} \times \Delta W$). Mean daily production for the interval can be determined by dividing the interval production by the interval length in days (Δt).

D. Exercise 3: Estimating Secondary Production—Previously Collected Samples

If quantitative samples have been collected previously (e.g., by the research leader) over 1 full year at regular intervals (e.g., monthly) with adequate replication (≥ 3 samples/date), this provides the opportunity to calculate production over the life span of any population. Not only can production be calculated for animals with life spans of ≤ 1 year, production also can be determined for populations with life spans >1 year because

TABLE 26.6

Worksheet for Calculating Mean Biomass, Interval Production, and Daily Production using the Increment-Summation Method

Date	Density (No./m²)	Mean $\overline{N}$	Indiv. mass (W, mg)	ΔW	Biomass (g/m²)	Interval prod. ($\overline{N} \times \Delta W$)	Daily prod. $\overline{N}\Delta W/\Delta t$
	a	$b = (a_1 + a_2)/2$	c	$d = c_1 - c_2$	$e = a \times c$	$f = b \times d$	$g = f/\Delta t$
	____	____	____	____	____	____	____
	____	____	____	____	____	____	____
	____	____	____	____	____	____	____
	____	____	____	____	____	____	____
	____	____	____	____	____	____	____

Note. Only four dates are shown for illustration, but additional dates are easily added. To obtain daily production, $\overline{N}\Delta W$ should be divided by Δt, the number of days between sampling dates.

multiple year classes (e.g., first-year cohort, second-year cohort, etc.) can be treated separately and simultaneously. One or more research teams can select a specific taxon (e.g., species to order). If the life history of the population is not known from earlier work, it may be necessary to construct a series of size–frequency histograms to assess developmental synchrony and identify cohorts (e.g., Waters and Crawford 1973, Benke and Wallace 1980). The data can be summarized in a manner similar to that described in the first two options (Table 26.4). After knowledge of life history is obtained, then any appropriate method can be applied to the data (e.g., Table 26.5 and Table 26.6).

E. Exercise 4. Estimating Secondary Production—Experimental Approach

Several of the exercises in this book, including the measurement of secondary production, can be used in measuring responses to experimental treatments in the field or in artificial streams (see Hauer 1993). For example, light, temperature, food type, habitat heterogeneity, and species can be controlled and manipulated in stream mesocosms. To measure secondary production as a response variable, numerous small to medium-sized individuals of one or more species or species groups could be collected from the natural stream and introduced to the mesocosms. A subsample of collected animals should be retained for initial size determinations. The number of animals introduced should likewise be recorded. Animals should be sampled at regular intervals or at the end of the experiment to obtain interval or total production as measured from population mortality and individual growth. Given these circumstances, a cohort method should be used to estimate production (e.g., Table 26.6).

IV. QUESTIONS

1. Compare the production estimates among the different populations, species, habitats, or experimental treatments found in your study. How much variation did you find? Give several reasons why this variation is likely to have occurred.

2. Calculate either a daily or annual P/B ratio from your data. Do you consider this relatively high or low? What is the difference between annual P/B and cohort P/B? Which one is a rate and why?

3. Although dry mass is the most common unit of mass used in invertebrate production studies, some consider ash-free dry mass more appropriate. Why might this be so?

4. The "Allen paradox" is the situation where there does not appear to be enough invertebrate biomass to satisfy the energetic needs of a predator (either a fish or an invertebrate predator). How might high values of the *P/B* ratio help resolve this paradox? What other explanations might there be for the paradox?

5. Why might secondary production be a better response variable for comparative or experimental studies than density or biomass?

6. Why is it necessary to use a correction factor (CPI) in the size–frequency method if the development time is much less than 1 year? Should the correction factor be used for populations in which development time is greater than 1 year?

7. If you are interested in estimating the production of a population with asynchronous development, you need to obtain an independent estimate of CPI or growth rate. For a stream invertebrate, how would you design a growth study that will provide you with a realistic estimate of either value?

8. From the data in Table 26.1, or from your own data (Table 26.6), calculate daily production with the instantaneous growth rate method. How does it compare to the increment-summation estimate?

V. MATERIALS AND SUPPLIES

Field Materials

Buckets (sturdy, 20-liter, with lids)
Coarse brush
Forceps
Plastic garbage bags or comparable (size of bag depends on sampler size)
Preservative (ethanol or formalin) stained with Phloxine B or Rose Bengal
Sampler for quantitative sampling (see Chapter 16)
Sieves (if samples partially processed in field)

Laboratory Materials

2- to 4-Dram vials (for storage of sorted samples)
100- to 500-ml Jars (for storage of unsorted samples)
70% Ethanol
Fine dissecting forceps
Shallow dishes (petri dishes)

Laboratory Equipment/Supplies

Analytical balance (optional)
Desiccator (optional)
Dissecting binocular microscope, light source (fiber optic), and ocular micrometer
Drying oven (optional)
Sieves (500-μm mesh or smaller)

REFERENCES

Allen, K. R. 1951. The Horokiwi Stream. A study of a trout population. *New Zealand Department of Fisheries Bulletin* **10**:1–238.

Benke, A. C. 1979. A modification of the Hynes method for estimating secondary production with particular significance for multivoltine populations. *Limnology and Oceanography* **24**:168–174.

Benke, A. C. 1984. Secondary production of aquatic insects. Pages 289–322 *in* V. H. Resh and D. M. Rosenberg (Eds.) *Ecology of Aquatic Insects.* Praeger Scientific, New York, NY.

Benke, A. C. 1993. Concepts and patterns of invertebrate production in running waters. *Verhandlungen der Internationalen Vereinigung für Theoretische und Angewandte Limnologie* **25**:15–38.

Benke, A. C., C. A. S. Hall, C. P. Hawkins, R. H. Lowe-McConnell, J. A. Stanford, K. Suberkropp, and J. V. Ward. 1988. Bioenergetic considerations in the analysis of stream ecosystems. *Journal of the North American Benthological Society* **7**:456–479.

Benke, A. C., and D. I. Jacobi. 1994. Production dynamics and resource utilization of snag-dwelling mayflies in a blackwater river. *Ecology* **75**:1219–1232.

Benke, A. C., and K. A. Parsons. 1990. Modelling black fly production dynamics in blackwater streams. *Freshwater Biology* **24**:167–180.

Benke, A. C., T. C. Van Arsdall, D. M. Gillespie, and F. K. Parrish. 1984. Invertebrate productivity in a subtropical blackwater river: The importance of habitat and life history. *Ecological Monographs* **54**:25–63.

Benke, A. C., and J. B. Wallace. 1980. Trophic basis of production among net-spinning caddisflies in a southern Appalachian stream. *Ecology* **61**:108–118.

Calow, P. 1992. Energy budgets. Pages 370–378 *in* P. Calow and G. E. Petts (Eds.) *The Rivers Handbook: Hydrological and Ecological Principles.* Blackwell Scientific, Oxford, UK.

Downing, J. A. 1984. Assessment of secondary production: The first step. Pages 1–18 *in* J. A. Downing and F. H. Rigler (Eds.) *A Manual on Methods for the Assessment of Secondary Productivity in Fresh Waters,* 2nd ed. Blackwell Scientific, Oxford, UK.

Downing, J. A., and F. H. Rigler (Eds.). 1984. *A Manual on Methods for the Assessment of Secondary Productivity in Fresh Waters,* 2nd ed. Blackwell Scientific, Oxford, UK.

Dudley, T. L., C. M. D'Antonio, and S. D. Cooper. 1990. Mechanisms and consequences of interspecific competition between two stream insects. *Journal of Animal Ecology* **59:**849–866.

Edmondson, W. T., and G. G. Winberg. 1971. *A Manual on Methods for the Assessment of Secondary Productivity in Fresh Waters.* Blackwell Scientific, Oxford, UK.

Elliott, J. M. 1981. A quantitative study of the life cycle of the net-spinning caddis *Philopotamus montanus* (Trichoptera: Philopotamidae) in a Lake District stream. *Journal of Animal Ecology* **50:**867–883.

Freeman, M. C., and J. B. Wallace. 1984. Production of net-spinning caddisflies (Hydropsychidae) and black flies (Simuliidae) on rock outcrop substrate in a small southeastern Piedmont stream. *Hydrobiologia* **112:**3–15.

Georgian, T., and J. B. Wallace. 1983. Seasonal production dynamics in a guild of periphyton-grazing insects in a southern Appalachian stream. *Ecology* **64:**1236–1248.

Gillespie, D. M., and A. C. Benke. 1979. Methods of calculating cohort production from field data—Some relationships. *Limnology and Oceanography* **24:**171–176.

Gillespie, D. M., D. L. Stites, and A. C. Benke. 1985. An inexpensive core sampler for use in sandy substrata. *Freshwater Invertebrate Biology* **4:**147–151.

Hamilton, A. L. 1969. On estimating annual production. *Limnology and Oceanography* **14:**771–782.

Hauer, F. R. 1993. Artificial streams for the study of macroinvertebrate growth and bioenergetics. *Journal of the North American Benthological Society* **12:**333–337.

Hauer, F. R., and A. C. Benke. 1987. Influence of temperature and river hydrograph on black fly growth rates in a subtropical blackwater river. *Journal of the North American Benthological Society* **6:**251–261.

Hauer, F. R., and A. C. Benke. 1991. Rapid growth of snag-dwelling chironomids in a blackwater river: The influence of temperature and discharge. *Journal of the North American Benthological Society* **10:**154–164.

Huryn, A. D., and J. B. Wallace. 1986. A method for obtaining in situ growth rates of larval Chironomidae (Diptera) and its application to studies of secondary production. *Limnology and Oceanography* **31:**216–222.

Huryn, A. D., and J. B. Wallace. 1987. Local geomorphology as a determinant of macrofaunal production in a mountain stream. *Ecology* **68:**1932–1942.

Hynes, H. B. N., and M. J. Coleman. 1968. A simple method of assessing the annual production of stream benthos. *Limnology and Oceanography* **13:**569–573.

Krueger, C. C., and F. B. Martin. 1980. Computation of confidence intervals for the size-frequency (Hynes) method of estimating secondary production. *Limnology and Oceanography* **25:**773–777.

Lauzon, M., and P. P. Harper. 1988. Seasonal dynamics of a mayfly (Insecta: Ephemeroptera) community in a Laurentian stream. *Holarctic Ecology* **11:**220–234.

Lugthart, G. J., and J. B. Wallace. 1992. Effects of disturbance on benthic functional structure and production in mountain streams. *Journal of the North American Benthological Society* **11:**138–164.

Marchant, R. 1986. Estimates of annual production for some aquatic insects from the La Trobe River, Victoria. *Australian Journal of Marine and Freshwater Research* **37:**113–120.

Meyer, E. 1989. The relationship between body length parameters and dry mass in running water invertebrates. *Archiv für Hydrobiologie* **117:**191–203.

Newman, R. M., and F. B. Martin. 1983. Estimation of fish production rates and associated variances. *Canadian Journal of Fisheries and Aquatic Sciences* **40:**1729–1736.

Peterson, B. J., and 17 others. 1993. Biological responses of a tundra river to fertilization. *Ecology* **74:**653–672.

Plante, C., and J. A. Downing. 1989. Production of freshwater invertebrate populations in lakes. *Canadian Journal of Fisheries and Aquatic Sciences* **46:**1489–1498.

Ross, D. H., and J. B. Wallace. 1981. Production of *Brachycentrus spinae* Ross (Trichoptera: Brachycentridae) and its role in seston dynamics of a southern Appalachian stream (USA). *Environmental Entomology* **10:**240–246.

Ross, D. H., and J. B. Wallace. 1983. Longitudinal patterns of production, food consumption, and seston utilization by net-spinning caddisflies (Trichoptera) in a southern Appalachian stream (USA). *Holarctic Ecology* **6:**270–284.

Short, R. A., E. H. Stanley, J. W. Harrison, and C. R. Epperson. 1987. Production of *Corydalus cornutus* (Megaloptera) in four streams differing in size, flow, and temperature. *Journal of the North American Benthological Society* **6:**105–114.

Smock, L. A. 1980. Relationships between body size and biomass of aquatic insects. *Freshwater Biology* **10:**375–383.

Smock, L. A., J. E. Gladden, J. L. Riekenberg, L. C. Smith, and C. R. Black. 1992. Lotic macroinvertebrate production in three dimensions: Channel surface, hyporheic and floodplain environments. *Ecology* **73:**876–886.

Vaughn, C. C., F. P. Gelwick, and W. J. Matthews. 1993. Effects of algivorous minnows on production of grazing stream invertebrates. *Oikos* **66:**119–128.

Wallace, J. B., and M. E. Gurtz. 1986. Response of *Baetis* mayflies (Ephemeroptera) to catchment logging. *American Midland Naturalist* **115:**25–41.

Wallace, J. B., and J. O'Hop. 1985. Life on a fast pad: waterlily leaf beetle impact on water lilies. *Ecology* **66:**1534–1544.

Waters, T. F. 1977. Secondary production in inland waters. *Advances in Ecological Research* **10:**91–164.

Waters, T. F., and G. W. Crawford. 1973. Annual production of a stream mayfly population: A comparison of methods. *Limnology and Oceanography* **18:**286–296.

Winberg, G. G. 1971. *Methods for the Estimation of Production of Aquatic Animals.* Academic Press, New York, NY.

CHAPTER 27

Leaf Breakdown in Stream Ecosystems

E. F. BENFIELD

Department of Biology
Virginia Polytechnic Institute and State University

I. INTRODUCTION

There are two possible sources of primary energy for streams: instream photosynthesis by algae, mosses, and higher aquatic plants and imported organic matter from streamside vegetation (e.g., leaves and other parts of vegetation). In small, heavily shaded streams, there is normally insufficient light to support substantial instream photosynthesis so energy pathways are supported largely by imported (allochthonous) energy. The bulk of imported energy enters such streams as autumnal leaf fall although additional leaf material may slide or blow into the stream from riparian zones over the rest of the year. Leaves falling into streams may be transported short distances but usually are caught by structures in the streambed to form *leaf packs* (Petersen and Cummins 1974, see also Chapter 11). Leaf packs are then *processed* in place by components of the stream community in a series of well documented steps.

Dead leaves entering streams in autumn are nutrition-poor because trees resorb most of the soluble nutrients (e.g., sugars and amino acids) that were present in the green leaves (Suberkropp *et al.* 1976, Paul *et al.* 1978). Within 1 or 2 days of entering a stream, many of the remaining soluble nutrients in leaves leach into the water. After leaching, leaves are composed mostly of structural materials like cellulose and lignin, neither of which are very digestible by animals. Within a few days, fungi and bacteria begin to colonize the leaves leading to a process known as "microbial conditioning" (Bärlocher and Kendrick 1975). The microbes produce a suite of enzymes that can digest the remaining leaf constituents and begin the conversion of leaves to smaller particles (Suberkropp and Klug 1976). After approximately 2 weeks, microbial conditioning leads to structural softening of the leaf and, among some species, fragmentation. Laboratory studies have shown that some species of aquatic hyphomycete fungi can reduce whole leaves to small particles given sufficient time (Suberkropp and Klug 1980). However, reduction in particle size from whole leaves (coarse particulate organic matter, CPOM) to fine particulate organic matter (FPOM) is accomplished mainly through the feeding activities of a variety of aquatic invertebrates collectively known as "shredders" (Cummins 1974, Cummins and Klug 1979, see also Chapter 21). Through the production of *orts* (i.e., fragments shredded from leaves but not ingested) and fecal pellets, shredders help reduce the particle size of organic matter. The particles then serve as a food resource for a variety of micro- and macroconsumers. In streams with few shredders, leaves may also be fragmented by a combination of microbial activity and physical factors such as current and abrasion (Benfield *et al.* 1977, Paul *et al.* 1978).

All species of leaves do not break down at the same rate. There appears to be a "leaf processing continuum" in most streams (Petersen and Cummins 1974) in which some species disappear rapidly ("fast processors"), some disappear moderately rapidly ("medium processors"), and some very slowly ("slow processors"). The leaf processing continuum results in the stream community having food resources throughout much of the annual cycle (Petersen and Cummins 1974). Differences in the rates at which *fast, medium,* and *slow* species break down in a particular stream appear to be mostly a function of initial physical and chemical properties of leaves (Webster and Benfield 1986). Species-specific breakdown rates may vary with stream, location in the stream, time of year, activity of microbes, presence of shredders, and other stream-specific factors (Webster and Benfield 1986).

Leaf-pack processing is an integrative ecosystem-level process because it links all of the elements mentioned above (i.e., leaf species, microbial activity, shredders, physical and chemical features of the stream). The major result of this linkage is that whole leaves are converted into fine particles which are then distributed downstream and used as an energy source by various consumers (see Chapter 10). The objective of this chapter is to gain insight into an important component of ecosystem-level processes in streams through measuring leaf breakdown rates.

II. GENERAL DESIGN

A. Site Selection

Leaf breakdown studies can be performed in virtually any size stream but small, shallow streams present fewer problems than large rivers. Streams with gravel or cobble substrates are preferable to those with sandy bottoms because of the difficulty of anchorage and the likelihood of burial in sandy-bottomed streams. Remote sites are preferable to sites that receive regular human traffic.[1] When placing litter bags in the stream, avoid spots that are likely to be significantly deeper during higher flows, areas of excessive erosion (e.g., next to cut banks) or deposition (e.g., point bars), and areas that may be unstable under higher flows (e.g., debris dams). Best results are usually obtained when litter bags are placed in shallow riffles closer to the bank than the thalweg, thus avoiding increased stream power during high flows.

B. General Procedure

This chapter involves placing preweighed leaf packs in a stream, recovering samples periodically, and estimating the rate at which the packs break down (i.e., disappear in the stream). Specifically, a large number of leaf packs are constructed and placed in a stream. At regular intervals over the course of the study (perhaps 3–7 months) three to five of the leaf packs are retrieved at each sampling date. Remaining leaf material is cleaned of debris and invertebrates, dried to constant mass, and weighed. Species-specific breakdown rates (k) are then calculated by using an exponential decay model, which assumes that rate of loss from the packs is a constant fraction of the amount of material remaining. Operationally, k is the negative slope of the line produced by a linear regression of the natural log of percentage leaf material remaining plotted against time.

III. SPECIFIC EXERCISES

A. Exercise 1: Leaf Litter Breakdown

1. Collect leaves from trees just before they are ready to fall (i.e., at abscission).

---

[1]Strings of colored litter bags are attractive to the curious.

TABLE 27.1
Example Data Sheet for Leaf Breakdown Study (All Values in Grams)

Sample ID	LPDM	PM	P + DM	P + AM	DM	AM	% Organic	AFDM
Maple 1[a]	8.27	1.0000	1.2500	1.0250	0.2500	0.0250	90	7.44
Maple 2	——	——	——	——	——	——	—	——
Maple 3	——	——	——	——	——	——	—	——
Oak 1	——	——	——	——	——	——	—	——
Oak 2	——	——	——	——	——	——	—	——
Oak 3	——	——	——	——	——	——	—	——
Days of incubation ————								

[a]Example data.

Note. LPDM, dry mass of leaf packs; PM, pan mass; P + DM, pan mass plus preashed mass of milled sample; P + AM, pan mass plus postashed mass of milled sample; DM, (P + DM − PM); AM, (P + AM − PM); % Organic, (DM − AM/DM × 100); AFDM, (LPDM × % Organic).

2. Air dry by putting leaves into a large cardboard box(es) with many 2.5-cm diameter holes covered by plastic window screen. Turn the box(es) daily and gently "fluff" the leaves to promote drying. Continue 5–8 days until leaves reach relatively constant mass. Leaves can be spread out on the floor or tables to dry. Thick piles of leaves should be turned over frequently to promote drying.

3. Weigh out 3- to 10-g portions on a top-loading analytical balance, and fashion them into leaf packs by one of several techniques described below. It is advisable to use the same mass (weight) of leaves for each pack. Record initial dry mass (LPDM) on a data sheet (see Table 27.1).

4. Construct mesh bags from bridal netting or similar material, poultry fencing, hardware cloth, plastic screen, other material that is mesh-like, or purchase commercial mesh bags used to package food items such as oranges or peanuts (e.g., 5-lb. citrus bags[2]). Regardless of material used, mesh openings should be as large as possible to allow access to the leaves by consumers yet retain the leaf material (methods reviewed by Webster and Benfield 1986).

5. Prepare enough packs for the entire exposure period. Set up a retrieval schedule according to the leaf type used. Fast decomposing leaf

[2]Plain citrus bags can be purchased in lots of 1000 from the Cady Bag Company, Pearson, GA, 1-800-243-2451.

species may disappear in 1–3 months and packs should be collected weekly or every 2 weeks. Medium and slow decomposing leaves may require 4 to 12 months to disappear and may be collected at monthly intervals (Webster and Benfield 1986). Since variability in the amount of material lost from leaf packs is relatively high, especially in the latter stages of decomposition, a minimum of 3 packs per species per site should be retrieved on each date in order to calculate a mean and standard error (e.g., to study the breakdown rate for one species at one site for 7 months you need start with 21 packs plus 3 for *handling loss* correction). In experiments involving more than one species, color code[3] the bags as to species. It can be difficult to distinguish between species when the leaves are in the middle to latter stages of decay. By their very nature, dried leaves are easily broken in handling. Therefore, it is necessary to account for losses encountered in fashioning, transporting, and placing the packs in the stream. This can be accomplished by preparing extra leaf packs that go through the entire process but are not left in the stream to incubate. The extra packs are processed and used to correct for "handling losses" as described below.

6. Place leaves into mesh bags. To reduce breakage, it is advisable to dampen the leaves with deionized water from a misting bottle after they are weighed but before they are placed in the bags.

7. Transport all packs to the stream site, handling carefully to avoid unnecessary breakage. Again, the packs can be wetted a bit to make them less vulnerable to breakage.

8. Secure the packs to the streambed. Depending on size and flow rate, various restraint systems are recommended. In small, shallow streams, leaf packs in mesh bags can be tied singly or in groups to gutter nails[4] pushed or driven into the streambed (Webster and Waide 1982). In larger or faster flowing streams, steel rods or metal fence posts driven into the streambed may be necessary to anchor the leaf packs. Alternatively, tie the packs to heavy wire attached to a tree along one bank. If none of these techniques seem appropriate, stronger devices such as those described in Benfield and Webster (1985) may be necessary. Place all packs, except those designated for handling loss correction, in the stream, spreading them out as much as possible as governed by the restraint system chosen, available space, etc. If more than one leaf species is being used, each species should be spread randomly through the array.

9. On retrieval from the stream, place each leaf pack individually into

[3]Plastic "embossing" tape stapled to the bags works well.
[4]Gutter nails are 9″ nails used to attach guttering to houses and can be obtained at most lumber or home improvement centers.

a zip-lock bag. Include an internal label (pencil or permanent marker on stiff paper) identifying the sample with sufficient information (e.g., retrieval date, site, species, etc.). Write the same information on the outside of the zip-lock bag, using a permanent marker. Place the samples on ice in a cooler and return them to the laboratory. Keep the bags in the cooler or refrigerate until processed.

10. Processing in the laboratory should be completed within 24 h of retrieval. Processing of leaf packs may involve several options depending on whether only leaf-pack breakdown is measured or if breakdown plus additional work with macroinvertebrates (see Chapter 16), microbes (Chapter 12), leaf chemistry (e.g., Maloney and Lamberti 1995), or other components is conducted.

11. Remove the leaf material and gently rinse the leaves of silt and debris. Place the cleaned leaves in paper bags, carrying the inside labels along. Label the outside of the paper bags with the same information that appears on the inside labels. Hang the bags on a line stretched across the laboratory and allow the leaf material to air dry to constant mass. Alternatively, dry in a hot air oven at no higher than 50°C for at least 24 h. After drying, weigh leaf material and record the dry mass (DM). If the project calls for saving macroinvertebrates, perform the rinsing of the remaining leaf material over a 250-μm sieve and place the macroinvertebrates in 70% ethanol. If you plan to do microbial analyses, subsample the leaves before drying and keep the subsamples cold and moist or frozen depending on the protocol required for the particular analyses planned.

12. In many cases, mineral deposits are not readily washed off the leaves and may result in errors in final dry mass. This problem can be mostly overcome by converting dry mass to ash-free dry mass (AFDM) (see Chapter 14). Organic matter combusts at about 550°C and the remaining material is mineral ash. When the mass of mineral ash is subtracted from initial dry mass, the result is the dry mass of the organic fraction (AFDM) of the leaf material. Begin the process by milling or grinding (e.g., Wiley Mill or mortar and pestle) all or significant portions of the "handling-loss" leaves and the leaves from each retrieval (all packs combined for each date) to a fine powder. Determine AFDM for the leaf packs as described below in step 13. The AFDM of the handling-loss leaves serves as the initial AFDM, (i.e., the AFDM of the leaves before they were put into the stream).

13. Mark the underside of aluminum weighing pans by inverting them over the bottom of a beaker and impressing a code using a metal probe. Record the identification codes on the data sheet. Heat the coded pans at 550°C for 30 min in a muffle furnace. Then, handling with gloves and tongs or forceps as appropriate throughout the process, place the pans in

a desiccator to cool, then weigh and record mass (weight) of each pan at the appropriate place on the data sheet.[5] Weigh out at least two subsamples of about 250 mg DM of each milled sample to a pan, oven-dry overnight at 50°C, and place in a desiccator to cool. Weigh the pans plus milled samples and record on the data sheet. Place pans plus milled samples in a muffle furnace at 550°C for 20 min, remove and stir with dissecting needle, and heat for an additional 20 min (Gurtz *et al.* 1980). Allow to cool and then wet-down material with distilled water, oven dry at 50°C for 24 h, and desiccate until pan and ash return to room temperature. Weigh pan plus ash and record on the data sheet. Subtract the pan weights from the pre-ashed pan plus sample and postash pan plus sample. Compute percentage organic matter (%OM) of the milled samples,

$$\%OM = \left(\frac{DM_{sample} - AM_{sample}}{DM_{sample}} \right) \times 100\%, \qquad (27.1)$$

where DM_{sample} is sample dry mass and AM_{sample} is sample ash mass.

14. Convert DM values of leaf packs to AFDM [AFDM = (DM) × (% organic matter)]. The mean AFDM of the *handling-loss* leaf packs is used as an estimate of the *initial* AFDM of all the leaf packs. Convert AFDM for each leaf pack to percentage AFDM remaining [(Initial AFDM − Final AFDM)/(Initial AFDM) × 100%]. Compute mean percentage AFDM remaining for each retrieval.

15. Regress the natural log (ln) of mean percentage AFDM remaining (*y*-axis) on days of exposure (*x*-axis) using the AFDM of the handling-loss leaf packs as 100% remaining for Day 0. The negative slope of the regression line is equal to the processing coefficient (*k*). Linear regression can be run on most modern calculators or on statistical software packages. If neither is available, you can approximate the slope by plotting the natural log of percentage remaining (*y*-axis) against days of exposure (*x*-axis), drawing the *best fit* line, and calculating the slope ($\Delta y/\Delta x$).

B. Exercise 2: Leaf Breakdown for One or Several Species

Perform the leaf breakdown study using one or several leaf species at one site in a stream. Describe the leaf breakdown pattern using a linear plot of percentage remaining against time and also compute the *k* value (Fig.

[5]Premuffling the pan is necessary to eliminate weight loss of the pan due to heating.

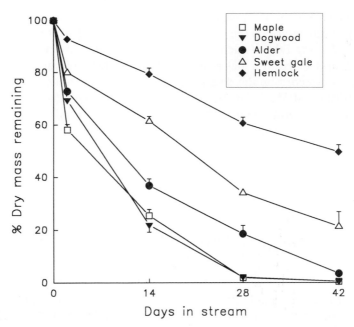

FIGURE 27.1 Processing of five different leaf species in a northern Michigan stream during summer (redrawn from Maloney and Lamberti 1995).

27.1). Processing coefficients (k) can also be computed using temperature (degree-days) as values on the x-axis in place of days (Petersen and Cummins 1974). Cumulative degree-days may be estimated by summing the average daily water temperatures (°C) over each incubation period and entering the appropriate values (i.e., the degree-days accumulated from Day 1 to the retrieval day) in place of days.

The simple single-species–single-site in one stream model can easily be expanded by including more species. For example, one could contrast the breakdown rates of presumed fast, moderate, and slow decomposing species. Another model could be to contrast riparian shrub or herbaceous leaves with tree leaves, or perhaps deciduous and evergreen shrub and/or tree leaves (e.g., Hauer *et al.* 1986, Maloney and Lamberti 1995). Species breakdown patterns can be compared by inspection or statistically. To compare by inspection, plot the percentage remaining on a linear scale versus days or degree days for each species on the same graph. Describe the differences in patterns of mass loss over time or degree day accumulation. Processing coefficients (k) may be compared statistically using analysis of covariance (ANCOVA) followed by a multiple comparison technique to determine whether the k values are significantly different (see Zar 1984 or Sokal and Rohlf 1995).

C. Exercise 3: Effects of Spatially Varying Stream Features on Leaf Breakdown Rates

Many designs are possible depending on the question(s) of interest. For example, use one or several species of leaves at single or multiple sites in one or several streams. Site differences could include riffles versus pools, high elevation versus low elevation, cobble substratum versus bedrock or sand, or canopied versus uncanopied. Stream differences could be based on stream order, gradient, geology, disturbance history, hardness, or nutrient level. Compare site–species combinations using the inspection or statistical methods described in Option 1.

D. Exercise 4: Effects of Anthropogenic Activities on Leaf Breakdown Rates

Investigate the impact of a municipal, industrial, or mining waste outfall on stream organic matter processes using leaf breakdown rate as an indicator (e.g., Paul *et al.* 1983). The usual protocol for evaluating the impact of a waste outfall on streams is to compare some value(s) upstream and downstream of the outfall at comparable sites (for a general discussion see Plafkin *et al.* 1989). Establish an upstream reference site(s) that is totally removed from any possible impact of the outfall in question. In wider streams, you may also use sites across the stream as reference sites. Establish a site just downstream from the outfall where impact is likely to be maximal, and then several additional sites further downstream including one at which you judge the impact of the outfall to be abated. Proceed with the study as outlined in the Section II above and analyze for site differences as described in Exercise 1.

IV. QUESTIONS

1. Are leaf pack breakdown and organic material decay essentially the same process? Why or why not?
2. What might be the impact to energy flow in a woodland stream if streamside (riparian) vegetation composition were simplified by removing all but one or two species? Can you think of examples where this has been done?
3. By what mechanisms might a pollution source alter the process of leaf breakdown in streams?
4. How might you attempt to experimentally separate the importance

of biological process (i.e., microbial conditioning and consumer feeding) from physical processes such as abrasion and fragmentation by currents in leaf breakdown in streams?

    5. What are some of the variables that make some leaves more resistant or susceptible to leaf breakdown processes in streams?

V. MATERIALS AND SUPPLIES

Field Supplies

Gutter nails, steel rebar, cable, as needed depending on leaf-pack design
Mesh bags
Zip-lock bags, paper bags, labels, tape, waterproof marker

Laboratory Supplies and Equipment

250-μm Mesh sieve
Aluminum weighing pans, tongs, gloves, forceps
Analytical balance accurate to ±0.1 mg
Drying oven
Grinding mill (or mortar and pestle)
Muffle furnace (optional)
Top loading analytical balance accurate to ±0.1 g

REFERENCES

Bärlocher, F., and B. Kendrick. 1975. Leaf conditioning by microorganisms. *Oecologia* **20:**359–362.

Benfield, E. F., D. R. Jones, and M. F. Patterson. 1977. Leaf pack processing in a pastureland stream. *Oikos* **29:**99–103.

Benfield, E. F., and J. R. Webster. 1985. Shredder abundance and leaf breakdown in an Appalachian mountain stream. *Freshwater Biology* **15:**113–120.

Cummins, K. W. 1974. Structure and function of stream ecosystems. *BioScience* **24:**631–641.

Cummins, K. W., and M. J. Klug. 1979. Feeding ecology of stream invertebrates. *Annual Review of Ecology and Systematics* **10:**147–172.

Gurtz, M. E., J. R. Webster, and J. B. Wallace. 1980. Seston dynamics in southern Appalachian streams: effects of clear-cutting. *Canadian Journal of Fisheries and Aquatic Sciences* **37:**624–631.

Hauer, F. R., N. L. Poff, and P. L. Firth. 1986. Leaf litter decomposition across broad thermal gradients in southeastern (USA) coastal plain streams and swamps. *Journal of Freshwater Ecology* **3:**545–552.

Maloney, D.C., and G. A. Lamberti. 1995. Rapid decomposition of summer-input leaves in a northern Michigan stream. *American Midland Naturalist* **133**:184–195.

Paul, R. W., Jr., E. F. Benfield, and J. Cairns, Jr. 1978. Effects of thermal discharge on leaf decomposition in a river ecosystem. *Verhandlungen der Internationalen Vereinigung für Theoretische und Angewandte Limnologie* **20**:1759–1766.

Paul, R. W., Jr., E. F. Benfield, and J. Cairns, Jr. 1983. Dynamics of leaf processing in a medium-sized river. Pages 403–423 *in* T. D. Fontaine and S. M. Bartell (Eds.) *Dynamics of Lotic Ecosystems.* Ann Arbor Press, Ann Arbor, MI.

Petersen, R. C., and K. W. Cummins. 1974. Leaf processing in a woodland stream. *Freshwater Biology* **4**:345–368.

Plafkin, J. L., M. T. Barbour, K. D. Porter, S. K. Gross, and R. M. Hughes. 1989. *Rapid Bioassessment Protocols for Use in Streams and Rivers: Benthic Macroinvertebrates and Fish.* Document/444/4-89-001, United States Environmental Protection Agency, Washington, DC.

Sokal, R. R., and F. J. Rohlf. 1995. *Biometry,* 3rd ed. Freeman, New York, NY.

Suberkropp, K., G. L. Godshalk, and M. J. Klug. 1976. Changes in the chemical composition of leaves during processing in a woodland stream. *Ecology* **57**:720–727.

Suberkropp, K., and M. J. Klug. 1976. Fungi and bacteria associated with leaves during processing in a woodland stream. *Ecology* **57**:707–719.

Suberkropp, K., and M. J. Klug. 1980. The maceration of deciduous leaf litter by aquatic hyphomycetes. *Canadian Journal of Botany* **58**:1025–1031.

Webster, J. R., and J. B. Waide. 1982. Effects of forest clearcutting on leaf breakdown in a southern Appalachian stream. *Freshwater Biology* **12**:331–344.

Webster, J. R., and E. F. Benfield. 1986. Vascular plant breakdown in freshwater ecosystems. *Annual Review of Ecology and Systematics* **17**:567–594.

Zar, J. H. 1984. *Biostatistical Analysis,* 2nd ed. Prentice–Hall, Engelwood Cliffs, NJ.

CHAPTER 28

Organic Matter Budgets

G. WAYNE MINSHALL

Stream Ecology Center
Department of Biological Sciences
Idaho State University

I. INTRODUCTION

A budget is a systematic accounting of inputs (income) and outputs (losses). In stream ecosystems *inputs* are interpreted broadly to include *autotrophic production* as well as that brought in by various physical vectors. *Outputs* are considered to incorporate community respiration (e.g., metabolism, decomposition) in addition to losses by downstream transport. Inputs retained within a particular stream reach over a given time interval are designated as *storage*. In an ecological context, budgets serve to identify the potential sources and fates of energy and materials in ecosystems and quantify their magnitude. Materials budgets of interest in the study of ecosystems usually are either specific chemical elements, particularly major macro- or micronutrients, or organic matter, either as a mass or as elemental carbon.

In stream ecology, organic matter budgets are important in determining the sources and fates of organic matter, in understanding the internal dynamics of these systems, and in making comparisons at different locations within a catchment or river basin or among different regions or stream types. The organic matter in a stream may originate from the surrounding terrestrial environment and be transported to the stream by wind, water, or direct deposition. Because of its origin outside of the stream boundaries, this material is referred to as *allochthonous* and is primarily of plant origin. Since much of this allochthonous plant material, in the form of leaves, twigs, and other parts, is dead by the time it reaches the stream, it also is often referred to as detritus (i.e., allochthonous detritus). The other main source of organic matter originates within the stream itself in the form of algae and higher aquatic plants and hence is termed *autochthonous*. Terrestrial insects, carcasses of various vertebrates, and the like also represent sources of organic input into streams. However, in terms of sheer mass, these are often insignificant compared to the magnitude of plant and dissolved sources. Therefore, in many budgets this component is ignored, even though it may be important qualitatively or, in some cases, quantitatively.

A major question in the study of streams is the relative importance of externally derived (allochthonous) versus internally derived (autochthonous) organic matter within and among streams and at different times. In many cases, particularly in shallow (wadable) streams in forest or shrublands, organic matter of terrestrial origin plays a major, often overriding, role in establishing stream ecosystem structure and function (e.g., Fisher and Likens 1972, Minshall *et al.* 1983). In other cases, particularly in meadow, grassland, and desert streams and in large rivers, algae or aquatic vascular plants are the principal source of organic matter, which alter lotic structure and function accordingly (Minshall 1978, Cushing and Wolf 1982, Minshall *et al.* 1983, 1992). New insights into the dynamics of flowing water ecosystems and terrestrial–aquatic linkages are based on the changing terrestrial dependence of these systems with different biogeographical areas, increasing channel size, varying types and amounts of streamside vegetation, different land-use practices, and the dynamics of input, storage, processing, and output of organic matter (e.g., De La Cruz and Post 1977, Cummins *et al.* 1983, Minshall *et al.* 1983, 1992, Duncan and Brusven 1985a,b, Naiman *et al.* 1987, Duncan *et al.* 1989, Meyer 1990).

The early use of materials budgets in streams was to determine the nutrient relationships in forested watersheds (Bormann *et al.* 1969, 1974, Bormann and Likens 1979). In these studies, the stream was considered to be a major pathway by which elements and products of the forest could be exported from a watershed. In this early view, streams were seen as nonreactive conduits that transported materials washed or drained from the forest to a weir or other convenient point for measurement by terrestrial ecologists. Subsequently, streams were shown to be complex, dynamic entities in which materials released from the forest and produced within the stream were intermixed, utilized, transformed, and released (often many times over) while being variously propelled and retarded along the channel before reaching a point at which a sample might be collected. More recently, the relationship has been shown to be even more complex with the demonstration that additional delays, transformations, and inputs may occur in streamwater entering the channel bed (e.g., hyporheic zone; see Chapters

6 and 30) and interchanging with both ground- and surface water (Minshall 1988, Triska *et al.* 1989).

Studies of organic matter budgets in streams often are inseparable from efforts to evaluate energy flow in these ecosystems. This is readily seen for some of the earliest studies on flowing waters (Odum 1957, Teal 1957, Tilly 1968), which also used spring habitats (rheocrenes) to simplify measurement. A subsequent generation of studies focused more directly on organic matter in streams draining small watersheds or in segments of larger streams (Nelson and Scott 1962, Fisher and Likens 1972, 1973, Hall 1972, Westlake *et al.* 1972, Sedell *et al.* 1974, Fisher 1977). Other, more recent, studies using comparable approaches include those by Minshall (1978), Mulholland and Kuenzler (1979), Mulholland (1981), Cushing and Wolf (1982), and Newbold *et al.* (1982). The transformation from thinking of streams as nonreactive conduits to that of reactive sticky traps was accompanied by the realization that streams were not simply a disjunct accumulation of erosional and depositional areas or a disconnected, open system in which nutrient cycling could not occur. Instead, they came to be viewed as continuously interconnected "systems" in which cycling was completed along a linear (upstream–downstream) gradient. The processing and resultant partial release of organic and inorganic nutrients from one part of a stream to another has come to be characterized as processing along a continuum (Vannote *et al.* 1980, Minshall *et al.* 1983, 1985) or spiraling (Webster and Patten 1979, Newbold *et al.* 1982). Modern studies combine the two concepts and permit the additional determination of such features as cycling distance, turnover time, and retention capacity (e.g., Minshall *et al.* 1992).

The purpose of this chapter is to provide a straightforward set of procedures for the construction of an organic matter budget. The basic components of a comprehensive annual organic matter budget are described and the different exercises provide different levels of intensity, from one type of material (Exercise 1) to many (Exercise 2) and for short time intervals (e.g., 1 day to a few weeks) to an annual budget or longer (Exercise 3). Exercise 4 provides an opportunity for dealing with some of the conceptual issues prior to or instead of collecting field samples. For ease in presentation, focus is on a short segment of a wadable stream in which there are no tributaries or diversions; however, the procedures may be expanded readily to address more complex situations. The methods are suitable for describing stream ecosystem functioning under a variety of natural and human-induced conditions, including the effects of fire, mining, livestock grazing, logging, and channelization. Little or no information is available on carbon budgets for areas outside of North America or in large rivers. In large rivers, water-column processes are expected to override benthic ones (Vannote *et al.* 1980) but few data are available to test this assumption (see e.g., Minshall *et al.* 1992). Accurate organic matter budgets for streams, especially as they enter the oceans, are needed for determination of global carbon budgets and effects on global climate. Particularly fruitful would be studies in areas of environmental extremes, such as tropical rainforests, tundra, and deserts, and under different types of resource management conditions ranging from national parks and wilderness areas to agricultural lands of various kinds and urban waterways. In addition, most of the budgets to date are rather crude and further investigation is needed to carefully define the effects of temporal and spatial variation at various levels of resolution (e.g., see Minshall 1988). This information would be valuable in improving past estimates (or defining their level of accuracy) and future measurements.

II. GENERAL DESIGN

The construction of a materials budget is based on a general mass balance approach in which the inputs (import), outputs (export), and standing stocks (storage) within a stream or segment are quantified. In the form of a mass balance equation, import (I) plus the change in storage (ΔS) is taken to be equal to export (E). Changes in storage may be either positive, in the case of accumulating conditions (e.g., aggrading channels) or negative, in the case of erosive conditions (e.g., degrading channels). With respect to organic matter (and energy), production ($+$) or decomposition ($-$) of materials within the study reach (e.g., community metabolism, ΔM) also may need to be considered. Thus,

$$E = I + \Delta S + \Delta M. \tag{28.1}$$

Alternatively, production (P) may be considered a type of import and respiration (R) a type of export, in which case

$$E + R = I + P + \Delta S. \tag{28.2}$$

Under equilibrium conditions (i.e., a "balanced" budget), the two sides of the equation will be equal and the net balance is zero. If the sum of the right-hand side of the equation exceeds the sum of the left-hand side, the system is losing organic matter and, if the reverse is true, the system is accumulating organic matter.

The description will be given for a complete organic matter budget but the specific exercises followed should reflect the actual component or components of interest (e.g., the simplest case would be restriction of the analysis to coarse particulate allochthonous detritus, as in Exercise 1). The major components of an organic matter budget for a stream segment are shown in Fig. 28.1. The input and output pathways are sometimes referred to as vectors or fluxes. Commonly, research effort is apportioned to focus on the main pathways, with minor fluxes being estimated less precisely, determined from literature values, or ignored.

A. Import

Allochthonous inputs occur in the form of dissolved and particulate organic matter. The dissolved portion (DOM, <0.45 μm) may enter via transport from upstream, groundwater, surface runoff, precipitation (through-fall), or *in situ* leachate from leaves. For each of these pathways except the last, the quantities involved can be determined from the product of the concentration in the particulate category of water involved times

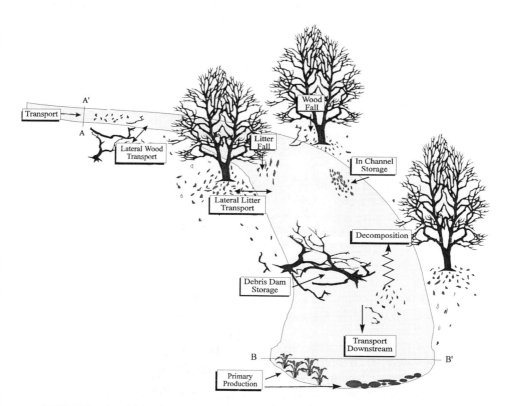

FIGURE 28.1 Major components of an organic matter budget for a forested stream segment defined by an upstream (A–A′) and a downstream (B–B′) transect.

the total volume or discharge of water. The particulate component (POM) vectors include stream transport, direct deposit (fall-in), wind borne, and gravitational movement down valley sideslopes. Frequently, POM is subdivided into ultrafine (UPOM, 0.45 to 50 μm), fine (FPOM, >50 to 1000 μm), and coarse (CPOM, >1000 μm) components, due to functional differences in importance (also see Chapters 10 and 11 for other size-fraction separation). Commonly, CPOM consists of twigs and whole leaves.

Autochthonous inputs correspond in general to the size categories given above, but arise as exudates from aquatic organisms (DOM) or from the scouring, sloughing, breakage, or uprooting of micro- and macroscopic organisms, especially algae and vascular hydrophytes (POM). This material is believed to enter a particular stream reach exclusively by means of waterborne transport from upstream. To date, no effort has been made to distinguish autochthonous versus allochthonous DOM fractions. However, the two DOM sources combined can easily account for 50% or more of the total OM input. In the case where the stream ecosystem or study-site boundaries encompass the headwaters, the terrestrial/aquatic origins of the material are easily discerned for delineating the input part of a budget. However, in the case of a stream segment, an added source of input will be the stream itself as the current transports organic materials across the upstream boundary (transect line) and into the segment of interest. Since this material is a composite of terrestrial and aquatic materials, it should be subdivided (partitioned) according to autochthonous and allochthonous origin, and often also by particle size, for the purpose of compiling a budget.

B. Storage

Material whose loss from the study boundaries is impeded by physical or biological means (see Chapters 10 and 11) is said to be in storage. DOM is assumed to move with the flow of water and hence not be stored or to be adsorbed or flocculated and thus converted to POM. Most organic matter budgets are computed for annual time periods, with changes in storage pools assumed to be negligible over that time frame. However, longer time periods may be required to adequately account for variations in channel aggradation or erosion (Cummins et al. 1983). For shorter time periods, such as a day to a few weeks, changes in storage can be considered imperceptible, unless major variations in stream discharge occur. Organic matter is stored in streams primarily through deposition and admixture with inorganic sediments in quiescent, depositional areas; entrapment around rocks; or accumulation in piles of woody debris. Lesser amounts may be captured by the nets and feeding appendages of filter-feeding invertebrates. Standing stocks of algae, vascular plants, macroinvertebrates, and fish (or other vertebrates) also may constitute important forms of storage.

C. Metabolism

Within the study segment or reach, organic matter may be elaborated by primary production and growth and consumed through respiration, including microbial decomposition. Net accumulation ($+$) or loss ($-$) is measured by assessing changes in standing crop or community metabolism (see Chapters 14 and 25).

Pathways of utilization of organic matter, availability, and rates of incorporation or decomposition, vary in response to the relative amounts of allochthonous and autochthonous components, as well as the particular makeup of each. For example, leaves of alder, dogwood, and willow are utilized more rapidly than those of maple, oak, or pine (see Chapter 27) and diatoms generally are consumed more readily than the green alga *Cladophora* or aquatic vascular plants. Autochthonous or allochthonous inputs that are not utilized through ingestion and assimilation or decomposition may be retained within a stream section (i.e., stored) or exported.

D. Export

Downstream transport is the most common means of export for all forms of organic matter in streams. During periods of overbank flow, lateral transport to the floodplain occurs also. Some loss to the atmosphere may occur as carbon dioxide, especially in poorly buffered waters, but this is generally regarded to be minor. Downstream transport is the product of the concentration in the streamwater and discharge (see Chapters 3, 10, and 11).

E. Site Selection

This procedure is applicable to a wide range of stream sizes. Generally, any safely wadable stream, which is accessible at various points along its banks, is suitable without additional precautions or logistical considerations. Larger streams require extra effort to safely access sampling sites and to adequately account for spatial variation in the distribution of organic matter in each of the horizontal, vertical, and longitudinal dimensions of a stream segment. The width of stream selected will affect sampling intensity and size of research team needed. Wider streams will require multiple samples across a transect. Narrow streams will facilitate sampling of all or most of the water passing a point. The entire height of the water column can be sampled in shallow waters, whereas deeper streams most likely will be partitioned into several strata for study. In small, headwater catchments or rheocrenes (i.e., flowing springs), the entire stream may be chosen for study. However, in most other streams, a segment, delineated by arbitrary upstream and downstream boundaries, will be the focus. A minimum stream length for study is several hundred meters; for example, in pioneering studies of Bear Brook and the Fort River, Fisher used 1700 m (Fisher and Likens 1973, Fisher 1977). In streams larger than third- or fourth-order, a length equal to 7× the mean width commonly is accepted as a minimum length of a segment for determining carbon budgets.

III. SPECIFIC EXERCISES

To properly frame the exercises in this chapter it is advisable to document basic stream conditions. Refer to Chapters 1–6 to obtain methods of placing the study stream in context of land forms, channel and bed forms, discharge and hydraulic properties, and water chemistry.

A. Exercise 1: Constructing an Organic Matter Budget

This exercise illustrates the general process of building an OM budget but CPOM is easier to deal with as an introductory exercise than are the other particle sizes or an entire budget.

1. Well in advance of this study (preferably in conjunction with step 2), visit a nearby stream and select a representative study segment several hundred meters in length, containing no tributaries. Establish a transect at both the upper and the lower ends. The lower transect should be located so that the same habitat type precedes it (e.g., a riffle) as is found immediately upstream of the upper transect. This will help assure that the measurements of import and export are not influenced by different hydraulic conditions.

2. Set out litter-fall baskets and blow-in/lateral-transport collectors for 30 days beginning approximately 1 week before leaf abscission. Litter-fall baskets can be produced by suspending plastic laundry baskets above the stream using nylon cord or rope. Line the baskets with plastic garbage bags for easy collection of material and transfer to the laboratory. The bags should be pierced with small holes at the time of installation to permit drainage of moisture. Place one or more collection baskets every 30 m, starting at one end of the segment.

3. Composite collectors to capture wind-blown, laterally transported material can be made by setting plastic kitchenware containers with small drain holes into the ground alongside the stream. Collectors are buried so the upper edge is flush with the surface of the ground. Plastic window screening, suspended about 25 cm above the top of the containers by means

of stiff wires stuck into the ground, will prevent entry of material from direct fall in. Place trays along the stream in the vicinity of each litter-fall basket, either one on each side of the stream or alternating from side to side.

4. Determine mass (g) of the collected plant matter after drying overnight at 100°C. Standardize the results on a per square meter per day basis by dividing the mass by the area of the basket opening (in m^2) and the number of days the collectors were in place, respectively. Ash a subsample of the material you collect and determine a conversion factor (see Chapter 10 for AFDM protocols). Refinements include weighing sticks and leaves separately and separating the leaves according to plant species. Express the results as mean AFDM per day per meter of stream length (based on the dimension of the trays parallel with the stream). Multiplying by segment length will yield the total amount imported into the segment per day. Calculate the total amount received over the entire segment per day (= mean AFDM/day × segment area as obtained in step 5).

5. Determine the cross-section profiles (widths and depths) for the upper and lower transects established in step 1 and for sufficient points in between to characterize the mean width and depth of the study segment (see Chapter 2). Segment area = mean width × length (m). Determine discharge at one of the transects (see Chapter 3). This measurement should be made carefully as it will strongly influence the upstream and downstream transport values. Alternatively, discharge could be determined at both transects and the values averaged. If more than one period will be devoted to this study, discharge should be measured each time transport samples are collected.

6. Collect and process CPOM in transport at a minimum of three points across each of the upper and lower transects. This may be done by setting paired nets (10-cm diameter opening and 1-mm mesh net) for a sufficient time to collect measurable quantities of CPOM (e.g., 1 h). However, the length of time that the nets are set depends largely on the concentration of CPOM in the stream being sampled. Current velocity must be determined to calculate the volume of water filtered by the sample nets. This may be done using a good quality current meter and then multiplying the area of the net opening × the current velocity × the time duration the nets were set. A single set of collections may suffice but, if possible, sampling should be repeated at least every 4 h for a 24-h cycle. Multiply stream discharge in m^3/24 h by the mean concentration of CPOM (i.e., g AFDM/m^3) to obtain total load per day for both import and export. At least separate the material into allochthonous and autochthonous components. Refinements include weighing sticks and leaves separately, and separating algae, moss, and the leaves of aquatic and terrestrial plants according to species.

7. For periods of short duration (up to a few weeks), in which no major changes in discharge or other disturbance to the streambed have occurred, it may be acceptable to assume no net change in storage (i.e., $\Delta S = 0$) for the interval. However, for longer periods, following major flooding, or simply to confirm the assumption of no net change, storage should be determined directly following the general approach described here. On two or more occasions, separated by several weeks (e.g., at the time the litter-fall baskets are set out and taken up), collect and process CPOM in storage from 5 to 10 locations within the study reach, using a stratified random design with respect to habitat types. An ordinary Surber or Hess benthic sampler fitted with a 1-mm mesh net (or whatever the minimum particle size of interest is) will suffice for sample collection. Determine ash-free dry mass (see Chapter 10 for AFDM protocols) of CPOM in storage, which has been gently washed free of inorganic sediments prior to drying and ashing. Express the results as g/m^2 and determine the net change during the period [i.e., time 2 results minus time 1 results = net change (+ or −)]. An additional improvement would be to determine empirically the number of samples needed to adequately represent the mean values for each habitat type or for the entire study reach (e.g., see Elliott 1977).

8. Net production is zero for allochthonous CPOM and is included in the storage calculation for aquatic macrophytes. Respiration losses may be measured directly using recirculating metabolism chambers (see Chapter 25) or estimated from published rates.

9. Summarize the data from your measurements as the means and standard deviations and enter them into the budget sheet (Table 28.1). Make

TABLE 28.1
Worksheet for Estimating the Daily (24-h) Coarse
Particulate Organic Matter Budget of a Stream Segment
(g Segment^{-1} day^{-1})

	Autochthonous	Allochthonous
Import		
Direct fall		
Wind blow/lateral transport		
Transport from upstream		
Storage		
All sources		
Metabolism		
Respiration		
Export		
Transport downstream		

any additional calculations needed and complete the budget by applying the mass balance equation.

B. Exercise 2: Incorporating Other Particle Sizes of Organic Matter and Community Metabolism

1. Sample fine and ultrafine organic matter using sets of nested nets of the appropriate mesh size(s) (see Chapter 10).

2. Measure primary productivity and community respiration (Chapter 25). Whole-system measures, such as obtained from the upstream–downstream method, may suffice. However, it is often desirable, particularly in larger streams, to partition metabolism measurements according to water column and benthic values using recirculating metabolism chambers.

C. Exercise 3: Extending Analyses to Periods of One or More Years (= Annual Budgets)

1. The simplest procedure is to obtain at least one set of measurements per season, multiply the results by the number of days in each season,

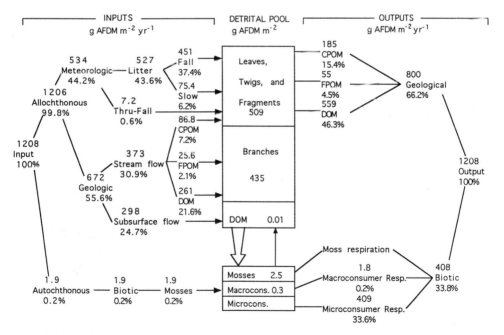

FIGURE 28.2 Sources and fates of organic matter in a small headwater stream (Bear Brook, New Hampshire (Fisher and Likens 1972)). Standing stocks of storage pools in g AFDM/m^2; input and output fluxes in g AFDM m^{-2} year^{-1}. Geologic fluxes are primarily the result of water movement; meteorologic fluxes are driven by gravity, wind, and precipitation. In this analysis, the stream is treated as a "segment" with organic matter both entering and leaving with stream flow.

and sum the results from each season to obtain an annual total. Further improvements might include expansion of the sampling to better represent seasonal variations, such as through monthly sampling. Day-to-day variations (particularly in transport) may be substantial, even during periods of stable hydrologic conditions, and are worthy of further study in their own right.

2. A more detailed analysis would lead one to isolate and better represent ecologically significant events within or among years. For example, an ice storm may change the loading of organic matter to the stream, its transport through a stream segment, the level of primary productivity, and the extent of scouring. Some of the most significant differences are associated with changes in the hydrograph (see Chapter 3). Effort should be made to sample conditions on each side of major changes in the hydrograph; for transport in particular, this should include measurements during the rising and falling limbs of the hydrograph.

D. Exercise 4: Sample Data for "Mythical Creek" Stream Segment

1. Use the data given in Fig. 28.2 to construct a budget (e.g., as in Table 28.1) for either CPOM only or for all size fractions and other components.

IV. QUESTIONS

1. Based on your solution of the mass balance equation, is the study segment in equilibrium, accumulating, or losing organic matter?

2. Which compartment(s) was most responsible for this response? How do you feel about the adequacy of the measurements of the OM in these compartments? How might you increase your level of "certainty" of these values in the future? What are other ways in which the study could be improved?

3. From your answers to Questions 1 and 2, what are the ecological implications of your results (a) for the study segment and (b) with respect to what you can infer about the sections of stream immediately upstream and downstream of the study segment?

4. How might you apply the approach used in this study to determine an organic matter budget for the entire mainstem river in which your study segment lies? For all of the comparable catchments in a, real or hypothetical, adjacent resource management unit?

5. What are some practical uses or "management insights" that could be derived from your measurements and findings?

6. How might changes in land use practices (e.g., logging) change the results?

7. How might different kinds of natural disturbances (e.g., an ice storm, an abrupt change in the hydrograph, a forest fire) affect the budget?

V. MATERIALS AND SUPPLIES

Field Materials

1-mm mesh seston nets (six plus a similar number of other mesh sizes (e.g., 50 μm) as needed)

1-mm mesh Tyler sieves or equivalent (e.g., embroidery hoops fitted with Nitex or other monofilament mesh cloth; plus other mesh sizes as needed)

Current meter, meter sticks, and 100-m measuring tapes

Equipment for primary productivity and community respiration measurement (see Chapter 25)

Nylon cord or rope

Plastic laundry baskets and plastic garbage bag liners ($\geq$10)

Plastic kitchenware trays ($\geq$10)

Plastic window screen cut to size 5 cm or more larger (l and w) than the plastic trays and coat-hanger wire supports

Surber or Hess sampler (fitted with 1-mm mesh net or other suitable mesh size if smaller particle sizes also are being investigated; see Chapters 10 and 11)

Laboratory Materials and Equipment

Drying oven (100°C, well vented for moisture release)

Muffle furnace (550°C), tongs, insulated gloves (not needed if AFDM is estimated from published conversion factors)

Top-loading balance

Weighing boats and crucibles

REFERENCES

Bormann, F. H., and G. E. Likens. 1979. *Pattern and Process in a Terrestrial Ecosystem.* Springer-Verlag, New York, NY.

Bormann, F. H., G. E. Likens, and J. S. Eaton. 1969. Biotic regulation of particulate and solution losses from a forest ecosystem. *BioScience* **19:**600–610.

Bormann, F. H., G. E. Likens, T. G. Siccama, R. S. Pierce, and J. S. Eaton. 1974. The export of nutrients and recovery of stable conditions following deforestation at Hubbard Brook. *Ecological Monographs* **44:**255–277.

Cummins, K. W., J. R. Sedell, F. J. Swanson, G. W. Minshall, S. C. Fisher, C. E.

Cushing, R. C. Petersen, and R. L. Vannote. 1983. Organic matter budgets for stream ecosystems: Problems in their evaluation. Pages 299–353 in J. R. Barnes and G. W. Minshall (Eds.) *Stream Ecology: Application and Testing of General Ecological Theory.* Plenum, New York, NY.

Cushing, C. E., and E. G. Wolf. 1982. Organic energy budget of Rattlesnake Springs, Washington. *American Midland Naturalist* 107:404–407.

De La Cruz, A. A., and H. A. Post. 1977. Production and transport of organic matter in a woodland stream. *Archives für Hydrobiologie* 80:227–238.

Duncan, W. F. A., and M. A. Brusven. 1985a. Energy dynamics of three low-order southeast Alaskan streams: autochthonous production. *Journal of Freshwater Ecology* 3:155–166.

Duncan, W. F. A., and M. A. Brusven. 1985b. Energy dynamics of three low-order southeast Alaskan streams: Allochthonous production. *Journal of Freshwater Ecology* 3:233–248.

Duncan, W. F. A., M. A. Brusven, and T. C. Bjornn. 1989. Energy-flow response models for evaluation of altered riparian vegetation in three southeast Alaskan streams. *Water Research* 23:965–974.

Elliott, J. M. 1977. *Some Methods for the Statistical Analysis of Samples of Benthic Invertebrates.* Freshwater Biological Association Scientific Publication 25, Ambleside, Cumbria, UK.

Fisher, S. G. 1977. Organic matter processing by a stream-segment ecosystem: Fort River, Massachusetts, USA. *Internationale Revue gesamten Hydrobiologie* 62:701–727.

Fisher, S. G., and G. E. Likens. 1972. Stream ecosystem: organic energy budget. *BioScience* 22:33–35.

Fisher, S. G., and G. E. Likens. 1973. Energy flow in Bear Brook, New Hampshire: an integrative approach to stream ecosystem metabolism. *Ecological Monographs* 43:421–439.

Hall, C. A. S. 1972. Migration and metabolism in a temperate stream ecosystem. *Ecology* 53:585–604.

Meyer, J. L. 1990. A blackwater perspective on riverine ecosystems. *BioScience* 40:643–650.

Minshall, G. W. 1978. Autotrophy in stream ecosystems. *BioScience* 28:767–771.

Minshall, G. W. 1988. Stream ecosystem theory: A global perspective. *Journal of the North American Benthological Society* 7:263–288.

Minshall, G. W., R. C. Petersen, K. W. Cummins, T. L. Bott, J. R. Sedell, C. E. Cushing, and R. L. Vannote. 1983. Interbiome comparison of stream ecosystem dynamics. *Ecological Monographs* 53:1–25.

Minshall, G. W., K. W. Cummins, R. C. Petersen, C. E. Cushing, D. A. Bruns, J. R. Sedell, and R. L. Vannote. 1985. Developments in stream ecosystem theory. *Canadian Journal of Fisheries and Aquatic Sciences* 42:1045–1055.

Minshall, G. W., R. C. Petersen, T. L. Bott, C. E. Cushing, K. W. Cummins, R. L. Vannote, and J. R. Sedell. 1992. Stream ecosystem dynamics of the Salmon River, Idaho: An 8th-order system. *Journal of the North American Benthological Society* 11:111–137.

Mulholland, P. J. 1981. Organic carbon flow in a swamp-stream ecosystem. *Ecological Monographs* **51**:307–322.

Mulholland, P. J., and E. J. Kuenzler. 1979. Organic carbon export from upland and forested wetland watersheds. *Limnology and Oceanography* **24**:960–966.

Naiman, R. J., J. M. Melillo, M. A. Lock, T. E. Ford, and S. R. Reice. 1987. Longitudinal patterns of ecosystem processes and community structure in a subarctic river continuum. *Ecology* **68**:1139–1156.

Nelson, P. J., and D. C. Scott. 1962. Role of detritus in the productivity of a rock outcrop community in a Piedmont stream. *Limnology and Oceanography* **7**:396–413.

Newbold, J. D., P. J. Mulholland, J. W. Elwood, and R. V. O'Neill. 1982. Organic carbon spiralling in stream ecosystems. *Oikos* **39**:266–272.

Odum, H. T. 1957. Trophic structure and productivity of Silver Springs. *Ecological Monographs* **27**:55–112.

Sedell, J. R., F. J. Triska, J. D. Hall, and N. H. Anderson. 1974. Sources and fates of organic inputs in coniferous forest streams. Pages 57–69 *in* R. H. Waring and R. L. Edmonds (Eds.) *Integrated Research in the Coniferous Forest Biome*, Bulletin 5. Coniferous Forest Biome, US-IBP.

Teal, J. M. 1957. Community metabolism in a temperate cold spring. *Ecological Monographs* **27**:293–302.

Tilly, L. J. 1968. Structure and dynamics of Cone Spring. *Ecological Monographs* **38**:169–197.

Triska, F. J., V. C. Kennedy, R. J. Avanzino, G. W. Zellweger, and K. E. Bencala. 1989. Retention and transport of nutrients in a third-order stream in northwestern California: Hyporheic processes. *Ecology* **70**:1893–1905.

Vannote, R. L., G. W. Minshall, K. W. Cummins, J. R. Sedell, and C. E. Cushing. 1980. The river continuum concept. *Canadian Journal of Fisheries and Aquatic Sciences* **37**:130–137.

Webster, J. R., and B. C. Patten. 1979. Effects of watershed perturbation: Stream potassium and calcium dynamics. *Ecological Monographs* **49**:51–72.

Westlake, D. F., H. Casey, H. Dawson, M. Ladle, R. H. K. Mann, and A. F. H. Marker. 1972. The chalk-stream ecosystem. Pages 615–637 *in* Z. Kajak and A. Hillbricht-Ilkowska (Eds.) *Productivity Problems of Freshwaters*. Proceedings IBP-UNESCO Symposium, Poland.

CHAPTER 29

Effects of Nutrient Enrichment on Periphyton

CATHERINE M. PRINGLE[*] AND FRANK J. TRISKA[†]

*Institute of Ecology
University of Georgia,
†Water Resources Division
United States Geological Survey

I. INTRODUCTION

Definitions of *periphyton* vary within the literature. Here we refer to it as microfloral growth upon substrata (Wetzel 1983). Stream periphyton communities are affected by a complex array of interacting factors including nutrient and toxicant loading, light, temperature, water velocity, and grazing pressure. Experimental hypothesis testing is essential for understanding periphyton development and production in natural systems. Control of environmental variables and experimental isolation of regulating mechanisms can be a difficult task in the field, particularly in streams. The more control exerted by the investigator, the more replicable the result, but the less applicable to natural systems. On the other hand, less-controlled field experiments and observations may accurately describe the current periphyton community at a particular site, but yield little insight into what factors control community development. Furthermore, *in situ* manipulations are difficult to replicate under temporally and spatially varying background conditions.

Factors controlling periphyton growth in streams are poorly known compared to our knowledge of production limits in lake. As pointed out by Peterson *et al.* (1983), the unambiguous demonstration of nutrient limitation of algal periphyton production in a stream ecosystem is difficult for several reasons. A very low concentration of dissolved nutrient in overlying water may meet periphyton requirements due to the large volume of constantly renewed water. Also, nutrient levels may be high at certain times of the day, during storms, or seasonally when allochthonous materials (e.g., autumnal leaf fall) release nutrients in high pulses.

Nutrient regulation of periphyton growth and production has been addressed using several different approaches listed here in order of increasing scale: (1) point source manipulation of nutrients via nutrient-diffusing substrata, (2) enrichment of the water using flow-through enclosures, (3) whole stream manipulations, and (4) integrated bioassays that combine bioassay techniques, allowing comparison of algal growth response across different scales. Each approach has advantages and disadvantages, so the most appropriate method will depend on the scale at which the investigator is addressing nutrient limitation (Pringle *et al.* 1988), the nature of the stream system under study, and the tractability of using a specific approach in that system. Below we provide important background information on each of the above approaches.

A. Point Source *in Situ* Nutrient Manipulations

Point source nutrient manipulations allow testing of periphyton response *in situ* without the artificiality of enclosures. Such manipulations sacrifice the control of an enclosure for the more natural interaction with the total aquatic environment. Point-source manipulations such as nutrient-diffusing substrata can be advantageous in that they introduce minimal solutes to the environment, an important consideration when the same habitat is simultaneously utilized by numerous investigators. Such manipulations simulate natural nutrient-rich subtrata which act as point sources of nutrients for attached algae. For instance, many larval Chironomidae consolidate sand grains into tubular-shaped retreats. The excreta of retreat-dwelling larvae provide a direct nutrient source to tube colonizing periphyton (Pringle 1985). Vascular macrophytes, wood debris, and other particulate inputs also constitute natural nutrient-diffusing substrata. Several types of nutrient-diffusing substrata, generally using an agar matrix, have been developed to test periphyton response to *in situ* point sources of specific elements or compounds.

Pringle and Bowers (1984) consolidated, washed, and sterilized sand from the streambed into petri dishes with agar solutions enriched with phosphate and nitrate to simulate the structure and texture of nutrient-rich chironomid tubes that are composed of sand grains. The growth response of algae colonizing these circular "bricks" was assessed. A different approach uses clay flower pots filled with nutrient-rich agar and sealed with plastic petri dishes (Fairchild and Lowe 1984, Fairchild *et al.* 1985). Clay pots provide a hard surface similar to an epilithic (rock) habitat, as opposed to sand–agar substrata, which are more representative of an epipelic (silty) habitat. Tate (1990) modified the clay pot technique for use in stream systems to minimize variable current regimes: clay flowerpot saucers were filled with agar, sealed with Plexiglas, and installed *in situ* with their bottom surface parallel with the current flow to provide a horizontal surface for algal periphyton colonization.

Major considerations in using nutrient-diffusing substrata include: the direction of nutrient transport from the substrata to the attached algal assemblage, the variability in the rate of nutrient release (i.e., released in a high initial pulse with rates decreasing in an exponential fashion), and that nutrient-diffusing substrata constitute a finite amount of nutrient (Pringle 1987). These factors can result in algal growth responses that are different from responses to nutrient additions to the whole system or larger mesocosms (Pringle 1987).

Nutrient-diffusing substrata have proven to be an effective tool to: (1) evaluate local nutrient recycling processes (Pringle and Bowers 1984, Pringle 1990), (2) validate theoretical models (Fairchild *et al.* 1985), (3) detect nutrient-limiting factors in a "shot-gun" approach (Pringle *et al.* 1988), and (4) supplement another bioassay technique (Grimm and Fisher 1986, Pringle 1987, Tate 1990).

B. Flow-Through Enclosures

Flow-through enclosures (i.e., flumes) can provide much flexibility for experimental manipulation, particularly for long-term studies. They serve as a valuable tool for isolating and modifying aspects of the physical, chemical, and biological environment of running water systems and for providing within-system replication. Such partially open enclosures are essential for long-term manipulation to minimize enclosure effects that could result in development of a unique and unrepresentative community. However, even partial enclosure of a natural system with walls may create a different environment than the unenclosed, natural system.

Partially enclosured flumes with a sealed or open bottom may be placed within the stream channel or may be located stream-side and receive stream water pumped from the channel. Nutrients are added to the water at the head of the flume. Flume channels located within the stream channel (Triska *et al.* 1983) or streamside (Rosemond 1993, Rosemond *et al.* 1993) have been used successfully to examine algal growth response to nutrient additions on either natural or artificial substrata. A classic study that used flumes *in situ* in a temperate, coastal-rainforest stream was conducted by Stockner and

Shortreed (1978), who found that $3\times$ increase of both NO_3 and PO_4 over ambient concentrations resulted in higher algal growth than when PO_4 alone was added.

C. Whole Stream Manipulations

The least controlled *in situ* manipulation uses the total stream as an experimental "arena." The major dispersal mechanism of an introduced solute is the hydrology of the channel, which adds enormous complexity (e.g., Bencala *et al.* 1984, Solute Transport Workshop 1990) to an otherwise straightforward experiment. While the flow-through enclosures described above control most hydrologic factors by simplifying or eliminating them, such control allows replication but sacrifices realism. In contrast, whole system manipulations allow a realistic response, but sacrifice control and replication. For logistical reasons, field experiments have typically been conducted in low-order streams at base flow.

Peterson *et al.* (1985, 1993) found that whole-stream enrichment of phosphorus in a tundra river resulted in a switch from heterotrophic to autotrophic production within the enriched stream reach. Furthermore, long-term phosphorus fertilization of the same pristine river for four consecutive summers illustrated cascading trophic effects: algal growth and standing crop increased during the first 2 years with subsequent increase in grazing insects and fish growth during the following 2 years (Peterson *et al.* 1993).

D. Integrated Bioassay

Integrated bioassays, which combine two or more different techniques, allow comparison of nutrient effects on algal periphyton across different scales. For example, Pringle (1987, 1990) combined the nutrient-diffusing substratum technique with flow-through bioassays to: (1) experimentally differentiate between algal growth response to nutrients derived from the substratum versus those introduced from the water column, (2) examine the responses of specific algal taxa to enrichment of the water versus the substratum, and (3) compare the effectiveness of flow-through systems and nutrient-diffusing substrata as *in situ* bioassay methods. Periphyton growth responded to combined influences of water and substratum enrichment in an additive or synergistic manner, depending on the types of nutrients (N, P) added from each source. When NO_3 was added to the substratum and PO_4 to the water, algal growth response was synergistic (Pringle 1987). Furthermore, specific algal taxa exhibited different responses to enrichment of substratum versus water.

This chapter describes four *in situ* field methods for quantitatively assessing algal growth response to nutrient additions. Three exercises assess nutrient limitation of periphyton growth over different spatial and temporal scales: (1) point-source nutrient enrichment using a nutrient-diffusing substratum technique, (2) nutrient additions to the water using a flow-through mesocosm, and (3) whole stream nutrient enrichment. These exercises can be combined in various permutations to examine nutrient effects on algal periphyton across different scales.

In choosing an exercise, researchers are encouraged to analyze the advantages and disadvantages of each approach in terms of: (1) the scale at which they wish to evaluate nutrient limitation of algal growth and (2) the tractability of the methods in light of the field sites available. The first two exercises are most easily applied to third- to fourth-order streams, while the third exercise is best applied to first- and second-order streams.

The specific objectives of this exercise are to: (1) introduce the concept of nutrient limitation of algal periphyton growth; (2) demonstrate how to assess effects of nutrient enrichment on algal growth, as measured by the accrual of algal standing crop (i.e., chlorophyll *a*, ash-free dry mass) through time; and (3) to illustrate the advantages and disadvantages of different techniques applied at different spatial scales.

II. GENERAL DESIGN

In this chapter we will evaluate the response of algal standing crop accrual, as measured by chlorophyll *a* and ash-free dry mass (AFDM), to nutrient manipulations using one (or more) of three exercises. Each of the exercises are designed to span a 5-week time frame, with the first (bioassay apparatus construction) and second (bioassay installation) sessions exclusively devoted to the exercise and the third, fourth, and fifth sessions requiring only 1–2 hr of work on the exercise, allowing for other modules within this book to be explored. All exercises will require regular visits to the stream study site (i.e., every 2–3 days for Exercise 1; every day for Exercises 2 and 3) for collection of water samples for background nutrients, removal of debris from artificial substrata, refilling nutrient reservoirs, and calibrating nutrient release rates. Analyses of nutrients (i.e., NO_3 and PO_4) can be facilitated by running any of the exercises in this chapter in conjunction with the exercises in Chapters 8 and 9.

A. Site Selection

Wadeable second- to fourth-order streams with relatively low ambient levels of nitrate ($NO_2 + NO_3 - N$) and phosphate (SRP) are ideal for this exercise. Within the study stream, a sunny reach (preferably <40% canopy cover) with a relatively simple channel (straight, uniform depth, and substratum type, sparse woody debris), should be selected. The length and width of the experimental reach should be chosen to accommodate the bioassay apparatus/materials associated with a given exercise. For instance, if Exercise 1 is pursued, the ideal site will provide suitable space for the installation of 48 clay saucers in 12 groups of four treatments.

B. General Procedures

Procedures for determining algal response to nutrient treatments will be assessed through analysis of chlorophyll a and AFDM. Chlorophyll a samples should be vacuum filtered (Whatman GF/F filters), ground in 9 ml buffered acetone, and analyzed spectrophotometrically (APHA *et al.* 1992). Ash-free dry mass samples should also be filtered onto preweighed Whatman GF/F filters (or equivalent), dried for 24 h, weighed, ashed, rehydrated, dried for 24 h, and reweighed (see Chapter 14 for detailed protocols).

Background stream nutrient chemistry must be collected every 2–3 days in acid-washed (10% HCl) and thoroughly rinsed 125-ml polyethylene bottles for analyses of NO_3 and soluble reactive phosphorus (SRP).

III. SPECIFIC EXERCISES

A. Exercise 1: Nutrient Limitation of Algal Growth Using Nutrient-Diffusing Clay Saucers

This exercise utilizes a nutrient-diffusing substratum technique designed by Tate (1990). The clay flowerpot-saucer technique has been chosen because of its minimal expense, the ease and facility with which it can be constructed and installed, and the "standardized" horizontal surface, relative to current flow, that the bottom of the saucer provides for algal periphyton colonization (Fig. 29.1). Nutrient-diffusing substrata are incubated within the stream in standardized conditions of current velocity, depth, and canopy cover. Saucers are then retrieved after 1-, 2-, and 3-week time periods and then processed for assessment of chlorophyll a and AFDM.

1. Construct nutrient-diffusing artificial substrata out of 10.2-cm-diameter clay flowerpot saucers glued to 12 × 12-cm Plexiglas plates. Fill four sets of 12 clay saucers with 225 ml of 2% agar solutions. Four batches of agar should be prepared for the following treatments, with each treatment represented by 9 saucers: (1) N enrichment, 0.5 mol/liter $NaNO_3$; (2) P enrichment, 0.1 mol/liter KH_2PO_4; (3) N+P enrichment, 0.5 mol/liter $NaNO_3$ and 0.1 mol/liter KH_2PO_4; (4) C, control with no nutrients (i.e., unenriched agar only). Drainage holes within the saucers should be closed with No. 00 neoprene stoppers. Data sheets should be prepared for data

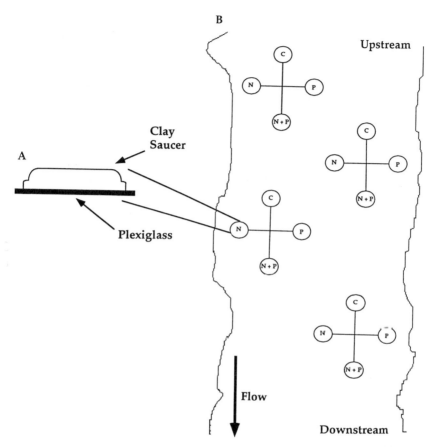

FIGURE 29.1 (A) Nutrient-diffusing clay saucer. (B) Orientation of diamond-shaped wooden platforms and nutrient-diffusing clay saucer treatments with respect to the current direction in stream. C, control; N, nitrate; P, phosphate; and N+P, nitrate and phosphate.

collection associated with field installation and should include columns for recording saucers number and treatment, depth and current velocity for each saucer location, extent of canopy cover at field site, retrieval dates, and chlorophyll *a* and AFDM values, which will be expressed in units of mg/m^2 surface area of artificial substratum surface (see Chapters 10, 11, and 14 for details of determining AFDM values).

    2. Saucers should be attached to wooden frames constructed in a diamond shape configuration, with a C saucer placed upstream, N and P saucers placed side by side downstream from C, and an N + P saucer at the downstream end, to minimize contamination among treatments (Figure

29.1, after Tate 1990). The wooden frames with saucers should be mounted in the stream bottom with stakes and positioned in a randomized block design within standardized conditions of current velocity, depth, and canopy cover. The latter may be measured with a spherical densiometer.

    3. Collect four sets of saucers after 1, 2, and 3 weeks. Place each saucer into a separate zip-lock bag and carefully transport to the laboratory. Scrape algae from the exposed top flat surface of each saucer (using toothbrushes and/or razor blades), dilute in a known volume of distilled water (e.g., 100–400 ml) in a 500-ml beaker, mix with a magnetic stirrer, and subsample (e.g., 20–50 ml) for chlorophyll *a* and AFDM.[1] Chlorophyll *a* and AFDM samples should be processed according to methods described above and in detail in Chapter 14.

B. Exercise 2: Nutrient Limitation of Algal Growth Using Flow-through Enclosures

In this exercise a flow-through flume system, based on the method of Peterson *et al.* (1983), has been selected because of its compact nature (<1 m^2), low expense, and ease of installation and maintenance in high-discharge situations. This compact system is composed of a bank of Plexiglas cylinders attached to a flotation device that allows the apparatus to rise and fall with variations in stream discharge (Fig. 29.2). Nutrients are dripped into upstream ends of the cylinders via Mariotte bottles. Banks of glass slides installed in downstream ends of cylinders (where nutrient concentrations are homogeneous), serve as substrata for algal colonization. Glass slides are retrieved through time to assess accrual of algal standing crop.

    1. Construct a bioassay apparatus consisting of five 1.2-m sections of clear plastic tube of 9.2-cm diameter attached by U-bolts to the top of a sheet of Plexiglas (see Fig. 29.2 for construction details). Each tube represents a treatment (e.g., tube 1, N enrichment; tube 2, P enrichment; tube 3, N + P enrichment; tubes 4 and 5, nonenriched controls). The apparatus is suspended from wood and styrofoam lateral floats. The upstream end of each tube should contain Lexan baffles to ensure turbulent mixing of the water and nutrient drip. A set of five microscope slide holders, each holding six slides attached to a strip of Plexiglas with rubber bands, is installed at the downstream end of each cylinder. Nutrients are introduced into the upstream end at a constant rate by siphoning concentrated solutions of $NaNO_3$ and KH_2PO_4 from 1-liter polyethylene Mariotte bottles through Teflon minibore tubing (0.56-mm-diameter) at a rate of ~25 ml/h (see Chapter 8

---

[1]The exact amount of the chlorophyll *a* or AFDM subsamples be determined by the researcher in the laboratory based on the density of algae within the algal homogenate.

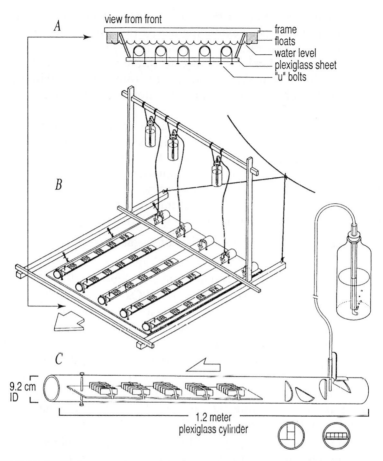

FIGURE 29.2 (A) Cross-sectional diagram of continuous-flow periphyton bioassay system. Bioassay tubes were suspended beneath the surface from a pair of outrigger floats. (B) Diagram of bioassay apparatus with Mariotte bottle frame tethered to a rope across the stream. Arrow denotes direction of water flow. (C) Detailed drawing of a single bioassay tube showing the Mariotte bottle, mixing baffles, and slide-mounting apparatus (modified from Peterson *et al.* 1983).

for details on making and use of Mariotte bottle). In the experiments of Peterson *et al.* (1983), this resulted in soluble reactive phosphorus (SRP) levels 5–15 μg P/L and nitrate values 50–150 μg N/L above ambient stream water concentrations. Nutrient drip rates from Mariotte bottles should be calibrated daily. Water samples should be collected at the downstream ends of each tube every 2–3 days for analyses of NO_3 and PO_4.

    2. Install the bioassay apparatus in an area of relatively uniform current velocity. The wood and styrofoam lateral floats should be adjusted so that

they are submerged just below the water surface. The apparatus should be securely tethered to a rope extending across the stream. This installation method allows the apparatus to be relatively flood-tolerant. Additions of food coloring or Rhodamine–WT dye to the upstream ends of the tubes and timing of dye movement will indicate the rate of flow within the tubes. These should be compared to similar water movement measurements taken outside the tubes.

3. After 1, 2, and 3 weeks, retrieve a set of six slides from each tube and discard the two on either side. Each slide should be placed into a sealed plastic bag and carefully transported to the laboratory. Two slides from each tube can be used for chlorophyll *a* analyses and two slides for AFDM analyses. All algae should be carefully removed from both sides of each glass slide using a razor blade, rinsed into a beaker with distilled water, mixed with a magnetic stirrer, and subsampled for chlorophyll *a* and AFDM.

C. Exercise 3: Nutrient Limitation of Algal Growth Using Whole-Stream Enrichment

In this exercise whole-stream nutrient enrichment will be employed by dripping concentrated $NaNO_3$ and/or KH_2PO_4 (or phosphoric acid) into the stream channel from a carboy fitted with a drip system. The decision regarding the nature of the nutrient addition will be determined by premeasurement of ambient nutrient levels (e.g., if the stream has low background phosphorus levels, then P enrichment may be selected; alternatively N enrichment may be desirable if ambient N levels indicate that the system is N-limited). Effects of whole-stream enrichment will be assessed by algal growth on artificial substrata (i.e., unglazed ceramic tiles) placed in both a pool and a riffle above and below the point of enrichment.

1. Use a large carboy (e.g., 50 liter), fitted with a spigot and vented by a narrow tube (to prevent changes in head pressure from affecting the flow rate) as a nutrient reservoir. Mount the carboy on a sturdy wooden stand or on the trunk of a riparian tree near a section of the stream where the channel narrows and is somewhat turbulent to enhance mixing of the added nutrients. Attach a length of tygon tubing sufficient to reach from the spigot to the stream surface and culminating in a micropipet tip. The micropipet tip will drip nutrients from the carboy reservoir into the stream. Attach it to a ringstand for stability and regulate the rate of nutrient addition by adjusting the height of the ringstand relative to the carboy. Calculate the solute concentration needed to attain a desired stream concentration (see Appendix 29.1).[2]

---

[2]Using this example (Appendix 29.1), the nutrient reservoir must be replenished approximately every 3 days (72–80 h). Make sure to record the time that the injection is stopped and restarted, between carboys. The rate of nutrient input can be checked by recording the

2. As the injectate solution is being prepared, place 15 unglazed ceramic clay tiles randomly within both a riffle and a pool habitat, both above and below the source of nutrient enrichment (total of 60 clay tiles, with 30 upstream controls and 30 downstream enriched). Treatment tiles should be placed far enough downstream so that the injectate solution is well mixed with streamwater by the time it passes over the tiles. Mixing can be checked by injecting a few milliliters of Rhodamine WT (or other dye) at the drip site. Care should be taken to minimize the variability in current and depth between sites selected for tile placement within a given habitat so that tiles are exposed to a relatively narrow range of current velocities (e.g., 25–30 cm/s for riffle and 0–5 cm/s in pool) and depths.

3. Retrieve from both the pool and the riffle habitat five replicate clay tiles after 1, 2, and 3 weeks. Process and analyze algal periphyton for both chlorophyll *a* and AFDM as in above exercises and in Chapter 14. Determine nutrient concentrations at control and treatment sites. Determine the mass of nutrient added from drip rates and residual solution.

D. Optional Activities for Additional Study

Optional activities, involving additional time and facilities include: (1) running integrated bioassay experiments (see illustration Fig. 29.3) that combine two of the protocols presented above (e.g., nutrient-diffusing substrata and flow through system) to experimentally separate effects of nutrients added from different sources and to evaluate algal response to nutrient perturbations at different scales (e.g., see Pringle 1987, 1990), (2) combining nutrient enrichment with ecosystem-level process studies (e.g., primary production and community respiration; see Chapter 25), and (3) assessing algal community composition response (see Chapter 13) to different nutrient treatments in each of the above exercises.

E. Data Analysis

At a minimum, chlorophyll *a* and AFDM should be graphed for each treatment as a function of time. Rates of algal standing crop accrual should be compared between treatments and/or habitats (e.g., pool versus riffle; Exercise 3). Inferential statistics (e.g., analysis of variance; see Zar 1984)

---

amount remaining in the carboy. The input rate should be calibrated daily by recording the time it takes to fill a volumetric flask (20–50 ml). Streamwater can be used to mix the reagents on-site. The flow rate with this type of apparatus will vary slightly as the liquid level in the carboy changes and also with changing viscosity due to temperature variations. The head variation can be minimized by mounting the carboy higher, thereby creating a large head and reducing the relative head variation caused by the difference in a full and a nearly empty carboy. If there are very large diel temperature shifts, however, some variation in flow rate will still occur.

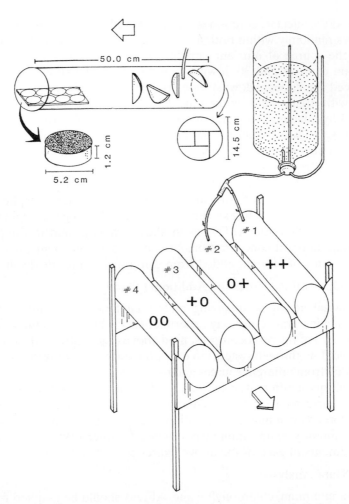

FIGURE 29.3 Schematic diagram of integrated bioassay system. Unshaded arrows denote current direction of current. (Top left) Single Plexiglas cylinder showing placement of nutrient-diffusing substrata, glass slides, baffles, and nutrient feeder line. Enlargement of nutrient-diffusing substrata (sand–agar in petri dish) and inside view down length of cylinder (illustrating baffle arrangement) are shown directly below. (Bottom) Bank of Plexiglas cylinders in wooden frame. Nutrients are added to upstream ends of cylinders 3 and 4 from a 20-liter carboy. Notation on cylinders refers to the treatment combination for substratum-water enrichment experiments where the first character refers to the presence (+) or absence (0) of substratum enrichment and the second character refers to the presence or absence of water enrichment (modified from Pringle 1987).

can be used to examine differences among various treatments in specific experiments.

IV. QUESTIONS

1. How might the physical/chemical nature of nutrient-diffusing substrata interfere with nutrient treatment effects? Consider the following possible complications:

  a. Agar is a strongly gelling seaweed hydrocolloid that is primarily composed of cell wall polysaccharides extracted from two types of marine red algae (*Gelidium* spp. and *Gracilaria* spp.) Given that the structure and composition of agar reflect the nutritional state of the algae from which it was derived, along with variable manufacturing and processing techniques that may introduce impurities, what types of growth-promoting or -inhibiting substances might agar contain and could these potentially interfere with treatment effects? How can an investigator minimize the unwanted effects of potential growth-promoting or -inhibiting substances? (i.e., Why is it important to use refined brands of commercial agar? Why should the same batch of agar be used in preparing treatments?)

  b. Flowerpot saucers are composed of clay, which is fine particles of hydrous aluminum silicates and other minerals that can act as a sink for certain elements such as phosphorus. Discuss the implications of this with respect to your experimental results (i.e., from the perspective of algal response to internal macronutrient and micronutrient stores (within nutrient-diffusing substratum) and to nutrients in overlying waters).

2. Contrast the effect of algal flora *on* ambient nutrient chemistry with respect to point-source nutrient amendments (nutrient diffusing substrata) and whole-stream nutrient enrichment. What effect would the relatively small biomass of algae on a phosphorus-diffusing substratum have on ambient nutrient chemistry versus the effect of algal flora within an entire stream reach enriched with phosphorus?

3. Why might algal response to nutrient-diffusing substrata not reflect water enrichment assays?

4. How might community composition and physiognomy (three dimensional structure) of an algal community be affected by nutrients introduced from different sources (e.g., substratum versus water)?

5. Why must caution be exerted in extrapolating experimental results from one spatial scale to another?

6. Compare and contrast the advantages of Exercises 1–3 in terms of replicability of treatments. Which of the experimental designs employ true replication of treatments? Which experimental design(s) is *pseudoreplicated* (sensu Hurlbert 1984)? How might one avoid pseudoreplication?

7. How might grazing benthic invertebrates have affected your experimental results? Is there any evidence that grazing insects might have obscured measurable treatment effects in your experiment by reducing algal standing crop? How might you quantify this effect?

V. MATERIALS AND SUPPLIES

Laboratory Materials

0.45-μm Millipore filters (for filtration of water samples)
125-ml Polyethylene bottles (for collection of water samples)
5, 10, and 20 ml Pipets
Acetone (for chlorophyll *a* analyses)
Duco cement
$NaNO_3$
KH_2PO_4 (or phosphoric acid—Exercise 3)
Polyethylene squeeze bottles (for distilled water washes of artificial substrata)
Whatman GF/F filters or equivalent (for chlorophyll *a* and AFDM samples)

Exercise 1

50 Clay flowerpot saucers (10.2 cm diameter)
50 Neoprene stoppers (00)
Agar
Plexiglas 1/8″ thick (sufficient amount for 36 12 × 12-cm squares)
Wood for construction of installation "diamonds" for saucers

Exercise 2

Glass slides (for periphyton colonization)
Plexiglas cylinders and plates, U-bolts
Wood and styrofoam (for lateral floats)

Exercise 3

2 Carboys (50 liter)
20 Liter bucket
60 Unglazed ceramic clay tiles (e.g., 7.3 × 15.3 cm)

Large funnel
Tygon tubing
Paddle (for stirring)
Pipet tip
Plastic spigot
Plastic tarp

Field Materials

Beakers (for algal homogenate)
Current velocity meter
Flagging
Meter sticks
Rhodamine WT dye or food coloring (Exercise 2)
Spherical densiometer
Toothbrushes and razor blades (for scraping algae)
Zip-lock bags (gallon size) (for saucer (Exercise 1) or tile (Exercise 3)
    retrieval; sandwich bag size for glass slide retrieval (Exercise 2))

Laboratory Equipment

Autoclave
Drying oven
Electronic balance (± 0.1 mg)
Filtering apparatus
Magnetic stirrer and hot plate
Muffle furnace
Spectrophotometer or fluorometer
Vacuum pump

APPENDIX 29.1 CALCULATIONS FOR DETERMINING SOLUTE INJECTION RATE FOR SPECIFIC STREAM CONCENTRATION

The solute, its concentration in the injectate solution, and the volume of the carboy reservoir will vary among sites due to factors such as discharge, target concentrations above background, the drip rate that can be reliably sustained, and the length of the experiment. Consider the following example:

Stream discharge =   10 liters/s
Nutrient solute to be added = $NaNO_3$
Target concentration =   100 μg N above ambient levels

Length of experiment =	14 days
Drip rate =	10 ml/min
Carboy size =	50 liters

1. Amount of water to be enriched (liters/hr): 10 liters/s $\times$ 60 s/min $\times$ 60 min/h = 36,000 liters/h

2. Amount of $NaNO_3$ amendment (g/h): 100 μg N/liter $\times$ 36,000 liters/h/0.16 (proportion of $NaNO_3$ that is N) = 22,500,000 μg $NaNO_3$/h = 22.5 g $NaNO_3$/h

3. Drip Rate = 10 ml/min $\times$ 60 min = 600 ml/h; a 50-liter carboy will allow 50 liter/0.6 liter/h or 83 h of constant nutrient addition

4. Concentration of $NaNO_3$ in the carboy: 22.5 g $NaNO_3$/h/0.6 liter = 37.5 g $NaNO_3$/liter

5. Mix into carboy 37.5 g $NaNO_3$/liter $\times$ 50 liter = 1875 g $NaNO_3$; fill to 50 liters

Tip. Place $NaNO_3$ in a separate container and add a known volume of water while stirring. Decant the dissolved nitrate solution into the carboy. Repeat until all $NaNO_3$ is dissolved. Then bring carboy up to 50 liters and stir. This will ensure that no nutrient salt remains in crystalline form in the carboy. Before mixing, cover the area below the carboy with a plastic tarp to contain accidental spills.

REFERENCES

APHA, AWWA, and WEF. 1992. *Standard Methods for the Examination of Water and Wastewater,* 18th ed. American Public Health Association, Washington, DC.

Bencala, K. E., V. C. Kennedy, G. W. Zellweger, A. P. Jackman, and R. J. Avanzino. 1984. Interactions of solutes and streambed sediment. 2. A dynamic analysis of coupled hydrologic and chemical processes that determine solute transport. *Water Resources Research* 20:1804–1814.

Fairchild, G. W., and R. L. Lowe. 1984. Artificial substrates which release nutrients: Effects on periphyton and invertebrate succession. *Hydrobiologia* 114:29–37.

Fairchild, G. W., R. L. Lowe, and W. B. Richardson. 1985. Algal periphyton growth on nutrient-diffusing substrates: An in situ bioassay. *Ecology* 66:465–472.

Grimm, N. B., and S. G. Fisher. 1986. Nitrogen limitation in a Sonoran desert stream. *Journal of the North American Benthological Society* 5:2–15.

Hurlbert, S. H. 1984. Pseudoreplication and the design of ecological field experiments. *Ecological Monographs* 54:187–211.

Peterson, B. J., J. E. Hobbie, T. L. Corliss, and K. Kriet. 1983. A continuous flow periphyton bioassay: Tests of nutrient limitation in a tundra stream. *Limnology and Oceanography* 28:583–591.

Peterson, B. J., J. E. Hobbie, A. E. Hershey, M. A. Lock, T. E. Ford, J. R. Vestal, V. L. McKinley, M. A. Hullar, M. C. Miller, R. M. Ventullo, and G. S. Volk. 1985. Transformation of a tundra river from heterotrophy to autotrophy by addition of phosphorus. *Science* 229:1383–1386.

Peterson, B. J., L. Deegan, J. Helfrich, J. E. Hobbie, M. Hullar, B. Moller, T. E.

Ford, A. Hershey, A. Hiltner, G. Kipphut, M. A. Lock, D. M. Fiebig, V. McKinley, M. C. Miller, and J. R. Vestal, R. Ventullo, and G. Volk. 1993. Biological responses of a tundra river to fertilization. *Ecology* **74:**653–672.

Pringle, C. M. 1985. Effects of chironomid (Insecta: Diptera) tube-building activities on stream diatom communities. *Journal of Phycology* **21:**185–194.

Pringle, C. M. 1987. Effects of water and substratum nutrient supplies on lotic periphyton growth: An integrated bioassay. *Canadian Journal of Fisheries and Aquatic Sciences* **44:**619–629.

Pringle, C. M. 1990. Nutrient spatial heterogeneity: Effects on the community structure, diversity and physiognomy of lotic algal communities. *Ecology* **71:**905–920.

Pringle, C. M., and J. Bowers. 1984. An in situ substratum fertilization technique: Diatom colonization on nutrient-enriched sand substrata. *Canadian Journal of Fisheries and Aquatic Sciences* **41:**1247–1251.

Pringle, C. M., R. Naiman, G. Bretchko, J. Karr, M. Oswood, J. Webster, R. Welcomme, and M. J. Winterbourn. 1988. Patch dynamics in lotic systems: The stream as a mosaic. *Journal of the North American Benthological Society* **7:**503–524.

Rosemond, A. D. 1993. Interactions among irradiance, nutrients, and herbivores constrain a stream algal community. *Oecologia* **94:**585–594.

Rosemond, A. D., P. J. Mulholland, and J. W. Elwood. 1993. Top–down and bottom–up control of stream periphyton: Effects of nutrients and herbivores. *Ecology* **74:**1264–1280.

Solute Transport Workshop. 1990. Concepts and methods for assessing solute dynamics in stream ecosystems. *Journal of the North American Benthological Society* **9:**95–119.

Stockner, J. G., and K. R. S. Shortreed. 1978. Enhancement of autotrophic production by nutrient addition in a coastal rainforest stream on Vancouver Island. *Journal of the Fisheries Research Board of Canada* **35:**28–34.

Tate, C. M. 1990. Patterns and controls of nitrogen in tallgrass prairie streams. *Ecology* **71:**2007–2018.

Triska, F. J., V. C. Kennedy, R. J. Avanzino, and B. N. Reilly. 1983. Effect of simulated canopy cover on regulation of nitrate uptake and primary production by natural periphyton assemblages. Pages 129–159 *in* T. D. Fontaine and S. M. Bartell (Eds.) *Dynamics of Lotic Ecosystems.* Ann Arbor Science Publishers, Ann Arbor, MI.

Wetzel, R. G. 1983. *Limnology,* 2nd ed. Saunders College Publishing, Philadelphia, PA.

Zar, J. H. 1984. *Biostatistical Analysis,* 2nd ed. Prentice–Hall, Englewood Cliffs, NJ.

CHAPTER 30

Surface–Subsurface Interactions in Streams

NANCY B. GRIMM

Department of Zoology
Arizona State University

I. INTRODUCTION

Stream–riparian ecosystems may be viewed as an amalgamation of several distinct subsystems. In two dimensions, there are longitudinally arranged units of the wetted stream, subsystems created by variations in current within the wetted stream (e.g., edge and center subsystems, thalweg, and backwaters), and lateral subsystems across the stream–riparian corridor defined by baseflow, annual floods (active channel zone), and vegetation type (e.g., see Triska *et al.* 1989, Gregory *et al.* 1991). A third, vertical, dimension encompasses the *hyporheic* (underflow) subsystem, which exists below the sediment surface of each of the longitudinal or lateral subsystems in most streams. The hyporheic zone is defined broadly as the region of saturated sediments and interstitial water directly beneath and lateral to the surface stream, which interacts via exchange of water and materials with the surface stream (see Chapter 6).

Adjacent vertical, lateral, and longitudinal subsystems in streams interact via the movement of water and its load of dissolved and suspended materials. Biological movement across subsystem boundaries may also occur; for example, as benthic invertebrates seek refuge from spates in the hyporheic zone (Palmer *et al.* 1992). If we take a subsystem approach to studying stream ecosystems, then we can describe the properties of each component, its linkage with other subsystems (i.e., what kinds of transfers take place), and, perhaps most interesting, the *consequences* of such interactions for biotic communities in each subsystem. Fisheries biologists, hydrologists, and stream ecologists long have known that surface and subsurface waters interact (e.g., Vaux 1962, 1968, Hynes 1983, Bencala *et al.* 1984, Grimm and Fisher 1984), but only recently have the consequences of this exchange been studied.

Exchange or interaction among subsystems occurs primarily via water movement. In many streams, exchange between surface and hyporheic waters is dictated by variation in the form of the bed (Savant *et al.* 1987, Thibodeaux and Boyle 1987, Harvey and Bencala 1993), proximity of underlying bedrock to the sediment surface (Vaux 1962, Grimm and Fisher 1984), or the influence of structural features of the channel (including biotic features such as macrophyte beds or fish nests; e.g., White 1990). These features cause variation in hydraulic head, a measure of the potential for water to move into or out of bed sediments. Although flow through the bed is slower than that of the surface stream, streams vary in the rate of water movement through bed materials. This variation is dictated by sediment porosity; in general, coarser sediments have high hydraulic conductivity and transmit water much more readily than finer sediments (Wroblicky *et al.* 1992). Thus, there are two primary features of stream beds that dictate exchange: hydraulic head and hydraulic conductivity. This exercise is specifically designed for streams with relatively high hydraulic conductivity; we focus here on variation in hydraulic head.

In addition to water movement, materials carried in flowing water may be exchanged among subsystems. This includes both dissolved and suspended materials. Although many dissolved materials behave as water does, suspended materials may be "filtered out" to varying extents by bed sediments. Thus, particulate materials in transport may be left on the sediment surface where water enters the bed (*downwells*) while high oxygen and low dissolved nutrients characteristic of surface water are transported to the hyporheic zone. Conversely, subsurface waters may be low in oxygen but exhibit elevated concentrations of dissolved nutrients, which can supply nutrient-limited surface communities at sites of subsurface discharge (*upwells*). Upwelling water usually represents a small input to a large flux (surface discharge), whereas downwelling water represents a large input to a smaller flux (hyporheic discharge).

Studies of the interaction of surface and subsurface subsystems can lend insight to biotic distributions and community composition. For example, in Sycamore Creek, Arizona, nitrogen is the limiting nutrient for primary production (Grimm and Fisher 1986). Hyporheic water that is enriched in nitrate because of nitrification in hyporheic sediments enters the surface stream at discrete upwelling zones (Valett *et al.* 1990, Grimm *et al.* 1991,

Jones *et al.* 1995a). At these sites algal biomass is higher, communities are dominated by filamentous green algae, and algae recover faster after disturbance than at downwelling sites (Valett *et al.* 1994). Blooms of algae associated with upwelling water also have been observed by others in mountain streams in New Mexico (Coleman and Dahm 1990) and large gravel-bed rivers in Montana (Stanford and Ward 1993). In Sycamore Creek, algal uptake results in longitudinal depletion of nitrate (Grimm 1987); downstream algal communities have lower biomass and recover slowly following disturbance (Valett *et al.* 1994) and often are dominated by nitrogen-fixing cyanobacteria (Fisher *et al.* 1982). A parallel consequence of surface–subsurface interaction for the hyporheic biota also has been discovered in Sycamore Creek. High dissolved oxygen and labile organic carbon in surface water are supplied to hyporheic microbial communities, which show higher respiration rates at downwelling zones than at upwelling zones (Jones *et al.* 1995b).

The objectives of the exercises in this chapter are to describe the physical, chemical, and biological properties of surface and subsurface environments at sites of exchange. This description will permit inference regarding the effect of surface–subsurface interactions on biotic distributions and physicochemical factors that potentially influence them. For a useful summary of techniques for measuring hydrologic exchange, see also Boulton (1993) or Chapter 6.

II. GENERAL DESIGN

The basic design of this exercise is a comparison of properties of surface and subsurface environments at points of exchange (upwelling and downwelling zones). An overall system map provides a context for the comparisons. The extent of replication is dependent upon time constraints; while the techniques described are relatively simple to perform, they may be time-consuming. Researchers may choose a relatively simple, descriptive design where one upwelling and one downwelling zone are compared; however, a more rigorous comparison would replicate ($n = 3$, at least) the hydrologic exchange zones. If this exercise is combined with other exercises (for example, those described in Chapters 5 and 6), more can be accomplished by a single group of researchers. Field experiments and measurements should be carried out by teams of from two to four researchers.

The mapping team is responsible for creating a map of the system, the hydrology team for hydrologic measurements, the biota team maps biotic distributions and collects some biological samples, the DO-temp team measures dissolved oxygen and temperature, and the chem-inverts team collects water and hyporheic invertebrate samples. This last team can be split into two teams if a sufficient number of researchers participate. Exercise 1 involves mapping, hydrologic measurements, and biotic sampling (mapping, hydrology, and biota teams). Exercise 2 additionally requires installation and sampling of wells (DO-temp and chem-inverts teams); this exercise may be more easily completed if wells are installed before the exercise begins. If Optional Exercise 3 is selected, an additional "process team" can be formed, and if Optional Exercise 4 is selected, a team can be created to set up experiments. Alternatively, the optional exercises provide follow-up projects that can be done by individual researchers.

A. Site Selection

The exercises in this chapter are designed for streams with beds dominated by fine- to medium-grained sand or gravel sediments of high hydraulic conductivity at low or baseflow conditions. Difficulties in installing wells or minipiezometers in coarse sediments and in obtaining interstitial samples from very fine silt or clay sediments usually preclude application of these techniques to such systems (see also Boulton 1993). In the optimum stream for these exercises, it should be possible to move sediments around with a shovel or insert minipiezometers or other sampling devices with a sledge hammer and steel rods. A single reach or two reaches (20–100 m length) in close proximity should be selected for study. The site selected could be one in which the hyporheic zone exercise (Chapter 6) had been performed previously, as it would be helpful to have some advance knowledge of the location of upwelling and downwelling zones. An open canopy permitting growth of abundant algae or macrophytes is another desirable (but not required) site characteristic.

B. Overview

The mapping team is responsible for generating an overall map of the reach, on which data gathered by the other teams can be replaced. The mapping team will locate and mark likely upwelling and downwelling zones, based on bedform variations (Fig. 30.1A), and measure stream gradient and discharge. A spatially explicit description of morphometric parameters such as stream width and depth, stream gradient, active channel width, riparian zone(s) width, and location of geologic or structural features (e.g., bedrock outcrops, rocky riffles, or woody debris accumulations) will help in understanding variables that influence subsystem exchange and its consequences for the biota.

The hydrology team will measure potential for exchange between hyporheic and surface waters. *Vertical hydraulic gradient* (VHG) is a measure of the pressure difference between surface and subsurface waters; its sign

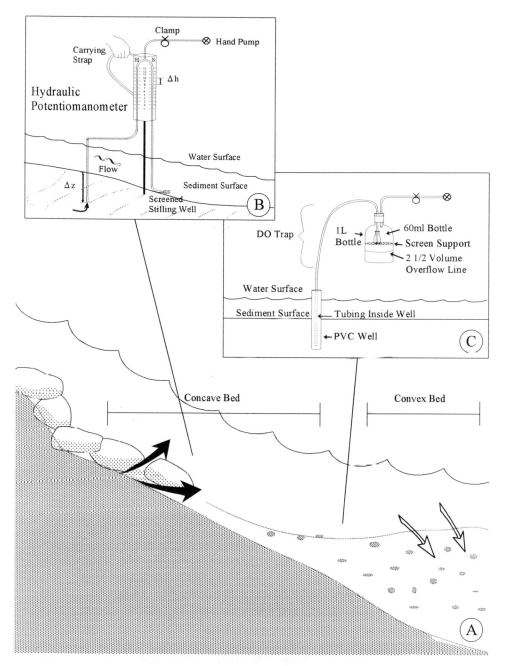

FIGURE 30.1 Methods and terms used in the study of surface–subsurface exchange. (A) Illustration of the base of a riffle and head of a sand/gravel run, showing position of upwelling zone (in a concave bed site) and downwelling zone (in a convex bed site). (B) Mini-piezometer and hydraulic potentiomanometer installed at an upwelling site. Note positive Δh, denoting upwelling. (C) Sampling well and DO trap installed at a downwelling site.

indicates whether water is entering (negative VHG; *downwelling*) or discharging from (positive VHG; *upwelling*) the hyporheic zone. Vertical hydraulic gradient is measured using a *minipiezometer* (Lee and Cherry 1979) and *hydraulic potentiomanometer* (Winter *et al.* 1988). The minipiezometer is inserted to a standard depth (Δz) in the sediments using a T-bar (Fig. 30.1B). Water is then drawn from the piezometer and the surface water through flexible tubing to the hydraulic potentiomanometer, where differences in height between subsurface and surface water levels (Δh) can be more easily read (Fig. 30.1B). Vertical hydraulic gradient is calculated as

$$VGH = \frac{\Delta h}{\Delta z}. \tag{30.1}$$

While measurement of VHG using minipiezometers is simple and provides useful information concerning exchange, inference that exchange is actually occurring must be made with caution. Subsurface water may be at pressure but prevented from upwelling by low permeability sediments. Similarly, algal mats may clog sediments and obstruct penetration of the bed (Kuznetsov 1968). Other methods for measuring VHG are described in Chapter 6.

The hydrology team will locate upwelling and downwelling zones and mark them for chemical and biological sampling. In addition to measurement of VHG, the hydrology team may determine subsurface flow velocity and locate lateral exchange sites using dye injections. These activities should be coordinated with the biota team to identify potential hydrologic influence on biotic distribution (in addition to upwelling and downwelling).

The biota team will identify and record locations of major patches of algae, macrophytes, and sessile invertebrates in the stream reach, based on information on dominant taxa provided by the coordinator. Nondestructive methods of counting and measuring patch size should be used where possible. The biota team also will collect benthic core samples or small cobbles at each upwelling and downwelling zone identified by the hydrology team for measurement of chlorophyll *a*, algal biomass, and benthic invertebrates. Finally, this team will determine hydrologic characteristics associated with major algal or macrophyte patches, if applicable (in collaboration with hydrology team). Because sampling of hyporheic invertebrates requires installation of wells, Exercise 2 must be included if subsurface and surface invertebrates are to be compared.

The DO-temp team will install sampling wells at upwelling and downwelling zones and collect water samples for determination of dissolved oxygen and temperature of surface and subsurface water. Dissolved oxygen may be measured either by the Winkler method or with an oxygen meter (see Chapter 5). Although water samples may be withdrawn through the minipiezometers, it is often easier to sample subsurface water from wells. Wells are most simply constructed from lengths of PVC pipe; a heavy-walled PVC with an internal diameter of 15–20 mm has been used successfully in fine-gravel beds. Wells are fitted over a steel T-bar that is slightly longer than the well and inserted by pounding the T-bar into the bed to a depth marked on the well. The T-bar is then carefully withdrawn, leaving the well in place. A simple pump and trap apparatus can be made to collect water samples for dissolved oxygen analysis (Fig. 30.1C). Temperature is measured using a thermometer suspended on a thin line into the well. If a large-diameter well is used, DO probes also may be inserted directly into the well (see Chapters 5 and 6).

The chem-inverts team will sample the wells installed by the DO-temp team for water chemistry and invertebrates and collect surface water chemistry samples. Since a large volume of water from wells is required to collect sufficient numbers of hyporheic invertebrates, a small diaphragm pump is used and water (for chemical analysis) is collected after sieving for invertebrates. This method samples invertebrates that are loosely associated with sediments or are swimming through interstitial space, and would tend to underestimate more sessile or sedentary forms. For this exercise, comparisons between sites using comparable methods are of primary interest. Numbers and kinds of hyporheic invertebrates might be expected to differ between upwelling and downwelling zones, because of differential species' tolerance of oxygen, temperature, and other factors. Differences between chemistry of hyporheic and surface waters in the two zones may be used to interpret differences in biota.

Optional Exercise 3 requires measures of surface metabolism (net production and respiration) and subsurface respiration. Metabolism measurements utilize some sort of chamber in which changes in dissolved oxygen concentration in both dark and light (surface) or dark only (subsurface) are measured. The problems, assumptions, and calculations associated with use of such chambers are described in Chapter 25. Protocol of this exercise employs short incubations of small cobbles or trays of gravel (surface) or sediments obtained by coring (subsurface) from each exchange zone. It is important to use water from the site at which the substrata were collected (i.e., interstitial water obtained from wells should be used for hyporheic sediment incubations).

Optional Exercise 4 involves experimental manipulations that must be followed over a longer time period. Experiments test the hypothesis that biotic distributions are influenced by hydrologic exchange between surface and subsurface water. If true, then biota moved from, for example, an upwelling zone to a downwelling zone should exhibit lower growth or poorer condition or should be eliminated when compared with biota that have not been moved. These experiments must be planned carefully and should utilize a biotic "patch" that shows a clear difference in abundance between upwelling and downwelling zones, has a relatively rapid turnover time (resulting in a quick response to manipulation), and is sufficiently common to allow replication.

III. SPECIFIC EXERCISES

A. Exercise 1: VHG and Biotic Distributions

Field Protocol: Mapping Team

1. Run a meter tape along the length of the reach and flag at evenly spaced intervals. Depending upon size of reach, flagging should delimit transects that will be 4–10 m apart (aim for five to seven transects per reach).

2. In the subreaches between each transect, roughly map out bedform variations, looking for likely upwelling (concave) or downwelling (convex) sites (Fig. 30.1A). If any are found, mark these for the hydrology team.

3. At each transect, measure and record stream width, active channel width, and width of riparian zone. Measure depth at evenly spaced intervals across the stream. Measure stream velocity at these same intervals (see Chapters 3 and 4).

4. Map substratum type: in the subreaches between transects, record major substratum type. Record location of any variations in substratum that occur between transects (e.g., a short riffle dominated by cobble within a 20-m subreach dominated by sand). These should be accurately located on the map so that comparison can be made with upwelling/downwelling location.

5. Fill out the map: map location and size of other physical features.

2. Field Protocol: Hydrology Team

1. Set up a data sheet as in Table 30.1.

2. Beginning at the top of the reach, measure VHG at mid-channel (or nearer the banks in swiftly flowing water) at evenly spaced intervals (aim for approximately 30–50 measurements per reach).

    a. At each measurement point, use a small sledge hammer to insert a T-bar fitted with a slightly shorter well marked at 25 or 30 cm into the sediment. Carefully withdraw the T-bar and drop a minipiezometer into the well. Push the minipiezometer to the bottom of the well using the T-bar, and, holding the T-bar in place, carefully withdraw the well, leaving the minipiezometer in the streambed.

TABLE 30.1
Example Data Sheet for Hydrologic Parameters

Distance along reach (m)	Δh (cm)	Δz (cm)	Notes
0	+2.5	25	Base of riffle
5	−0.5	25	Right bank Higher elevation
5	+1.0	25	Mid-channel

Note. Data sheet contains sample data.

b. Attach the tubing from the hydraulic potentiomanometer to the minipiezometer, and set the tubing with the stilling well on the streambed (out of direct current). Slowly suck water from both ends into the hydraulic potentiomanometer, being careful to avoid air bubbles. When there is a continuous column of water in both tubes, clamp the tubing above the Y. Slowly release the clamp, allowing the water levels to drop down to the measurement scale.

c. Read and record the potential difference (*hydraulic head;* Δh). Be sure to record the *sign:* if the hyporheic column is higher, the sign is +; if the surface column is higher, the sign is − (see also Fig. 30.1B). Also record the depth of the piezometer (Δz). Enter these values in the table next to the value for distance along reach. Record any other information on location.

3. If the mapping team locates a possible strong upwelling or downwelling zone, measure and record VHG at that location so other teams can begin sampling. Otherwise, continue measurements downstream, marking strong upwelling or downwelling zones (with flagging on wire stakes) for sampling.

4. At sites where the stream is braided or appears to vary laterally in elevation, flow velocity, or substratum, make additional lateral measurements of VHG.

5. After VHG has been measured for the entire reach, return to a site of upwelling for measurement of interstitial flow rate. Inject a small bolus of dye (fluorescein is highly visible and thus useful for such qualitative sampling) 5 cm beneath the sediment surface either with a syringe fitted with a long cannula or through a minipiezometer. Record injection time. Wait for dye to appear at the sediment surface; record time of first appear-

ance. Measure distance from injection point to emergence point and calculate distance travelled by the dye. Repeat at another site to verify rate. If dye does not appear, dig at the point of injection to locate dye bolus. If dye cannot be found, the site is not an upwelling site!

6. Collaborate with biota team to identify sites where water enters the stream from lateral interstitial flow (parafluvial zone). Use dye injections (described above) to trace water movement.

Field Protocol: Biota Team

1. Walk along the reach and identify major patch types of algae, macrophytes, and/or sessile invertebrates. If you do not know what they are, give each type a code name (e.g., green alga, 1; invertebrate, 3, etc.).[1]

2. Draw a rough map of the reach, recording the locations as marked by the mapping team. Later, transfer patch locations to the more detailed map generated by the mapping team.

3. Draw major patches on the map. Measure the variables found on Table 30.2 (select those that apply to study system from the following list) and record on a data sheet.

4. Collect a benthic invertebrate core sample at each upwelling and downwelling zone selected for sampling (see Chapters 15 and 16 for additional ideas on sampling invertebrates).

   a. Core to standard depth (5–10 cm) using a section of PVC pipe (5–10 cm diameter), and place core contents into a small bucket.
   b. Add water; elutriate sample by swirling bucket and pouring off water plus invertebrates (plus detritus and algae) through a 62-μm mesh sieve or net. Repeat at least five to seven times (or until no additional invertebrates are collected).
   c. Rinse contents of sieve/net into a labeled bag or vial with 70% ethanol (final concentration) to preserve.
   d. Replicate benthic invertebrate cores if sites (upwelling, downwelling) are not replicated.

5. Collect an algal core sample at each upwelling and downwelling zone selected for sampling (see Chapters 13 and 14 for additional information on sampling algae).

   a. Core to standard depth (2–3 cm) using a 3- to 6-cm diameter section of tubing, or a cut-off syringe.
   b. Drain water from core without losing sample.

---

[1]Focus on the most abundant or easily recognized patch types.

TABLE 30.2
Biotic Variables to Be Measured and Recorded

Patch type	Patch characteristics to measure	Location (reference to detailed maps)
Algae—mats or filaments	Mat size (length × width)	Edge or center
	Algal height or mat thickness	Upwelling or downwelling
	Algal condition	Distance along reach
Algae–microscopic communities on sediments	Color	Major substratum types
	Continuity of cover	Major physical features
		(These notes should be taken to aid in placing patches on map)
Macrophytes	Bed size (length × width)	
	Plant height	
	Plant condition	
	Plant density (low, medium, high)	
Invertebrates—large sessile (attached) or conspicuous grazers	Density (quantitative; number per unit area)	
	Diversity (number of taxa)	
	Apparent feeding mode (see Chapter 21)	

    c. Place sample into a labeled bag and store on ice.
    d. Replicate algal cores if sites are not replicated.

    6. Alternative to coring: If substrata are too large for coring, collect cobbles for determination of algal biomass and species composition. Place two to three cobbles from each site into a labeled bag, and store on ice.

    7. Work with hydrology team to determine hydrologic characteristics of particularly striking biotic patches, if not located in sampling sites. For example, algal mats or macrophyte beds may be especially abundant at points where lateral interstitial flow enters the stream or at the heads of small side channels formed where water upwells from the sediments. As before, record relevant patch characteristics as well as hydrologic characteristics (VHG, flow direction determined using dye injections).

Laboratory Protocol

    1. Calculate VHG ($=\Delta h/\Delta z$).
    2. Calculate interstitial flow velocity (distance/time).

3. Place VHG, flow path direction (if any), and biotic patch data on the overall system map generated by the mapping team.

4. Process algal core samples.

    a. Obtain a small "grab sample" (a few bits of gravel or 1 ml of slurry) and place in a vial with algal preservative (see Chapter 13) for later identification of species.

    b. Divide the gravel/sand sample in half and analyze subsamples for ash-free dry mass (AFDM) and chlorophyll *a* (use a modification of protocols in Chapter 14 wherein subsamples are extracted (chlorophyll *a*) or dried and ashed (AFDM) directly).

    c. Calculations: use area of corer to express chlorophyll *a* and AFDM on an areal basis; multiply by 2 to correct for subsampling.

5. Process algal cobble samples.

    a. Scrape and brush material from cobbles into a small volume of distilled water. Add water to this to make up to standard volume (e.g., 50 or 100 ml).

    b. Homogenize the scraped material plus water.

    c. Subsample by pipetting a known volume of the homogenate.

    d. Filter the chlorophyll *a* subsample and analyze filters for chlorophyll *a*. Place AFDM subsample into drying vessel. Preserve a third subsample for algal identification. Measure area of cobbles (see protocols in Chapter 14).

    e. Calculations: use cobble area to express chlorophyll *a* and AFDM on an areal basis; multiply by total homogenate volume/subsample volume to correct for subsampling.

6. Identify and count invertebrates in preserved samples (see protocol in Chapter 16).

B. Exercise 2: VHG, Biotic Distribution, and Chemistry

Mapping team, hydrology team, and biota team: same as for Exercise 1 above.

Field Protocol: DO-Temp Team

1. Install wells at upwelling and downwelling sites selected for sampling.

    a. Insert a 50-cm T-bar fitted with a 48-cm well to a sediment depth of 30 cm using a sledge hammer.[2]

---

[2]Top of well must be above water surface level, so use a longer T-bar and well in deeper water.

    b. Carefully remove the T-bar while holding the well, leaving the well in place.

    c. Bail the well: insert long tubing connected to DO trap into well (see Fig. 30.1C). Push tubing to the bottom of the well, then pull up slightly (1 cm) to avoid aspirating sand. Pump ca. 300 ml water from the well and discard.[3]

2. Collect hyporheic oxygen samples.

    a. Insert a DO bottle (with stopper removed) into the trap. Replace and tighten top of trap, ensuring that long tubing piece inside the cap is inserted into the DO bottle.

    b. Pump water from the well to overflow the DO bottle 2.5 times.

    c. Important: clamp the tubing to stop flow of water from the well, then remove the DO bottle.

    d. Fix sample using Winkler reagents (see Chapter 5), and store in darkness.

    e. Alternatives (if difficulty in obtaining 60-ml sample is encountered): (1) Pump a sample of hyporheic water into an empty bottle or jar and measure its oxygen content using a dissolved oxygen electrode (see Chapter 5); (2) drop electrode into wide-bore well (see Chapter 6); (3) fill a 10-ml syringe from a continuous column of water pumped from the hyporheic zone; analyze dissolved oxygen using a micro-Winkler technique (see Grimm 1987 or Chapter 5).

    3. Measure and record hyporheic water temperature. Suspend thermometer into well, withdraw and read temperature, or read directly from DO probe (wide-bore wells).

    4. Collect surface oxygen samples using the DO trap employed in step 2, pumping from the surface stream. Fix and store in dark. Alternative: use DO electrode.

    5. Measure and record surface water temperature.

Field Protocol: Chem-inverts Team

    1. Working at the same site as the DO-temp team, obtain samples of surface water. First rinse the bottle(s) three times with streamwater; then fill from the thalweg.

    2. After oxygen samples have been obtained from the well, collect hyporheic water and invertebrate samples.

---

[3]Temperature can be measured on this water.

a. Insert long end of tubing connected to the diaphragm pump setup into the well. Push tubing to the bottom of the well, then pull up slightly (1 cm) to avoid aspirating sand.

b. Ensure that cap is tightened onto the collection bottle; then begin pumping slowly. Diaphragm pumps can generate quite a vacuum, so be sure you do not pump so fast as to empty the well or aspirate stones. Pump until 2, 3, or 4 liters of water is collected.[4]

c. Remove bottle cap and slowly pour contents *through* a 62-μm mesh net *into* (1) rinse bottle (for rinsing invertebrate net) and (2) water sample bottles, first rinsing bottles and then filling.

d. Rinse net contents into a prelabeled zip-lock bag or jar, using the filtered water in the rinse bottle. Store on ice.[5]

Laboratory Protocol

1. Mapping team, hydrology team, and biota team: follow steps for Exercise 1.

2. Acidify and titrate oxygen samples (see Chapter 5).

3. Filter water samples and analyze for nutrients (nitrate-N, ammonium-N, soluble reactive P, others as suggested by prior data or coordinator) using standard methods (APHA *et al.* 1992; see also Chapter 9).

4. Identify and count hyporheic invertebrates in samples from wells. Since many species of hyporheic invertebrates are poorly known, the level of taxonomic resolution may be relatively coarse. Try, however, to recognize abundant species (even if you do not know their names) to determine whether differences in species composition exist between upwelling and downwelling zones. See Chapter 15 for protocols on identifying meiofaunal invertebrates.

C. Exercise 4: VHG, Biotic Distribution, and Metabolism (Optional)

Mapping, hydrology, biota, DO-temp, and chem-inverts teams: same as for Exercises 1 and 2 above.

Field Protocol: Process Team

1. For especially fine sediments, it may be necessary to place trays filled with precleaned sediments in the stream for colonization by periphyton several weeks before measurements are made. Chambers as simple as

---

[4]In finer sediments, volumes may have to be reduced.

[5]If samples cannot be processed within 2–3 days, rinse net contents into zip-lock bag or jar with 70% ethanol to preserve the sample. It is best, however, to examine specimens live if at all possible.

zip-lock bags or stoppered cores, or as sophisticated as those described in Chapter 25, may be used.

2. Collect material to be incubated and water for incubations. Label chamber with hydrologic type (up- or downwelling), subsystem (surface or hyporheic), and replicate number (suggested minimum number of replicates is five; giving total $N = 5 \times 4$ treatments $= 20$).

3. A second team member will collect a reservoir of "initial" stream water from each of four sites: surface/upwelling, hyporheic/upwelling, surface/downwelling, hyporheic/downwelling. Use wells installed in Exercise 2 to obtain hyporheic water. After thoroughly mixing each of the 4 reservoirs of initial stream water, collect triplicate dissolved oxygen samples from each using a hand pump-DO trap apparatus (from Exercise 2) *or* insert dissolved oxygen electrode into reservoir and measure *initial* dissolved oxygen.

4. Carefully fill chambers with water from the appropriate reservoir and add material to be incubated.

5. Incubate chambers in light (surface) or dark (bury in sediments or cover with black plastic) *under stream water* (to maintain realistic temperature) for 1–1.5 h.

8. Carefully measure "final" dissolved oxygen (Winkler or dissolved oxygen meter).

9. Return contents of chambers to laboratory for analyses of chlorophyll *a* and AFDM.

Laboratory Protocol: Process Team

1. Titrate dissolved oxygen.
2. Calculate surface net primary production (P_N) in mg O_2 m^{-2} h^{-1},

$$P_N = \Delta DO \cdot \frac{V}{A \cdot t}, \tag{30.2}$$

where ΔDO is change in oxygen during light incubation (mg/liter); V, volume of water in chamber (liters); A, area of sediment sampled (or of cobbles used) (m^2 = cm^2/10,000); and t, incubation time (h).

3. Calculate hyporheic respiration (R) in mg O_2 m^{-2} h^{-1} or in mg g sediment^{-1} h^{-1},

$$R = \Delta DO \cdot \frac{V}{w \cdot t} \tag{30.3}$$

$$R = \Delta DO \cdot \frac{V}{A \cdot t}, \tag{30.4}$$

where w is dry mass of sediment (g). Expressing respiration per unit sediment mass will allow comparison with published values.

D. Exercise 4: Optional Experiment

Field Protocol: Relocation Experiments

1. If mapping of biotic distributions in relation to upwelling and downwelling zones has not been done, generate a stream reach map following procedures outlined in Exercise 1 above.

2. Locate one to three sites each of upwelling and downwelling (treatments may be replicated at the level of cores or sites, or both). Using information on algal or macrophyte patch type distribution, select one or two dominant patch types from each surface environment for experimental manipulation. Remove 6 to 10 cores (50–100 cm^2) from each patch at each upwelling site. Replace half of the cores in the same location, and move the remaining cores to a downwelling site. Conduct the reciprocal experimental relocation (downwelling to upwelling) for dominant downwelling zone patches. Mark the experimental cores with flagging and leave in place for at least 2 weeks. Prior to sampling, record condition (see Experiment A) of control (removed and replaced) and treatment (removed and relocated) algae/macrophytes. Recore the experimental units and return to the laboratory (on ice) for analysis of chlorophyll a, biomass, and community structure.

3. Epilithic algae relocation experiments (cobbles): follow the design described in step 2, but relocate/replace cobbles rather than cores. For laboratory analysis, pay special attention to possible changes in microscopic algal community structure.

4. Sessile invertebrate relocation experiments (cobbles): if sessile, epilithic invertebrates are abundant, an experiment following the design in Step 3 could be analyzed for invertebrate abundance as well as algal variables.

Field Protocol: Nutrient Enrichment Experiments

1. If differences in biota between upwelling and downwelling zones are caused by enrichment of surface biota by nutrient-rich, upwelling water, then addition of a suspected limiting nutrient should have a smaller effect on algal growth/accrual at an upwelling zone than at a downwelling zone. Experiments using substrata that release nutrients (see Chapter 29) can be set up to test this prediction. Substrata can be prepared with and without the potentially limiting nutrient and incubated in the stream at upwelling and downwelling zones. Data from Exercise 2 should be examined to determine whether differences between subsurface and surface water chemistry indicate a nutrient effect. Atomic ratios of nitrogen to phosphorus also should be calculated to identify which of these elements is potentially

limiting (Grimm 1992, see also Chapter 9). Alternatively, both nutrients could be added so that any nutrient effect associated with upwelling might be detected.

Data Analysis

1. Compare upwelling and downwelling zones (surface stream) in terms of (a) vertical hydraulic gradient; (b) channel features; (c) primary producer abundance, condition, and types; (d) invertebrate abundance and types; (e) temperature and chemical parameters; and (f) metabolism. Some of these differences may be examined graphically (e.g., with bar graphs) and, if replicate sites were sampled, tested for significance using t tests (Sokal and Rohlf 1995).

2. Compare upwelling and downwelling zones (hyporheic zone) in terms of (a) hyporheic invertebrate abundance and types; (b) temperature and chemical parameters; and (c) metabolism, again using graphical and statistical methods as in step 1.

3. For parameters relevant to and measured at all sampling points (temperature, chemical parameters, dissolved oxygen), a two-way analysis of variance (Sokal and Rohlf 1995) with subsystem (surface, hyporheic) and hydrology (upwelling, downwelling) as main factors may be used to detect significant differences. Hint: compare upwelling and downwelling zones in terms of the *difference* between surface and subsurface environments.

4. Compare treatment (removal–relocation) to control (removal–replacement) experimental units (algae/macrophyte cores, cobbles) in terms of condition, biomass, invertebrate abundance, or change in community structure, using graphical analysis and t tests.

5. Compare upwelling zones with downwelling zones in terms of response to nutrient enrichment (difference between treatment and control nutrient-diffusing substrata) using t test.

IV. QUESTIONS

1. What channel features are related to location of upwelling and downwelling zones?

2. Is there evidence that exchange influences the distribution of periphyton, macrophytes, riparian vegetation, or invertebrates in the surface stream or channel?

3. What factors might explain difference in water chemistry or metabolism between upwelling and downwelling zones in the surface stream?

4. To what factors or processes would you attribute differences in community structure or abundance of hyporheic invertebrates between upwelling and downwelling sites?

5. How do hyporheic water temperature and chemistry vary between upwelling and downwelling sites? What might cause these differences (if any)?

6. Does hyporheic metabolism at downwelling sites differ from that at upwelling sites? Why?

7. If you found a large difference in some physical or chemical parameters between surface and subsurface waters at upwelling zones, but a small difference in the same parameters at downwelling zones, how would you explain this? How would you explain the reverse?

8. What prediction is tested by the experiment relocating algae/macrophytes/invertebrates? Of what hypothesis is this prediction a logical consequence? Was the hypothesis supported by the data or not? If not, what can be concluded about the hypothesis? (Hint: use this structure: *If ...* [hypothesis] *and I move algae growing at a downwelling site to an upwelling site* [experimental manipulation] *then ...* [prediction].)

9. What prediction is tested by the nutrient enrichment experiment? Of what hypothesis is this prediction a logical consequence? Was the hypothesis supported by the data or not? If not, what can be concluded about the hypothesis? (Hint: use this structure: *If ...* [hypothesis] *and ...* [experimental manipulation] *then ...* [prediction].)

V. MATERIALS AND SUPPLIES

Mapping Team

Clipboard with gridded paper
Flagging and markers (for labeling flagging)
Items for flotation (for measuring discharge)
Meter stick(s)
Meter tape(s)
Stakes
Stopwatch or wristwatch with timer

Hydrology Team

40-cm Minipiezometers (5–10)
48-cm Well
50-cm T-bar
Biodegradable, nontoxic dye (e.g., fluorescein) in syringe

Hand pump
Hydraulic potentiomanometer (Fig. 30.1B)
Small sledge hammer
Stopwatch

Biota Team

62-μm Mesh nets or sieves
70% Ethanol and squirt bottle
Corers (5–10 cm diameter; one or more each for algae and invertebrates)
Cooler with ice (for unpreserved samples)
Sample bags (for algae)
Sample vials or jars (for invertebrates)
Small bucket or plastic pitcher
Small ruler for measuring plant/algal heights

DO-temp Team

48-cm (or longer) wells (enough to leave in place during sampling)
50-cm (or longer) T-bar
Dissolved oxygen bottles (60-ml size, or smaller for low hydraulic conductivity sites)
Dissolved oxygen reagents (see Chapter 5)
Dissolved oxygen trap with hand-operated vacuum pump
Noniced cooler or box for oxygen samples (to keep dark)
Small sledge hammer
Thermometers (see Chapter 5)
Alternative: dissolved oxygen meter, probe, and collection vessel
Alternative: micro-Winkler collection vessels (10-ml syringes)

Chem-inverts Team

4-Liter heavy-walled polyethylene bottle (or other heavy plastic; thin-walled materials will collapse
62-μm mesh net or sieve
Cooler with ice
Diaphragm pump mounted on board or bucket (to keep out of water; if mounted on bucket, may invert bucket to sit while pumping)
Hoses to pump and well
Lid for polyethylene bottle fitted with two hose connectors (for pump and well hoses)
Plastic bags or jars (for invertebrates)
Squirt bottle (field wash bottle)
Water chemistry bottles

Process Team

Black plastic
Buckets or carboys for reservoir of "initial" water
Buckets to transfer materials to be incubated
Chambers (20)
Dissolved oxygen measurement equipment
Well sampling equipment

Experiments Team

Coring devices
Flagging and stakes
Nutrient-diffusing substrata (see Chapter 29)

REFERENCES

APHA, AWWA, and WEF. 1992. *Standard Methods for the Examination of Water and Wastewater,* 18th ed. American Public Health Association, Washington, DC.

Bencala, K. E., V. C. Kennedy, G. W. Zellweger, A. P. Jackman, and R. J. Avanzino. 1984. Interactions of solute and streambed sediment. 1. An experimental analysis of cation and anion transport in a mountain stream. *Water Resources Research* **20:**1797–1803.

Boulton, A. J. 1993. Stream ecology and surface-hyporheic hydrologic exchange: Implications, techniques, and limitations. *Australian Journal of Marine and Freshwater Research* **44:**553–564.

Coleman, R. L., and C. N. Dahm. 1990. Stream geomorphology: Effects on periphyton standing crop and primary production. *Journal of the North American Benthological Society* **9:**293–302.

Fisher, S. G., L. J. Gray, N. B. Grimm, and D. E. Busch. 1982. Temporal succession in a desert stream ecosystem following flash flooding. *Ecological Monographs* **52:**93–110.

Gregory, S. V., F. J. Swanson, W. A. McKee, and K. W. Cummins. 1991. An ecosystem perspective of riparian zones. *Bioscience* **41:**540–551.

Grimm, N. B. 1987. Nitrogen dynamics during succession in a desert stream. *Ecology* **68:**1157–1170.

Grimm, N. B. 1992. Biogeochemistry of nitrogen in arid-land stream ecosystems. *Journal of the Arizona–Nevada Academy of Science* **26:**130–146.

Grimm, N. B., and S. G. Fisher. 1984. Exchange between interstitial and surface water: Implications for stream metabolism and nutrient cycling. *Hydrobiologia* **111:**219–228.

Grimm, N. B., and S. G. Fisher. 1986. Nitrogen limitation in a Sonoran Desert stream. *Journal of the North American Benthological Society* **5:**2–15.

Grimm, N. B., H. M. Valett, E. H. Stanley, and S. G. Fisher. 1991. Contribution of the hyporheic zone to stability of an arid-land stream. *Verhandlungen der Internationalen Vereinigung für Theoretische und Angewandte Limnologie* **24:**1595–1599.

Harvey, J. W., and K. E. Bencala. 1993. The effect of streambed topography on surface-subsurface water exchange in mountain catchments. *Water Resources Research* **29:**89–98.

Hynes, H. B. N. 1983. Groundwater and stream ecology. *Hydrobiologia* **100:**93–99.

Jones, J. B., Jr., S. G. Fisher, and N. B. Grimm. 1995a. Nitrification in the hyporheic zone of a desert stream ecosystem. *Journal of the North American Benthological Society* **14:**249–258.

Jones, J. B., Jr., S. G. Fisher, and N. B. Grimm. 1995b. Vertical hydrologic exchange and ecosystem metabolism in a Sonoran Desert stream. *Ecology* **76:**942–952.

Kuznetsov, S. L. 1968. Recent studies on the role of microorganisms in the cycling of substances in lakes. *Limnology and Oceanography* **13:**211–224.

Lee, D. R., and J. A. Cherry. 1979. A field exercise on groundwater flow using seepage meters and mini-piezometers. *Journal of Geology Education* **27:**6–10.

Palmer, M. A., A. E. Bely, and K. E. Berg. 1992. Response of invertebrates to lotic disturbance: A test of the hyporheic refuge hypothesis. *Oecologia* **89:**182–194.

Savant, S. A., D. D. Reible, and L. J. Thibodeaux. 1987. Convective transport within stable river sediments. *Water Resources Research* **23:**1763–1768.

Sokal, R. R., and F. J. Rohlf. 1995. *Biometry,* 3rd ed. Freeman, New York, NY.

Stanford, J. A., and J. V. Ward. 1993. An ecosystem perspective of alluvial rivers: Connectivity and the hyporheic corridor. *Journal of the North American Benthological Society* **12:**48–60.

Thibodeaux, L. J., and J. D. Boyle. 1987. Bedform-generated convective transport in bottom sediment. *Nature* **325:**341–343.

Triska, F. J., V. C. Kennedy, R. J. Avanzino, G. W. Zellweger, and K. E. Bencala. 1989. Retention and transport of nutrients in a 3rd-order stream in northwestern California: Hyporheic processes. *Ecology* **70:**1893–1905.

Valett, H. M., S. G. Fisher, N. B. Grimm, and P. Camill. 1994. Vertical hydrologic exchange and ecological stability of a desert stream ecosystem. *Ecology* **75:**548–560.

Valett, H. M., S. G. Fisher, and E. H. Stanley. 1990. Physical and chemical characteristics of the hyporheic zone of a Sonoran desert stream. *Journal of the North American Benthological Society* **9:**201–215.

Vaux, W. G. 1962. *Interchange of Stream and Intergravel Water in a Salmon Spawning Riffle.* United States Fish and Wildlife Service, Special Scientific Report-Fisheries 405. Washington, DC.

Vaux, W. G. 1968. Intragravel flow and interchange of water in a streambed. *United States Fish and Wildlife Services Fisheries Bulletin* **66:**479–489.

White, D. S. 1990. Biological relationships to convective flow patterns within stream beds. *Hydrobiologia* **196:**149–158.

Winter, T. C., J. W. LaBaugh, and D. O. Rosenberry. 1988. The design and use of a hydraulic potentiomanometer for direct measurement of differences in hydraulic head between groundwater and surface water. *Limnology and Oceanography* **33:**1209–1214.

Wroblicky, G. J., M. E. Campana, H. M. Valett, J. A. Morrice, K. S. Henry, C. N. Dahm, J. V. Hurley, and J. M. Noe. 1992. Remote monitoring of stream hyporheic zones with inexpensive pressure transducer-data acquisition systems. Pages 267–277 *in* J. A. Stanford and J. J. Simons (Eds.) *Proceedings of the First International Conference on Ground Water Ecology.* American Water Resource Association Technical Publication Series, Tampa, FL.

CHAPTER 31

Macroinvertebrates as Biotic Indicators of Environmental Quality

VINCENT H. RESH,
MARILYN J. MYERS, AND
MORGAN J. HANNAFORD

Department of Environmental Science, Policy and Management
University of California, Berkeley

I. INTRODUCTION

The use of living organisms to assess water quality is a century-old approach (Cairns and Pratt 1993), but its widespread use is much more recent in North America and the United Kingdom than in Continental Europe (Metcalf 1989). The former countries have relied mainly on chemical and physical measures, even though water pollution is essentially a biological problem. One problem of relying solely on chemical and physical measurements to evaluate water quality is that they provide data that primarily reflect conditions that exist when the sample is taken. In essence, a physicochemical approach provides a "snapshot" of water quality conditions. In contrast, biological monitoring gives an indication of past conditions as well as current conditions. So rather than merely being an instantaneous measure, the biological data provide an integrated "moving picture" of the recent past.

Of all the potential groups of freshwater organisms that have been considered for use in biological monitoring, benthic macroinvertebrates (which mainly consist of aquatic insects, mites, molluscs, crustaceans, and annelids) are most often recommended (Hellawell 1986). There are many advantages in using these organisms in water quality monitoring (Table 31.1).

TABLE 31.1

Advantages and Difficulties to Consider in Using Benthic Macroinvertebrates for Biological Monitoring (Summarized from Rosenberg and Resh 1993a, Who Also Discuss How to Overcome the Difficulties Mentioned)

Advantages	Difficulties to consider
(1) Being ubiquitous, they are affected by perturbations in all types of waters and habitats	(1) Quantitative sampling requires large numbers of samples, which can be costly
(2) Large numbers of species offer a spectrum of responses to perturbations	(2) Factors other than water quality can affect distribution and abundance of organisms
(3) The sedentary nature of many species allows spatial analysis of disturbance effects	(3) Seasonal variation may complicate interpretations or comparisons
(4) Their long life cycles allow effects of regular or intermittent perturbations, variable concentrations, etc., to be examined temporally	(4) Propensity of some macroinvertebrates to drift may offset the advantage gained by the sedentary nature of many species
(5) Qualitative sampling and analysis are well developed, and can be done using simple, inexpensive equipment	(5) Perhaps too many methods for analysis available
(6) Taxonomy of many groups is well known and identification keys are available	(6) Certain groups are not well known taxonomically
(7) Many methods of data analysis have been developed for macroinvertebrate communities	(7) Benthic macroinvertebrates are not sensitive to some perturbations, such as human pathogens and trace amounts of some pollutants
(8) Responses of many common species to different types of pollution have been established	
(9) Macroinvertebrates are well suited to experimental studies of perturbation	
(10) Biochemical and physiological measures of the response of individual organisms to perturbations are being developed	

In this chapter, we introduce procedures to: (1) conduct physical habitat assessments of streams and their surrounding riparian zones and (2) use benthic macroinvertebrates in evaluating water quality. This is done by assessing selected physical features of a stream habitat, collecting and identifying benthic macroinvertebrates, and using this latter information to calculate a *biotic index score*. The idea of combining information from habitat assessments and biotic index scores is a fairly new one, but the linkage is related to the inescapable fact that benthic macroinvertebrates may be influenced by habitat quality (e.g., bank stability) as well as water quality (e.g., a pollutant present in the water).

Various measurements of the physical habitat of the channel, floodplain, and water constituents can provide important information to stream ecologists. Such information is important because when the physical habitat is in poor condition we would expect that the *biological health* of the stream would be affected adversely as well. For example, if a stream's banks are unstable and regularly sloughing, the input of fine sediment likely would negatively impact sensitive fish and macroinvertebrates. Sediment-intolerant species likely would decline and the species composition in the stream might change to species that are tolerant to sedimentation. A quantitative assessment of the stream's habitat may indicate impairment, and provide insight into mitigation measures to correct the problem. For example, a common stream-improvement project is to replant riparian vegetation next to the stream to reduce the source of sedimentation.

Biotic indices are based on the premise that *pollution tolerance* differs among various benthic organisms. *Tolerance scores* for each taxon are intended for a single type of pollution; usually these are for organic pollution but recently tolerance scores for acidification have been developed (Johnson *et al.* 1993). In most biotic indices, the tolerance of the taxa that comprise a macroinvertebrate community and the numbers (or their proportion of the total, or some other categorical measure of abundance) of each taxon are used to calculate a single score. In some of the European biotic indices, an indicator value for the different taxa is also added (e.g., the Saprobien Index used in Germany). Various biotic indices and their formulae are discussed by Metcalfe (1989) and Resh and Jackson (1993).

What level of taxonomy is used in assigning tolerance scores? In some citizens' monitoring programs where volunteer participants depend on "picture keys" to name the organisms present, identification is usually to a mixture of the order- and family-levels. In this exercise we will use family-level identifications, but most State, Provincial, and Federal agencies of the United States and Canada involved in water quality monitoring usually use a mixture of generic- and species-level identifications. Because the tolerance of many benthic macroinvertebrates differs within a family and may differ even within a genus (Resh and Unzicker 1975), the finer the taxonomic resolution the more reliable the assignment of *tolerance values.*

Biotic indices are one of several types of measures that are routinely used in biological monitoring. Most contemporary survey approaches rely on multiple measures, also referred to as *metrics,* of community structure and function. These measures can be grouped into several categories: (1) taxa richness (e.g., family-level, generic-level, species-level), (2) enumerations (e.g., number of all macroinvertebrates collected, proportions of selected orders such as Ephemeroptera, Plecoptera, and Trichoptera (EPT)), (3) community diversity indices (e.g., Shannon's index), (4) community similarity indices (e.g., the Pinkham–Pearson index), or (5) functional feeding-groups ratios (e.g., percentage of the "shredder" functional group; see also Chapter 21). Thus, although calculation of a biotic index is an important and commonly used component of a biomonitoring program, an approach that uses several different measures is likely to be of greater value (e.g., Plafkin *et al.* 1989).

The objective of this chapter is to describe how to conduct physical habitat assessments and to evaluate macroinvertebrate community structure and function within the context of biotic indices. This chapter is intended for students and also for researchers who have not conducted these types of evaluations. For a broader perspective on biological monitoring, consult the book edited by Rosenberg and Resh (1993b). For regulatory agency personnel and experienced researchers, Resh and Jackson (1993) provide a summary of rapid assessment approaches to benthic macroinvertebrate monitoring, Resh and McElravy (1993) give an overview of quantitative studies in this regard, and Resh *et al.* (1995) discuss the design and implementation of rapid bioassessment approaches, including the methods described in this chapter. Students may find the coverage of biological monitoring in the chapter of Rosenberg and Resh (1995) sufficient for a broad understanding of concepts, principles, and future information needs.

II. GENERAL DESIGN

The process of conducting a biological assessment, or a physical habitat assessment of a stream and its surrounding riparian area, can range from detailed measurements made over a long period of time (months or years) to a relatively quick survey of present conditions. In Exercise 1, we present methods for rapid biological and physical habitat assessment that can be completed in 1 day. These methods are particularly appropriate for a survey or reconnaissance study when little is known about a stream, when a spill or pollution problem needs to be evaluated quickly, or for comparisons over a geographic area with similar topographic features. The exercise is based on techniques developed for aquatic biologists in consulting firms and government regulatory agencies. With modification, they can also be applied in establishing pollution control programs in newly industrialized and developing countries (Resh 1995).

A. Site Selection

Prior to beginning this study, appropriate sample sites must be selected. The optimal design for this study is to have an impacted reach of stream to compare with an unimpacted or "reference" reach. Often reaches are chosen in the same stream, for instance above and below a point source of pollution. Alternatively, sample sites could be in two different streams.[1] The principle that should be remembered when two different streams are compared, or if different reaches in the same stream are compared, is that physical characteristics of the sites selected should be as similar as possible. A few examples of important characteristics are: (1) the gradient of the compared reaches should be very similar (within 1%) (e.g., comparison of a high gradient (5% slope) reach with a meandering (0.5% slope) reach is not meaningful), (2) the substrata composition or at least the dominate substratum size of each reach should be similar (e.g., a sand-dominated channel will have different macroinvertebrates than a cobble-dominated channel), (3) the stream order should be the same so that similar-size streams are compared, and (4) the streams should be either permanent *or* intermittent (c.g., an intermittent stream may appear perennial in winter and spring, but its invertebrate fauna will be very different than the fauna of a perennial stream).

Once the stream reaches have been chosen, individual sampling sites within them must be selected. Riffles often are selected as sites for macroinvertebrate sampling because of the abundance and diversity of organisms found in such habitats. Decide whether to sample riffles only, a combination of riffles and pools, or to sample all habitats in proportion to their occurrence.

B. General Procedures

Physical Habitat Assessment   In any stream study, certain habitat characteristics are nearly always measured and recorded: water temperature,

---

[1] In conducting these procedures within a class setting, if an impacted stream cannot be compared with a reference stream, participants could sample macroinvertebrates in distinctly different habitats within one stream (e.g., pools, riffles, vegetated stream margins) or repeatedly sample the same type of habitat within a single stream. Thus, the participants could examine variability in sampling while still learning the concepts and techniques used in biological assessment.

air temperature, stream width, stream depth, and discharge (see Chapters 2–5). These measurements provide information about the size of the stream and what types of organisms might be living there (e.g., some organisms typically live only in cold water, some only in warm water).

Other characteristics that are often measured are the stability of the stream banks, the gradient of the stream, the amount that the stream is shaded by riparian vegetation, the composition of the bottom substrata, the complexity of microhabitats within the stream, and the number of pieces of large wood present (Platt *et al.* 1983, see also Chapters 2, 5, and 11). Water quality measurements such as the amount of dissolved oxygen, pH, the levels of nutrients, or the number of enteric bacteria (*E. coli*) present are also often measured (see Chapters 5 and 9).

Choice of which measurements to be taken depends on the questions being asked. For example, an investigator might not need to measure dissolved oxygen in a clean, high-gradient mountain stream because it is likely that the water is saturated with dissolved oxygen. However, in a slow-moving, warm stream that may be subject to organic pollution (e.g., sewage), measures of dissolved oxygen content (both during the day and during the night) are crucial.

Biological Assessment   Many measures and indices have been developed to quantify and evaluate the biological characteristics of a stream; a biotic index is one of the oldest approaches. A biotic index is simply a weighted average (by numerical abundance) of the tolerance scores of the different organisms collected.

Tolerance scores for individual taxa are derived in a number of ways. One approach is to use collections of benthic invertebrates from streams of varying water quality and relate abundance and distribution of organisms to these conditions. These types of surveys have been published for Wisconsin (Hilsenhoff 1988) and the southeastern United States (Lenat 1993). The more common method of assigning tolerance scores is based on "expert opinion," either modifying previously published tolerance scores for use in different regions or estimating scores based on experience. An alternative approach has been developed in the United Kingdom (Wright *et al.* 1984, 1988), which involves the prediction of macroinvertebrate occurrence at a given site from a number of environmental variables.

III. SPECIFIC EXERCISES

A. Exercise 1: Field Option

Physical Habitat Assessment   The first four measurements are all related to discharge (see also Chapter 3 for detailed discussion and methods).

These four measurements are taken together in a reach having few obstructions and a uniform flow. Record the measurements of physical habitat as they are taken.

1. *Mean stream width.* Measure the width of the stream in meters, from water's edge to water's edge and perpendicular to the flow, for three different transects across the stream.

2. *Mean stream depth.* Along the same transects as above, measure the depth (in cm) at 1/4 the distance from the water's edge, again at 1/2 the distance (midstream), and at 3/4 of the way across. Add the three values and divide by 4 (divide by 4 to account for the shallow water from the bank edge to the 1/4 distance mark). Record the average depth in meters for each transect.

3. *Current velocity.* Follow the protocols described in Chapter 3. Alternatively, lay a tape measure along the edge of the stream (5–10 m is sufficient). Drop a neutrally bouyant float (e.g., orange or lemon) into the water several meters upstream of where the tape measure begins and measure the amount of time it takes for the float to pass the length of the tape (if the water is very shallow a twig or cork can be used). Repeat five times and calculate the average velocity (m/s).

4. *Discharge.* Discharge is the product of width × mean depth × velocity (all measured in meters). Following the protocol described in Chapter 3, calculate discharge (m³/s) for each of the three cross sections. Calculate the mean of the three discharge measurements.

5. *Air temperature.* Measure air temperature before you measure water temperature so that the bulb of the thermometer is dry. Air temperature is measured because it provides information about the expected water temperature.[2]

6. *Water temperature.* Using either an electronic thermocouple or mercury thermometer (see Chapter 5), read the temperature while the probe is still in the water and after a reasonable equilibration period. Always record the time when temperatures are taken.

7. *Gradient.* Gradient equals rise (upstream increase) in stream height divided by the length of stream run examined. Measure stream gradient using standard surveying techniques, autolevel and staff, or hand-held Abney-level and staff. Alternatively, lay a tape measure along the edge of the stream for about 30 m. Hold a meter stick straight up and down at the water's edge at the beginning of the tape, and hold another meter stick at the other end of the tape. Stretch a string between the two meter sticks,

---

[2]For example, if the air temperature is 30°C and the study streamwater temperature expected at 20–30°C, but the water temperature is only 10°C, we might look for an underwater spring source. The consequence of such conditions would lead to different expectations about the stream fauna.

and place the end of the string at the 50-cm mark on the downstream stick. Attach a string level to the string (midway between the two meter sticks) and move the string up or down along the upstream meter stick until the bubble on the level is centered. Record the difference between the heights on the two meter sticks (rise) and divide by the length between the two sticks (run). Example: The height on the downstream stick is 50 cm and that on the upstream stick is 47 cm; $50 - 47 = 3$ cm. If the distance between the two sticks is 30 m, then the gradient equals 0.03 m $\div$ 30 m = $0.001 = 0.1\%$.

8. *Water clarity.* Collect a 500-ml water sample from mid-stream in a prewashed nalgene bottle. Return the sample to the laboratory and analyze for turbidity (APHA *et al.* 1992). Alternatively, record whether the water is clear, slightly turbid, or muddy (e.g., whether you can or cannot see the bottom of the stream). Try to note any source of sediment (e.g., storm runoff, construction activities).

9. *Habitat parameters.* The habitat parameters describe components of the stream channel and the surrounding riparian area (Table 31.2). Judge the condition of these parameters as optimal, suboptimal, marginal, or poor. These habitat parameters are only a few of many possibilities that have been proposed and used. For each parameter, read the description given for each of the four conditions, select the one that most clearly identifies what you see, and circle the number that corresponds to the condition class. When you have gone through all the parameters, sum the circled values. Normally, this value is compared to the number obtained when the same habitat parameters are evaluated for a "reference" stream (i.e., a stream in excellent condition in the same geographical area). (Also see the variability discussion in the data analysis section.)

Biological Assessment

1. *Sampling benthic invertebrates.* Although many devices and techniques are available for collecting aquatic invertebrates (see Chapter 16), a D-frame net or a kick screen (a window screen stretched between two poles) is recommended for this study. The number of samples and habitats to be sampled are based on study objectives and design.

2. *Field sample processing.* Once the macroinvertebrates are collected, samples can be either "picked" (the macroinvertebrates are removed from the substrata and debris in the sample) in the field or returned to the laboratory for processing (see below). If the samples are to be field picked, place the contents of the net or screen into a large, white enamel pan with enough water from the stream to cover the invertebrates. Using forceps and an eyedropper, pick out at least 200 invertebrates and place them in a jar with 70% alcohol. Although the tendency will be to pick out the

TABLE 31.2

Form for Recording the Physical Habitat Assessment (Based on Barbour and Stribling 1991)

Habitat parameter	Optimal		Suboptimal		Marginal		Poor	
Bottom substrate	More than 60% of bottom is gravel, cobble, and boulders. Even mix of substratum size classes	20	30–60% of bottom is cobble or boulder substrata. Substrate may be dominated by one size class.	15	10–30% of substrata is large materials. Silt or sand accounts for 70–90% of bottom.	10	Substrate dominated by silt and sand. Gravel cobble and larger sizes <10%.	5
Habitat complexity	A variety of types (logs, branches, boulders, aquatic vegetation, undercut banks) and sizes of material form a diverse habitat.	20	Structural types or sizes of material is less than optimum but adequate cover still provided.	15	Habitat dominated by only one or two structural components. Amount of cover is limited.	10	Monotonous habitat with little diversity. Silt and sand dominate and reduce habitat diversity and complexity.	5
Pool quality	25% of the pools are as wide or wider than the mean stream width and are >1 m deep.	20	<5% of the pools are >1 m deep and wider than mean stream width. Majority of pools are < mean width and <1 m deep.	15	<1% of the pools are >1 m deep and wider than stream width. Pools present may be very deep or very shallow. Variety of pools or quality is fair.	10	Majority of pools are small and shallow. Pools may be absent.	5
Bank stability	Little evidence of past bank failure and little potential for future mass wasting into channel.	20	Infrequent or very small slides, mostly healed over. Low future potential.	15	Mass wasting moderate in frequency and size. Raw spots eroded during high flows.	10	Frequent or large slides. Banks unstable and contributing sediment to stream.	5
Bank protection	Over 80% of streambank surfaces are covered by vegetation, boulders, bedrock, or other stable materials.	20	50–80% of the streambanks covered with vegetation, cobble, or larger material.	15	25–50% of the streambank is covered by vegetation.	10	<25% of the streambank is covered by vegetation or stable materials.	5
Canopy	Vegetation of various heights provides a mix of shade and filtered light to water surface.	20	Discontinuous vegetation provides areas of shade alternating with areas of full exposure. Or filtered shade occurs <6 h/day	15	Shading is complete and dense. Or filtered shade occurs <3 h/day.	10	Water surface is exposed to full sun nearly all day long.	5
Score:								

largest organisms, it is essential that all species and size classes are sampled proportionately. Faster-moving organisms will be harder to catch but every effort should be made to sample all taxa evenly.[3] Be aware of organisms in cryptic cases that resemble pieces of substratum or debris; some insects, such as caddisflies, build cases out of natural materials. If one sample does not provide at least 200 individuals, take one or more additional samples until the total number of invertebrates collected reaches this number.

3. *Subsampling.* If a sample contains far more individuals than needed (e.g., >400) then subsampling is advisable. This can be done by marking a grid in the bottom of the pan, using random numbers to select individual squares, and then picking out all the macroinvertebrates from the selected squares until you have a total of 200 individuals. Another option is to use dividers (like a metal ice-cube tray) designed as the grid for your pan. Dividers will prevent the macroinvertebrates from moving from one square to another. The important thing to remember about subsampling is that the procedure you use should not over- or underrepresent any particular group.

4. *Laboratory sample processing.* If the sample is to be taken back to the laboratory (for sorting), place the contents of the sample into an appropriate sized bottle, plastic container, or a zip-lock bag. Add enough 95% ethanol to cover the contents and, with the water in the sample, produce a final concentration of 70% ethanol. Label the sample as described below. In the laboratory, make sure there is good ventilation, and then use forceps to pick out all the macroinvertebrates from the organic and inorganic material in the sample (see Chapter 16 for sorting tips). If large numbers of macroinvertebrates were collected, subsample in the same manner as described above. Place the invertebrates in a vial(s) or jar with 70% ethanol and the label(s).

5. *Sample labeling.* Label the sample clearly, giving the date, a clear description of the location including the county and state, a brief description of the habitat type (e.g., pool, riffle), and the collector's name(s). Write the information on a paper tag using a pencil, and place the tag in the jar with the sample.

6. *Invertebrate identification.* Once the sample has been picked, sort the macroinvertebrates into groups of similar-looking organisms (i.e., those you think represent a single species or taxon). Use the general key in Chapter 16 to identify an individual from each group to the family-level. Record the information on the data sheet provided (Table 31.3). Good general keys for more detailed identifications are available for all benthic macroinvertebrate groups (e.g., Pennak 1989, Thorp and Covich 1991), specific groups such as the insects (e.g., Lehmkuhl 1979, Merritt and Cum-

---

[3]Soda water or "club soda" can be added to the pan to anesthetize the animals.

TABLE 31.3
Form for Recording Macroinvertebrate Data

Date: _____

Name: _____

Site: _____

A	B	C	D
	No. of	Biotic	
Order/family	organisms	score	Total

1. _____ _____ × _____ = _____
2. _____ _____ × _____ = _____
3. _____ _____ × _____ = _____
4. _____ _____ × _____ = _____
5. _____ _____ × _____ = _____
6. _____ _____ × _____ = _____
7. _____ _____ × _____ = _____
8. _____ _____ × _____ = _____
9. _____ _____ × _____ = _____
10. _____ _____ × _____ = _____
11. _____ _____ × _____ = _____
12. _____ _____ × _____ = _____
13. _____ _____ × _____ = _____
14. _____ _____ × _____ = _____
15. _____ _____ × _____ = _____
16. _____ _____ × _____ = _____
17. _____ _____ × _____ = _____
18. _____ _____ × _____ = _____
19. _____ _____ × _____ = _____
20. _____ _____ × _____ = _____

Family biotic index = total of Column D divided by total of Column B = _____

% EPT = total Ephemeroptera, Trichoptera, and Plecoptera divided by total of Column B = _____

Taxa richness = total No. of taxa = _____

mins 1996), macroinvertebrates of specific regions (e.g., Clifford 1991), and insects of specific regions (e.g., Usinger 1956, Peckarsky *et al.* 1990).

B. Exercise 2: Laboratory Option

1. *Laboratory preparation.* This exercise is intended for classroom instruction only. The instructor must assemble macroinvertebrates from previous collections or make special collections prior to the laboratory session. Plan to provide at least 100 macroinvertebrates per student. Assemble two "invertebrate soups," one that represents a reference stream and one that represents an impacted stream. To make the "soup" place all the invertebrates that represent the reference site together in a bowl and cover with 70% ethanol. Do the same for the macroinvertebrates from the impacted site.

2. *Sampling.* To take a sample from the invertebrate soup, swirl the soup to evenly distribute the organisms. Then using a tea strainer or small aquarium net, dip into the soup to obtain a sample. The purpose of this is to obtain a random sample of about 100 invertebrates. The sample is then placed in a petri dish with ethanol, sorted, and identified as outlined in Exercise 1.

C. Data Analysis

1. *Tolerance scores.* On the worksheet provided (Table 31.3), list the names of the macroinvertebrate families collected and the number of individuals in each family in the sample. Look up the tolerance score (Appendix 31.1) for each family and write it in the next column; multiply the values in the number column by the tolerance score for that row. Sum the resulting numbers, and then divide this sum by the total number of individuals. Equation (31.1) is used to calculate the family biotic index (FBI) (Hilsenhoff 1988),

$$FBI = 1/N \sum n_i t_i, \qquad (31.1)$$

where n_i is the number of individuals in a family; t_i, the tolerance score for that family; and N, the total number of individuals in the sample.

The information recorded in Table 31.3 can also be used to calculate several other useful measures (e.g., the total number of families; percentage of total organisms that are Ephemeroptera, Plecoptera, and Trichoptera (EPT); percentage of total organisms of a particular functional feeding group). Resh and Jackson (1993) provide many examples of the various measures that can be calculated. These measures can then be analyzed in the same way outlined below for FBI. Analysis of habitat scores and FBI values are conducted in two steps: (1) Individual participants evaluate their own data from the habitat assessment, calculate a biotic index, and determine the water quality category of the sites selected; (2) within-site and between-site variability is examined by using the data from all participants.

2. *Biotic index—Analysis of individual data.* Water quality can be evaluated using the FBI by comparing the index value calculated from a benthic sample with a predetermined scale of "biological condition." A scale developed for use in Wisconsin to determine the degree of organic pollution is given in Table 31.4 (Hilsenhoff 1988). For example, find Group A in the sample data set (Table 31.5). An index of 4.5 was calculated from a test stream sample, which would indicate a "good" water quality rating on a scale of "very poor" to "excellent." Find the water quality rating that describes the FBI scores you calculated.

Because thresholds may differ regionally, you can also assess water quality by comparing a test site to a regional reference site, or to a "paired" unimpacted site. Some protocols (e.g., Plafkin *et al.* 1989) recommend calculating the percentage similarity between the test site and the regional or paired reference-site (Table 31.6). Calculate this as follows:

$$\% \text{ similarity} = (\text{reference FBI/test FBI}) \times 100. \qquad (31.2)$$

In the example provided, Group A would show a 76% similarity between sites, indicating that the test site is "slightly impaired" relative to the reference. Calculate a percentage similarity of your reference site and test site using Eq. (31.2) and then match this to the water quality thresholds provided in Table 31.4; these thresholds are recommended by the U.S.

TABLE 31.4
Water Quality Based on Family
Biotic Index Values

Family biotic index	Water quality
0.00–3.75	Excellent
3.76–4.25	Very good
4.26–5.00	Good
5.01–5.75	Fair
5.76–6.50	Fairly poor
6.51–7.25	Poor
7.26–10.00	Very poor

Note. From Hilsenhoff (1988).

TABLE 31.5
Sample Data Set for *t*-Test

Group	Reference site	Test site
A	3.4	4.5
B	3.2	5.2
C	3.9	6.1
D	5.6	7.9
E	3.1	5.2
F	5.3	5.7
G	4.3	6.5
H	4.3	5.4
I	5.1	6.3
J	3.2	4.7

Summary statistics

Reference site:	$n_1 = 10$	$x_1 = 4.14$	$s_1^2 = 0.89$	$\Sigma\, x_1 = 41.4$	$\Sigma\, (x^2)_1 = 179.3$
Test site:	$n_2 = 10$	$x_2 = 5.75$	$s_2^2 = 1.00$	$\Sigma\, x_2 = 57.5$	$\Sigma\, (x^2)_2 = 339.6$

Environmental Protection Agency (Plafkin *et al.* 1989) but may require regional adjustment.

3. *Habitat rating.* Analysis of habitat scores calculated for the two stream sites is basically the same as the percentage similarity example above. In this case, the test site score is divided by the paired site (or reference) score. If no reference score is available, use the maximum score possible as a substitute. Table 31.7 is a scale recommended by Barbour and Stribling (1991) and the U.S. Environmental Protection Agency (Plafkin *et*

TABLE 31.6
Biological Condition Using Percentage
Similarity of FBI Calculated Between
Test-Site and Reference-Site Samples

% Similarity	Biological condition
≥85%	Unimpaired
84–70%	Slightly impaired
69–50%	Moderately impaired
<50%	Severely impaired

Note. Modified from Plafkin *et al.* (1989).

TABLE 31.7
Comparison of Habitat Assessment Scores
(Percentage Similarity) from the Test Site
and the Reference Site

% Similarity	Habitat evaluation
≥ 90%	Comparable
75–89%	Supporting
60–74%	Partially supporting
≤59%	Not supporting

Note. Modified from Plafkin *et al.* (1989).

al. 1989) to assess whether a test site is comparable (i.e., degree of agreement with conditions at reference site) to the habitat of the reference site. Compare your results to this scale.

4. *Graphical analysis of group data.* Examine the variability of the rapid assessment data. The easiest way to view variability is to graph it. With each point representing a single collection, place each of the sites on the x-axis and the calculated measure (e.g., FBI or habitat score) on the y-axis. Mark the predetermined water quality thresholds directly on this graph and observe the range of water quality assessments obtained within the same site.

5. *Measure of variability.* Another way to compare the variability of metrics (e.g., FBI) is to express the standard deviation of the sample as a percent of the sample mean. This is called the coefficient of variation (CV) and is computed as follows:

$$CV = (\text{standard deviation/mean}) \times 100. \qquad (31.3)$$

The advantage of calculating the CV is that it has no units. Therefore, it can be used to compare the variability of different types of measures (e.g., %EPT, taxa richness). Barbour *et al.* (1992) provide an example of its use.

6. *Comparison of replicate samples.* Finally, we will use the FBI values obtained to determine if there is a statistically significant difference between the two sites. A *t* test can be used to indicate if the two sample means are equal, or if they are significantly different. To do this test, calculate the mean ($\bar{x}$), variance (s^2), sum of measures (Σx), sum of squared measures (Σx^2), and the number of observations for each site (n) using samples as replicate observations (see Zar 1984 or Sokal and Rohlf 1995). Also, an example is provided in Table 31.5 (see Narf 1985).

The *t* test is used to choose between two hypotheses regarding the two populations that were sampled. The null hypotheses (H_0) is that the two population means are equal. The alternative hypothesis (H_A) is that the two means come from different populations. What the t test actually tells us is the probability (P) that the null hypothesis (H_0) is true (i.e., the probability that $\text{mean}_1 = \text{mean}_2$). By convention, when the probability is less than 1 in 20 ($P < 0.05$) the two means are considered to be significantly different.

To calculate the t statistic, take the difference between the sample means and divide by the standard error of this difference, as shown by

$$t = \frac{\bar{x}_1 - \bar{x}_2}{\sqrt{s_{n_1+n_2}^2((1/n_1) + (1/n_2))}}, \tag{31.4}$$

where the means and sample sizes are taken directly from the summary statistics. The pooled variance ($S_{n_1+n_2}^2$) is calculated as follows:

$$s_{n_1+n_2}^2 = \frac{(\Sigma x^2 - \bar{x}\Sigma x)_1 + (\Sigma x^2 - \bar{x}\Sigma x)_2}{n_1 + n_2 - 2}. \tag{31.5}$$

A demonstration of a t test using the sample data above is provided below:

$$t = \frac{5.75 - 4.14}{\sqrt{0.94\,((1/10) + (1/10))}} = \frac{1.61}{0.43} = 3.74. \tag{31.6}$$

The critical value for t (t_{crit}) is then looked up in a T table. The t_{crit} depends on the number of samples used to calculate t. The data above has a total of 20 samples, therefore 18 degrees of freedom (df); where $df = n_1 + n_2 - 2$. For 18 df, the $t_{crit(0.05)} = 2.10$ (Table 31.8). Our calculated t is greater than this critical t value, therefore we conclude that the two means are significantly different ($P < 0.05$). (If the calculated t is negative, use the absolute value). A T table should be consulted for critical values (e.g., Zar 1984, Sokal and Rohlf 1995) but Table 31.6 contains some two-tailed $t_{crit(0.05)}$ values for sample sizes ($N = n_1 + n_2$) ranging from 7 to 37 (df, 5–35). Determine if the two sites that the class or group sampled are significantly different using the FBI.

The above procedure can be done with many different types of measures (e.g., those listed earlier). Note that the t test assumes that both

TABLE 31.8
List of Two-Tailed Critical t-Values
($P < 0.05$) for Various Degrees of Freedom (df)

N	df	Critical t
7	5	2.57
12	10	2.23
17	15	2.13
22	20	2.09
27	25	2.06
32	30	2.04
42	40	2.02

Note. Use the next smaller N if your sample size is between the values provided, or consult a more extensive T Table (e.g., Zar 1984).

samples come from a normally distributed population and that the variances (s^2) of both samples are equal. When sample sizes are equal, these assumptions may be "relaxed." However, it is always a good idea to know if these assumptions are being met. A quick rule-of-thumb to check for equal variances is to make sure the ratio of the sample variances is less than 2. If this ratio is greater than 2, then you may be able to transform the data a number of different ways, or use a test that does not assume equal variance. Refer to a statistical text (e.g., Zar 1984, Sokal and Rohlf 1995) for further information on these alternatives. Always be confident that the various assumptions for parametric statistics are met before drawing a conclusion based on the test.

IV. QUESTIONS

1. Were the measurements of physical habitat used in this exercise good indicators of stream condition? What other measurements of the physical habitat could be assessed? Choose two and describe what the optimal condition would be and what the poor condition would be. How would the measurements you selected influence the abundance and distribution of macroinvertebrates? What about other stream organisms such as fish or amphibians?

2. What might you conclude if the physical habitat looked optimal but there were few types or numbers of macroinvertebrates in the stream? How would you proceed to examine this discrepancy in more detail?

3. What are the strengths of using a subjective evaluation of the physical habitat? What are the weaknesses?

4. If you were setting up a study to compare a pristine stream with an impacted stream, list three variables that you would try to keep constant between the two sites. Why?

5. Did the study streams have different FBI scores? If several samples were taken from each stream, how similar were the FBI scores from different samples? What about different habitats? How could the variability among samples have been reduced?

6. Assume that you are in charge of setting up a long-term monitoring program on an urban stream. What characteristics do you think would be important to monitor? If the stream were in a wild land forest, how would this change your monitoring program?

7. How important is the history of the stream in terms of the stream condition that you observed (e.g., what past events might still influence the study stream)?

8. We used an index based on invertebrate families. Do you think an index based on generic or species tolerances would have been better? Why or why not? What might limit the use of species-level tolerances?

9. Single measurements of water quality are often taken at one site. Do you think water quality measures such as pH, temperature, and dissolved oxygen are more or less variable than biological measures such as the FBI or %EPT?

10. How do you think collections made during different seasons might influence the FBI, %EPT, and total number of individuals collected?

V. MATERIALS AND SUPPLIES

Accurate maps showing collecting sites
Biotic scores (Appendix 31.1)
Calculator
Dissecting microscopes
Enamel pans
Equipment for measuring current velocity (see Chapter 3)
Equipment for measuring gradient
Equipment for collection of macroinvertebrates (see Chapter 16)
Ethanol (95 and 70%)
Keys for identifying macroinvertebrates (e.g., Appendix 16.1, Merritt
     and Cummins 1996)
Petri dishes for sorting samples

Thermometer
Vials, jars, or plastic zip-lock bags

APPENDIX 31.1

Tolerance Values for Macroinvertebrates (Data from Hilsenhoff 1988, except *Adapted from Lenat 1993, **Adapted from Bode 1988)

Plecoptera		Calamoceratidae*	3	Empididae	6		
Capniidae	1	Glossosomatidae	0	Ephydridae	6		
Chloroperlidae	1	Helicopsychidae	3	Psychodidae	10		
Leuctridae	0	Hydropsychidae	4	Simuliidae	6		
Nemouridae	2	Hydroptilidae	4	Muscidae	6		
Perlidae	1	Lepidostomatidae	1	Syrphidae	10		
Perlodidae	2	Leptoceridae	4	Tabanidae	6		
Pteronarcyidae	0	Limnephilidae	4	Tipulidae	3		
Taeniopterygidae	2	Molannidae	6				
		Odontoceridae	0	**Amphipoda****			
Ephemeroptera		Philpotamidae	3	Gammaridae	4		
Baetidae	4	Phryganeidae	4	Talitridae	8		
Baetiscidae	3	Polycentropodidae	6				
Caenidae	7	Psychomyiidae	2	**Isopoda****			
Ephemerellidae	1	Rhyacophilidae	0	Asellidae	8		
Ephemeridae	4	Sericostomatidae	3				
Heptageniidae	4	Uenoidae	3	**Acariformes****	4		
Leptophlebiidae	2						
Metretopodidae	2	**Megaloptera**		Decapoda**	6		
Oligoneuridae	2	Corydalidae	0				
Polymitarcyidae	2	Sialidae	4	**Mollusca****			
Potomanthidae	4			Lymnaeidae	6		
Siphlonuridae	7	**Lepidoptera**		Physidae	8		
Trichorythidae	4	Pyralidae	5	Sphaeridae	8		
Odonata		**Coleoptera**		**Oligochaeta****	8		
Aeshnidae	3	Dryopidae	5				
Calopterygidae	5	Elmidae	4	**Hirudinea****			
Coenagrionidae	9	Psephenidae	4	Bdellidae	10		
Cordulegastridae	3						
Corduliidae	5	**Diptera**		**Turbellaria****			
Gomphidae	1	Anthericidae	2	Platyhelminthidae	4		
Lestidae	9	Blepharoceridae	0				
Libellulidae	9	Ceratopogonidae	6				
Macromiidae	3	Blood-red	8				
		Chironomidae					
Trichoptera		Other Chironomidae	6				
Brachycentridae	1	Dolochopodidae	4				

REFERENCES

APHA, AWWA, and WEF. 1992. *Standard Methods for the Examination of Water and Wastewater,* 18th ed. American Public Health Association, Washington, DC, USA.

Barbour, M. T., J. L. Plafkin, B. P. Bradley, C. G. Graves, and R. W. Wisseman. 1992. Evaluation of EPA's rapid bioassessment benthic metrics: Metric redundancy and variability among reference stream sites. *Journal of Environmental Toxicology and Chemistry* **11:**437–449.

Barbour, M. T., and J. B. Stribling. 1991. Use of habitat assessment in evaluating the biological integrity of stream communities. Pages 25–38 *in Biological Criteria: Research and Regulation.* EPA-440-5-91-005, United States Environmental Protection Agency, Office of Water, Washington, DC.

Cairns, J., Jr., and J. R. Pratt. 1993. A history of biological monitoring using benthic macroinvertebrates. Pages 10–27 *in* D. M. Rosenberg and V. H. Resh (Eds.) *Freshwater Biomonitoring and Benthic Macroinvertebrates.* Chapman & Hall, New York, NY.

Clifford, H. F. 1991. *Aquatic Invertebrates of Alberta.* The Univ. of Alberta Press, Edmonton, Alberta, Canada.

Hellawell, J. M. 1986. *Biological Indicators of Freshwater Pollution and Environmental Management.* Elsevier, New York, NY.

Hilsenhoff, W. L. 1988. Rapid field assessment of organic pollution with a family level biotic index. *Journal of the North American Benthological Society* **7:**65–68.

Johnson, R. K., T. Wiederholm, and D. M. Rosenberg. 1993. Freshwater biomonitoring using individual organisms, populations, and species assemblages of benthic macroinvertebrates. Pages 40–158 *in* D. M. Rosenberg and V. H. Resh (Eds.) *Freshwater Biomonitoring and Benthic Macroinvertebrates.* Chapman & Hall, New York, NY.

Lehmkuhl, D. M. 1979. *How to Know the Aquatic Insects.* Brown, Dubuque, IA.

Lenat, D. R. 1993. A biotic index for the southeastern United States: Derivation and list of tolerance values, with criteria for assigning water-quality ratings. *Journal of the North American Benthological Society* **12:**279–290.

Merritt, R. W., and K. W. Cummins (Eds.). 1996. *An Introduction to the Aquatic Insects of North America,* 3rd ed. Kendall/Hunt, Dubuque, IA.

Metcalfe, J. L. 1989. Biological water quality assessment of running waters based on macroinvertebrate communities: History and present states in Europe. *Environmental Pollution* **60:**101–139.

Narf, R. P., E. L. Lange, and R. C. Wildman. 1984. Statistical procedures for applying Hilsenhoff's biotic index. *Journal of Freshwater Ecology* **2:**441–448.

Peckarsky, B. L., P. R. Fraissinet, M. A. Penton, and D. J. Conklin. 1990. *Freshwater Macroinvertebrates of Northeastern North America.* Cornell Univ. Press, Ithaca, NY.

Pennak, R. W. 1989. *Freshwater Invertebrates of the United States,* 3rd ed. Wiley, New York, NY.

Plafkin, J. L., M. T. Barbour, K. D. Porter, S. K. Gross, and R. M. Hughes. 1989. *Rapid Bioassessment Protocols for Use in Streams and Rivers. Benthic Macroin-*

vertebrates and Fish. EPA/444/4-89/0001. Office of Water Regulations and Standards, United States Environmental Protection Agency, Washington, DC.

Platts, W. S., W. F. Megahan, and G. W. Minshall. 1983. *Methods for Evaluating Stream, Riparian, and Biotic Conditions.* General Technical Report INT-138. United States Department of Agriculture, Forest Service, Ogden, UT.

Resh, V. H. 1995. Freshwater benthic macroinvertebrates and rapid assessment procedures for water quality monitoring in developing and newly industrialized countries. Pages 167–177 *in* W. S. Davis and T. P. Simon (Eds.) *Biological Assessment and Criteria.* Lewis Publishers, Boca Raton, FL.

Resh, V. H., and J. K. Jackson. 1993. Rapid assessment approaches to biomonitoring using benthic macroinvertebrates. Pages 195–223 *in* D. M. Rosenberg and V. H. Resh (Eds.) *Freshwater Biomonitoring and Benthic Macroinvertebrates.* Chapman & Hall, New York, NY.

Resh, V. H., and E. P. McElravy. 1993. Contemporary quantitative approaches to biomonitoring using benthic macroinvertebrates. Pages 159–194 *in* D. M. Rosenberg and V. H. Resh (Eds.) *Freshwater Biomonitoring and Benthic Macroinvertebrates.* Chapman & Hall, New York, NY.

Resh, V. H., and J. D. Unzicker. 1975. Water quality monitoring and aquatic organisms: The importance of species identification. *Journal of the Water Pollution Control Federation* **47:**9–19.

Resh, V. H., R. H. Norris, and M. T. Barbour. 1995. Design and implementation of rapid assessment approaches for water resource monitoring using benthic macroinvertebrates. *Australian Journal of Ecology* **20:**108–121.

Rosenberg, D. M., and V. H. Resh. 1993a. Introduction to freshwater biomonitoring and benthic macroinvertebrates. Pages 1–9 *in* D. M. Rosenberg and V. H. Resh (Eds.) *Freshwater Biomonitoring and Benthic Macroinvertebrates.* Chapman & Hall, New York, NY.

Rosenberg, D. M., and V. H. Resh (Eds.). 1993b. *Freshwater Biomonitoring and Benthic Macroinvertebrates.* Chapman & Hall, New York, NY.

Rosenberg, D. M., and V. H. Resh. 1996. Use of aquatic insects in biomonitoring. Pages 87–97 *in* R. W. Merritt and K. W. Cummins (Eds.) *An Introduction to the Aquatic Insects of North America,* 3rd ed. Kendall/Hunt, Dubuque, IA.

Sokal, R. R., and F. J. Rohlf. 1995. *Biometry,* 3rd ed. Freeman, New York, NY.

Thorp, J. H., and A. P. Covich. 1991. *Ecology and Classification of North American Freshwater Invertebrates.* Academic Press, New York, NY.

Usinger, R. L. (Ed.). 1956. *Aquatic Insects of California.* Univ. of California Press, Berkeley, CA.

Wright, J. F., P. D. Armitage, M. T. Furse, and D. Moss. 1988. A new approach to the biological surveillance of river quality using macroinvertebrates. *Verhandlungen der Internationalen Vereinigung für Theoretische und Angewandte Limnologie* **23:**1548–1552.

Wright, J. F., D. Moss, P. D. Armitage, and M. T. Furse. 1984. A preliminary classification of running water sites in Great Britain based on macroinvertebrate species and prediction of community type using environmental data. *Freshwater Biology* **14:**221–256.

Zar, J. H. 1984. *Biostatistical Analysis,* 2nd ed. Prentice–Hall, Englewood Cliffs, NJ.

Index